More information about this series at https://link.springer.com/bookseries/558

Lecture Notes in Computer Science 13493

Étienne Baudrier · Benoît Naegel ·
Adrien Krähenbühl · Mohamed Tajine (Eds.)

Discrete Geometry
and Mathematical Morphology

Second International Joint Conference, DGMM 2022
Strasbourg, France, October 24–27, 2022
Proceedings

 Springer

Editors
Étienne Baudrier 🆔
University of Strasbourg
Strasbourg, France

Benoît Naegel 🆔
University of Strasbourg
Strasbourg, France

Adrien Krähenbühl 🆔
University of Strasbourg
Strasbourg, France

Mohamed Tajine
University of Strasbourg
Strasbourg, France

ISSN 0302-9743   ISSN 1611-3349 (electronic)
Lecture Notes in Computer Science
ISBN 978-3-031-19896-0   ISBN 978-3-031-19897-7 (eBook)
https://doi.org/10.1007/978-3-031-19897-7

This Springer imprint is published by the registered company Springer Nature Switzerland AG
The registered company address is: Gewerbestrasse 11, 6330 Cham, Switzerland

Preface

This volume contains the papers presented at DGMM 2022: the IAPR International Conference on Digital Geometry and Mathematical Morphology held during October 24–27, 2022, in Strasbourg.

DGMM is sponsored by the International Association of Pattern Recognition (IAPR), and is associated with the IAPR Technical Committee on Discrete Geometry and Mathematical Morphology (TC18).

This is the second joint event between the two main conference series of IAPR TC18, the International Conference on Discrete Geometry for Computer Imagery (DGCI), with 21 previous editions, and the International Symposium on Mathematical Morphology (ISMM), with 14 previous editions.

There were 45 submissions. Each submission was single-blind reviewed by at least two reviewers. Based on 128 detailed reviews, we accepted 33 papers. The authors of these 33 articles are from 10 different countries: Austria, Australia, Brazil, France, Germany, Hungary, Italy, the Netherlands, Spain, and the UK.

The DGMM 2022 papers highlight the current trends and advances in discrete geometry and mathematical morphology, be they purely theoretical contributions, algorithmic developments, or novel applications in image processing, computer vision, and pattern recognition.

In addition, four internationally well-known researchers were invited for keynote lectures:

- Isabelle Bloch, on "Hybrid AI for knowledge representation and model-based medical image understanding - Towards explainability"
- Nicolas Courty, on "Sliced Wasserstein on Manifolds: Spherical and Hyperbolical cases"
- Christer Kiselman, on "Digital Geometry, Mathematical Morphology, and Discrete Optimization: a survey"
- Christian Ronse, on "Reflections on a scientific career and its possible legacy"

Three of these speakers proposed an article that can be found in this volume. Following the tradition of both DGCI and ISMM, the DGMM 2022 proceedings appear in Springer's LNCS series and a special issue of the Journal of Mathematical Imaging and Vision, with extended versions of selected outstanding contributions, is planned.

We wish to thank the members of the Program Committee and all the volunteer reviewers for their efforts in reviewing all submissions on time and giving extensive feedback. We would like to thank all the other people involved in this conference: first, the Steering Committee for giving us the chance to organize DGMM 2022; second, the four invited speakers, Isabelle Bloch, Nicolas Courty, Christer Kiselman and Christian Ronse, for accepting to share their recognized expertise; and finally, the most important component of any scientific conference, the authors for producing the high-quality and original contributions.

We are thankful to the IAPR for its sponsorship and we acknowledge the EasyChair conference management system that was invaluable in handling the paper submission, the review process, and putting this volume together. We also acknowledge Springer for making possible the publication of these proceedings in the LNCS series.

September 2022

Étienne Baudrier
Benoît Naegel
Adrien Krähenbühl
Mohamed Tajine

Organization

DGMM 2022 was organized by the IMAGeS team of the ICube Laboratory, University of Strasbourg, France.

Organizing Committee

Program Chairs

Étienne Baudrier	University of Strasbourg, France
Benoît Naegel	University of Strasbourg, France
Adrien Krähenbühl	University of Strasbourg, France
Mohamed Tajine	University of Strasbourg, France

Local Organization Chair

Étienne Le Quentrec	University of Strasbourg, France

Steering Committee

Jesús Angulo-Lopez	Ecole des Mines de Paris, France
Isabelle Bloch	Sorbonne Université - LIP6, France
Gunilla Borgefors	Uppsala University, Sweden
Srećko Brlek	Université du Québec à Montréal, Canada
Bernhard Burgeth	Universität des Saarlandes, Germany
David Coeurjolly	CNRS - LIRIS, France
Isabelle Debled-Rennesson	Université de Lorraine - LORIA, France
Andrea Frosini	Università di Firenze, Italy
Marie Jos Jiménez	University of Seville, Spain
Yukiko Kenmochi (Chair)	CNRS - GREYC, France
Walter Kropatsch	TU Wien, Austria
Jacques-Olivier Lachaud	University Savoie Mont Blanc - LAMA, France
Cris Luengo Hendriks	Flagship Biosciences, USA
Petros Maragos	National Technical University of Athens, Greece
Laurent Najman	ESIEE Paris - LIGM, France
Benjamin Perret (TC18)	ESIEE Paris - LIGM, France
Dan Schonfeld	University of Illinois, USA
Hugues Talbot	CentralSupelec - Université Paris-Saclay, France
Michael Wilkinson (Vice Chair)	University of Groningen, Netherlands

Program Committee

Eric Andres	Université de Poitiers, France
Jesùs Angulo-Lopez	Ecole des Mines de Paris, France
Peter Balazs	University of Szeged, Hungary
Samy Blusseau	CMM, Mines Paris, PSL Research University, France
Petra Bosilj	Lincoln Centre for Autonomous Systems, UK
Nicolas Boutry	LRDE, EPITA, France
Michael Breuß	Brandenburgische Technische Universität, Germany
Srecko Brlek	Université du Québec à Montréal, France
Sara Brunetti	University of Siena, Italy
Jean Cousty	LIGM, Université Gustave Eiffel, ESIEE, France
Yan Gerard	Université Clermont Auvergne, LIMOS, France
Rocio Gonzalez-Diaz	University of Seville, Spain
Bertrand Kerautret	LIRIS, France
Walter G. Kropatsch	TU Wien, Austria
Jacques-Olivier Lachaud	LAMA, University of Savoie Mont Blanc, France
Paulo Miranda	University of São Paulo, Brazil
Laurent Najman	LIGM, Université Gustave Eiffel, ESIEE, France
Phuc Ngo	Lorraine University, LORIA, France
Akihiro Sugimoto	National Institute of Informatics, Japan
Michael H. F. Wilkinson	University of Groningen, Netherlands

Additional Reviewers

Erchan Aptoula	Andrea Frosini
Isabelle Bloch	Simon Gazagnes
Gunilla Borgefors	Gaëlle Largeteau-Skapin
Luc Brun	Thierry Geraud
Bernhard Burgeth	Silvio Guimaraes
Joseph Chazalon	Lajos Hajdu
David Coeurjolly	Ronaldo Fumio Hashimoto
Lidija Comic	Atsushi Imiya
Loïc Crombez	Damien Jamet
Guillaume Damiand	María José Jiménez
Isabelle Debled-Rennesson	Marvin Kahra
Étienne Decencière	Péter Kardos
Eva Dokladalova	Yukiko Kenmochi
Éric Domenjoud	Christer Kiselman
Jonathan Fabrizio	Bastien Laboureix
Attila Fazekas	Cris L. Luengo Hendriks
Fabien Feschet	Filip Malmberg

Daniel Martins Antunes
Arnold Meijster
Benedek Nagy
Guillaume Noyel
Georgios Ouzounis
Kalman Palagyi
Eduardo Paluzo-Hidalgo
Nicolas Passat
Samuel Peltier
Kacper Pluta
Xavier Provençal
Bastien Rivier
Jos Roerdink
Tristan Roussillon
Philippe Salembier

Deise Santana Maia
Karthik Seemakurthy
Isabelle Sivignon
Ryan Slechta
Robin Strand
Hugues Talbot
Lama Tarsissi
Édouard Thiel
Guillaume Tochon
Yiying Tong
Álvaro Torras-Casas
Marc Van Droogenbroeck
Santiago Velasco-Forero
Martin Welk
François Willot

Contents

Discrete Geometry - Models, Transforms, and Visualization

Invited Papers

Reflections on a Scientific Career and Its Possible Legacy

Christian Ronse[(✉)] [ID]

ICube, Université de Strasbourg, CNRS, 300 Bd Sébastien Brant,
CS 10413, 67412 Illkirch Cedex, France
`cronse@unistra.fr`

Abstract. I give the history of my research career, its evolving scientific topics, my main results, and how the computer science and image processing community reacted to them. I briefly describe my current research on generalized flat morphology based on threshold summation. I finally discuss possible future developments arising from my works, to be pursued by a new generation.

Keywords: Discrete geometry and topology · Mathematical morphology · Connections and partitions · Poset and lattice theory

1 Introduction

Benoît Naegel suggested me to present a summary of my research career in Discrete Geometry and Mathematical Morphology, and the possible lessons to be drawn from it. Then I can also describe my current work on generalized flat morphology.

I will give here only the most important ideas and results in my long career. For more details, the reader can consult the complete list of my scientific publications:

https://christianronse.github.io/publist.html

and the one of my unpublished research reports (with PDF scans attached):

https://christianronse.github.io/wdrep.html

My career has known several changes in geographical localization, employment and scientific field of research, some quite drastic, leading to numerous topics. I first summarize various things that I did before turning to image processing (Sect. 2). Then I describe my work in DGMM, divided into four topics (Sects. 3, 4, 5, and 6). Next, I present my current research topic, generalized flat morphology based on threshold summation (Sect. 7). Finally, I discuss possible research arising from my works, notably in poset and lattice theory, to be pursued by the younger generation (Sect. 8).

© Springer Nature Switzerland AG 2022
É. Baudrier et al. (Eds.): DGMM 2022, LNCS 13493, pp. 3–16, 2022.
https://doi.org/10.1007/978-3-031-19897-7_1

2   Before Image Processing

I was trained in pure mathematics at the Université Libre de Bruxelles (1972–76), where I got interested in group theory, finite geometries and graph theory. Then I went to the University of Oxford to do my M.Sc. (1977) and D.Phil. (1979), with a thesis on finite permutation groups.

Although lattice-theoretical methods frequently intervene in group theory (cf. the Jordan-Hölder theorem), in combinatorics, and in geometry (cf. matroid theory, and the lattice-theoretical characterization of projective spaces [1]), throughout my studies I never heard of lattice theory. Indeed, this topic is generally spurned within the community of pure mathematics, and I give a few examples of that rejection in the following webpage:

https://christianronse.github.io/lt.html

With the development of computer science, the usefulness of lattice theory was recognized in programming semantics [3], formal concept analysis [4], fuzzy set theory, mathematical morphology [40], spatial logic [2], etc.

After my thesis, I was hired at the Philips Research Laboratory Brussels, where I was immediately told that unless I find a new simple group, I would not do any more group theory. I was first assigned to work with Marc Davio on the design of "switching networks", that is, how to efficiently construct a correspondence between n inputs and n outputs by combining similar correspondences for 2 inputs and 2 outputs; for instance, how to obtain all permutations by a succession of involutions, see Fig. 1. This involves a mixture of combinatorics and boolean design. In my view, it was the most boring period of my research career, and it only led to banal results.

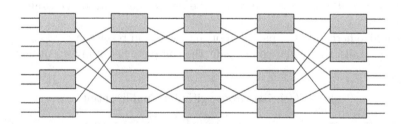

Fig. 1. Decomposition of permutations of 8 inputs into 5 stages of involutions, using the Beneš network, a recursive form of the Clos network. From Wikipedia.

My first notable work at Philips was my study [9] of feedback shift registers, that is, sequences $(x_n)_{n \in \mathbb{N}}$ given by a recurrence of the form $x_{n+m} = f(x_n, \ldots, x_{n+m-1})$. When f is linear, the sequence can be described in terms of the roots of the associated polynomial (cf. the Fibonacci sequence for the polynomial $X^2 - X - 1$). In the case where the terms x_n are in a finite field, I gave an algebraic theory of such recurrences which bears some similarity with the theory of polynomials and ideals.

3   Discrete Geometry and Topology

In the early 1980's, I was advised to turn to digital image processing. I started with 2D binary images on a square grid, with a special emphasis on their topology. I gave an algorithm (implemented in Pascal) for detecting connected components and holes in a 2D binary image [32].

My best work in this domain was my topological analysis of 2D thinning: my Philips internal report R470 and the two journal articles derived from it. The first one [10] characterized the topological validity of a 2D thinning or reduction by a "strongly deletable set" of removed pixels, and showed that it can be achieved by a succession of removals of simple pixels (i.e., a digital homotopy). The second one [11] applied the theory to parallel algorithms for thinning or reduction: when topology is not preserved, one must necessarily remove a "minimal non-deletable set", which will be either a pixel, a pair of adjacent pixels, or in the case of 8-adjacency on the foreground, an isolated 2 × 2 triangle or square.

In order to publish these results, I had to "swim against the current" of utilitarian image processing. My colleague Pierre Devijver expressed his reservations about such topological abstractions. I submitted my first paper to IEEE PAMI, where it was rejected; one referee was positive, the two others negative, in the opinion that it represents only an elaboration on the work of Azriel Rosenfeld, and advising me to study the preservation of geometry instead of topology. The editor Theodosios Pavlidis seemed to have some regrets in his decision, and asked me if I wanted a 4th referee. I then submitted it to PRL, where it was also rejected, the referee complained that it was "difficult to read". Finally I submitted it to TCS, and after one year of silence, the editor in chief Maurice Nivat sent me an postcard saying that the referee was enthusiastic, so it would be the first article on image analysis to appear in TCS.

The second paper, submitted to DAM, was almost immediately accepted, without revision. A few years later, these two papers would be cited in numerous works on thinning. There is a lesson to be drawn from this episode: if you believe in your work, you must pursue it, even if it is shunned by the mainstream.

Another interesting subject that I investigated at that time is the application of convexity to discrete geometry, in particular characterizations of digital convexity and of digital straight line segments. The two problems are related: a straight line segment is the prototype of a convex set, and given a digital set X, the set of all pairs $(a, b) \in \mathbb{R}^2$ such that X is in the digitization of the line $y = ax + b$ is convex. The methodology can also be applied to the quantization of linear and affine functions on $\mathbb{R}^n$. See my Philips internal report R485 and my publications on these topics between 1985 and 1990.

I must say that all these results were obtained in isolation, I had almost no contacts with the discrete geometry community. I first became acquainted with it on the occasion of the 1st DGCI conference held in Strasbourg in 1991, where I was an invited speaker.

Much later, in Strasbourg at the end of the 1990's, I started investigating *Hausdorff discretization* in collaboration with my colleague Mohamed Tajine. Given a metric space (E, d), the metric d leads to the Hausdorff metric H_d

on compact subsets of E. Given a discrete subspace D of E, for any compact subset K of E, we can choose as discretization of K any finite subset S of D which minimizes the Hausdorff distance $H_d(K, S)$ [38]. It includes as particular case the covering and supercover discretizations [39]. The theory can be generalized to closed subsets of E instead of compact ones. Under minimal topological conditions on E and D, the discretization of a connected closed subset of E is connected for a specific adjacency graph on D [35]. More problems must be investigated in this theory, in particular its topological aspects [44].

4   Algebraic Framework for Mathematical Morphology

In the early 1980's, at Philips Research Laboratory Brussels, I investigated some practical problems in image analysis, in particular digital subtraction angiography for extracting blood vessels from X-ray images. I studied non-linear filters, such as the median and rank filters, their weighted counterparts, the Kramer-Bruckner filter, etc. This led me to the concept of *order-configuration* filters [12], where the new value on a pixel is the value of one pixel in its window, chosen according to the ordering of the values of all pixels in that window. In fact, this corresponds to what one calls a *flat operator* in mathematical morphology.

Having invented in 1986 a new flat operator, the parametric (or rank-max) opening, and applied it to digital subtraction angiography (see my Philips internal document WD47), I was introduced to the mathematical morphology community and I met Henk Heijmans. We immediately agreed to formalize the lattice-theoretical basis of morphological operators, relying on classical lattice-theoretical references, in particular the 1980 edition of [3]. Our two papers [6,34] immediately became classical. Later we collaborated again on the topic of annular filters [5,33].

I studied various types of idempotent operators, in particular a generalization of openings, where for each structuring element we take the composition of a hit-or-miss transform followed by the dilation by the foreground structuring element [13]. In fact, the grey-level hit-or-miss transform [7] (and the derived idempotent operator) has been very useful in the analysis of angiographic images [8].

Mathematical morphology was put to practical use by several of my doctoral students: Vincent Agnus (motion measurement), Benoît Naegel (segmentation of 3D hepatic images), Julien Lamy (segmentation of 3D images of the colon), Nicolas Passat (segmentation of 3D cerebral vascular images), Erchan Aptoula (multivariate image analysis), Bessem Bouraoui (segmentation of 3D coronary images), and Alice Dufour (cerebrovascular atlas).

Since the 1980's, I never saw any problem switching between discrete geometry and mathematical morphology, or combining both in practical applications. This was not obvious for many others. When I came to Strasbourg in 1992, the discrete geometry group headed by Jean Françon belonged to the computer graphics team, and I once heard Françon complain that there were too many image analysis talks at DGCI. Reciprocally, most researchers in mathematical morphology had only the most basic notions in discrete geometry (dating from the early 1970's).

In Strasbourg, the two topics of discrete geometry and mathematical morphology progressively joined in the early 2000's, becoming an official research axis within the image analysis team in 2007. At that time, the two topics were also combined in the A3SI team of the LIGM at Marne-la-Vallée; indeed, its leader Gilles Bertrand had been working on discrete geometry, discrete topology and mathematical morphology since 1984. However, in most of France, DG and MM remained separated. The grouping of the two within the IAPR TC18 and the GDR IM of the CNRS was achieved much later, in 2016 and 2017 respectively.

5 Connectivity and Partitioning (Also Partial)

I come now to what I consider the most original part of my research: the theory of connections and partial connections, and its relations with the lattices of partitions and of partial partitions. Indeed, apart Jean Serra and myself, few people obtained significant fundamental results on this topic, and some of my works in this domain have become rather famous.

Serra [40] unified various types of connectivity into the notion of *connection*: a family C of subsets of the space E to which belong the empty set and all singletons, and such that for any subset B of C having a non-empty intersection, the union of B belongs to C. The elements of C are called *connected*. He also characterized connections in terms of the *point openings* γ_p, $p \in E$: for any $p \in E$ and $X \in \mathcal{P}(E)$, $\gamma_p(X)$ is the connected component of X containing p when $p \in X$, and the empty set when $p \notin X$.

In [14] I gave a characterization of connections in terms of a family of *separators*. I also studied the lattice of connections and the construction of connections, in particular what one calls *second-generation connections*, for instance Serra's *connection by dilation* [40], and the *connection by opening* first suggested by Heijmans: we intersect a connection with the invariance domain of an opening, and we add the singletons, see Fig. 2. In this latter example, adding the singletons is an artificial operation necessary for satisfying the axiom that all singletons belong to the connection. If we omit that, we have then a *partial connection*: $\emptyset \in C$ and for $B \subseteq C$ with $\bigcap B \neq \emptyset$, we must have $\bigcup B \in C$; here the singletons do not necessarily belong to C.

I studied thus partial connections in [17]. For a connection, the connected components of a set according to a connection from a partition of that set [40]; in the case of a partial connection, they make a *partial partition* of that set, equivalently, a partition of a subset of it. Thus I studied also partial partitions. They constitute a complete lattice for the standard order ($\pi_1 \leq \pi_2$ iff each block of π_1 is included in a block of π_2). Serra [42] had shown that a family C of subsets of E comprising the empty set is a connection if and only if the set of partitions of E with blocks in C is sup-closed; I extended this result, with more equivalent conditions, and obtained the same result in the case of a partial connection and partial partitions [17].

Given an operator σ that associates to each $X \in \mathcal{P}(E)$ a partial partition $\sigma(X)$ of X, we obtain an anti-extensive operator $\beta(\sigma)$ on partial partitions which

Fig. 2. Left: a structuring element B in $\mathbb{Z}^2$, connected for a connection $\mathcal{C}$; the set of all $\mathcal{C}$-connected subsets of $\mathbb{Z}^2$ invariant under the opening by B constitutes a partial connection $\mathcal{C}_B$. Right: the connected components of X under that partial connection $\mathcal{C}_B$ are the two $\mathcal{C}$-connected components of the opening $X \circ B$ (in grey); in the connection made of $\mathcal{C}_B$ plus all singletons, all singletons of $X \setminus (X \circ B)$ (in black) will be connected components.

acts on a partial partition by applying σ to each block separately: $\beta(\sigma)(\pi) = \bigcup_{B \in \pi} \sigma(B)$. Then $\beta(\sigma)$ is an opening on partitions if and only if σ decomposes a set into its connected components for some connection; we have the same result for partial partitions and partial connections [21]. Non-increasing idempotent operators of the form $\beta(\sigma)$ were studied in [22], in particular those involved in the iterative segmentation scheme of Serra or the "constrained segmentation" according to Soille.

I analysed other operations on partitions and partial partitions, in particular adjunctions [19], and closures obtained from closures on sets [24].

Serra generalized connections to arbitrary complete lattices (for instance, the one of numerical images) [41], and together [36] we generalized the related operation of *geodesic reconstruction* to complete lattices where the binary infimum distributes arbitrary suprema. I have generalized the geodesic reconstruction to some non-distributive lattices, in particular the one of label images [31] and those of partitions and of grey-level images with Keshet's reference order [18].

6 New Orders on Partial Partitions

In [25–27] I introduced a dozen orders on partial partitions. Given a partial partition π, the *support* of π, supp(π), is the union of all blocks of π; thus π is a partition of supp(π); the *background* of π, back(π), is the complement of the support, so back$(\pi) = E \setminus$ supp(π).

In [25], I first noticed that the growth of a partial partition for the standard order combines three basic operations: creating new blocks (from background points), inflating individual blocks (adding to them background points), and merging blocks. The three inverse operations are removing blocks, deflating blocks (removing some but not all of their points), and splitting blocks. These six operations can be represented by a triangle, where the operation corresponding to any arrow can be obtained by composing two operations corresponding to the two arrows forming a path from its origin to its goal, see Fig. 3, left:

- Inflating blocks B_i into the blocks C_i $(i \in I)$ can be done by creating the blocks $C_i \setminus B_i$ $(i \in I)$, then merging the two blocks B_i and $C_i \setminus B_i$ for all $i \in I$.
- Merging blocks B_i $(i \in I)$ can be done by removing all blocks B_i except one, B_j, then inflating the block B_j into $\bigcup_{i \in I} B_i$.
- In a *non-empty* partial partition, creating blocks B_i $(i \in I)$ can be done by inflating another block C into $C \cup \bigcup_{i \in I} B_i$, then splitting the block $C \cup \bigcup_{i \in I} B_i$ into C and all B_i $(i \in I)$.

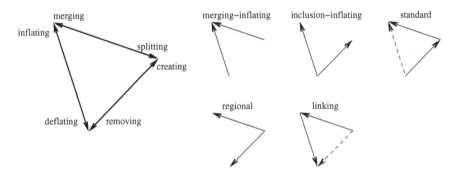

Fig. 3. Left: the diagram made by the 3 basic operations on blocks (upward arrows) and their 3 inverses (downward arrows). Right: orders obtained by combining two types of operations; here a dashed arrow indicates a third operation that can be obtained by combining the two others. Top right: the 3 orders obtained by combining two types of basic operations (upward arrows). Bottom right: 2 orders obtained by combining one type of basic operation (upward arrow) with the inverse of another type of basic operation (downward arrow).

Each type of basic growth operation taken alone gives rise to an order: the *inclusion* order $\subseteq$, the *inflating* and *merging* orders. Combining two of them gives rise to the *merging-inflating, inclusion-inflating* and *standard* orders, see Fig. 3, top right; note that in the standard order growth can be obtained by creating blocks followed by merging blocks, from which inflating blocks also arises. In these six orders, the growth of a partial partition can only increase $|\mathsf{supp}(\pi)|$ and $|\mathsf{supp}(\pi)| - |\pi|$, so these orders are appropriate to describe the bottom-up growth of a segmentation.

Elaborating on the "building order" suggested by Serra [43], I introduced in [26] a generalization of block merging, called *apportioning*: some blocks will disappear, and their contents is apportioned between one or several remaining blocks, that is, each disappearing block either is merged with one remaining block, or is split into several parts, and each part is merged with one remaining block, see Fig. 4. The resulting *apportioning order* contains the merging order. As suggested by Serra, this order can be used to describe the elimination of "parasitic" segmentation classes, for instance thick borders separating regions.

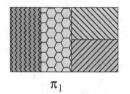 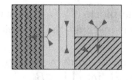 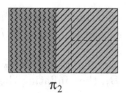

π_1 π_2

Fig. 4. The apportioning order. Left: the initial partial partition π_1; the pink and yellow block will be apportioned. Middle: the pink block is split into two parts, one is merged with the blue block, and the other with the green block; the yellow block is merged with the blue block. Right: the resulting greater partial partition π_2. (Color figure online)

Combining apportioning with inflating blocks, one obtains the *apportioning-inflating order*, which contains the merging-inflating order; combining apportioning with creating blocks lead to the *extended order*, which contains the standard order. These orders are again appropriate to describe the bottom-up growth of a segmentation, with the added possibility of eliminating "parasitic" blocks.

In [27] I studied some orders where the growth of a partial partition can only decrease $|\mathsf{supp}(\pi)|$ and $|\pi|$, so these orders can be relevant to image reduction (or simplification), for instance in skeletonization and skeleton pruning. First, I considered two orders where the growth of a partial partition combines one of the three basic operations with the inverse of another one, see Fig. 3, bottom right. In the *regional order*, a partial partition grows by merging or removing blocks, and it constitutes a complete lattice isomorphic to the one of partitions of $E \cup \{\wp\}$ for an additional point $\wp$. In the *linking order*, a partial partition grows by merging and deflating blocks, from which removing blocks can be achieved; it contains thus the regional order. Both orders seem relevant to some topics in image processing.

I also considered the *partial merging* and *partial apportioning* orders, where a "disappearing" block to be merged or apportioned can lose some points before being merged or apportioned; finally there is the *joining order*, which contains the five previous ones: here a partial partition grows by apportioning and deflating blocks (we can thus remove blocks). However the interpretation of these three orders is more difficult, and their practical relevance is questionable.

There is one combination of upward and downward arrows that I did not give in Fig. 3, bottom right: removing blocks combined with inflating blocks, from which merging blocks can be obtained. Unlike the orders given above, the order generated in this way does not have good properties.

I have not yet pursued some topics briefly investigated in [23], such as maximal partial partitions in compound segmentation, or the selection of an optimal partial partition with respect to some criteria.

7   Current Research: Generalized Flat Morphology

In [15] I gave a general lattice-theoretical theory of flat morphological operators derived from increasing operators on binary images. Let E be the space of points and V the set of image values (for instance grey-levels, vectors, labels, etc.); in fact, V can be any *complete lattice* with at least two elements. Consider the set V^E of images $E \to V$.

For an image $F : E \to V$ and $v \in V$, define the *threshold set* $\mathsf{X}_v(F) = \{p \in E \mid F(p) \geq v\}$; these sets form a *stack*, in other words $\mathsf{X}_v(F)$ is decreasing in v: $w > v \Rightarrow \mathsf{X}_w(F) \subseteq \mathsf{X}_v(F)$. For $B \subseteq E$ and $v \in V$, the *cylinder of base B and level v* is the function $C_{B,v}$ given by setting for $p \in E$: $C_{B,v}(p) = v$ if $p \in B$, and $C_{B,v}(p) = \bot$ if $p \notin B$. Then every function $F : E \to V$ is the upper envelope of the sets $\{v\} \times \mathsf{X}_v(F)$, in other words, $F = \bigvee_{v \in V} C_{\mathsf{X}_v(F),v}$. Thus, F can be recovered by superposing its thresholdings at all values $v \in V$.

Let the operator $\psi : \mathcal{P}(E) \to \mathcal{P}(E)$ on binary images be *increasing*: $X \subseteq Y \Rightarrow \psi(X) \subseteq \psi(Y)$. Then for any $F : E \to V$, the sets $\psi(\mathsf{X}_v(F))$ form a stack, they decrease with v. Now take the upper envelope of the sets $\{v\} \times \psi(\mathsf{X}_v(F))$, in other words $\bigvee_{v \in V} C_{\psi(\mathsf{X}_v(F)),v}$, and this defines $\psi^V(F)$, the result of the operator ψ^V on the image F. For every point $p \in E$ we have:

$$\psi^V(F)(p) = \bigvee \{v \in V \mid p \in \psi(\mathsf{X}_v(F))\}. \tag{1}$$

Then $\psi^V : V^E \to V^E : F \mapsto \psi^V(F)$ is the *flat operator corresponding to* ψ, or the *flat extension of* ψ [15].

However, when the operator ψ is not increasing, this does not work, as the sets $\psi(\mathsf{X}_v(F))$ do not form a stack. In [29] I proposed to replace the lattice-theoretical threshold superposition by a *threshold summation*. Indeed, in the case where V is a chain and ψ is increasing, in (1) the set of all $v \in V$ such that $p \in \psi(\mathsf{X}_v(F))$ forms an interval bounded by $\min V$ and $\psi^V(F)(p)$, and the summation will measure its length. Writing χ for the characteristic function $\mathcal{P}(E) \to \{0,1\}^E$, we get

- in the *discrete case* $V = \{t_0, \ldots, t_n\}$, where $t_0 < \cdots < t_n$: $\psi^V(F)(p) = t_0 + \sum_{i=1}^{n} (t_i - t_{i-1}) \chi\psi(\mathsf{X}_{t_i}(F))(p)$.
- in the *continuous case* $V = [\bot, \top]$: $\psi^V(F)(p) = \bot + \int_\bot^\top \chi\psi(\mathsf{X}_t(F))(p)\, dt$.

This method was given in [45] to construct flat operators in the case where $V = \{0, 1, \ldots, n\}$. In [29] I extended this approach to operators which are not necessarily increasing, and whose result is not necessarily a binary image (for instance, the morphological Laplacian). Moreover, I did not restrict myself to the case where V is a chain, my theory works also for multivalued images.

My first step was to define, for a poset $P \subseteq \mathbb{R}^m$, the summation $\mathcal{S}_{[a,b]}(f)$ of a function $f : P \to \mathbb{R}$ over an interval $[a, b]$ in P. I defined it in the case where the function f is *bounded, non-negative and decreasing*. Assume now that the poset P is *bounded*, that is, it has least and greatest elements; then every function $f : P \to \mathbb{R}$ *of bounded variation* is the difference between two bounded, non-negative and decreasing functions. So, for a decomposition $f = g - h$, where g and h are bounded, non-negative and decreasing, we can define $\mathcal{S}_{[a,b]}(f) = \mathcal{S}_{[a,b]}(g) - \mathcal{S}_{[a,b]}(h)$; but for this definition to be independent of the choice of the decomposition $g - h$, the summation must be *additive on* P, that is, for all bounded, non-negative and decreasing functions $g, h : P \to \mathbb{R}$, and all $a, b \in P$ with $a < b$, we must have $\mathcal{S}_{[a,b]}(g + h) = \mathcal{S}_{[a,b]}(g) + \mathcal{S}_{[a,b]}(h)$. Fortunately, the summation is additive on any bounded chain, cartesian product of bounded chains, or any complete sublattice of a cartesian product of complete chains.

Compared to the classical method of [15], my new approach to flat operators is restricted to grey-level and vector images. Let $U = \mathbb{R}^m$ or $U = u_1 \mathbb{Z} \times \cdots \times u_m \mathbb{Z}$ $(u_1, \ldots, u_m > 0)$; let V be a bounded subset of U, having least and greatest elements $\bot, \top$: $\{\bot, \top\} \subseteq V \subseteq [\bot, \top] \subseteq U$. We will define operators acting on images $E \to V$, and the resulting images will be $E \to U$; in other words, we will have operators $V^E \to U^E$. Assuming that $V = [\bot, \top]$ or V is a complete sublattice of $[\bot, \top]$, then the summation $\mathcal{S}$ is additive on V.

Recall the characteristic function $\chi : \mathcal{P}(E) \to \{0,1\}^E$. Call a *binary image transformation* a map $\mathcal{P}(E) \to \mathcal{P}(E)$, for instance a dilation or erosion; now call a *binary image measurement* a map $\mathcal{P}(E) \to K^E$, where K is a *finite* interval in $\mathbb{Z}$; for instance the morphological Laplacian $\chi\delta + \chi\varepsilon - 2\chi\mathbf{id} : \mathcal{P}(E) \to \{-1, 0, 1\}^E$, where $\mathbf{id}$ is the identity operator. Obviously, every binary image transformation ψ corresponds to the binary image measurement $\chi\psi$, with $K = \{0, 1\}$.

Given a binary image measurement μ, define the *no-shift flat extension* μ^{-V} of μ by setting for any image $F : E \to V$ and point $p \in E$:

$$\mu^{-V}(F)(p) = \mathcal{S}\big(\mu(\mathsf{X}_v(F))(p) \mid v \in V\big), \tag{2}$$

in other words we take the summation over the interval $[\bot, \top]$ of the function $V \to \mathbb{R} : v \mapsto \mu(\mathsf{X}_v(F))(p)$; for this definition to make sense, this function $v \mapsto \mu(\mathsf{X}_v(F))(p)$ must be of bounded variation.

Given a binary image transformation ψ, define the *shifted flat extension* ψ^{+V} of ψ by setting for any image $F : E \to V$ and point $p \in E$:

$$\psi^{+V}(F)(p) = \bot + (\chi\psi)^{-V}(F)(p) = \bot + \mathcal{S}\big(\chi\psi(\mathsf{X}_v(F))(p) \mid v \in V\big); \tag{3}$$

again, the summation must be well-defined, in other words, the function $v \mapsto \chi\psi(\mathsf{X}_v(F))(p)$ must be of bounded variation. When ψ is increasing, the new definition coincides with the classical one of [15]: $\psi^{+V}(F) = \psi^V(F)$.

In [29] I showed that several non-increasing morphological operators on grey-level images, such as the Beucher gradient, the morphological Laplacian, the white and black top-hats, and Soille's "unconstrained" hit-or-miss transform are the no-shift flat extensions of their counterparts for binary images. Some general properties of flat operators shown in [15] also hold in the generalized framework. Moreover, the flat extension of a *connected* operator is connected.

In the submitted sequel [30], I studied further properties of generalized flat operators. In the case of increasing binary image transformations, it is known [15] that the flat extensions of the union and intersection of operators are the supremum and infimum of their flat extensions for the componentwise order on V^E.

When the operators are not increasing or do not have binary values, this supremum and infimum takes another form. The flat extension of the composition of an increasing binary image transformation followed by a binary image measurement is the composition of their respective flat extensions, but this does not hold if the binary image transformation is not increasing. For a binary image transformation that is not increasing, commutation with contrast mappings holds only for linear plus constant contrast mappings, and commutation with thresholding always fails. I have introduced the dual form of the flat extension, and then the dual flat extension of an operator coincides with the dual by inversion of its flat extension; under some conditions, it is also equal to the flat extension of the dual by complementation of the operator. Finally, in the case of images with *non-negative* values ($\perp = 0$ or $(0, \ldots, 0)$), convolution by a finite mask of values is the flat extension of the same convolution operator on binary images.

8 Legacy for Future Generations

I write with the perspective that mathematical theory will still remain useful in image processing research, in particular that investigators will not restrict themselves to "deep learning" approaches to problem solving.

I stress that in mathematical morphology, one must have specialists knowing the long mathematical heritage of poset and lattice theory [4]. Matheron, Serra and myself have all sometimes proposed concepts and proved theorems that were already given by mathematicians long ago. So, in [37] Serra and myself strived to rely as much as possible on old references for definitions and theorems. Already in the late 1980's, Heijmans and myself had relied to a great extent on the 1980 edition of [3] for our study of the lattice-theoretical basis of morphological operators [6, 34].

Moreover, old results in lattice theory can be useful in the analysis of some problems. In my investigation of flat operators [15, 16], I relied on two different characterizations of completely distributive complete lattices; in particular the second one [16] had been given by Raney in 1960 and much later rediscovered by Matheron under the name of "monoseparation".

In my study of orders on partial partitions, I found that for a finite space E, the growth from an initial partial partition π_1 to a final one π_2 involves indecomposable elementary steps of different types (creating a singleton block, increasing a block by one point, merging two blocks, etc.), and each type of step appears a constant number of times in the growth. This is similar to the Jordan-Hölder theorem in group theory. So in [28] I gave the most general definition of upper or lower semimodularity in a poset, and proved a Jordan-Hölder theorem on maximal chains, from which one can derive as examples the same theorem in group theory, the above property of my orders on partial partitions, and also a previous similar result in the poset of all closure ranges on a poset (dual of the poset of closure operators on that poset) [20].

Many other concepts and results from old papers on lattice theory could find their use in mathematical morphology, and possibly in discrete topology. Some young researchers should investigate these links in depth.

References

1. Birkhoff, G.: Lattice Theory, American Mathematical Society Colloquium Publications, vol. 25, 3rd edn., 8th printing. American Mathematical Society, Providence (1995)
2. Bloch, I., Heijmans, H., Ronse, C.: Mathematical morphology. In: Aiello, M., Pratt-Hartmann, I., van Benthem, J. (eds.) Handbook of Spatial Logics, pp. 857–944. Springer, Heidelberg (2007). https://doi.org/10.1007/978-1-4020-5587-4_14
3. Gierz, G., Hofmann, K., Keimel, K., Lawson, J., Mislove, M., Scott, D.: Continuous Lattices and Domains, Encyclopedia of Mathematics and its Applications, vol. 93. Cambridge University Press (2003)
4. Grätzer, G.: General Lattice Theory, 2nd edn. Birkhäuser, Basel (2003)
5. Heijmans, H., Ronse, C.: Annular filters for binary images. IEEE Trans. Image Process. **8**(10), 1330–1340 (1999). https://doi.org/10.1109/83.791959
6. Heijmans, H., Ronse, C.: The algebraic basis of mathematical morphology - Part I: Dilations and erosions. Comput. Vis. Graph. Image Process. **50**, 245–295 (1990)
7. Naegel, B., Passat, N., Ronse, C.: Grey-level hit-or-miss transforms - Part I: Unified theory. Pattern Recogn. **40**(2), 635–647 (2007)
8. Naegel, B., Passat, N., Ronse, C.: Grey-level hit-or-miss transforms- Part II: Application to angiographic image processing. Pattern Recogn. **40**(2), 648–658 (2007)
9. Ronse, C.: Feedback Shift Registers. LNCS, vol. 169. Springer, Heidelberg (1984). https://doi.org/10.1007/BFb0023974
10. Ronse, C.: A topological characterization of thinning. Theor. Comput. Sci. **43**, 31–41 (1986)
11. Ronse, C.: Minimal test patterns for connectivity preservation in parallel thinning algorithms for binary digital images. Discret. Appl. Math. **21**(1), 67–79 (1988)
12. Ronse, C.: Order-configuration functions: mathematical characterizations and applications to digital signal and image processing. Inf. Sci. **50**(3), 275–327 (1990)
13. Ronse, C.: A lattice-theoretical morphological view on template extraction in images. J. Vis. Commun. Image Represent. **7**(3), 273–295 (1996). https://doi.org/10.1006/jvci.1996.0024
14. Ronse, C.: Set-theoretical algebraic approaches to connectivity in continuous or digital spaces. J. Math. Imaging Vis. **8**(1), 41–58 (1998)
15. Ronse, C.: Flat morphology on power lattices. J. Math. Imaging Vis. **26**(1/2), 185–216 (2006). https://doi.org/10.1007/s10851-006-8304-1
16. Ronse, C.: Anamorphoses and flat morphological operators on power lattices. Acta Appl. Math. **103**(1), 59–85 (2008). https://doi.org/10.1007/s10440-008-9219-1
17. Ronse, C.: Partial partitions, partial connections and connective segmentation. J. Math. Imaging Vis. **32**(2), 97–125 (2008). https://doi.org/10.1007/s10851-008-0090-5
18. Ronse, C.: Reconstructing masks from markers in non-distributive lattices. Appl. Algebra Eng. Commun. Comput. **19**(1), 51–85 (2008). https://doi.org/10.1007/s00200-008-0064-2
19. Ronse, C.: Adjunctions on the lattices of partitions and of partial partitions. Appl. Algebra Eng. Commun. Comput. **21**(5), 343–396 (2010). https://doi.org/10.1007/s00200-010-0129-x
20. Ronse, C.: The poset of closure systems on an infinite poset: detachability and semi-modularity. Port. Math. **67**(4), 437–452 (2010). https://doi.org/10.4171/PM/1872
21. Ronse, C.: Idempotent block splitting on partial partitions, I: isotone operators. Order **28**(2), 273–306 (2011). https://doi.org/10.1007/s11083-010-9171-3

22. Ronse, C.: Idempotent block splitting on partial partitions, II: non-isotone operators. Order **28**(2), 307–339 (2011). https://doi.org/10.1007/s11083-010-9190-0
23. Ronse, C.: Orders on partial partitions and maximal partitioning of sets. In: Soille, P., Pesaresi, M., Ouzounis, G.K. (eds.) ISMM 2011. LNCS, vol. 6671, pp. 49–60. Springer, Heidelberg (2011). https://doi.org/10.1007/978-3-642-21569-8_5
24. Ronse, C.: Closures on partial partitions from closures on sets. Math. Slovaca **63**(5), 959–978 (2013). https://doi.org/10.2478/s12175-013-0147-9
25. Ronse, C.: Ordering partial partitions for image segmentation and filtering: merging, creating and inflating blocks. J. Math. Imaging Vis. **49**(1), 202–233 (2014). https://doi.org/10.1007/s10851-013-0455-2
26. Ronse, C.: Orders on partial partitions based on block apportioning. Acta Appl. Math. **141**(1), 69–105 (2016). https://doi.org/10.1007/s10440-014-0004-z
27. Ronse, C.: Orders for simplifying partial partitions. J. Math. Imaging Vis. **58**(3), 382–410 (2017). https://doi.org/10.1007/s10851-017-0717-5
28. Ronse, C.: Semimodularity and the Jordan-Hölder theorem in posets, with applications to partial partitions. J. Algebraic Combin. **50**(3), 255–280 (2019). https://doi.org/10.1007/s10801-018-0852-0
29. Ronse, C.: Flat morphological operators from non-increasing set operators, I: general theory. Math. Morphol. Theory Appl. **5**(1), 70–107 (2021). https://doi.org/10.1515/mathm-2020-0109
30. Ronse, C.: Generalised flat morphology, II: more properties and hybrid operators. J. Math. Imaging Vis. (2022, submitted, under review)
31. Ronse, C., Agnus, V.: Geodesy on label images, and applications to video sequence processing. J. Vis. Commun. Image Represent. **19**(6), 392–408 (2008). https://doi.org/10.1016/j.jvcir.2008.04.002
32. Ronse, C., Devijver, P.: Connected Components in Binary Images: The Detection Problem. Research Studies Press (1984)
33. Ronse, C., Heijmans, H.: A lattice-theoretical framework for annular filters in morphological image processing. Appl. Algebra Eng. Commun. Comput. **9**(1), 45–89 (1998). https://doi.org/10.1007/s002000050095
34. Ronse, C., Heijmans, H.: The algebraic basis of mathematical morphology - Part II: openings and closings. Comput. Vis. Graph. Image Process.: Image Underst. **54**, 74–97 (1991)
35. Ronse, C., Mazo, L., Tajine, M.: Correspondence between topological and discrete connectivities in Hausdorff discretization. Math. Morphol. Theory Appl. **3**(1), 1–28 (2019). https://doi.org/10.1515/mathm-2019-0001
36. Ronse, C., Serra, J.: Geodesy and connectivity in lattices. Fund. Inform. **46**(4), 349–395 (2001)
37. Ronse, C., Serra, J.: Algebraic foundations of morphology. In: Najman, L., Talbot, H. (eds.) Mathematical Morphology: From Theory to Applications, chap. 2, pp. 35–80. ISTE/Wiley (2010)
38. Ronse, C., Tajine, M.: Discretization in Hausdorff space. J. Math. Imaging Vis. **12**(3), 219–242 (2000). https://doi.org/10.1023/A:1008366032284
39. Ronse, C., Tajine, M.: Hausdorff discretization for cellular distances, and its relation to cover and supercover discretizations. J. Vis. Commun. Image Represent. **12**(2), 169–200 (2001). https://doi.org/10.1006/jvci.2000.0458
40. Serra, J. (ed.): Image Analysis and Mathematical Morphology, II: Theoretical Advances. Academic Press, London (1988)
41. Serra, J.: Connectivity on complete lattices. J. Math. Imaging Vis. **9**(3), 231–251 (1998)

42. Serra, J.: A lattice approach to image segmentation. J. Math. Imaging Vis. **24**(1), 83–130 (2006)
43. Serra, J.: Grain building ordering. In: Soille, P., Pesaresi, M., Ouzounis, G.K. (eds.) ISMM 2011. LNCS, vol. 6671, pp. 37–48. Springer, Heidelberg (2011). https://doi.org/10.1007/978-3-642-21569-8_4
44. Tajine, M., Ronse, C.: Topological properties of Hausdorff discretizations. In: Goutsias, J., Vincent, L., Bloomberg, D.S. (ed.) Mathematical Morphology and its Applications to Image and Signal Processing, ISMM 2000, pp. 41–50. Kluwer Academic Publishers, Palo Alto (2000)
45. Wendt, P., Coyle, E., Gallagher, N.: Stack filters. IEEE Trans. Acoust. Speech Signal Process. **34**(4), 898–911 (1986)

Hybrid Artificial Intelligence for Knowledge Representation and Model-Based Medical Image Understanding - Towards Explainability

Isabelle Bloch[(✉)]

Sorbonne Université, CNRS, LIP6, Paris, France
isabelle.bloch@sorbonne-universite.fr

Abstract. In this paper, we advocate that combining several frameworks in artificial intelligence, adopting a hybrid point of view for both knowledge data representation and reasoning, offers opportunities towards explainability. This idea is illustrated on the example of image understanding, in particular in medical imaging, formulated as a spatial reasoning problem.

Keywords: Hybrid artificial intelligence · Explainable Artificial Intelligence (XAI) · Spatial reasoning · Image understanding

1 Introduction: Hybrid Artificial Intelligence

While symbolic and statistical machine learning methods for artificial intelligence (AI) have been developed rather independently for decades, with alternated predominance of one or the other along time, a trend is to merge both types of approaches. Examples include neuro-symbolic approaches (see e.g. [18,25,26, 32,34]), among others. Here hybrid artificial intelligence is intended in a broader sense, as the combination of several AI methods, whatever their type. These methods may belong to the domains of abstract knowledge representation and formal reasoning, based on logics, structural representations (such as graphs and hypergraphs, ontologies, concept lattices...), fuzzy sets, machine learning.

Such combinations take inspiration from cognitive functions. Roughly speaking, according to Kahneman [31] who distinguished two systems for thinking, named system 1 and system 2, we may consider, from an AI point of view, modeling system 1 by deep learning and system 2 by symbolic reasoning. Developing neuro-symbolic approaches is a new trend to combine the two systems (see e.g. [32]). But again, more theories will be committed in our view of hybrid AI.

This work was partly supported by the author's chair in Artificial Intelligence (Sorbonne Université and SCAI). A part of the work mentioned in this paper was performed while the author was with LTCI, Télécom Paris, Institut Polytechnique de Paris.

The aim of this paper is not to propose new methods for hybrid AI, but rather highlight how this way of thinking and designing AI systems offers opportunities towards explainability, in the field of explainable AI (XAI), and as a mean to maintain the link between knowledge and data. In that domain too, the two main branches are developed quite independently, with early work (e.g. Peirce at the end of the 19th century) focusing on logical reasoning based on abduction on the one hand, versus recent methods focusing on features or data most involved in a decision on the other hand (to name but a few). In the first paradigm, knowledge is represented by symbols, in a given logic, and the reasoning power of this logic plays then a major role. Reasoning is based on axioms, theories and inference rules, leading to provable, non-refutable conclusions. In the second paradigm, where data and experience play the major role, statistical guarantees can be achieved, but conclusions are potentially refutable.

As an example, these ideas are illustrated in the field of image understanding, formulated as a spatial reasoning problem, as described in Sect. 2. Examples of combinations of different AI methods are given both for knowledge and data representation in Sect. 3, and for reasoning in Sect. 4. These methods find concrete applications in medical imaging (only briefly mentioned in this paper). Finally a short discussion on open research directions concludes the paper (Sect. 5).

Although no technical details are given in this paper, it is noticeable that mathematical morphology is a useful theory for knowledge representation (in particular spatial relations, in conjunction with fuzzy set theory) and for reasoning (e.g. abductive reasoning in various logics), as shown in our previous work (see e.g. [2,6–8,11]). For example, defining the region of space, in an image, satisfying some spatial relation with respect to a reference object can be obtained by dilating the reference object (whether crisp or fuzzy) with a structuring element modeling the semantics of the spatial relation of interest. Another example is the use of erosion or derived operators to provide explanations to observations according to a knowledge base (i.e. abductive reasoning), by applying these operators to the set of models of logical formulas or to a concept lattice.

2   Image Understanding and Spatial Reasoning

Image understanding refers, at the simplest level, to the problem of recognizing an object or structure, or several objects in an image, either real, as an observation of a part of the real world, or synthetic. More generally, relations between these objects should be considered, towards a global recognition of the scene and a higher level interpretation. The question of semantics is central since it is not directly in the image, but should be inferred based on visual features. We advocate that knowledge should be involved in this process. Indeed, while purely data driven approaches have proved to be powerful in image and computer vision problems, with sometimes impressive results, they still require a good accessibility to numerous and annotated data, which is not always possible and which induces high costs (in terms of both human interactions and computation). Knowledge and models have then an important role to play. Image understanding is then formulated as a spatial reasoning problem, combining representations of data and knowledge, pertaining to both objects and relations between objects (in particular spatial relations), and reasoning on them.

Spatial reasoning has been largely developed in symbolic AI, based mostly on logics and benefiting from the reasoning apparatus of these logics [1]. It has been much less developed for image understanding, where purely symbolic approaches are limited to account for numerical information. This again votes for hybrid approaches. Spatial reasoning evolved from purely qualitative and symbolic approaches to more and more hybrid methods, involving methods from mathematical morphology, fuzzy sets, graphs, machine learning, etc. to gain in expressivity (sometimes at the price of increased complexity). As an example, let us mention region connection calculus (RCC) that was first proposed in logical frameworks (first order, modal), and then augmented with fuzzy sets to handle imprecision, with mathematical morphology, with lattice-based reasoning, etc. [1, 9, 33, 39–41]. The main ingredients in spatial reasoning include knowledge representation, imprecision representation and management, fusion of heterogeneous information (whether knowledge or data), reasoning and decision making. Approaches for spatial reasoning take a lot of inspiration from work in philosophy, linguistics, human perception, cognition, neuro-imaging, art, etc. (see e.g. a related discussion for the case of spatial distances in [5]).

Models for image understanding are particularly useful to represent, in a formal way, knowledge (about the domain, the scene content and in particular its structure), image information (type of acquisition, geometry, characteristics of signal and noise...), the potential imperfections of knowledge and data (imprecision, uncertainty, incompleteness...), as well as the combination of knowledge and image information. These models are then included in algorithms to guide image understanding in concrete applications. Conversely, models can be built from data, for instance to infer knowledge, or to provide a digital twin of a patient as a 3D model, useful to plan a surgery or a therapy, as well as to explain the plan.

An important issue is the semantic gap [42], with the following question: how to link visual percepts from the images to symbolic descriptions? In artificial intelligence, this is close to the notions known as the anchoring or symbol grounding problem [15, 29]. Solving the semantic gap issue has bidirectional consequences: on the one hand, it allows moving from a concept to its instantiation in the image (or feature) space, as a guide during spatial reasoning. On the other hand, it is part of the explainability, since it links results inferred from the image to concepts related to prior knowledge. For instance, anatomical knowledge says that the heart is between the lungs. Since the heart might be difficult to recognize directly in a medical image (e.g. a non-enhanced CT image), we may rely on its relative position with respect to the lungs (which are easier to detect in such images) to perform the task. Conversely, we can explain the recognition of an image region as the heart *because* it is between the lungs.

3 Representations

Representations of spatial entities can take various forms, either in the spatial domain (region, key points, bounding box...), or abstractly, as in region connection calculus (RCC), as formulas in a given logic. Semi-quantitative (or semi-qualitative) representations as fuzzy sets (in either domain) constitute a good midway and can accommodate both numerical and symbolic representations [46]. This becomes even more significant when considering representations of spatial relations. The usefulness of fuzzy representations of spatial relations was already advocated in the 1970's [24], to account for their intrinsic imprecision in concepts such as "close to", "to the right of", that are nevertheless perfectly understandable in a given context. In our previous work, we designed mathematical models of several relations (set theoretical, topological, distances, directional relations, more complex relations such as between, along, parallel...) by combining formalisms from mathematical morphology and fuzzy sets. The common underlying structure is the one of complete lattices, that allows instantiating the definitions, with the very same formalism, in different frameworks: sets, fuzzy sets, graphs and hypergraphs, formal concept lattices, conceptual graphs, ontologies... Note that most of these frameworks carry structural information, useful for instance to represent the spatial arrangement of objects in a scene and in an image. To take a simple example, a graph can represent this structure, where vertices correspond to objects (e.g. anatomical structures in medical images) and edges correspond to relations between objects (e.g. contrast between two structures in a given imaging modality, relative position between objects...), this graph being enhanced with the (fuzzy) representations of objects and their properties, and of relations. For instance, the representation of a spatial relation can be abstract, as extracted from an ontology for example, or linked to the concrete domain of an image (degree of satisfaction of the relation, region of space where the relation to some object is satisfied...). Other structured representations of knowledge (including spatial knowledge) may rely on grammars, decision trees, relational algebras on temporal or spatial configurations, or graphical models. They can also benefit from a fuzzy modeling layer, to cope with imprecision.

The relevance of fuzzy sets for knowledge representation relies in their capability to capture linguistic as well as quantitative knowledge and information. A useful notion is the one of linguistic variable [47], where symbolic values (defined at an ontological level) have semantics defined by membership functions on a concrete domain (at the image or features level). The membership functions and their parameters can be handcrafted, according to some expert knowledge on the application domain. They can also be learned, for instance from annotated data [4]. The advantage of such representations is that linguistic characterizations may be less specific than numerical ones (and therefore need less information). Their two levels (syntactic and semantic) allow on the one hand for approximate modeling of vague concepts and reasoning on them, and allow on the other hand solving the semantic gap issue by providing semantics in concrete domains, according to each specific context.

4   Reasoning

Based on the previous representations, the reasoning part takes various forms, separately or in combination, again in the spirit of hybrid AI. Let us mention a few, mostly from our previous work, which led to applications in medical imaging, in particular for brain structure recognition[1]: matching between a model and an image based on graph representations [3,12,22,38]; sequential spatial reasoning mimicking the usual cognitive process where one may focus on an object that is easy to detect and to recognize, and then move progressively to more and more difficult objects by exploring the space based on the spatial relations with respect to previously recognized objects [10,14,19,23]; exploration of the whole space and reducing progressively the potential region for each object, again mimicking a type of cognitive process, for instance by expressing the task as a constraint satisfaction problem [21,37]; logical reasoning based on abduction, to find the best explanations to the observations according to the available knowledge [45]; logical reasoning driven by an ontology [30].

In all these methods, an important feature is the combination of several approaches within the framework of hybrid AI, with the aim of explainability. Abstract knowledge representation and formal reasoning (typically using logics) allow building a knowledge base representing prior information (on anatomy for the considered examples), and to reason on it. Structural representations (graphs and hypergraphs, ontologies, conceptual graphs, concept lattices...) are frameworks to convert expert knowledge on the spatial organization of organs into operational computational models. Medical images may suffer from imprecision, due to their discrete nature, to the potential partial volume effect, to reconstruction algorithms or to image processing steps. Knowledge is also often expressed in vague terms, although perfectly intelligible by a human. All these imperfections are modeled by resorting to fuzzy sets theory. In particular, the semantic gap issue is solved by using linguistic variables, which link concepts to visual percepts in the images, or more generally representations as fuzzy sets in a concrete domain. This is a very crucial point to maintain the links between data and knowledge, and is indeed key to explainability. Considering the example of structure recognition based on spatial reasoning, explanations become natural by identifying the spatial relations that actually play a role in the recognition, as mentioned above. Furthermore, from a knowledge base on anatomy, expressed in some logics, and from segmentation and recognition results, higher level interpretations of an image can be derived using abductive reasoning [45]. The language in which the knowledge is expressed should be defined according to the granularity level expected for the interpretation and to whom the description is dedicated (the explainee). For instance the description of the content of a pathological brain image will depend on whether the explainee is anyone (without assuming any particular expertise), the patient itself, or a medical expert

---

[1] These are only examples and similar approaches have been developed in other application domains, such as satellite imaging, video, music representations, etc.

who wants to make a decision guided by this description to interact with other experts.

Now, considering the huge recent developments in machine learning, and in particular deep learning, a recent trend is to combine such approaches with knowledge driven methods. This can be done at several levels (see e.g. [44]): to enhance the input (e.g. by including in the input of a neural network a result of some image processing method as in [17]); as regularization terms in the loss function (e.g. to force the satisfaction of some relations) or to focus attention on specific patches based on geometric or topological information (e.g. vessel tree [43]); or as post-processing to improve results (e.g. [13]). Again one of the advantages of such hybrid approaches is to improve interpretability and explainability. This is particularly important in medical imaging to increase the confidence the user may have in an approach based on deep learning, and therefore to increase the adoption of such techniques.

5 Discussion

To go further in the field of hybrid AI and XAI for image understanding, principles expressed and discussed more generally in AI could be instantiated in this particular domain of application, and pave the way for new research directions.

This starts with the definition of interpretability and explainability. An interesting distinction is proposed in [20], where interpretability is defined via the composition of elements that are meaningful for humans, while explanation is strongly related to causality, and understanding is linked to unifying diversity under a commun principle (this is maybe somewhat different when interpreting an individual image as in medical imaging).

Seeing explanations as causality has been widely addressed, in particular by Halpern and Pearl [27,28], and by Miller [35,36], where structural models play a major role. Notions such as contrast and relevance are put to the fore, and would be also important to consider in image understanding. For instance, explaining why a decision was made by an algorithm and not another decision is a way to make explanations more convincing. The level of explanation should depend on the explainee, as mentioned before, and a deeper study of this aspect could take inspiration from the work on intelligibility in [16] (for instance based on projections on a given vocabulary). This goes with the idea of human-centered evaluation of AI systems.

It has been advocated in [34] that new research should aim at developing *a hybrid, knowledge driven, reasoning based approach, centered around cognitive models, that could provide the substrate for a richer, more robust AI than is currently possible.* This is exactly what research in image understanding based on hybrid AI is trying to do, but still at a modest level.

Finally it would be interesting to investigate more deeply to which extent hybrid AI and XAI could help answering questions related to ethics, in particular in radiology.

Acknowledgements. The author would like to emphasize that the ideas summarized in this paper benefited from many joint works with post-doctoral researchers and PhD candidates, with colleagues in universities and research centers in several countries, with university hospitals, and with industrial partners. Thanks to all of them!

References

1. Aiello, M., Pratt-Hartmann, I., van Benthem, J. (ed.).: Handbook of Spatial Logic. Springer, Cham (2007). https://doi.org/10.1007/978-1-4020-5587-4
2. Aiguier, M., Atif, J., Bloch, I., Pino Pérez, R.: Explanatory relations in arbitrary logics based on satisfaction systems, cutting and retraction. Int. J. Approximate Reasoning **102**, 1–20 (2018)
3. Aldea, E., Bloch, I.: Toward a better integration of spatial relations in learning with graphical models. In: H. Briand, F. Guillet, G.R., Zighed, D. (eds.) Advances in Knowledge Discovery and Management. Studies in Computational Intelligence, vol. 292. pp. 77–94. Springer, Heidelberg (2010). https://doi.org/10.1007/978-3-642-00580-0_5
4. Atif, J., Hudelot, C., Fouquier, G., Bloch, I., Angelini, E.: From Generic Knowledge to Specific Reasoning for Medical Image Interpretation using Graph-based Representations. In: International Joint Conference on Artificial Intelligence IJCAI 2007, pp. 224–229. Hyderabad, India (2007)
5. Bloch, I.: On fuzzy spatial distances. In: Hawkes, P. (ed.) Advances in Imaging and Electron Physics, vol. 128, pp. 51–122. Elsevier, Amsterdam (2003)
6. Bloch, I.: Fuzzy spatial relationships for image processing and interpretation: a review. Image Vision Comput. **23**(2), 89–110 (2005)
7. Bloch, I.: Spatial reasoning under imprecision using fuzzy set theory, formal logics and mathematical morphology. Int. J. Approximate Reasoning **41**(2), 77–95 (2006)
8. Bloch, I.: Mathematical morphology and spatial reasoning: fuzzy and bipolar setting. TWMS J. Pure Appl. Math. **12**(1), 104–125 (2021). Special Issue on Fuzzy Sets in Dealing with Imprecision and Uncertainty: Past and Future Dedicated to the memory of Lotfi A. Zadeh
9. Bloch, I.: Modeling imprecise and bipolar algebraic and topological relations using morphological dilations. Math. Morphol. Theory Appl. **5**(1), 1–20 (2021)
10. Bloch, I., Géraud, T., Maître, H.: Representation and fusion of heterogeneous fuzzy information in the 3D space for model-based structural recognition - application to 3D brain imaging. Artif. Intell. **148**, 141–175 (2003)
11. Bloch, I., Lang, J., Pino Pérez, R., Uzcátegui, C.: Morphologic for knowledge dynamics: revision, fusion, abduction. Technical report arXiv:1802.05142, arXiv cs.AI (2018)
12. Cesar, R., Bengoetxea, E., Bloch, I., Larranaga, P.: Inexact graph matching for model-based recognition: evaluation and comparison of optimization algorithms. Pattern Recognit. **38**, 2099–2113 (2005)
13. Chopin, J., Fasquel, J.B., Mouchère, H., Dahyot, R., Bloch, I.: Improving semantic segmentation with graph-based structural knowledge. In: El Yacoubi, M., Granger, E., Yuen, P.C., Pal, U., Vincent, N. (eds.) International Conference on Pattern Recognition and Artificial Intelligence. ICPRAI 2022. Lecture Notes in Computer Science, vol. 13363, pp. 173–184. Springer, Cham (2022). https://doi.org/10.1007/978-3-031-09037-0_15

14. Colliot, O., Camara, O., Bloch, I.: Integration of fuzzy spatial relations in deformable models - application to brain MRI segmentation. Pattern Recognit. **39**, 1401–1414 (2006)
15. Coradeschi, S., Saffiotti, A.: Anchoring symbols to vision data by fuzzy logic. In: Hunter, A., Parsons, S. (eds.) ECSQARU'99. LNCS, vol. 1638, pp. 104–115. Springer, London (1999)
16. Coste-Marquis, S., Marquis, P.: From explanations to intelligible explanations. In: 1st International Workshop on Explainable Logic-Based Knowledge Representation (XLoKR 2020) (2020)
17. Couteaux, V., et al.: Automatic knee meniscus tear detection and orientation classification with Mask-RCNN. Diagn. Interv. Imaging **100**, 235–242 (2019)
18. De Raedt, L., Dumancic, S., Manhaeve, R., Marra, G.: From statistical relational to neuro-symbolic artificial intelligence. In: Bessiere, C. (ed.) Twenty-Ninth International Joint Conference on Artificial Intelligence, IJCAI-20, pp. 4943–4950 (2020)
19. Delmonte, A., Mercier, C., Pallud, J., Bloch, I., Gori, P.: White matter multi-resolution segmentation using fuzzy set theory. In: IEEE International Symposium on Biomedical Imaging (ISBI), pp. 459–462. Venice, Italy (2019)
20. Denis, C., Varenne, F.: Interprétabilité et explicabilité de phénomènes prédits par de l'apprentissage machine. Rev. Ouverte Intell. Artif. **3**, 287–310 (2022)
21. Deruyver, A., Hodé, Y.: Constraint satisfaction problem with bilevel constraint: application to interpretation of over-segmented images. Artif. Intell. **93**(1–2), 321–335 (1997)
22. Fasquel, J., Delanoue, N.: A graph based image interpretation method using a priori qualitative inclusion and photometric relationships. IEEE Trans. Pattern Anal. Mach. Intell. **41**(5), 1043–1055 (2019)
23. Fouquier, G., Atif, J., Bloch, I.: Sequential model-based segmentation and recognition of image structures driven by visual features and spatial relations. Comput. Vis. Image Underst. **116**(1), 146–165 (2012)
24. Freeman, J.: The modelling of spatial relations. Comput. Graph. Image Process. **4**(2), 156–171 (1975)
25. d'Avila Garcez, A., Lamb, L.C.: Neurosymbolic AI: the 3rd wave. CoRR abs/2012.05876 (2020)
26. Garnelo, M., Shanahan, M.: Reconciling deep learning with symbolic artificial intelligence: representing objects and relations. Curr. Opin. Behav. Sci. **29**, 17–23 (2019)
27. Halpern, J.Y., Pearl, J.: Causes and explanations: a structural-model approach. Part I: causes. Br. J. Philos. Sci. **56**(4), 843–887 (2005)
28. Halpern, J.Y., Pearl, J.: Causes and explanations: a structural-model approach. Part II: explanations. Br. J. Philos. Sci. **56**(4), 889–911 (2005)
29. Harnad, S.: The symbol grounding problem. Physica **42**, 335–346 (1990)
30. Hudelot, C., Atif, J., Bloch, I.: Fuzzy spatial relation ontology for image interpretation. Fuzzy Sets Syst. **159**, 1929–1951 (2008)
31. Kahneman, D.: Thinking. Fast and Slow, Penguin, New York (2012)
32. Kautz, H.: The third AI summer: AAAI Robert S. Engelmore Memorial Lect. AI Mag. **43**(1), 93–104 (2022)
33. Landini, G., Galton, A., Randell, D., Fouad., S.: Novel applications of discrete mereotopology to mathematical morphology: signal processing: image communications **76**, 109–117 (2019)
34. Marcus, G.: The next decade in AI: four steps towards robust artificial intelligence. CoRR abs/2002.06177 (2020)

35. Miller, T.: Explanation in artificial intelligence: insights from the social sciences. Artif. Intell. **267**, 1–38 (2019)
36. Miller, T.: Contrastive explanation: a structural-model approach. Knowl. Eng. Rev. **36**, E14 (2021)
37. Nempont, O., Atif, J., Bloch, I.: A constraint propagation approach to structural model based image segmentation and recognition. Inf. Sci. **246**, 1–27 (2013)
38. Perchant, A., Bloch, I.: Fuzzy morphisms between graphs. Fuzzy Sets Syst. **128**(2), 149–168 (2002)
39. Randell, D., Cui, Z., Cohn, A.: A Spatial Logic based on Regions and Connection. In: Nebel, B., Rich, C., Swartout, W. (eds.) Principles of Knowledge Representation and Reasoning KR'92, pp. 165–176. Kaufmann, San Mateo, CA (1992)
40. Schockaert, S., De Cock, M., Cornelis, C., Kerre, E.E.: Fuzzy region connection calculus: representing vague topological information. Int. J. Approximate Reasoning **48**(1), 314–331 (2008)
41. Schockaert, S., De Cock, M., Kerre, E.E.: Spatial reasoning in a fuzzy region connection calculus. Artif. Intell. **173**(2), 258–298 (2009)
42. Smeulders, A., Worring, M., Santini, S., Gupta, A., Jain, R.: Content-based image retrieval at the end of the early years. IEEE Trans. Pattern Anal. Mach. Intell. **22**(12), 1349–1380 (2000)
43. Virzi, A., et al.: Segmentation of pelvic vessels in pediatric MRI using a patch-based deep learning approach. In: Melbourne, A., et al. (eds.) PIPPI/DATRA -2018. LNCS, vol. 11076, pp. 97–106. Springer, Cham (2018). https://doi.org/10.1007/978-3-030-00807-9_10
44. Xie, X., Niu, J., Liu, X., Chen, Z., Tang, S., Yu, S.: A survey on incorporating domain knowledge into deep learning for medical image analysis. Med. Image Anal. **69**, 101985 (2021)
45. Yang, Y., Atif, J., Bloch, I.: Abductive reasoning using tableau methods for high-level image interpretation. In: Hölldobler, S., Krötzsch, M., Peñaloza, R., Rudolph, S. (eds.) KI 2015. LNCS (LNAI), vol. 9324, pp. 356–365. Springer, Cham (2015). https://doi.org/10.1007/978-3-319-24489-1_34
46. Zadeh, L.A.: Fuzzy Sets. Inf. Control **8**, 338–353 (1965)
47. Zadeh, L.A.: The concept of a linguistic variable and its application to approximate reasoning. Inf. Sci. **8**, 199–249 (1975)

Digital Geometry, Mathematical Morphology, and Discrete Optimization: A Survey

Christer Oscar Kiselman[✉]

Department of Information Technology, Uppsala University,
P. O. Box 337, SE-751 05 Uppsala, Sweden
christer.kiselman@it.uu.se

Abstract. We study difficulties that appear when well-established definitions and results in Euclidean geometry, especially in the theory of convex sets and functions in vector spaces, are translated into a discrete setting. Solutions to these problems are sketched.

1 Introduction

The purpose of this note is to briefly present three distinct but related branches of science: digital geometry, mathematical morphology, and discrete optimization. They are related by a common mindset and also by the many fields of knowledge where they can be successfully applied and provide reliable solutions. They also have in common the fact that the advent of computers has made it possible to actually perform operations that have been studied for a long time.

Digital geometry is, in simple terms, the geometry of the computer screen. While Euclid's straight lines and planes have been studied for more than two millennia, the concept of a digital straight line was clarified as late as in 1974 by Azriel Rosenfeld (1931–2004), a pioneer in the field. So work in this area has, strictly speaking, not been going on for more than fifty years.

Mathematical morphology is, in equally simple words, the theory and practice of transformations of sets and functions with an emphasis on their shapes, among them the Boolean operations, but far from just these. In many cases, these transformations have been known for a long time, but they have come into focus for the same reason as digital geometry: the transformations can actually be performed on a computer. The concept of shape in geometry is fundamental: when do two figures have similar shapes? Approaching this seemingly innocent question will reveal deep problems.

Among the operations of great importance here, infimal convolution and discrete convolution on a group stand out.

Discrete optimization is a natural companion to digital geometry: we look for the best solution to a problem among several possibilites, but now in a discrete set.

My book (2022) might serve as an introduction to all three fields.

É. Baudrier et al. (Eds.): DGMM 2022, LNCS 13493, pp. 26–31, 2022.
https://doi.org/10.1007/978-3-031-19897-7_3

2   Digital Geometry

Discrete objects, like carpets and mosaics, have been around for thousands of years. A carpet can consist of many knots, a mosaic of many little stones, tessellas, but only finite in number. So this is the beginning of a study of locally finite spaces.

Between two distinct points in the plane there is a third, the midpoint, but on the computer screen two pixels can be neighbors without any pixel between them. This makes the logic of digital geometry quite different from Euclidean geometry. So we have to start from scratch in our thinking. A convenient model for the computer screen is $\mathbf{Z}^2$, the set of points in the plane with integer coordinates— although the screen is bounded, it is easier to work with a digital plane without borders.

The concept of a straight line, apart from the easiest ones (which are the horizontal, the vertical and the diagonal lines), is highly non-trivial. When we come to planes in three-space and in higher dimensions, the difficulties in finding useful and consistent definitions become even harder. There are several definitions of a two-plane in three-space, all useful and each with its merits.

The two lines defined in $\mathbf{Z}^2$ by $y = x$ and $y = -x+1$ do not intersect although they are not parallel. Is this disturbing?

The two distinct parallel lines defined by $y = x$ and $y = \lfloor x + 1/2 \rfloor$, where $\lfloor t \rfloor$ is defined by the floor function

$$\mathbf{R} \ni t \mapsto \lfloor t \rfloor \in \mathbf{Z}, \qquad t - 1 < \lfloor t \rfloor \leqslant t, \qquad t \in \mathbf{R},$$

intersect in infinitely many points.

References for digital geometry include the doctoral thesis by Reveillès (1991) and the book by Klette and Rosenfeld (2004).

3   Mathematical Morphology

The branch of science now called mathematical morphology started with the study of minerals: the task was to describe shapes of particles taken from mines in a more precise way than by just indicating their size. The theory was (quite logically) born at *l'École nationale supérieure des mines de Paris*, founded already in 1783 and now with campuses in Fontainebleau and Paris. So we are talking about descriptions of shape and comparisons of shapes.

The two creators of this most successful branch of science were Georges Matheron (1930–2000) and Jean Serra, both at Fontainebleau at the time—their institution is now called *le Centre de morphologie mathématique* to honor the new branch of science. The current director of the center is Jesús Angulo.

The most important early books on the subject are the books by Matheron & Serra (2002), Matheron (1975), Serra (1982) and the one edited by Serra (1988). Among more recent sources, the one edited by Najman and Talbot (2008a; 2008b) is important.

4   Discrete Optimization

To optimize means 'to find the best'. Is the lowest or the highest price the best? First we must decided which function to consider.

We can usually not buy half an apple or half a car; quite often, only entire numbers are allowed when we want to find an optimal solution to a problem.

Convex functions are indeed convenient when it comes to minimizing: a local minimum is automatically also a global minimum. But when we come to functions defined on the integers, this simple fact is no longer valid.

5   Discretizing Convexity

In no other field do the difficultics to discretize definitions and results come clearer to the surface than for convexity. First of all, the theory of convex functions in real vector spaces is an extremely well-developed branch of science with forceful methods and a lot of important applications. It is therefore of special interest to see what happens when we endeavour to make sense of discrete versions of convexity.

Let us look at three fundamental properties of convexity in vector spaces:

1. The image under a linear mapping of a convex set is convex. Equivalently, the marginal function of a convex function is convex.
2. A local minimum of a convex function is global.
3. Between two given disjoint convex sets there is a separating hyperplane.

The first two are easily seen to be valid; the third is the Hahn–Banach Theorem in finite dimension.

Let us note that these three important properties fail conspicuously in a discrete setting:

1. Look at the function $f\colon \mathbf{Z}^2 \to \mathbf{Z}$ defined by $f(x,y) = |x - my|$ for $(x,y) \in \mathbf{Z}^2$. Its marginal function $f_{\mathrm{marg}}\colon \mathbf{Z} \to \mathbf{Z}$, defined as

$$f_{\mathrm{marg}}(x) = \inf_{y \in \mathbf{Z}} f(x,y), \qquad x \in \mathbf{Z},$$

   is equal to $|x|$ for $-m \leqslant x \leqslant m$ and given for other values of x by the fact that it is periodic with period 2 m. Thus f_{marg} is a function with teeth as large as we please. It is not reasonable to call such a function convex.
   Still, we are tempted to accept f as a convex function since its extension to $\mathbf{R}$, given by the same formula, is convex. We conclude that it is in no way sufficient to assume that a function on a discrete set like $\mathbf{Z}^n$ has a convex extension to $\mathbf{R}^n$.
2. The function $\mathbf{Z} \ni x \mapsto \lfloor x/m \rfloor$, which is one of several digitalizations of a linear function, thus of a convex function, has a local minimum in quite a large set; indeed it is $\geqslant -1$ in the set $\{x \in \mathbf{Z};\ -m \leqslant x < +\infty\}$, where $m \in \mathbf{N}$ can be as large as we want, but it is not bounded from below.

3. Let us look at the two disjoint sets

$$A = \{(x, y, z) \in \mathbf{Z}^3;\ z \leqslant 1 - 2|x + y - 1|\} \text{ and}$$
$$B = \{(x, y, z) \in \mathbf{Z}^3;\ 2|x - y| \leqslant z\}.$$

A separating plane must have the equation $z = H(x, y)$ for some affine function H, which leads to

$$1 - 2|x + y - 1| \leqslant H(x, y) \leqslant 2|x - y|, \qquad (x, y) \in \mathbf{Z}^2.$$

Then these inequalities must hold also for $(x, y) \in \mathbf{R}^2$, in particular for $x = y = 1/2$, which is not true. (The point $(1/2, 1/2)$ exists in $\mathbf{R}^2$ but not in $\mathbf{Z}^2$.)

There are satisfactory solutions to these three problems—we shall return to these in Sect. 7.

6   Duality in Convexity and Discrete Convexity

By duality I understand a situation when two structures operate against each other. A well-known example is the duality between the Lebesgue spaces $L^p(\mathbf{R}^n)$ and $L^q(\mathbf{R}^n)$ given by the bilinear form

$$L^p(\mathbf{R}^n) \times L^q(\mathbf{R}^n) \ni (f, g) \mapsto \int_{\mathbf{R}^n} f(x)g(x)\mathrm{d}x \in \mathbf{C},$$

where $1 < p < +\infty$ with $1/p + 1/q = 1$ or (p, q) equal to $(1, \infty)$ or $(\infty, 1)$. This duality corresponds in the theory of distributions to the bilinear form

$$\mathscr{D}'(\mathbf{R}^n) \times \mathscr{D}(\mathbf{R}^n) \ni (u, \varphi) \mapsto u(\varphi) \in \mathbf{C}.$$

In convexity theory, the Fenchel transformation is a most successful example of duality.

Given $f \colon \mathbf{R} \to [-\infty, +\infty]$ we define its **Fenchel transform** $\tilde{f}$ by

$$\tilde{f}(\xi) = \sup_{x \in \mathbf{R}^n} (\xi \cdot x - f(x)), \qquad \xi \in \mathbf{R}^n.$$

This function is always convex and lower semicontinuous, and it takes the value $-\infty$ only if it is equal to $-\infty$ everywhere. Taking the transformation twice, we get $\tilde{\tilde{f}} \leqslant f$ with equality if and only if f has the three mentioned properties.

For functions defined on a discrete set like $\mathbf{Z}^n$ we can still define it just by restricting the variable x to $\mathbf{Z}^n$, but as we have seen, this does not suffice to guarantee convexity of marginal functions.

7   Solutions

We shall write $\mathscr{P}(X)$ for the family of all subsets of a given set X and shall denote by $\mathscr{F}(X,Y)$ the family of all mappings $f\colon X \to Y$. Finally, let $\mathscr{W}$ be the family of all finite compositions of difference operators D_a defined by $(D_a f)(x) = f(x+a) - f(x)$.

Definition 7.1. We define two mappings

$$\Phi\colon \mathscr{P}(\mathscr{W}) \to \mathscr{F}(X,Y) \text{ and } \Psi\colon \mathscr{F}(X,Y) \to \mathscr{P}(\mathscr{W})$$

by

$$\Phi(\mathscr{A}) = \{f \in \mathscr{F}(X,Y);\ R(f) \geqslant 0 \text{ for all } R \in \mathscr{A}\}, \qquad \mathscr{A} \in \mathscr{P}(\mathscr{W}).$$

and

$$\Psi(\mathscr{B}) = \{R \in \mathscr{W};\ R(f) \geqslant 0 \text{ for all } f \in \mathscr{B}\}, \qquad \mathscr{B} \in \mathscr{P}(\mathscr{F}(X,Y)).$$

These two mappings are obviously decreasing and the compositions

$$\Psi \circ \Phi\colon \mathscr{P}(\mathscr{W}) \to \mathscr{P}(\mathscr{W}) \text{ and } \Phi \circ \Psi\colon \mathscr{F}(X,Y) \to \mathscr{F}(X,Y)$$

are increasing and larger then the respective identity—we have a Galois correspondance.

Definition 7.2. We shall say that a function $f\colon \mathbf{Z}^2 \to \mathbf{R}$ is $\mathscr{C}$-*convex* if $R(f) \geqslant 0$ for all $R \in \mathscr{C}$, where $\mathscr{C}$ is any subset of $\mathscr{W}$.

Definition 7.3. Convexity in one integer variable. We define

$$(R_1(f))(x) = f(x-1) - 2f(x) + f(x+1), \qquad x \in \mathbf{Z},$$

and $\mathscr{W}_1 = \{R_1\}$.

To be $\mathscr{W}_1$-convex is equivalent to possessing a convex extension defined on all of $\mathbf{R}$. This is the only reasonable definition of convexity on $\mathbf{Z}$.

Definition 7.4. A convexity property in two integer variables. We define

$$(R_{2,b}(f))(x,y) = f(x-1,y+b) - 2f(x,y) + f(x+1,y), \qquad (x,y) \in \mathbf{Z}^2,\ b = -1,0,1,$$

and $\mathscr{W}_2 = \{R_{2,-1}, R_{2,0}, R_{2,1}\}$.

Theorem 7.5. *With $\mathscr{W}_1$ and $\mathscr{W}_2$ as just defined, the marginal function of a $\mathscr{W}_2$-convex function $f\colon \mathbf{Z}^2 \to \mathbf{R}$ is $\mathscr{W}_1$-convex.*

The proof is in my book (2022), Theorem 12.3.1, page 262.

So this gives a satisfactory solution to the problem about convexity of the marginal function of a function of two discrete variables. Similar results hold for functions of any number of variables; see Theorem 12.9.1 on page 280 in my book.

References

Kiselman, C.O.: Elements of Digital Geometry, Mathematical Morphology, and Discrete Optimization, xxiv + 461 p. World Scientific, Singapore (2022)

Matheron, G.: Random Sets and Integral Geometry, xxiii + 261 p. (1975)

Matheron, G., Serra, J.: The birth of mathematical morphology. In: Talbot, H., Beare, R. (eds.), Mathematical Morphology. Proceedings of the VIth International Symposium-ISMM 2002. Sydney, 3–5 April 2002, pp. xii + 441 9. Collingwood VIC CSIRO Publishing (2002)

Najman, L., Talbot, H.: Introduction à la morphologie mathématique. In: Najman, L., Talbot, H. (eds.), vol. 2008, pp. 19–45 (2008)

Najman, L., Talbot, H. (eds.) : Introduction à la morphologie mathématique. 1. Approches déterministes, 264 p. Paris, Lavoisier (2008)

Reveillès, J.P.: Géométrie discrète, calcul en nombres entiers et algorithmique. Strasbourg. Université Louis Pasteur. Thèse d'État presented on 1991 December 20 (1991)

Serra, J.: Image Analysis and Mathematical Morphology, xviii + 461 p. Academic Press, London et al. (1982)

Serra, J., (ed.): Image Analysis and Mathematical Morphology. Volume 2: Theoretical Advances, xvii + 411 p. London et al. Academic Press (1988)

Talbot, H., Beare, R.: Mathematical Morphology. In: Proceedings of the VIth International Symposium–ISMM 2002. Sydney, 3–5 April 2002, xii + 441 p. Collingwood VIC CSIRO Publishing (2002)

Discrete and Combinatorial Topology

Gradient Vector Fields of Discrete Morse Functions and Watershed-Cuts

Nicolas Boutry[1], Gilles Bertrand[2], and Laurent Najman[2]

[1] EPITA Research and Development Laboratory (LRDE),
Le Kremlin-Bicêtre, France
nicolas.boutry@lrde.epita.fr
[2] Univ Gustave Eiffel, CNRS, LIGM, 77454 Marne-la-Vallée, France
{gilles.bertrand,laurent.najman}@esiee.fr

Abstract. In this paper, we study a class of discrete Morse functions, coming from Discrete Morse Theory, that are equivalent to a class of simplicial stacks, coming from Mathematical Morphology. We show that, as in Discrete Morse Theory, we can see the gradient vector field of a simplicial stack (seen as a discrete Morse function) as the only relevant information we should consider. Last, but not the least, we also show that the Minimum Spanning Forest of the dual graph of a simplicial stack is induced by the gradient vector field of the initial function. This result allows computing a watershed-cut from a gradient vector field.

Keywords: Topological data analysis · Mathematical morphology · Discrete Morse Theory · Simplicial stacks · Minimum Spanning Forest

1 Introduction

We present here several results relating Mathematical Morphology [17] (MM) to Discrete Morse Theory [13] (DMT). This strengthens previous works highlighting links between MM and topology. In [6,7], it is demonstrated that watersheds are included in skeletons on pseudomanifolds of arbitrary dimension. Recently (see [1–3]), some relations between MM and Topological Data Analysis [16,20] (TDA) have been exhibited: the *dynamics* [14], used in MM to compute markers for watershed-based image-segmentation, is equivalent to the *persistence*, a fundamental tool from Persistent Homology [11].

In this paper, the first main result links the spaces used in MM and in TDA: the main mathematical spaces used in DMT, *discrete Morse functions* [19] (DMF), are equivalent, under some constraints, to spaces well-known in MM and called *simplicial stacks* [6–8]. Simplicial stacks are a class of weighted simplicial complexes whose upper threshold sets are also complexes. Indeed, in a DMF, the values locally increase when we increase the dimension of the face we are observing; in a simplicial stack, it is the opposite. Without surprise, we can then observe that, under some constraints, any DMF is the opposite of a simplicial stack, and conversely.

© Springer Nature Switzerland AG 2022
E. Baudrier et al. (Eds.): DGMM 2022, LNCS 13493, pp. 35–47, 2022.
https://doi.org/10.1007/978-3-031-19897-7_4

In TDA, it is a common practice to consider that the main information conveyed by a DMF is its gradient vector field (GVF), naturally obtained by pairing neighbor faces with the same value. Two DMFs with the same GVF are then considered to be equivalent. Using the very same principle on simplicial stacks, we can go further, and consider that a GVF encodes not only a class of DMFs, but also the corresponding class of simplicial stacks.

The relation between TDA and MM in the context of DMFs and stacks is not limited to the previous observations. In [8], the authors proved that a watershed-cut is a Minimum Spanning Forests (MSF) cut in the dual graph of a simplicial stack. We prove here that such a MSF can be extracted from the GVF of the simplicial stack (seen as a DMF). Relations between watersheds and Morse theory have long been informally known [10], but this is the first time that a link is presented in the discrete setting, relying on a precise definition of the watershed. Furthermore, as far as we know, this is the first time that a concept from Discrete Morse Theory is linked to a classical combinatorial optimization problem.

The plan of this paper is the following. Section 2 recalls the mathematical background necessary to our proofs. Section 3 shows the equivalence between DMF's and simplicial stacks. Section 4 studies the link between MSFs and GVFs. Section 5 concludes the paper.

2   Mathematical Background

2.1   Simplicial Complexes, Graphs and Pseudomanifolds

We call *(abstract) simplex* any finite nonempty set of arbitrary elements. The *dimension* of a simplex x, denoted by $\dim(x)$, is the number of its elements minus one. In the following, a simplex of dimension d will also be called a *d-simplex*. If x is a simplex, we set $\mathrm{Clo}(x) = \{y | y \subseteq x, y \neq \emptyset\}$. A finite set X of simplices is a *cell* if there exists $x \in X$ such that $X = \mathrm{Clo}(x)$.

If X is a finite set of simplices, we write $\mathrm{Clo}(X) = \cup\{\mathrm{Clo}(x) | x \in X\}$, the set $\mathrm{Clo}(X)$ is called the *(simplicial) closure* of X. A finite set X of simplices is a *(simplicial) complex* if $X = \mathrm{Clo}(X)$.

In the sequel of the paper, $\mathbb{K}$ denotes a simplicial complex. A subcomplex of $\mathbb{K}$ is a subset of $\mathbb{K}$ which is also a complex. Any element in $\mathbb{K}$ is a face of $\mathbb{K}$ and we call *d-face* of $\mathbb{K}$ any face of $\mathbb{K}$ whose dimension is d. If σ, τ are two faces of $\mathbb{K}$ with $\tau \subset \sigma$, we say that σ is a coface of τ. Any d-face of $\mathbb{K}$ that is not included in any $(d+1)$-face of $\mathbb{K}$ is called a $(d-)$*facet* of $\mathbb{K}$ or a *maximal face* of $\mathbb{K}$.

The dimension of $\mathbb{K}$, written $\dim(\mathbb{K})$, is the largest dimension of its faces: $\dim(\mathbb{K}) = \max\{\dim(x) | x \in \mathbb{K}\}$, with the convention that $\dim(\emptyset) = -1$. If d is the dimension of $\mathbb{K}$, we say that $\mathbb{K}$ is *pure* whenever the dimension of all its facets equals d.

Suppose that there is a pair of simplices[1] $(\sigma^{(p-1)}, \tau^{(p)})$ of $\mathbb{K}$ with $\sigma \subset \tau$ such that the only coface of σ is τ. Then $\mathbb{K} \setminus \{\sigma, \tau\}$ is a simplicial complex called *an*

---

[1] The superscripts correspond to the dimensions of the faces.

elementary collapse of $\mathbb{K}$. For an elementary collapse, such a pair $\{\sigma, \tau\}$ is called a *free pair*, and σ is called a *free face*. Note that elementary collapses preserve simple homotopy type [21]. A free pair $\{\sigma^{(d-1)}, \tau^{(d)}\}$ is called a *free d-pair*, and $\mathbb{K} \setminus \{\sigma^{(d-1)}, \tau^{(d)}\}$ is called an *elementary d-collapse*. If a complex $\mathbb{K}'$ is the result of a sequence of elementary d-collapses of $\mathbb{K}$, we say that $\mathbb{K}'$ is a *d-collapse of* $\mathbb{K}$. If, furthermore, there is no free d-pair for $\mathbb{K}'$, then $\mathbb{K}'$ is an *ultimate d-collapse of* $\mathbb{K}$.

In this paper, a *graph* $\mathcal{G}$ is a pure 1-dimensional simplicial complex. A subgraph is a subset of a graph which is also a graph. We denote the vertices (the 0-dimensional elements) of a graph $\mathcal{G}$ by $V(\mathcal{G})$, and the edges (the 1-dimensional elements) by $E(\mathcal{G})$.

Let X be a set of simplices, and let $d \in \mathbb{N}$. Let $\pi = \langle x_0, \dots, x_l \rangle$ be a sequence of d-simplices in X. The sequence π is a *d-path* from x_0 to x_l in X if $x_{i-1} \cap x_i$ is a $(d-1)$-simplex in X, for any $i \in \{1, \dots, l\}$. Two d-simplices x and y in X are said to be *d-linked* for X if there exists a d-path from x to y in X. We say that the set X is *d-connected* if any two d-simplices in X are d-linked for X. We say that the set $Y \subset X$ is a *d-connected component* (or simply, a *connected component*) of X if Y is d-connected and maximal for this property.

Let X be a set of simplices, and let $\pi = \langle x_0, \dots, x_l \rangle$ be a d-path in X. The d-path π is said *simple* if for any two distinct i and j in $\{0, \dots, l\}$, $x_i \neq x_j$. It can be easily seen that X is d-connected if and only if, for any two d-simplices x and y of X, there exists a simple d-path from x to y in X.

A complex $\mathbb{K}$ of dimension d is said to be a $d-pseudomanifold$ if

(1) $\mathbb{K}$ is pure,
(2) any $(d-1)-$face of $\mathbb{K}$ is included in exactly two $d-$faces of $\mathbb{K}$, and
(3) $\mathbb{K}$ is $d-$connected.

In the sequel of the paper, $d \geq 1$ is an integer, and $\mathbb{M}$ denotes a d-pseudomanifold.

Proposition 1 (Ultimate collapses [6]). *Let $\mathbb{K}$ be a proper subcomplex of the d-pseudomanifold $\mathbb{M}$. If the dimension of $\mathbb{K}$ is equal to d, then necessarily there exists a free d-pair for $\mathbb{K}$. In other words, the dimension of an ultimate d-collapse of $\mathbb{K}$ is necessarily $d - 1$.*

Following Proposition 1, we say that an ultimate d-collapse of $\mathbb{K} \subset \mathbb{M}$ is *thin*.

Let $x \in \mathbb{M}$, the star of x (in $\mathbb{M}$), denoted by $\text{St}(x)$, is the set of all simplices of $\mathbb{M}$ that include x, i.e., $\text{St}(x) = \{y \in \mathbb{M} \mid x \subseteq y\}$. If A is a subset of $\mathbb{M}$, the set $\text{St}(A) = \cup_{x \in A} \text{St}(x)$ is called the *star* of A (in $\mathbb{M}$). A set A of simplices of $\mathbb{M}$ is a star (in $\mathbb{M}$) if $A = star(A)$.

2.2   Simplicial Stacks

Let F be a mapping $\mathbb{M} \to \mathbb{Z}$. For any face σ of $\mathbb{M}$, the value $F(\sigma)$ is called the *altitude* of F at σ. For $k \in \mathbb{Z}$, the *k-section* of F, denoted by $[F \geq k]$ is equal to $\{\sigma \in \mathbb{M} \mid F(\sigma) \geq k\}$. A *simplicial stack* F on $\mathbb{M}$ is a map from $\mathbb{M}$ to $\mathbb{Z}$ which satisfies that any of its k-section is a (possibly empty) simplicial complex. In other words, a map F is a simplicial stack if, for any two faces σ and τ of $\mathbb{M}$ such that $\sigma \subseteq \tau$, $F(\sigma) \geq F(\tau)$.

We say that a subset A of $\mathbb{M}$ is a *minimum* of F at altitude $k \in \mathbb{Z}$ when A is a connected component of $[F \leq k] := \{\sigma \in \mathbb{M} \mid F(\sigma) \leq k\}$ and $A \cap [F \leq k-1] = \emptyset$. In the following, we denote by $M_-(F)$ the union of all minima of F. We note that, if F is a simplicial stack, then $M_-(F)$ is a star. The *divide* of a simplicial stack F is the set of all faces of $\mathbb{M}$ which do not belong to any minimum of F. Note that since $M_-(F)$, is a star, the divide is a simplicial complex.

Let σ be any face of $\mathbb{M}$. When σ is a free face for $[F \geq F(\sigma)]$, we say that σ is a *free face for* F. If σ is a free face for F, there exists a unique face τ in $[F \geq F(\sigma)]$ such that (σ, τ) is a free pair for $[F \geq F(\sigma)]$, and we say that (σ, τ) is a *free pair* for F. Let (σ, τ) be a free pair for F, then it is also a free pair for $[F \geq F(\sigma)]$. Thus, τ is a face of $[F \geq F(\sigma)]$, and we have $\sigma \subseteq \tau$. Therefore, we have $F(\tau) \geq F(\sigma)$ and $F(\tau) \leq F(\sigma)$ (since F is a stack), which imply that $F(\tau) = F(\sigma)$. Let $\mathbb{N} \subseteq \mathbb{M}$, the *indicator function* of $\mathbb{N}$, denoted by $1_{\mathbb{N}} : \mathbb{M} \to \{0,1\}$, is the mapping such that $1_{\mathbb{N}}(\sigma)$ is equal to 1 when σ belongs to $\mathbb{N}$ and is equal to 0 when σ belongs to $\mathbb{M} \setminus \mathbb{N}$. The *lowering* of F at $\mathbb{N}$ is the map $F - 1_{\mathbb{N}}$ from $\mathbb{M}$ into $\mathbb{Z}$. Let $(\sigma^{(d-1)}, \tau^{(d)})$ be a free pair for F. The map $F - 1_{\{\sigma,\tau\}}$ is called an *elementary d-collapse of* F. Thus, this elementary d-collapse is obtained by subtracting 1 to the values of F at σ and τ. Note that the obtained mapping is still a simplicial stack. If a simplicial stack F' is the result of a sequence of elementary d-collapses on F, then we say that F' is a *d-collapse* of F. If, furthermore, there is no free pair $(\sigma^{(d-1)}, \tau^d)$ for F', then F' is an *ultimate d-collapse* of F.

2.3   Watersheds of Simplicial Stacks

Let A and B be two nonempty stars in $\mathbb{M}$. We say that B is an extension of A if $A \subseteq B$, and if each connected component of B includes exactly one connected component of A. We also say that B is an *extension* of A if $A = B = \emptyset$. Let X be a subcomplex of the pseudomanifold $\mathbb{M}$ and let Y be a collapse of X, then the complement of Y in $\mathbb{M}$ is an extension of the complement of X in $\mathbb{M}$. Let A be a nonempty open set in a pseudomanifold $\mathbb{M}$ and let X be a subcomplex of $\mathbb{M}$. We say that X is a *cut* for A if the complement of X is an extension of A and if X is minimal for this property. Observe that there can be several distinct cuts for a same open set A and, in this case, these distinct cuts do not necessarily contain the same number of faces.

Let $\pi = \langle x_0, \dots, x_\ell \rangle$ be a d-path in $\mathbb{M}$, and let F be a function on $\mathbb{M}$. We say that the d-path π is *descending* for F if for any $i \in \{1, \dots, \ell\}$, $F(x_i) \leq F(x_{i-1})$.

Let X be a subcomplex of the pseudomanifold $\mathbb{M}$. We assume that X is a cut for $M_-(F)$. We say that X is a *watershed-cut* of F if for any $x \in X$, there exists two descending paths $\pi_1 = \langle x, x_0, \dots, x_\ell \rangle$ and $\pi_2 = \langle x, y_0, \dots, y_m \rangle$ such that (1) x_ℓ and y_m are simplices of two distinct minima of F; and (2) $x_i \notin X$, $y_j \notin X$, for any $i \in \{0, \dots, \ell\}$ and $j \in \{0, \dots, m\}$.

Several equivalent definitions of the watershed for pseudo-manifolds are given in [6,7]. Also, it was shown that a watershed-cut of F is necessarily included in an ultimate d-collapses of F. Thus, by Proposition 1, a watershed-cut is a thin divide.

In this paper, we focus on a definition relying on combinatorial optimization, more precisely on the minimum spanning tree. For that, we need a notion of "dual graph" of a pseudomanifold.

Starting from a d-pseudomanifold $\mathbb{M}$ valued by $F : \mathbb{M} \to \mathbb{Z}$, we define the *dual (edge-weighted) graph* of F as the 3-tuple $\mathcal{G}_F = (V, E, F_\mathcal{G})$ whose vertex set V is composed of the d-simplices of $\mathbb{M}$, whose edge set E is composed of the pairs $\{\sigma, \tau\}$ such that σ, τ are d-faces of $\mathbb{M}$ and $\sigma \cap \tau$ is a $(d-1)$-face of $\mathbb{M}$, and whose edge weighting $F_\mathcal{G}$ is made as follows: for two distinct d-faces σ, τ in $\mathbb{M}$ sharing a $(d-1)$-face of $\mathbb{M}$, $F_\mathcal{G}(\{\sigma, \tau\}) = F(\sigma \cap \tau)$.

Let A and B be two non-empty subgraphs of the dual graph $\mathcal{G}_F$ of F. We say that B is a forest relative to A when

(1) B is an extension of A; and
(2) for any extension $C \subseteq B$ of A, we have $C = B$ whenever B and C share the same vertices.

Informally speaking, the second condition imposes that we cannot remove any edge from B while keeping an extension of A that has the same vertex set as B. We say that B is a *spanning forest relative to* A for $\mathcal{G}_F$ if B is a forest relative to A and if B and $\mathcal{G}_F$ share the same vertices.

The *weight* of A is defined as: $F_\mathcal{G}(A) := \sum_{u \in E(A)} F_\mathcal{G}(u)$. We say that B is a *minimum spanning forest (MSF)* relative to A for $F_\mathcal{G}$ if B is a spanning forest relative to A for $F_\mathcal{G}$ and if the weight of B is less than or equal to the weight of any other spanning forest relative to A for $F_\mathcal{G}$.

Let A be a subgraph of $\mathcal{G}_F$, and let X be a set of edges of $\mathcal{G}_F$. We say that X is an *MSF cut* for A if there exists an MSF B relative to A such that X is the set of all edges of $\mathcal{G}_F$ adjacent to two distinct connected components of B.

In the following, if S is a set of $(d-1)$-faces of $\mathbb{M}$, we set $\text{Edges}(S) = \{\{\sigma, \tau\} \in E(\mathcal{G}_F) \mid \sigma \cap \tau \in S\}$. The *dual graph of the minima* of F is the graph whose vertex set is the set M of d-faces of the minima of F and whose edge set is composed of the edges of $\mathcal{G}_F$ linking two elements of M.

Theorem 2 (Theorem 16 p. 10 [6]). *Let X be a set of $(d-1)$-faces of $\mathbb{M}$, and let $F : \mathbb{M} \to \mathbb{Z}^+$ be a simplicial stack. The complex resulting from the closure of X is a watershed-cut of F if, and only if, $\text{Edges}(X)$ is a MSF cut for the dual graph of the minima of F.*

In other words, to compute the watershed of a stack F, it is sufficient to compute in $\mathcal{G}_F$ a MSF cut relative to the graph associated with the minima of F. Different algorithms for computing MSF cuts are detailed in [8,9].

2.4 Basic Discrete Morse Functions

We rely here on the formalism presented in [19], with results from Forman and Benedetti. A function $F : A \to B$ is said to be $2-1$ when, for every $b \in B$, there exist at most two values $a_1, a_2 \in A$ such that $F(a_1) = F(a_2) = b$. Let $\mathbb{K}$ be a simplicial complex. A function $F : \mathbb{K} \to \mathbb{Z}$ is called *weakly increasing* if $F(\sigma) \leq F(\tau)$ whenever the two faces σ, τ of $\mathbb{K}$ satisfy $\sigma \subseteq \tau$.

A *basic discrete Morse function* $F : \mathbb{K} \to \mathbb{Z}$ is a weakly increasing function which is $2 - 1$ and satisfies the property that if $F(\sigma) = F(\tau)$, then $\sigma \subseteq \tau$ or $\tau \subseteq \sigma$.

Let $F : \mathbb{K} \to \mathbb{Z}$ be a basic discrete Morse function. A simplex σ of $\mathbb{K}$ is said to be *critical* when F is injective on σ. Otherwise, σ is called *regular*. When σ is a critical simplex, $F(\sigma)$ is called a *critical value*. If σ is a regular simplex, $F(\sigma)$ is called a *regular value*.

Discrete Morse functions are more general than basic discrete Morse functions. A discrete Morse function (DMF) F on $\mathbb{K}$ is a function from $F : \mathbb{K} \to \mathbb{Z}$ such that for every p-simplex $\sigma \in \mathbb{K}$, we have

$$|\{\tau^{(p-1)} \subset \sigma \mid F(\tau) \geq F(\sigma)\}| \leq 1 \tag{1}$$

and

$$|\{\tau^{(p-1)} \supset \sigma \mid F(\tau) \leq F(\sigma)\}| \leq 1. \tag{2}$$

However, to each discrete Morse function, there exists a basic discrete Morse function which is equivalent in the following sense (see Theorem 4 and Proposition 5). Two discrete Morse functions F, F' defined on the same simplicial complex $\mathbb{K}$ are said to be *Forman-equivalent* when for any two faces $\sigma^{(p)}, \tau^{(p+1)} \in \mathbb{K}$ satisfying $\sigma \subset \tau$, $F(\sigma) < F(\tau)$ if and only if $F'(\sigma) < F'(\tau)$. Hence, in this paper, we focus on basic discrete Morse functions.

Let F be a basic discrete Morse function on $\mathbb{K}$. The *(induced) gradient vector field* (GVF) $\overrightarrow{\mathrm{grad}}$ of F is defined by

$$\overrightarrow{\mathrm{grad}}(F) := \left\{ (\sigma^{(p)}, \tau^{(p+1)}) \mid \sigma, \tau \in \mathbb{K} , \sigma \subset \tau , F(\sigma) \geq F(\tau) \right\}. \tag{3}$$

If (σ, τ) belongs to $\overrightarrow{\mathrm{grad}}(F)$, then it is called a *vector* (for F) whose σ is the *tail* and τ is the *head*. The vector (σ, τ) is sometimes denoted by $\overrightarrow{\sigma\tau}$.

Let $\mathbb{K}$ be a simplicial complex. A *discrete vector field* V on $\mathbb{K}$ is defined by

$$V := \{(\sigma^{(p)}, \tau^{(p+1)}) \mid \sigma \subset \tau, \text{each simplex of } \mathbb{K} \text{ is in at most one pair}\} \tag{4}$$

Naturally, every GVF is a discrete vector field.

Let V be a discrete vector field on a simplicial complex $\mathbb{K}$. A *gradient path* is a sequence of simplices: $(\tau_{-1}^{(p+1)},)\sigma_0^{(p)}, \tau_0^{(p+1)}, \sigma_1^{(p)}, \tau_1^{(p+1)}, \ldots, \sigma_{k-1}^{(p)}, \tau_{k-1}^{(p+1)}, \sigma_k^{(p)},$ of $\mathbb{K}$, beginning at either a critical simplex $\tau_{-1}^{(p+1)}$ or a regular simplex $\sigma_0^{(p)}$, such that $(\sigma_\ell^{(p)}, \tau_\ell^{(p+1)})$ belongs to V and $\tau_{\ell-1}^{(p+1)} \supset \sigma_\ell^{(p)}$ for $0 \leq \ell \leq k - 1$. If $k \neq 0$, then this path is said to be *non-trivial*. Note that the last simplex does not need to be in a pair in V. A gradient path is said to be *closed* if $\sigma_k^{(p)} = \sigma_0^{(p)}$.

Theorem 3 (Theorem 2.51 p.61 of [19]). *A discrete vector field is the GVF of a discrete Morse function if, and only if, this discrete vector field contains no non-trivial closed gradient paths.*

Theorem 4 (Theorem 2.53 p.62 of [19]). *Two discrete Morse functions defined on a same complex $\mathbb{K}$ are Forman-equivalent if, and only if, they induce the same GVF. A consequence is that any two Forman-equivalent discrete Morse functions defined on a simplicial complex have the same critical simplices.*

Proposition 5. *If F is a discrete Morse function, there exists F' a basic discrete Morse function that is Forman-equivalent to F.*

Proposition 5 is a consequence of [19, Proposition 4.16]. Starting from a DMF and computing its GVF, it is possible (by correctly ordering all the simplices) to compute a basic DMF Forman-equivalent to it; such an algorithm preserves the GVF. A precise algorithm, together with a proof of Proposition 5 relying on this algorithm, will be provided in an extended version of this paper.

3   A Class of Simplicial Stack Equivalent to Morse Functions

Simplicial stacks are weakly decreasing. We call *basic simplicial stack*, a simplicial stack F that is $2-1$ and satisfies the property that if $F(\sigma) = F(\tau)$, then $\sigma \subseteq \tau$ or $\tau \subseteq \sigma$. The proof of the following is straightforward.

Proposition 6. *Let F be a function defined on $\mathbb{M}$. Then F is a basic simplicial stack if and only if $-F$ is a basic discrete Morse function.*

Hence, all properties of basic discrete Morse functions hold true for basic simplicial stacks, and conversely. In the sequel of this paper, we exemplify that fact with gradient vector fields.

Relying on Proposition 6, we define the *gradient vector field of a basic simplicial stack F* as the GVF of the DMF $-F$ it corresponds to.

As stated in Theorem 4, two basic DMF's are Forman-equivalent if, and only if, they induce the same GVF. In other words, at each GVF corresponds a class of DMF's. Using Proposition 6, we have a bijection between the space of basic DMF's and the space of basic simplicial stacks. This leads to the following corollary:

Corollary 7. *If F is a basic DMF defined on $\mathbb{M}$, there exists a class $\mathbb{BD}$ of basic DMF's and a class $\mathbb{SS}$ of basic simplicial stacks, bijective to $\mathbb{BD}$, such that each F' in one of those classes has the same gradient vector field as the one of F.*

4   The Minimum Spanning Forest of a Stack and the GVF

4.1   The Forest Induced by a GVF

Let F a basic simplicial stack. As any k-section of F is a simplicial complex, and as F is 2-1, we have the following proposition:

Proposition 8. *Let F be a basic simplicial stack. We have:*

1. $M_-(F)$ *is a set of simplices of dimension d.*
2. *Each minimum of F is made of a single simplex of* $M_-(F)$.
3. *The set of edges of the dual graph of the minima is empty.*

Let $\overrightarrow{\text{grad}}$ be the GVF of F. Let $\overrightarrow{ab}$ be a vector of $\overrightarrow{\text{grad}}$ such that $dim(a) = d-1$. Since $\mathbb{M}$ is a pseudomanifold, the face a is included in two d-faces, the face b and another d-face c. We write $[\overrightarrow{ab}] = \{\{b\}, \{c\}, \{b, c\}\}$ and we consider the graph:

$$\mathcal{G}(\overrightarrow{\text{grad}}) = \cup\{[\overrightarrow{ab}] \mid \overrightarrow{ab} \in \overrightarrow{\text{grad}}, dim(a) = d - 1\}. \tag{5}$$

Let $\mathcal{G}^+(\overrightarrow{\text{grad}})$ be the union of $\mathcal{G}(\overrightarrow{\text{grad}})$ and of $\mathcal{G}(M_-(F))$, where $\mathcal{G}(M_-(F))$ is the dual graph of the minima of F.

Proposition 0. *The graph* $\mathcal{G}^+(\overrightarrow{\text{grad}})$ *is a spanning forest relative to the dual graph of the minima of F.*

The proof of this proposition relies on the following fact: any critical simplices of F that is not a minimum of F is of dimension strictly lower than d.

Proof. We first show that $\mathcal{G}^+(\overrightarrow{\text{grad}})$ spans all vertices of the dual graph $\mathcal{G}_F$: as $\mathcal{G}^+(\overrightarrow{\text{grad}})$ contains the dual graph of the minima, we only need to show that for any d-face σ of $\mathbb{M}$, $\sigma \notin M_-(F)$, there is a pair $(\tau^{(d-1)}, \sigma)$ of simplices in $\overrightarrow{\text{grad}}$.

[19, Remark 2.42] states that, for any simplex τ, exactly one of the following holds true:

(i) τ is the tail of exactly one vector
(ii) τ is the head of exactly one vector
(iii) τ is neither the tail nor the head of a vector; that is τ is critical

By Proposition 8, item 2, each minimum of F is made of a single simplex of $M_-(F)$. By remark [19, Remark 2.42] above, it remains to show that, if σ is not a minimum of F, it is regular, and hence the head of exactly one vector. As F is a simplicial stack, its k-section for $k = F(\sigma)$ contains all the simplices ν such that $\nu \subset \sigma$. We have $F(\nu) \geq F(\sigma)$. Because σ is not a minimum, there exists a simplex $\tau^{(d-1)}$ such that $F(\tau^{(d-1)}) = F(\sigma)$ with $\tau^{(d-1)} \subset \sigma$. This implies that $(\tau^{(d-1)}, \sigma) \in \overrightarrow{\text{grad}}$.

By Theorem 3, $\mathcal{G}(\overrightarrow{\text{grad}})$ does not contain any closed 1-path. Hence, $\mathcal{G}^+(\overrightarrow{\text{grad}})$ is a forest relative to the dual graph of the minima of F. □

Following Proposition 9, we say in the sequel that $\mathcal{G}^+(\overrightarrow{\text{grad}})$ is the *forest induced* by the GVF $\overrightarrow{\text{grad}}$.

4.2   The Forest Induced by a GVF Is the MSF

Proposition 10. *Let $F : \mathbb{M} \to \mathbb{Z}$ be a basic simplicial stack, and let $\overrightarrow{grad}$ be the GVF of F. Then, any gradient-path $\pi = (\pi(k))_{k \in [0,N]}$ of the GVF is increasing, that is, for any $k \in [0, N-1]$, $F(\pi(k)) \leq F(\pi(k+1))$.*

Proof. Let π some gradient-path of $\overrightarrow{grad}$, and let us assume without loss of generality, that $\pi(0)$ is a d-face of $\mathbb{M}$. We know that the $(d-1)$-face $\pi(2k+1)$ is paired with the n-face $\pi(2k+2)$ in $\overrightarrow{grad}$ for any $k \in [0, (N-1)/2 - 1]$ (N is odd), which means that $F(\pi(2k+1)) = F(\pi(2k+2))$. We also know that F is a stack, and then F decreases when we increase the dimension of the face, so for any $k \in [0, (N-1)/2 - 1]$, $F(\pi(2k)) \leq F(\pi(2k+1))$. $\qquad\square$

Lemma 11. *(MST Lemma [4,15]). Let $\mathcal{G} = (V, E, F)$ be some edge-weighted graph. Let $v \in V$ be any vertex in $\mathcal{G}$. A minimum spanning tree for $\mathcal{G}$ must contain an edge vw that is a minimum weighted edge incident on v.*

Theorem 12. *Let $F : \mathbb{M} \to \mathbb{Z}^+$ be a basic simplicial stack, and let $\overrightarrow{grad}$ be the GVF of F. The forest induced by $\overrightarrow{grad}$ is the unique MSF relative to $M_-(F)$ of the dual graph of F.*

Proof. By Proposition 9, $\mathcal{G}^+(\overrightarrow{grad})$ is a spanning forest relative to $M_-(F)$, the minima of F. As F is a basic simplicial stack, hence 2-1, all edges of the dual graph $\mathcal{G}_F = (V, E, F_{\mathcal{G}})$ of F have a unique weight, and the MSF of the dual graph $\mathcal{G}_F$ is unique. It remains to prove that the induced forest is of minimum cost.

Since gradients do not exist on minima, let us consider a d-simplex $\sigma \in V \setminus M_-(F)$. Then, by Proposition 9, there exists exactly one vector in $\overrightarrow{grad}$, which can be written $(\tau \cap \sigma, \sigma)$, with $\tau \in V$. By the definition of $\overrightarrow{grad}$, we have $F(\tau \cap \sigma) = F(\sigma)$.

Let $\theta \in V \setminus \{\tau\}$ some d-simplex such that $\{\tau, \theta\}$ belongs to E. Since F is a simplicial stack, either the $(d-1)$-face $\tau \cap \theta$ is critical (and $F(\tau \cap \theta) > \max(F(\tau), F(\theta))$), or it is regular and $\tau \cap \theta$ is paired with θ in $\overrightarrow{grad}$ (and $F(\theta) = F(\tau \cap \theta) > F(\tau)$ by Proposition 10). Therefore, $\{\sigma, \tau\}$ is the lowest cost edge incident to τ:

$$F_{\mathcal{G}}(\{\sigma, \tau\}) = F(\tau \cap \sigma) = F(\sigma) < \min\{F_{\mathcal{G}}(\{\theta, \tau\}) \; ; \; \{\theta, \tau\} \in E, \; \theta \neq \sigma\} \quad (6)$$

and thus belongs to the MST of F by Lemma 11.

As by Proposition 9, the induced forest is a spanning forest relative to the dual graph of the minima of F, it is then the minimum spanning tree of the dual graph relative to the minima of F, which concludes the proof. $\qquad\square$

A summary of this result is depicted in Fig. 1, which shows a piece of a pseudomanifold of dimension 2.

Using Theorem 2 and Theorem 12, we can conclude that the cut of the forest induced by the GVF is also a watershed-cut. This leads to the following corollary.

Corollary 13. *Let $F : \mathbb{M} \to \mathbb{Z}^+$ be a basic simplicial stack. Then, the watershed-cut of F is provided equivalently by the MSF of F or by the GVF of F.*

Figure 2 illustrates this corollary: each tree of the induced forest is a connected component of the dual graph, called a catchment basin of the watershed-cut.

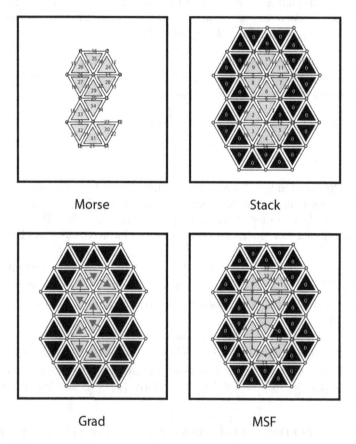

Fig. 1. Starting from a Morse function, we obtain its equivalent simplicial stack up to the minus sign. For simplicity, the simplicial stack is valued by 0 on all d-simplices at the border. Then, we deduce the GVF of the initial Morse function and its MSF. This illustrates that **the MSF is the forest induced by the GVF** of both a discrete Morse function and the corresponding simplicial stack.

Fig. 2. Illustration of the watershed-cut obtained from the GVF of a discrete Morse function: we have a partition of the pseudo-manifold, such that each tree in the forest is a basin of the watershed-cut. We also highlight the equality between the forest induced by the GVF (in blue) and the MSF of the dual graph (in red). (Color figure online)

5   Conclusion

In this paper, we highlight some links between several notions that exist in Discrete Topology and in Mathematical Morphology:

- discrete Morse functions are equivalent, under some constraints, to simplicial stacks;
- gradient vector fields in the Morse sense are applicable to simplicial stacks;
- and the gradient vector field of a simplicial stack induces the Minimum Spanning Forest of its dual graph, leading to watershed-cuts.

In the extended version of this paper, we will relax the constraints for the equivalence between discrete Morse function and simplicial stacks, and we will show how to use the watershed to define a purely discrete version of the well-known Morse-Smale complex [12].

In the future, we will continue looking for strong relations linking Discrete Morse Theory and Mathematical Morphology, with the goal of using morphological tools for topological data analysis. We also aim at making clearer the relation between discrete topology and discrete Morse theory, following [18] that was inspired by [5].

Acknowledgements. The authors would like to thank both Julien Tierny and Thierry Géraud, for many insightful discussions.

References

1. Boutry, N., Géraud, T., Najman, L.: An equivalence relation between morphological dynamics and persistent homology in 1D. In: Burgeth, B., Kleefeld, A., Naegel, B., Passat, N., Perret, B. (eds.) ISMM 2019. LNCS, vol. 11564, pp. 57–68. Springer, Cham (2019). https://doi.org/10.1007/978-3-030-20867-7_5
2. Boutry, N., Géraud, T., Najman, L.: An equivalence relation between morphological dynamics and persistent homology in n-D. In: Lindblad, J., Malmberg, F., Sladoje, N. (eds.) DGMM 2021. LNCS, vol. 12708, pp. 525–537. Springer, Cham (2021). https://doi.org/10.1007/978-3-030-76657-3_38
3. Boutry, N., Najman, L., Géraud, T.: Some equivalence relation between persistent homology and morphological dynamics. J. Math. Imaging Vis. (2022). https://doi.org/10.1007/s10851-022-01104-z, https://hal.archives-ouvertes.fr/hal-03676854
4. Cormen, T.H., Leiserson, C.E., Rivest, R.L.: Introduction to Algorithms. 23rd printing (1999)
5. Couprie, M., Bertrand, G.: New characterizations of simple points in 2D, 3D, and 4D discrete spaces. IEEE Trans. Pattern Anal. Mach. Intell. **31**(4), 637–648 (2008)
6. Cousty, J., Bertrand, G., Couprie, M., Najman, L.: Collapses and watersheds in pseudomanifolds. In: Wiederhold, P., Barneva, R.P. (eds.) IWCIA 2009. LNCS, vol. 5852, pp. 397–410. Springer, Heidelberg (2009). https://doi.org/10.1007/978-3-642-10210-3_31
7. Cousty, J., Bertrand, G., Couprie, M., Najman, L.: Collapses and watersheds in pseudomanifolds of arbitrary dimension. J. Math. Imaging Vis. **50**(3), 261–285 (2014). https://doi.org/10.1007/s10851-014-0498-z
8. Cousty, J., Bertrand, G., Najman, L., Couprie, M.: Watershed cuts: minimum spanning forests and the drop of water principle. IEEE Trans. Pattern Anal. Mach. Intell. **31**(8), 1362–1374 (2009)
9. Cousty, J., Bertrand, G., Najman, L., Couprie, M.: Watershed cuts: thinnings, shortest path forests, and topological watersheds. IEEE Trans. Pattern Anal. Mach. Intell. **32**(5), 925–939 (2010)
10. De Floriani, L., Fugacci, U., Iuricich, F., Magillo, P.: Morse complexes for shape segmentation and homological analysis: discrete models and algorithms. Comput. Graph. Forum **34**(2), 761–785 (2015)
11. Edelsbrunner, H., Harer, J.: Persistent homology - a survey. Contemp. Math. **453**, 257–282 (2008)
12. Edelsbrunner, H., Harer, J., Natarajan, V., Pascucci, V.: Morse-Smale complexes for piecewise linear 3-manifolds. In: Proceedings of the Nineteenth Annual Symposium on Computational Geometry, pp. 361–370 (2003)
13. Forman, R.: A Discrete Morse Theory for cell complexes. In: Yau, S.T. (ed.) Geometry. Topology for Raoul Bott. International Press, Somerville MA (1995)
14. Grimaud, M.: New measure of contrast: the dynamics. In: Image Algebra and Morphological Image Processing III, vol. 1769, pp. 292–306. International Society for Optics and Photonics (1992)
15. Motwani, R., Raghavan, P.: Randomized Algorithms. Cambridge University Press, Cambridge (1995)
16. Munch, E.: A user's guide to topological data analysis. J. Learn. Anal. **4**(2), 47–61 (2017)
17. Najman, L., Talbot, H.: Mathematical Morphology: From Theory to Applications. Wiley, Hoboken (2013)

18. Robins, V., Wood, P.J., Sheppard, A.P.: Theory and algorithms for constructing discrete Morse complexes from grayscale digital images. IEEE Trans. Pattern Anal. Mach. Intell. **33**(8), 1646–1658 (2011)
19. Scoville, N.A.: Discrete Morse Theory, vol. 90. American Mathematical Soc. (2019)
20. Tierny, J.: Introduction to Topological Data Analysis. Technical report, Sorbonne University, LIP6, APR team, France (2017). https://hal.archives-ouvertes.fr/cel-01581941
21. Whitehead, J.H.C.: Simplicial spaces, nuclei and m-groups. Proc. Lond. Math. Soc. **2**(1), 243–327 (1939)

Towards Topological Analysis of Non-symmetric Tensor Fields via Complexification

Bernhard Burgeth[1], Andreas Kleefeld[2,3]([✉]), Eugene Zhang[4], and Yue Zhang[4]

[1] Faculty of Mathematics and Computer Science, Saarland University, 66041 Saarbrücken, Germany
burgeth@math.uni-sb.de

[2] Jülich Supercomputing Centre, Forschungszentrum Jülich GmbH, Wilhelm-Johnen-Straße, 52425 Jülich, Germany
a.kleefeld@fz-juelich.de

[3] Faculty of Medical Engineering and Technomathematics, University of Applied Sciences Aachen, Heinrich-Mußmann-Str. 1, 52428 Jülich, Germany

[4] School of Electrical Engineering and Computer Science, Oregon State University, Corvallis, OR 97331, USA
zhange@engr.orst.edu, zhangyue@oregonstate.edu

Abstract. Fields of asymmetric tensors play an important role in many applications such as medical imaging (diffusion tensor magnetic resonance imaging), physics, and civil engineering (for example Cauchy-Green-deformation tensor, strain tensor with local rotations, etc.). However, such asymmetric tensors are usually symmetrized and then further processed. Using this procedure results in a loss of information. A new method for the processing of asymmetric tensor fields is proposed restricting our attention to tensors of second-order given by a 2×2 array or matrix with real entries. This is achieved by a transformation resulting in Hermitian matrices that have an eigendecomposition similar to symmetric matrices. With this new idea numerical results for real-world data arising from a deformation of an object by external forces are given. It is shown that the asymmetric part indeed contains valuable information.

Keywords: Asymmetric tensor fields · Spectral decomposition · Line integral convolution

1 Introduction

Fields of tensors are an essential notion in many applications such as medical imaging, physics, and civil engineering. Tensors make their natural appearance as Cauchy-Green-deformation tensor, strain tensor with local rotations, permittivity tensor, etc., or as structure tensor in image processing itself. Although the

© Springer Nature Switzerland AG 2022
É. Baudrier et al. (Eds.): DGMM 2022, LNCS 13493, pp. 48–59, 2022.
https://doi.org/10.1007/978-3-031-19897-7_5

notion of a tensor is quite sophisticated especially in mathematical literature, in the context of this article we consider them simply as 2×2-arrays of complex numbers subjected to the standard computational rules of matrix calculus. Despite this fact, we refer to any mapping from a suitable set $\Omega \subset \mathbb{R}^2$ into the set of matrices as a tensor field as it is common.

Symmetric matrices or second order tensors possess an eigenvalue decomposition with real eigenvalues and mutually orthogonal eigenvectors. Hence, a decomposition of a symmetric matrix S as

$$S = Q \cdot D \cdot Q^\top$$

with a diagonal matrix D and an orthogonal matrix Q is at our disposal. As a consequence the functional calculus is sufficiently rich to pave the way to transfer algorithms designed for the processing of real-valued data (functions) to the setting of matrix-valued data (functions), let us refer to [4].

The visualization of symmetric tensors often makes use of the corresponding quadratic form resulting in ellipses ($n = 2$) or ellipsoids ($n = 3$) (see for example [8]), casting the information about eigenvalues and eigenvectors in an appealing visual form. Particular visualization methodologies focus on the overall appearance of the tensor field, its topological, global structure, and its connectivity. Prominent is the line integral convolution (LIC) procedure that relies on the dominant eigenvector, that is the (normalized) eigenvector belonging to the largest eigenvalue of a symmetric matrix [5]. Clearly this concepts is no longer applicable if the existence of a real-valued eigenvector cannot be guaranteed as it is the case for general non-symmetric, hence, mostly non-diagonalizable matrices. However, tensors of the latter type are of particular interest in many applications. As a remedy a symmetrization is used leading to a manageable tensor field, but at the price of a loss of the information captured in the asymmetric part.

This is the reason why existing research [1,7,9–11] even in the 2D-case is based on their visualization of asymmetric tensor fields relying on the decomposition of an asymmetric tensor into the product of three matrix components, whose corresponding physical concepts in civil engineering are respectively expansion/contraction, rotation, and pure shear. But these approaches have their intricacies and their generalization to dimension $n = 3$ does not seem to be straightforward.

In this article, our response to this dilemma is "complexifying" the asymmetric tensors by applying a mapping $\varXi$ from the set of real square matrices $\mathbb{R}^{n \times n}$ into the set of Hermitian tensors $\mathrm{Herm}(n) := \{K \in \mathbb{C}^{n \times n} : K = K^*\}$ with $n \geq 2$ defined by (see also [2,3])

$$\varXi : \begin{cases} \mathbb{R}^{n \times n} \longrightarrow & \mathrm{Herm}(n) \\ A \longmapsto \frac{1}{2}(A + A^\top) + \frac{i}{2}(A - A^\top). \end{cases} \tag{1}$$

We will elaborate more on this mapping $\varXi$ in Sect. 2, while we report on the application of hermitization together with the LIC-procedure to real data sets in Sect. 3. A short summary and an outlook is given in Sect. 4.

2   The Conversion Process

Since tensors describing general deformation fields are usually not symmetric, they are symmetrized leading to a loss of information captured in the asymmetric part. In this article, those tensors are pre-processed by applying a mapping Ξ given by (1) in the special case $n = 2$. Indeed, for any $A \in \mathbb{R}^{2 \times 2}$ we have $(\Xi(A))^* = \Xi(A)$, where $^\top$ and * means transpose and conjugate transpose, respectively. The reason for this pre-processing step is the fact that Hermitian tensors allow for a rich tensor (or matrix) calculus almost as amenable as in the case of real symmetric matrices since in both cases the matrices form a real vector space and are unitarily diagonalizable with real eigenvalues. We point out that some properties of this mapping and Hermitian tensors also hold for $n \geq 2$.

Proposition 1. Ξ maps $\mathbb{R}^{n \times n}$ bijectivly into the real vector space $\mathrm{Herm}(n)$.

Proof. The mapping Ξ is linear on a finite dimensional space and has a trivial kernel. Hence, it is an isomorphism and therefore invertible.

For the sake of brevity we set $H = (H_{ij})_{i,j=1,2} := \Xi(T)$ and assume

$$H = U \Lambda U^*, \qquad \Lambda = \mathrm{diag}(\kappa_1, \kappa_2)$$

with real eigenvalues $\kappa_1 \geq \kappa_2$. A straightforward reckoning reveals

$$\kappa_1 = (H_{11} + H_{22})/2 + \sqrt{(H_{11} + H_{22})^2/4 - (H_{11}H_{22} - H_{12}H_{21})} \qquad (2)$$

and for the associated major eigenvector (not normalized)

$$u_1 = (H_{12}, -(H_{11} - \kappa_1))^\top.$$

The second component of u_1 is real, that is, $\mathrm{Im}(-(H_{11} - \kappa_1)) = 0$. That means the imaginary part of this major eigenvector is aligned in the real x-direction.

Remark 1. Note that the eigenvector u_1 can be multiplied by an arbitrary constant, say $c \in \mathbb{C}$, and still it will be an eigenvector. In fact, we can specifically choose $c = e^{i\beta}$ with $\beta \in [0, 2\pi]$ without changing the original length of the eigenvector. That means we could theoretically align the major eigenvector in any direction (if desired).

3   Numerical Results

Note that all the following figures are created with Matlab 2018a. The line integral convolution algorithm has been downloaded from the web page https://itp.tugraz.at/~ahi/Uni/AppSoft/LIC/ which has been implemented in Matlab by A. Hirczy.

   Further, note that the data are generated from two real experiments. Precisely, the 2D tensor fields we are dealing with in this section are derived from the deformation of an object by external forces. The strain tensor is derived from the gradient of the displacement occurring when external forces to an object are applied (see [6, Section 2.2.3]). Here, we consider two configurations of forces, indicated by the arrows in Fig. 1, acting on a 2D object of a square-like shape with coordinates $[-0.675, 0.675] \times [-0.675, 0.675]$. Non-zero forces are acting on the right side of the square-shaped body, resembling a parabolic force pattern in the first case (data set named "*parabolic*"), while the pattern is more sinusoidal in the second case (data set named "*sinusoidal*").

First, we will visualize the non-symmetric stress tensor field obtained from the deformation gradient. Precisely, taking the partial derivative of the displacement vector field with respect to the material coordinates gives the material displacement gradient tensor which can be written as $F - I$ where F is the deformation gradient tensor field (see [6, Section 2.1.6 and Section 2.2.3]).

The stress tensor under consideration is a multiplication between a symmetric stress tensor and the deformation gradient tensor. It is an asymmetric tensor, thus it might give us some new insight.

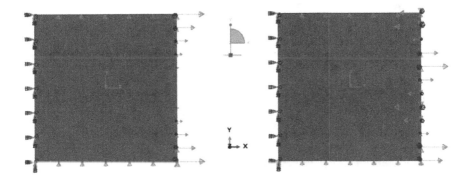

Fig. 1. The two experiments *parabolic* and *sinusoidal*.

3.1 Non-symmetric Stress Tensor

Each of the resulting data sets are stress tensor fields with a (pixel) resolution of 217×217. We transform the non-symmetric matrix into a Hermitian matrix for each pixel and compute the real and imaginary part of the major eigenvector corresponding to κ_1 of (2). In sum, this results in two different vector fields. From these vector fields integral lines are derived through convolution and visualized by the LIC procedure. In addition, we plot the major eigenvector fields in quiver plots as an alternative representation method for better visual comparison. Note that we only visualize every fifth vector resulting in a 43×43 resolution to avoid cluttering. We begin with the *parabolic* data set.

When the imaginary part is aligned in positive x-direction, then we color the arrow in green. Otherwise, it is colored red, as shown in Fig. 2. Both the LIC-representation and the quiver plot of the real part of the major eigenvector field clearly indicates that the *parabolic* pattern of forces applied on the right hand side of the shape creates a corresponding response inside. The anti-symmetric "left-right" pattern in the imaginary part of the major eigenvector field is a striking feature in the quiver-plot while it is not so clearly discernible in the LIC-picture. The later is no surprise since by its very construction the LIC-procedure is not predisposed to capture the discontinuous behavior of the imaginary part of the vector components.

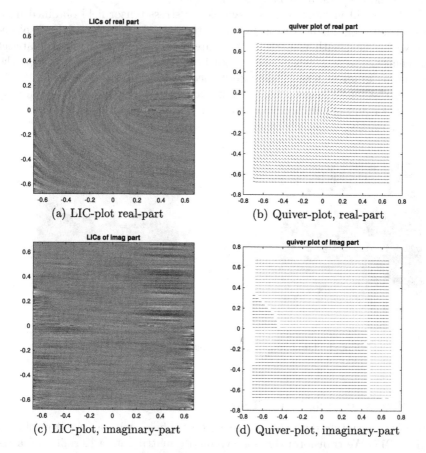

(a) LIC-plot real-part   (b) Quiver-plot, real-part

(c) LIC-plot, imaginary-part   (d) Quiver-plot, imaginary-part

Fig. 2. First column: Graphical representation of the line integral convolution (LIC) of the real and imaginary part of the eigenvector corresponding to κ_1 of (2) for the data set *parabolic* having resolution 217×217. Second column: Thinned out quiver plots (43×43 resolution) of the real and imaginary part of the same eigenvector fields.

In some applications, where tensorial quantities are derived from gradients, for example, indefinite matrices may play a role. Using polar decomposition for symmetrization inevitably causes positive definiteness of the resulting tensor, which means an additional loss of information. Nevertheless, even in the case of positivity, the asymmetry captured in the imaginary part of the Hermitian tensor reveals discontinuity properties of the data (visible in the quiver plots) that are independent of the rotational ambiguity mentioned in Remark 1, and hence they should not be discarded. Whether these discontinuities indicate possible locations of emerging fractures in materials or real anomalies in flow patterns is not yet clear. At this stage of our research a reasonable and authoritative explanation still eludes the authors.

For the sake of comparison, we show in Fig. 3 the major eigenvector field in its LIC-representation after the non-symmetric stress tensor data have been symmetrized in each pixel simply by means of $A \mapsto (A + A^\top)/2$.

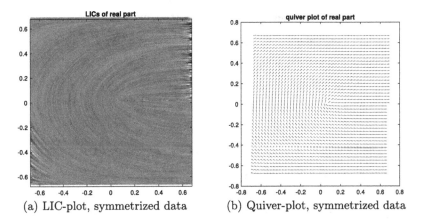

(a) LIC-plot, symmetrized data   (b) Quiver-plot, symmetrized data

Fig. 3. Left: Graphical representation of the line integral convolution (LIC) of the eigenvector corresponding to κ_1 of the symmetrization $(A + A^\top)/2$ for the data set *parabolic*. Right: Thinned out quiver plots (43×43 resolution) of the same symmetrized real eigenvector fields.

As expected, the imaginary part is zero and does not contain any information, hence we refrain from a graphical representation. A comparison of the major eigenvector field stemming from the real part of the hermitization versus the symmetrized tensor field reveals very little differences in the LIC-representation as well as in the quiver plot. The reason might be that the non-symmetry in the original data is not very pronounced and hence the imaginary part is rather small. Nevertheless, the similarity between hermitization and symmetrized version speaks for the reliability of the proposed approach.

Next, we process and visualize the stress tensor fields stemming from the *sinusoidal* data set as illustrated in Fig. 4.

The LIC-representation and the quiver plot of the real part of the major eigenvector field reveal an eddy-like structure inside the shape as a response to the *sinusoidal* pattern of forces applied on the right border. The LIC image of the imaginary part of the major eigenvector field indicates a complicated pattern inside the object, and as before, the quiver plot capable of capturing

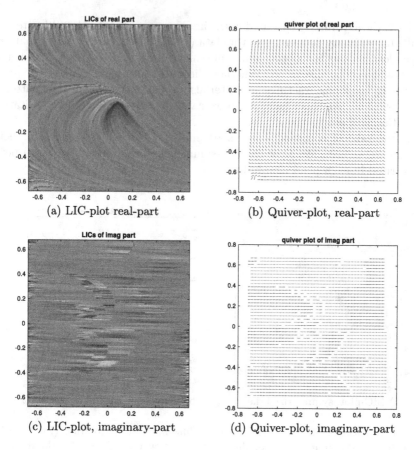

(a) LIC-plot real-part

(b) Quiver-plot, real-part

(c) LIC-plot, imaginary-part

(d) Quiver-plot, imaginary-part

Fig. 4. First column: Graphical representation of the line integral convolution (LIC) of the real and imaginary part of the eigenvector corresponding to κ_1 of (2) for the data set *sinusoidal* having resolution 217×217. Second column: Thinned out quiver plots (43×43 resolution) of the real and imaginary part of the same eigenvector fields.

the discontinuous "left-right" pattern, provides a similar but more discernible internal structure. A thorough interpretation of such newly discovered pattern will be the subject of future research.

For comparison, we show in Fig. 5 the major eigenvector after applying line integral convolution to data being symmetrized via $(A + A^\top)/2$ in each pixel.

Again, the vanishing imaginary part is not graphically represented. In contrast to the findings for the previous data set a comparison of the major eigenvector field stemming from the real part of the hermitization versus the symmetrized tensor field reveals significant differences in the LIC-representation as well as in the quiver plot. We attribute this to a more pronounced non-symmetry and a larger imaginary component if compared with the first data set. The data set *sinusoidal* suggests that symmetrization indeed destroys a significant portion of information and that this information not only might be preserved by hermitiza-

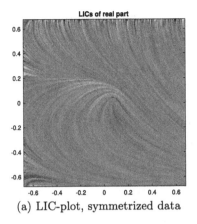

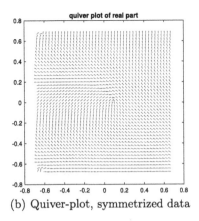

(a) LIC-plot, symmetrized data   (b) Quiver-plot, symmetrized data

Fig. 5. Left: Graphical representation of the line integral convolution (LIC) of the eigenvector corresponding to κ_1 of the symmetrization $(A + A^\top)/2$ for the data set *sinusoidal*. Right: Thinned out quiver plots (43×43 resolution) of the same symmetrized real eigenvector fields.

tion but also, via its imaginary part, might eventually lead to new interpretations and insights.

3.2   Deformation Gradient Tensor

The deformation gradient tensor as a non-symmetric tensor is a meaningful quantity, but it is not very much studied in the literature. We proceed almost as before. We extract the deformation gradient tensor fields for both force configurations (*parabolic* and *sinusoidal*), we use hermitization and then produce graphical LIC and quiver representations as shown in Figs. 6 and 7.

In case of the data set *parabolic* the LIC representations of the deformation gradient tensor field and the stress tensor field (refer to Fig. 2) look very much the same if we look at the real part of the major eigenvector fields. Seemingly, the same holds true for the quiver plots. However, the imaginary part exhibits some differences between the deformation gradient field and the stress tensor field both in the LIC and the quiver plot, albeit more pronounced in the latter one, as expected.

The situation is different for the *sinusoidal* data set. We note some discrepancies between the deformation gradient field, see Fig. 7, and the corresponding stress tensor field, see Fig. 4, in all four representations, namely, LIC- and quiver plots, real and imaginary parts.

The experiments show, that significant information about a field of non-symmetric tensors is captured in the imaginary parts of its hermitization form, which solely results from the non-symmetry. The outcome suggests that symmetrization of the tensors indeed eliminates information to some extent. The imaginary part of the major eigenvector field displayed in quiver plots reveals

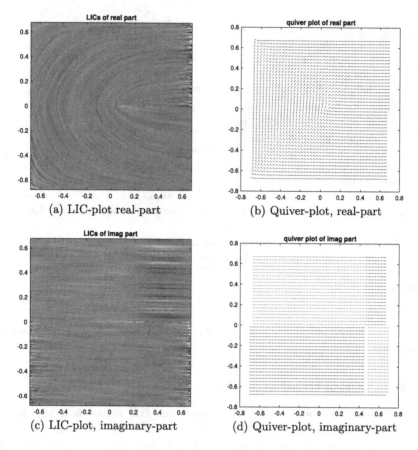

(a) LIC-plot real-part

(b) Quiver-plot, real-part

(c) LIC-plot, imaginary-part

(d) Quiver-plot, imaginary-part

Fig. 6. First column: Graphical representation of the line integral convolution (LIC) of the real and imaginary part of the eigenvector corresponding to κ_1 of (2) for the data set *parabolic* having resolution 217×217. Second column: Thinned out quiver plots (43×43 resolution) of the real and imaginary part of the same eigenvector fields.

this information, however, those plots by no means lend themselves to a straightforward interpretation; hence more research is needed in this direction.

4   Summary and Outlook

There are numerous examples of real second-order tensors in medical imaging or civil engineering that are symmetric, hence they have an eigendecomposition, which allows for a rather straightforward processing and analysis. However, the original tensors encountered in applications might not be symmetric. Their analysis and processing is much more cumbersome, since SVD and Jordan decomposition are complicated and sometimes insufficient substitutes for the lack of an eigendecomposition. In order to circumvent this difficulty, the tensors are often

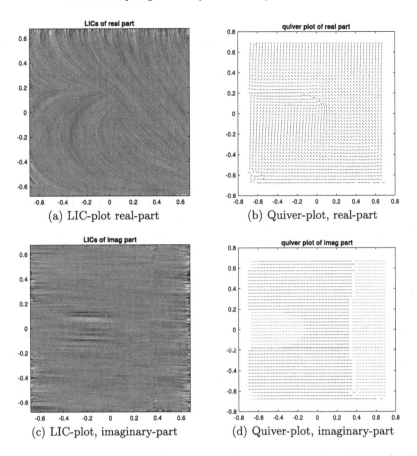

(a) LIC-plot real-part (b) Quiver-plot, real-part

(c) LIC-plot, imaginary-part (d) Quiver-plot, imaginary-part

Fig. 7. First column: Graphical representation of the line integral convolution (LIC) of the real and imaginary part of the eigenvector corresponding to κ_1 of (2) for the data set *sinusoidal* having resolution 217×217. Second column: Thinned out quiver plots (43×43 resolution) of the real and imaginary part of the same eigenvector fields.

symmetrized, and the asymmetric part is discarded, but this possibly comes with the price of loosing the information captured in the asymmetry. Rewriting a non-symmetric tensor as a Hermitian tensor, hermitization for short, allows to preserve this information since Hermitian matrices possess an eigendecomposition as well. However, the eigenvectors have complex-valued components, which means, that the non-symmetry information is mainly cast into the imaginary part of the eigenvectors while its real part is very close to the one stemming from the symmetrized version. Although the complex eigenvector may be shifted in phase (i.e. multiplied by $e^{i\beta}$ with $\beta \in [0, 2\pi]$) the abrupt discontinuous behavior of the imaginary part of the major eigenvector identifies edges in the tensor field that clearly stem from the asymmetric parts of the tensors.

The numerical results on non-symmetric 2D stress and deformation gradient tensor fields clearly indicate that this part of information is indeed relevant, albeit difficult to interpret from the quiver and LIC-plots.

In fact, we hope to find a better connection of the presented subject to image processing and mathematical morphology in the future. The discovered discontinuity represents a boundary of a region. If the external force pattern changes, then this region will change as well leading to a moving boundary. Maybe this movement can be related to some morphological operation. Knowledge of this connection might enable us to predict the boundary of a yet unknown force field. Hence, the extensive simulation of the mechanical response of external forces applied to materials can be replaced by simple morphological operations to obtain the location of such a discontinuity. This is especially relevant in three dimensions.

More tests and research efforts in this direction will be a topic of our future research.

References

1. Auer, C., Kasten, J., Kratz, A., Zhang, E., Hotz, I.: Automatic, tensor-guided illustrative vector field visualization. In: 2013 IEEE Pacific Visualization Symposium, Sydney, pp. 265–272. IEEE (2013). https://doi.org/10.1109/PacificVis.2013.6596154
2. Burgeth, B., Kleefeld, A.: Towards processing fields of general real-valued square matrices. In: Schultz, T., Özarslan, E., Hotz, I. (eds.) Modeling, Analysis, and Visualization of Anisotropy. MV, pp. 115–144. Springer, Cham (2017). https://doi.org/10.1007/978-3-319-61358-1_6
3. Burgeth, B., Kleefeld, A.: A unified approach to PDE-driven morphology for fields of orthogonal and generalized doubly-stochastic matrices. In: Angulo, J., Velasco-Forero, S., Meyer, F. (eds.) ISMM 2017. LNCS, vol. 10225, pp. 284–295. Springer, Cham (2017). https://doi.org/10.1007/978-3-319-57240-6_23
4. Burgeth, B., Pizarro, L., Breuß, M., Weickert, J.: Adaptive continuous-scale morphology for matrix fields. Int. J. Comput. Vis. **92**(2), 146–161 (2011). https://doi.org/10.1007/s11263-009-0311-4
5. Cabral, B., Leedom, L.C.: Imaging vector fields using line integral convolution. In: Proceedings of the 20th Annual Conference on Computer Graphics and Interactive Techniques, SIGGRAPH 1993, pp. 263–270. ACM, New York (1993). https://doi.org/10.1145/166117.166151
6. Dimitrienko, Y.I.: Nonlinear Continuum Mechanics and Large Inelastic Deformations, Solid Mechanics and Its Applications, vol. 174. Springer, New York (2011). https://doi.org/10.1007/978-94-007-0034-5
7. Khan, F., et al.: Multi-scale topological analysis of asymmetric tensor fields on surfaces. IEEE Trans. Vis. Comput. Graph. **26**(1), 270–279 (2020). https://doi.org/10.1109/TVCG.2019.2934314
8. O'Meara, O.: Introduction to Quadratic Forms. Classics in Mathematics. Springer, Berlin (2000). https://doi.org/10.1007/978-3-642-62031-7
9. Palke, D., et al.: Asymmetric tensor field visualization for surfaces. IEEE Trans. Vis. Comput. Graph. **17**(12), 1979–1988 (2011). https://doi.org/10.1109/TVCG.2011.170

10. Zhang, E., Yeh, H., Lin, Z., Laramee, R.S.: Asymmetric tensor analysis for flow visualization. IEEE Trans. Vis. Comput. Graph. **15**(1), 106–122 (2009). https://doi.org/10.1109/TVCG.2008.68

11. Zheng, X., Pang, A.T.: 2D asymmetric tensor analysis. In: IEEE Visualization, VIS 2005, pp. 3–10 (2005). https://doi.org/10.1109/VISUAL.2005.1532770

A Heuristic for Short Homology Basis of Digital Objects

Aldo Gonzalez-Lorenzo[✉][ID], Alexandra Bac, and Jean-Luc Mari[ID]

Aix Marseille Univ, CNRS, LIS, Marseille, France
{aldo.gonzalez-lorenzo,alexandra.bac,jean-luc.mari}@univ-amu.fr

Abstract. Finding the minimum homology basis of a simplicial complex is a hard problem unless one only considers the first homology group. In this paper, we introduce a general heuristic for finding a short homology basis of any dimension for digital objects (that is, for their associated cubical complexes) with complexity $\mathcal{O}(m^3 + \beta_q \cdot n^3)$, where m is the size of the bounding box of the object, n is the size of the object and β_q is the rank of its qth homology group. Our heuristic makes use of the thickness-breadth balls, a tool for visualizing and locating holes in digital objects.

We evaluate our algorithm with a data set of 3D digital objects and compare it with an adaptation of the best current algorithm for computing the minimum radius homology basis by Dey, Li and Wang [10].

Keywords: Digital object · Persistent homology · Short homology basis · Thickness-breadth

1 Introduction

A 3D digital object is a finite set of voxels, that is, a subset of the regular grid $\mathbb{Z}^3$. This is a fundamental data representation of shapes because it is well suited for data coming from volumetric acquisition devices, while its regular structure allows for efficient data structures and algorithms. A digital object can be endowed with a topological space by defining an adjacency relation between its voxels, and more generally, by defining an associated topological object called cubical complex. With a cubical complex we can start studying the topological properties of a shape such as its connectivity, its homology and its homotopy groups.

The homology groups are algebraic objects that rigorously define the notion of *hole* in a space. Let us give an informal presentation in dimension one. The elements of the first homology group are closed curves (called 1-cycles) under an equivalence relation that implies that two 1-cycles are related if one can be continuously transformed into the other. Hence, if an object has a tunnel, a 1-cycle around this tunnel is not equivalent to a small 1-cycle on a side and thus its homology group is not trivial. The first homology group is a quotient group with the structure of a vector space, so its dimension is the number of holes in

© Springer Nature Switzerland AG 2022
É. Baudrier et al. (Eds.): DGMM 2022, LNCS 13493, pp. 60–70, 2022.
https://doi.org/10.1007/978-3-031-19897-7_6

the object, and a basis can be used to visualize the holes of the object (see Fig. 1 for instance).

In order to enhance the visualization of the holes in an object, we look for a minimal homology basis, that is, a set of 1-cycles generating the homology group with total minimal length. Such a set is then tight around the holes and allows us to identify them separately. Short homology bases find application in contexts when one wishes to identify and locate the holes of an object such as in [13,23].

Finding a minimal homology basis made of 1-cycles can be solved in polynomial time using techniques from graph theory [11]. Unfortunately, the generalization of this problem to higher dimension (with closed manifolds instead of simply curves) has been proved to be NP-hard.

In this paper, we introduce a general heuristic for computing a short homology basis for digital objects and homology groups of any dimension. Our algorithm takes advantage of the *thickness-breadth balls* [16]. The thickness-breadth balls were conceived as a visualization and a quantification tool for holes in digital objects avoiding the use of homology groups because of the hardness of optimizing its size, but we show in this paper that they can indeed be used for computing a homology basis.

Our algorithm is easy to implement and runs in $\mathcal{O}(m^3 + \beta_q \cdot n^3)$, where m is the number of points in the bounding box of the digital object, n is its size and β_q is the rank of its qth homology group. We evaluate our heuristic on a set of 3D digital objects and compare it with a randomized version of the algorithm by Dey et al. [10] with similar time complexity. We show that our heuristic produces good results and we provide an open-source implementation to ensure the replicability of our results.

2   State of the Art

The problem of computing a minimal homology basis is well studied for simplicial complexes. Simplicial complexes are endowed with nonnegative weights on their simplices, which induce weights for every cycle. Thus, given a weighted simplicial complex K and a dimension $q > 0$, the problem is to find a set of cycles with minimum total weight such that their corresponding homology classes form a qth homology basis. This problem was proved to be NP-hard for general simplicial complexes and dimension $q > 1$ by Chen and Freedman [8], while Dey et al. [10] found an algorithm in $\mathcal{O}(n^\omega + \beta_1 \cdot n^2)$ for the case $q = 1$, where n is the number of simplices, $\omega < 2.38$ is the matrix multiplication exponent and β_1 is the rank of the first homology group $H_1(K)$. Indeed, the case $q = 1$ has brought particular attention and different specialized algorithms exist for specific subclasses of simplicial complexes [11,14,22].

Despite the hardness of the problem of minimal homology basis for dimension $q > 1$, there are works studying how to efficiently find a solution. Escolar and Hiraoka studied how to compute a solution using integer linear programming in [15]. Chen and Freedman introduced in [7] a notion of weight of a cycle called *radius* and showed that the minimum basis using this measure can be computed in $\mathcal{O}(\beta_q \cdot n^4)$. Their algorithm looks for the shortest non-trivial cycle in the complex by computing the persistent homology of one filtration per 0-dimensional simplex, seals this cycle, and repeats until the complex becomes acyclic. Dey et al. improved this result in [10] by using annotations and considering all the cycles at the same time with an algorithm that runs in $\mathcal{O}(n^{\omega+1})$. While both these algorithms are not guaranteed to find a good approximation for a minimal homology basis, they can be used as a heuristic.

3   Background

3.1   Digital Objects and Cubical Complexes

A d-dimensional *digital object* is a finite subset of $\mathbb{Z}^d$. Its elements are called pixels when $d = 2$, voxels when $d = 3$, or points in general.

We endow a digital object with an *adjacency relation*. Two points $x, y \in \mathbb{Z}^d$ are said to be $(3^d - 1)$-adjacent if $\|x - y\|_\infty \leq 1$. The transitive closure of this binary relation defines the connectivity of a digital object. We can obtain more topological information from a digital object by defining its associated cubical complex.

An *elementary interval* is an interval of the form $[k, k + 1]$ or a degenerate interval $[k, k]$, where $k \in \mathbb{Z}$. An *elementary cube* is the Cartesian product of d elementary intervals. The number of non-degenerate intervals in this product is the *dimension* of σ, which is denoted $\dim(\sigma)$. An elementary cube of dimension q will be called a q-*cube*. Given two elementary cubes σ and τ such that $\sigma \subsetneq \tau$, we say that σ is a *face* of τ.

A (finite) *cubical complex* K is a collection of elementary cubes such that for every $\sigma \in K$, its faces are also contained in K. For each $q \geq 0$, we denote by K_q the set of the q-cubes of K.

Given a digital object O, its associated cubical complex, $K[O]$, is defined as follows: for each point $x = (x_1, \cdots, x_d)$ of O, we add to the cubical complex the d-cube $[x_1, x_1 + 1] \times \cdots \times [x_d, x_d + 1]$ together with its faces.

Note that we can also consider the $(2d)$-adjacency relation for the digital object and define an associated cubical complex.

3.2   Homology

In this work, we consider only homology with coefficients in the field $\mathbb{Z}_2 := (\mathbb{Z}/2\mathbb{Z}, +, \cdot)$. For a more general presentation of homology theory, we refer the reader to [18,20].

A *chain complex* (C, ∂) is a sequence of $\mathbb{Z}_2$-vector spaces $\mathsf{C}_0, \mathsf{C}_1, \ldots$ (called *chain groups*) and linear maps $\partial_1 : \mathsf{C}_1 \to \mathsf{C}_0, \partial_2 : \mathsf{C}_2 \to \mathsf{C}_1, \ldots$ (called *boundary operators*) such that $\partial_{q-1} \circ \partial_q = 0$, for all $q > 0$.

A cubical complex K induces a chain complex. C_q is the $\mathbb{Z}_2$-vector space generated by the q-cubes of K. Its elements (called q-*chains* or chains of dimension q) are formal sums of q-cubes with coefficients in $\mathbb{Z}_2$, and thus, they can be interpreted as sets of q-cubes. The linear operator ∂_q maps each q-cube to the sum of its $(q-1)$-dimensional faces. For the 0-cubes, we define $\partial_0 = 0$.

A q-chain x is a *cycle* if $\partial_q(x) = 0$, and a *boundary* if $x = \partial_{q+1}(y)$ for some $(q+1)$-chain y. By the property $\partial_{q-1} \circ \partial_q = 0$, every boundary is a cycle, but the converse is not true: a cycle that is not a boundary contains a "hole". The qth *homology group* of the chain complex (C, ∂) is the quotient space $H_q(\mathsf{C}) := \ker(\partial_q)/\operatorname{im}(\partial_{q+1})$, whose elements are classes under the equivalence relation

$$\forall x, y \in \mathsf{C}_q, \quad x \sim y \Leftrightarrow x - y \in \operatorname{im}(\partial_{q+1}).$$

The dimension of the qth homology group is called the qth *Betti number* and is denoted by β_q, which is considered to be the number of holes of dimension q in the chain complex.

An *annotation* [5] is a homomorphism $a : \mathsf{C}_q \to H_q(\mathsf{C})$ such that for any two q-cycles $x, y \in \ker(\partial_q)$, $a(x) = a(y)$ if and only if $[x] = [y]$. Such a map can be encoded in a $\beta_q(K) \times |K_q|$ matrix and allows us to check if two cycles are homologous or if a cycle is a boundary.

3.3 Persistent Homology

Given a sequence of cubical complexes $K^1, \cdots, K^m$, (standard) homology allows us to obtain only the Betti numbers of each subcomplex K^i. However, if these complexes are nested, then *persistent homology* [12] provides global homological information.

Let K be a cubical complex. We say that a function $f : K \to \{a_1 < \cdots < a_m\} \subset \mathbb{R}$ is *monotone* if $f(\sigma) \leq f(\tau)$ for all $\sigma \subset \tau$. In such a case, $K^i := f^{-1}(] - \infty, a_i])$ is a cubical complex for $1 \leq i \leq m$, and $K^1 \subset \cdots \subset K^m = K$ is the *filtration* induced by f.

Persistent homology formalizes the idea that we can track the same hole in consecutive subcomplexes of a filtration, and hence detect the *lifespan* of the hole. The definition of persistent homology is based on the inclusion between the subcomplexes of the filtration, which induces a chain map between their chain groups and a homomorphism between their homology groups. We omit the details of this algebraic construction and directly point out that for each dimension q, persistent homology defines a *persistence diagram* $PD_q(f) \subset \mathbb{R}^2$, where each point $(a_i, a_j) \in PD_q(f)$ implies that there is a non-trivial homology class $[x]$ that appears in $f^{-1}(] - \infty, a_i])$ and becomes trivial in $f^{-1}(] - \infty, a_j])$.

Computing the persistence diagrams is a very active research field [21]. An algorithm in matrix multiplication time was introduced in [19]. Recent algorithms [2,4] have theoretical cubic worst-case complexity but near linear complexity in practice. These algorithms compute a partial matching of the cubes of the complex, $M = \bigcup_{i \in I}\{(\sigma_i, \tau_i)\}$, such that $\dim(\sigma_i) + 1 = \dim(\tau_i)$. The persistence diagram $PD_q(f)$ is derived from the matching M: its points are the pairs $(f(\sigma), f(\tau))$ for each $(\sigma, \tau) \in M, \dim(\sigma) = q$, together with the points $(f(\sigma), \infty)$ for each unmatched q-cube σ. Moreover, persistent homology algorithms also produce a representative cycle for each pair $(\sigma, \tau) \in M$ and for each unmatched cube while computing this matching M. Remark that the cycles associated to the unmatched q-cubes of M are a basis of $H_q(K)$.

4   Algorithm

In this section, we present our heuristic for obtaining a short homology basis using the thickness-breadth balls.

Let us first define precisely our problem. Let $O \subset \mathbb{Z}^d$ be a digital object and let us fix a dimension $q \geq 0$. The size of a chain $x \in C_q$ is the number of its q-cubes. We are looking for a set of cycles $\{x_1, \ldots, x_g\}$ such that their homology classes $\{[x_1], \ldots, [x_g]\}$ are a basis for the homology group $H_q(K[O])$ and the sum of the sizes of the cycles is small. From now on we call $g = \beta_q(K[O])$.

Our algorithm for computing a short homology basis consists of four steps: (1) compute the breadth balls; (2) compute one filtration for each breadth ball and extract a homology basis; (3) sort and annotate the cycles; (4) extract the earliest basis. In the following we detail each of these steps.

Step 1: Breadth balls. Given the digital object O with bounding box $BB = \prod_{i=1}^{d}[a_i, b_i] \subset \mathbb{Z}^d$, we define the filtration f_{sdt} over the cubical complex $K[BB]$ induced by the signed distance transform of O. The *signed distance transform* of O is the function that maps each point $x \notin O$ to its distance to O and each point $x \in O$ to its negative distance to $\mathbb{Z}^d \setminus O$. The persistent homology computation of this filtration f_{sdt} produces a matching M, from which we take the subset of q-cells $\mathcal{B} = \{\tau \in K_q[BB] : (\sigma, \tau) \in M, f_{sdt}(\sigma) < 0 < f_{sdt}(\tau)\}$. By the k-triangle Lemma [12], $\mathcal{B}$ contains g q-cells. Also, $\mathcal{B}$ is a subset of $K[BB] \setminus K[O]$. See [16] for more details about the definition, properties and computation of the thickness-breadth balls.

This step has time complexity $\mathcal{O}(m^3)$, where m is the number of cells in $K[BB]$, that is, $m = \prod_{i=1}^{d} 2(b_i - a_i + 1) + 1$.

Step 2: Compute g filtrations. For each cell $\tau \in \mathcal{B}$, we define a filtration f_τ over the complex $K[O]$ based on the discrete geodesic distance. The *discrete geodesic distance* between two 0-cubes σ_1, σ_2 is the size of the shortest path connecting σ_1 to σ_2 along the 1-cubes of $K[O]$. Let τ_0 be the closest 0-cube of $K[O]$ to τ (breaking ties arbitrarily), for each 0-cube σ of $K[O]$, $f_\tau(\sigma)$ is the minimum geodesic distance from σ to τ_0. This function f_τ is extended to the rest of the cubes of $K[O]$ by taking the maximum value over their 0-dimensional faces so that it is a filtration.

We compute the persistent homology of this filtration f_τ and extract the cycles associated to the unmatched q-cubes of $K[O]$. There are g such cycles, namely $\{x_{\tau,i}\}_{i=1}^{g}$, and their classes generate the qth homology group $H_q(K[O])$.

Taking the filtration associated to each cell of $\mathcal{B}$, we obtain a set of g^2 cycles $\bigcup_{\tau \in \mathcal{B}}\{x_{\tau,1}, \ldots, x_{\tau,g}\}$. Assuming that computing persistent homology has cubic complexity, the running time of this step is in $\mathcal{O}(g \cdot n^3)$, where n is the number of cells in $K[O]$.

Step 3: Sort and annotate. We sort the g^2 cycles by size and put them in a list $(x_1, \ldots, x_{g^2})$ in $\mathcal{O}(g^2 n + g^2 \log(g^2))$. Next, we need to annotate these cycles to obtain a list of elements in the homology group $H_q(K[O])$.

Busaryev et al. introduced in [5] an algorithm for computing an annotation in matrix multiplication time. This algorithm uses a LSP decomposition, which is difficult to implement [6]. We present here a simpler algorithm based on the notion of *homological discrete vector field* [17] with cubic complexity.

Let us fix an arbitrary filtration f over the cubical complex $K[O]$ and call $M = \bigcup_{i \in I} \{(\sigma_i, \tau_i)\}$ the matching of cells produced by the persistent homology computation of f. By fixing the dimension q, we define two subsets of cells, $P = \{\sigma : (\sigma, \tau) \in M, \dim(\sigma) = q\}$ and $S = \{\tau : (\sigma, \tau) \in M, \dim(\tau) = q+1\}$. Let i_S be the inclusion map from the chain group generated by the cells of S, $\mathbb{Z}_2[S]$, to C_{q+1} and j_P be the projection map from C_q to the chain group generated by the cells of P, $\mathbb{Z}_2[P]$. One can prove that the map $j_P \circ \partial_{q+1} \circ i_S$ is a bijection from $\mathbb{Z}_2[S]$ to $\mathbb{Z}_2[P]$ and that the map

$$a = (j_C \circ \partial_{q+1} \circ i_S) \circ (j_P \circ \partial_{q+1} \circ i_S)^{-1} \circ j_P + j_C$$

is an annotation, where C is the set of the unmatched q-cells of M. To compute this annotation, we only need the extract two submatrices of the boundary matrix ∂_{q+1}, $A = (j_C \circ \partial_{q+1} \circ i_S)$ and $B = (j_P \circ \partial_{q+1} \circ i_S)$, and compute $A \cdot B^{-1}$, which requires $\mathcal{O}(n^3)$ time.

Then, we apply the annotation map a to the g^2 sorted cycles and obtain a $g \times g^2$ matrix with columns $y_i = a(x_i)$. We compute each column of this matrix in $\mathcal{O}(g \cdot n)$.

Altogether, this step takes $\mathcal{O}(g^2 n + g^2 \log(g^2) + n^3 + n^3 + g^2 \cdot g \cdot n) = \mathcal{O}(g \cdot n^3)$ time.

Step 4: Earliest basis. Let Y be a $m \times n$ matrix with rank r, the set of columns $\{c_{j_1}, \ldots, c_{j_r}\}$ is an *earliest basis* if the indices $\{j_1, \ldots, j_r\}$ are the first (with respect to the lexicographical order) index set such that the corresponding columns of Y have full rank. Busaryev et al. introduced in [5] an algorithm in matrix multiplication time for computing an earliest basis using a LSP decomposition. However, its implementation is complex and thus we prefer to transform the matrix into column echelon form using Gaussian elimination, whose running time is in $\mathcal{O}(m^2 n)$.

Given the matrix Y with columns $a(x_1) \ldots a(x_{g^2})$, we perform Gaussian elimination in order to obtain its earliest basis $\{a(x_{j_1}), \ldots, a(x_{j_g})\}$. Then, the output of our algorithm is the corresponding set of cycles $\{x_{j_1}, \ldots, x_{j_g}\}$. Since the column vectors are all independent, the homology classes $a(x_{j_i})$ are a homology basis and their corresponding cycles x_j are representatives of these homology classes.

The matrix Y has dimensions $g \times g^2$ and thus this step requires $\mathcal{O}(g^4)$ time.

In all, our algorithm has complexity $\mathcal{O}(m^3 + g \cdot n^3)$, where m is the number of cells in $K[BB]$.

5   Results

Our heuristic can be seen as a special case of the algorithm for computing a minimal radius homology basis of Dey et al. [10], which we will refer to as MinRadiusBasis. This algorithm also computes filtrations based on the geodesic distance to a 0-cube and takes all the cycles to choose a minimal basis among them. The difference is that MinRadiusBasis computes a filtration for each 0-cube of the complex, which guarantees to find a minimum radius homology basis at the price of $\mathcal{O}(n^{\omega+1})$ time complexity.

To do a fair comparison, we adapt MinRadiusBasis to only compute the filtrations for a random subset of $\beta_q(K[O])$ 0-cubes.

Our software was developed in C++ using the library DGtal [1] for the distance transform computation and PHAT [3] for computing persistent homology. The source code is available under GPL at https://github.com/agonlor/tb-basis.

Our dataset consists of six 3D digital objects with one connected component and no cavities. These digital objects are available together with our implementation. For each object, we compute a first homology basis with our algorithm and with MinRadiusBasis with 10 different random subsets of 0-cubes. Table 1 presents the average total size of the homology bases computed with these two algorithms.

Table 1. Average sizes of the homology bases produced by our algorithm and MinRadiusBasis.

Object	Us	MinRadiusBasis
Amphora_200	930	1068.0
Dancing_200	902	1042.6
Eight_200	356	417.0
Fertility_200	962	999.8
Neptune_200	640	663.0
Pegasus_200	1032	1182.8

Our algorithm produces a shorter homology basis than MinRadiusBasis in average for all the digital objects, which shows that it is better to choose the subset of 0-cubes using the breadth balls than simply taking a random sampling. Surprisingly, the adaptation of MinRadiusBasis produces quite short homology basis for this dataset, and some executions of MinRadiusBasis succeed at finding a shorter basis than our algorithm.

We illustrate the homology bases computed by our algorithm in Fig. 1. Note that in a cubical complex, a minimal cycle is not necessarily tight around a hole.

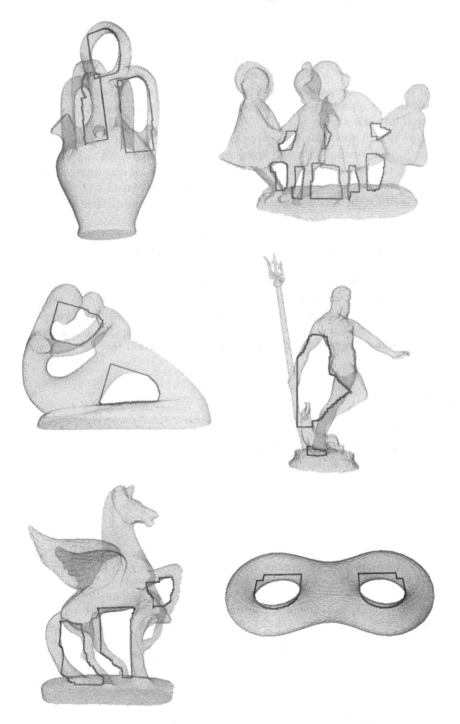

Fig. 1. Homology bases computed by our algorithm on the dataset. We have removed some cycles for the sake of clarity.

6   Conclusion

We have introduced a new algorithm for computing a short homology basis for digital objects which is based on the notion of breadth balls. We have made an experimental comparison between our algorithm and an adaptation of the state-of-the-art algorithm for computing minimum radius homology bases and displayed its output.

There is still room for improving these results. First, it seems possible to combine our algorithm with the linear programming approach in [15] for post-processing the cycles to further reduce their size. Also, if one wants to use the homology basis as a visualization tool, it would be convenient to transform the cycles to reduce their length estimation [9] without increasing their size, so that the cycles are *tight* around the holes.

The relation between the homology basis and the thickness-breadth balls can be exploited further. It may be possible to choose a subset of the breadth balls of an object and compute their corresponding short cycles. In this way, we can filter the holes of the object and compute a partial short homology basis faster.

Let us conclude by noting that our algorithm can be adapted to compute a short cohomology basis using the thickness balls. To the best of our knowledge, the problem of finding a minimum cohomology basis has not been studied even though a cohomology basis provides a useful visualization tool for the holes in an object.

References

1. DGtal: Digital geometry tools and algorithms library. http://dgtal.org
2. Bauer, U., Kerber, M., Reininghaus, J.: Clear and compress: computing persistent homology in chunks. In: Bremer, P.-T., Hotz, I., Pascucci, V., Peikert, R. (eds.) Topological Methods in Data Analysis and Visualization III. MV, pp. 103–117. Springer, Cham (2014). https://doi.org/10.1007/978-3-319-04099-8_7
3. Bauer, U., Kerber, M., Reininghaus, J., Wagner, H.: PHAT - persistent homology algorithms toolbox. J. Symb. Comput. **78**, 76–90 (2017). https://doi.org/10.1016/j.jsc.2016.03.008
4. Boissonnat, J., Dey, T.K., Maria, C.: The compressed annotation matrix: an efficient data structure for computing persistent cohomology. Algorithmica **73**(3), 607–619 (2015). https://doi.org/10.1007/s00453-015-9999-4
5. Busaryev, O., Cabello, S., Chen, C., Dey, T.K., Wang, Y.: Annotating simplices with a homology basis and its applications. In: Fomin, F.V., Kaski, P. (eds.) SWAT 2012. LNCS, vol. 7357, pp. 189–200. Springer, Heidelberg (2012). https://doi.org/10.1007/978-3-642-31155-0_17
6. Busaryev, O., Dey, T.K., Wang, Y.: Homology annotations via matrix reduction. Technical report, OSU-CISRC-4/12-TR04, Department of Computer Science and Engineering, the Ohio State University (2012). https://web.cse.ohio-state.edu/~dey.8/paper/annot/basis_TR.pdf
7. Chen, C., Freedman, D.: Measuring and computing natural generators for homology groups. Comput. Geom. **43**(2), 169–181 (2010). https://doi.org/10.1016/j.comgeo.2009.06.004

8. Chen, C., Freedman, D.: Hardness results for homology localization. Discret. Comput. Geom. **45**(3), 425–448 (2011). https://doi.org/10.1007/s00454-010-9322-8
9. Coeurjolly, D., Klette, R.: A comparative evaluation of length estimators of digital curves. IEEE Trans. Pattern Anal. Mach. Intell. **26**(2), 252–257 (2004). https://doi.org/10.1109/TPAMI.2004.1262194
10. Dey, T.K., Li, T., Wang, Y.: Efficient algorithms for computing a minimal homology basis. In: Bender, M.A., Farach-Colton, M., Mosteiro, M.A. (eds.) LATIN 2018. LNCS, vol. 10807, pp. 376–398. Springer, Cham (2018). https://doi.org/10.1007/978-3-319-77404-6_28
11. Dey, T.K., Sun, J., Wang, Y.: Approximating loops in a shortest homology basis from point data. In: Proceedings of the 26th ACM Symposium on Computational Geometry, Snowbird, Utah, USA, 13–16 June 2010, pp. 166–175 (2010). https://doi.org/10.1145/1810959.1810989
12. Edelsbrunner, H., Harer, J.: Computational Topology - An Introduction. American Mathematical Society (2010)
13. Emmett, K.J., Schweinhart, B., Rabadan, R.: Multiscale topology of chromatin folding. In: Suzuki, J., Nakano, T., Hess, H. (eds.) Proceedings of the 9th EAI International Conference on Bio-inspired Information and Communications Technologies (formerly BIONETICS), BICT 2015, New York City, USA, 3–5 December 2015, pp. 177–180. ICST/ACM (2015). http://dl.acm.org/citation.cfm?id=2954838
14. Erickson, J., Whittlesey, K.: Greedy optimal homotopy and homology generators. In: Proceedings of the Sixteenth Annual ACM-SIAM Symposium on Discrete Algorithms, SODA 2005, Vancouver, British Columbia, Canada, 23–25 January 2005, pp. 1038–1046 (2005). http://dl.acm.org/citation.cfm?id=1070432.1070581
15. Escolar, E.G., Hiraoka, Y.: Optimal cycles for persistent homology via linear programming. In: Fujisawa, K., Shinano, Y., Waki, H. (eds.) Optimization in the Real World. MI, vol. 13, pp. 79–96. Springer, Tokyo (2016). https://doi.org/10.1007/978-4-431-55420-2_5
16. Gonzalez-Lorenzo, A., Bac, A., Mari, J.-L., Real, P.: Two measures for the homology groups of binary volumes. In: Normand, N., Guédon, J., Autrusseau, F. (eds.) DGCI 2016. LNCS, vol. 9647, pp. 154–165. Springer, Cham (2016). https://doi.org/10.1007/978-3-319-32360-2_12
17. Gonzalez-Lorenzo, A., Bac, A., Mari, J.L., Real, P.: Allowing cycles in discrete Morse theory. Topol. Appl. **228**, 1–35 (2017). https://doi.org/10.1016/j.topol.2017.05.008
18. Kaczynski, T., Mischaikow, K., Mrozek, M.: Computational Homology, vol. 157, chap. 2, 7, pp. 255–258. Springer, Heidelberg (2004)
19. Milosavljevic, N., Morozov, D., Skraba, P.: Zigzag persistent homology in matrix multiplication time. In: Proceedings of the 27th ACM Symposium on Computational Geometry, Paris, France, 13–15 June 2011, pp. 216–225 (2011). https://doi.org/10.1145/1998196.1998229
20. Munkres, J.R.: Elements of Algebraic Topology. Addison-Wesley (1984)
21. Otter, N., Porter, M.A., Tillmann, U., Grindrod, P., Harrington, H.A.: A roadmap for the computation of persistent homology. EPJ Data Sci. **6**(1), 17 (2017). https://doi.org/10.1140/epjds/s13688-017-0109-5
22. Rathod, A.: Fast algorithms for minimum cycle basis and minimum homology basis. In: Cabello, S., Chen, D.Z. (eds.) 36th International Symposium on Computational Geometry, SoCG 2020, Zürich, Switzerland, 23–26 June 2020. LIPIcs, vol. 164, pp. 64:1–64:11. Schloss Dagstuhl - Leibniz-Zentrum für Informatik (2020). https://doi.org/10.4230/LIPIcs.SoCG.2020.64

23. Zhang, X., Wu, P., Yuan, C., Wang, Y., Metaxas, D.N., Chen, C.: Heuristic search for homology localization problem and its application in cardiac trabeculae reconstruction. In: Kraus, S. (ed.) Proceedings of the Twenty-Eighth International Joint Conference on Artificial Intelligence, IJCAI 2019, Macao, China, 10–16 August 2019, pp. 1312–1318. ijcai.org (2019). https://doi.org/10.24963/ijcai.2019/182

Completions and Ramifications

Gilles Bertrand$^{(\boxtimes)}$

Univ Gustave Eiffel, CNRS, LIGM, 77454 Marne-la-Vallée, France
gilles.bertrand@esiee.fr

Abstract. We investigate ramifications, which are simplicial complexes constructed with a very simple inductive property: if two complexes are ramifications, then their union is a ramification whenever their intersection is a ramification. We show that the collection of all ramifications properly contains the collection of all collapsible complexes and that it is properly contained in the collection of all contractible complexes. We introduce the notion of a ramification pair, which is a couple of complexes satisfying also an inductive property. We establish a strong relation between ramification pairs and ramifications. In particular, the collection of ramification pairs is uniquely determined by the collection of ramifications. Also we provide some relationships between ramification pairs, collapsible pairs, and contractible pairs.

Keywords: Combinatorial topology · Ramifications · Contractibility · Collapse · Completions

1 Introduction

Simple homotopy, introduced by J. H. C. Whitehead in the early 1930's, may be seen as a refinement of the concept of homotopy [1]. Two simplicial complexes are simple homotopy equivalent if one of them may be obtained from the other by a sequence of elementary collapses and expansions.

Simple homotopy plays a fundamental role in combinatorial topology [1–3]. Also, many notions relative to homotopy in the context of computer imagery rely on the collapse operation. In particular, this is the case for the notion of a simple point, which is crucial for all image transformations that preserve the topology of the objects [4–6].

In this paper, we investigate ramifications, which are simplicial complexes constructed with a very simple inductive property: if two complexes are ramifications, then their union is a ramification whenever their intersection is a ramification.

It could be seen that the collection of all trees satisfies the above property. Also, any complex of arbitrary dimension is a ramification whenever it is collapsible, *i.e.*, whenever it reduces to a single vertex with a sequence composed solely of collapses.

Our main results include the following:

- We show that the collection $\mathbb{R}$ of all ramifications properly contains the collection $\mathbb{E}$ of all collapsible complexes. Also we show that $\mathbb{R}$ is properly contained

© Springer Nature Switzerland AG 2022
E. Baudrier et al. (Eds.): DGMM 2022, LNCS 13493, pp. 71–83, 2022.
https://doi.org/10.1007/978-3-031-19897-7_7

in the collection $\mathbb{H}$ of all contractible complexes, *i.e.*, all complexes that are homotopy equivalent to a single vertex.

- We introduce the notion of a ramification pair, which is a couple of complexes satisfying also an inductive property. We show there is a strong relation between the collection of all ramification pairs $\ddot{\mathbb{R}}$ and $\mathbb{R}$. In particular, $\ddot{\mathbb{R}}$ is uniquely determined by $\mathbb{R}$.
- We show that $\ddot{\mathbb{R}}$ properly contains the collection of all collapsible pairs, and that $\ddot{\mathbb{R}}$ is properly contained in the collection of all contractible pairs.

The paper is organized as follows. First, we give some basic definitions for simplicial complexes (Sect. 2) and simple homotopy (Sect. 3). Then, we recall some facts relative to completions, which allow us to formulate inductive properties (Sect. 4). We investigate the containment relations between the collections $\mathbb{E}$, $\mathbb{R}$, and $\mathbb{H}$ in Sect. 5. Then, we introduce the collection $\ddot{\mathbb{R}}$ of ramification pairs and give the fundamental relation between $\ddot{\mathbb{R}}$ and $\mathbb{R}$ (Sect. 6). In Sect. 7, we make clear the relations between $\ddot{\mathbb{R}}$, collapsible pairs, and contractible pairs. Note that the paper is self contained. Nevertheless, for the sake of place, several proofs are not included, these proofs may be found in an online archive [11].

2 Basic Definitions for Simplicial Complexes

Let X be a finite family composed of finite sets. The *simplicial closure of X* is the complex $X^- = \{y \subseteq x \mid x \in X\}$. The family X is a *(simplicial) complex* if $X = X^-$. We write $\mathbb{S}$ for the collection of all finite simplicial complexes. Note that $\emptyset \in \mathbb{S}$ and $\{\emptyset\} \in \mathbb{S}$, $\emptyset$ is the *void complex*, and $\{\emptyset\}$ is the *empty complex*.

Let $X \in \mathbb{S}$. An element of X is *a simplex of X* or *a face of X*. A *facet of X* is a simplex of X that is maximal for inclusion. For example, the family $X = \{\emptyset, \{a\}, \{b\}, \{a, b\}\}$ is a simplicial complex with four faces and one facet. Note that the empty set is necessarily a face of X whenever $X \neq \emptyset$.

A *simplicial subcomplex* of $X \in \mathbb{S}$ is any subset Y of X that is a simplicial complex. If Y is a subcomplex of X, we write $Y \preceq X$.

Let $X \in \mathbb{S}$. The *dimension* of $x \in X$, written $dim(x)$, is the number of its elements minus one. The *dimension of X*, written $dim(X)$, is the largest dimension of its simplices, the *dimension of $\emptyset$*, the void complex, being defined to be -1. Observe that the dimension of the empty complex $\{\emptyset\}$ is also -1.

A complex $A \in \mathbb{S}$ is *a cell* if $A = \emptyset$ or if A has precisely one non-empty facet x. We set $A^\circ = A \setminus \{x\}$ and $\emptyset^\circ = \emptyset$. We write $\mathbb{C}$ for the collection of all cells. A cell $\alpha \in \mathbb{C}$ is *a vertex* if $dim(\alpha) = 0$.

The *ground set* of $X \in \mathbb{S}$ is the set $\underline{X} = \cup\{x \in X \mid dim(x) = 0\}$. Thus, if $A \in \mathbb{C}$, with $A \neq \emptyset$, then $\underline{A}$ is precisely the unique facet of A. In particular, if α is a vertex, we have $\alpha = \{\emptyset, \underline{\alpha}\}$.

We say that $X \in \mathbb{S}$ and $Y \in \mathbb{S}$ are *disjoint*, or that X is *disjoint from Y*, if $\underline{X} \cap \underline{Y} = \emptyset$. Thus, X and Y are disjoint if and only if $X \cap Y = \emptyset$ or $X \cap Y = \{\emptyset\}$.

If $X \in \mathbb{S}$ and $Y \in \mathbb{S}$ are disjoint, the *join of X and Y* is the simplicial complex XY such that $XY = \{x \cup y \mid x \in X, y \in Y\}$. Thus, $XY = \emptyset$ if $Y = \emptyset$ and $XY = X$ if $Y = \{\emptyset\}$. The join αX of a vertex α and a complex $X \in \mathbb{S}$ is a *cone*.

3   Simple Homotopy

We recall some basic definitions related to the collapse operator [1].

Let $X \in \mathbb{S}$ and let x, y be two distinct faces of X. The couple (x, y) is a *free pair for* X if y is the only face of X that contains x. Thus, the face y is necessarily a facet of X. If (x, y) is a free pair for X, then $Y = X \setminus \{x, y\}$ is *an elementary collapse of* X, and X is *an elementary expansion of* Y. We say that X *collapses onto* Y, or that Y *expands onto* X, if there exists a sequence $\langle X_0, ..., X_k \rangle$ such that $X_0 = X$, $X_k = Y$, and X_i is an elementary collapse of X_{i-1}, $i \in [1, k]$. The complex X is *collapsible* if X collapses onto $\emptyset$. We say that X is *(simply) homotopic to* Y, or that X and Y are *(simply) homotopic*, if there exists a sequence $\langle X_0, ..., X_k \rangle$ such that $X_0 = X$, $X_k = Y$, and X_i is an elementary collapse or an elementary expansion of X_{i-1}, $i \in [1, k]$. The complex X is *(simply) contractible* if X is simply homotopic to $\emptyset$.

Let $\alpha = \{\emptyset, \underline{\alpha}\}$ be an arbitrary vertex. We observe that $(\emptyset, \underline{\alpha})$ is a free face for α. Thus α collapses onto $\emptyset$, that is, the void complex. It follows that a complex is contractible if and only if it is homotopic to a single vertex. Also a non-void complex is collapsible if and only if it collapses onto a single vertex.

Remark 1. We observe that a complex $X \in \mathbb{S}$, $X \neq \emptyset$, is an elementary collapse of a complex Z, if and only if we have $Z - X \cup \gamma D$ and $X \cap \gamma D = \gamma D^\circ$, where D, $D \neq \emptyset$, is a cell, and γ is a vertex disjoint from D. See also [1], p. 247.

Let $X, Y \in \mathbb{S}$, and α be a vertex disjoint from $X \cup Y$. We can check that:

1) If $x, y \in X \setminus Y$, then (x, y) is a free pair for X iff (x, y) is a free pair for $X \cup Y$.
2) If $x \in X \setminus Y$ is a facet of X, then $(x, \underline{\alpha} \cup x)$ is a free pair for $\alpha X \cup Y$.
3) If $x, y \in X$, then the couple (x, y) is a free pair for X if and only if $(\underline{\alpha} \cup x, \underline{\alpha} \cup y)$ is a free pair for $\alpha X \cup Y$.

By induction, we have the following results which will be used in this paper.

Proposition 1. *Let* $X, Y \in \mathbb{S}$. *The complex* X *collapses onto* $X \cap Y$ *if and only if* $X \cup Y$ *collapses onto* Y.

Proposition 2. *Let* $X, Y \in \mathbb{S}$, *and let* α *be a vertex disjoint from* $X \cup Y$.
The complex $\alpha X \cup Y$ *collapses onto* $\alpha(X \cap Y) \cup Y$.
In particular, the complex αX *collapses onto* $\emptyset$. *Thus any cone is collapsible.*

Proposition 3. *Let* $X, Y \in \mathbb{S}$, $Z \preceq X$, *let* α *be a vertex disjoint from* $X \cup Y$.
The complex X *collapses onto* Z *if and only if* $\alpha X \cup Y$ *collapses onto* $\alpha Z \cup X \cup Y$.
In particular, if X *is collapsible, then* $\alpha X \cup Y$ *collapses onto* $X \cup Y$.

4   Completions

We give some basic definitions for completions. A completion may be seen as a rewriting rule that permits to derive collections of sets. See [7] for more details.

Let **S** be a given collection and let $\mathcal{K}$ be an arbitrary subcollection of **S**. Thus, we have $\mathcal{K} \subseteq \mathbf{S}$. In the sequel of the paper, the symbol $\mathcal{K}$, with possible superscripts, will be a dedicated symbol (a kind of variable).

Let κ be a binary relation on $2^{\mathbf{S}}$, thus $\kappa \subseteq 2^{\mathbf{S}} \times 2^{\mathbf{S}}$. We say that κ is *finitary*, if **F** is finite whenever $(\mathbf{F}, \mathbf{G}) \in \kappa$.

Let $\langle K \rangle$ be a property that depends on $\mathcal{K}$. We say that $\langle K \rangle$ is a *completion* (*on* **S**) if $\langle K \rangle$ may be expressed as the following property:

$->$ If $\mathbf{F} \subseteq \mathcal{K}$, then $\mathbf{G} \subseteq \mathcal{K}$ whenever $(\mathbf{F}, \mathbf{G}) \in \kappa$. $\qquad\qquad \langle \kappa \rangle$

where κ is a finitary binary relation on $2^{\mathbf{S}}$.

If $\langle K \rangle$ is a property that depends on $\mathcal{K}$, we say that a given collection $\mathbf{X} \subseteq \mathbf{S}$ satisfies $\langle K \rangle$ if the property $\langle K \rangle$ is true for $\mathcal{K} = \mathbf{X}$.

Theorem 1 (from [7]). *Let $\langle K \rangle$ be a completion on **S** and let $\mathbf{X} \subseteq \mathbf{S}$. There exists, under the subset ordering, a unique minimal collection that contains **X** and that satisfies $\langle K \rangle$.*

If $\langle K \rangle$ is a completion on **S** and if $\mathbf{X} \subseteq \mathbf{S}$, we write $\langle \mathbf{X}; K \rangle$ for the unique minimal collection that contains **X** and that satisfies $\langle K \rangle$. We say that $\langle \mathbf{X}; K \rangle$ is a *completion system* and that **X** is the *starting collection* of $\langle \mathbf{X}; K \rangle$.

Let $\langle K_1 \rangle, \langle K_2 \rangle, ..., \langle K_k \rangle$ be completions on **S**. We write $\wedge$ for the logical "and". It may be seen that $\langle K \rangle = \langle K_1 \rangle \wedge \langle K_2 \rangle ... \wedge \langle K_k \rangle$ is a completion. In the sequel, we write $\langle K_1, K_2, ..., K_k \rangle$ for $\langle K \rangle$. Thus, if $\mathbf{X} \subseteq \mathbf{S}$, the notation $\langle \mathbf{X}; K_1, K_2, ..., K_k \rangle$ stands for the smallest collection that contains **X** and that satisfies each of the properties $\langle K_1 \rangle, \langle K_2 \rangle, ..., \langle K_k \rangle$. We observe that, if $\langle K \rangle$ and $\langle Q \rangle$ are two completions on **S**, then we have $\langle \mathbf{X}; K \rangle \subseteq \langle \mathbf{X}; K, Q \rangle$ whenever $\mathbf{X} \subseteq \mathbf{S}$.

5   Ramifications

5.1   Definition

The notion of a dendrite was introduced in [7] as a way for defining a collection made of acyclic complexes. Let us consider the collection $\mathbf{S} = \mathbb{S}$, and let $\mathcal{K}$ denote an arbitrary collection of simplicial complexes.

We define the two completions $\langle R \rangle$ and $\langle D \rangle$ on $\mathbb{S}$: For any $S, T \in \mathbb{S}$,

$->$ If $S, T \in \mathcal{K}$, then $S \cup T \in \mathcal{K}$ whenever $S \cap T \in \mathcal{K}$. $\qquad\qquad \langle R \rangle$
$->$ If $S, T \in \mathcal{K}$, then $S \cap T \in \mathcal{K}$ whenever $S \cup T \in \mathcal{K}$. $\qquad\qquad \langle D \rangle$

Let $\mathbb{D} = \langle \mathbb{C}; R, D \rangle$. Each element of $\mathbb{D}$ is a *dendrite* or an *acyclic complex*. We have the general result [7]:

A non-void complex is a dendrite iff it is acyclic in the sense of homology.

We set $\mathbb{R} = \langle \mathbb{C}; R \rangle$. Each element of $\mathbb{R}$ is a *ramification*. Thus, the collection $\mathbb{R}$ is the unique minimal collection that contains $\mathbb{C}$ and that satisfies the property $\langle R \rangle$. Also, the collection $\mathbb{R}$ is the very collection that may be obtained by starting from $\mathcal{K} = \mathbb{C}$, and by iteratively adding to $\mathcal{K}$ all the sets $S \cup T$ such that $S, T \in \mathcal{K}$ and $S \cap T \in \mathcal{K}$.

Note that the notion of a ramification corresponds to the buildable complexes introduced by J. Jonsson [3]. Here, we have a formulation in terms of completions.

The collection of all cones provides a basic example of ramifications. If a cone αZ has more than one facet, then it may be split in two distinct cones αX and αY such that $\alpha Z = \alpha X \cup \alpha Y$. Since $\alpha X \cap \alpha Y$ is a cone, and αZ is a cell whenever αZ has a single facet, it follows by induction that any cone is a ramification.

5.2   Ramifications and Collapsible Complexes

Let us denote by $\mathbb{E}$ the collection of all complexes X such that $\emptyset$ expands onto X, *i.e.*, such that X is collapsible. This collection may be described by completions. See Sec. 6 of [7] and Sec. 8 of [10]. Now let us consider the alternative definition of an elementary collapse given in Remark 1. If $X \in \mathbb{S}$, $X \neq \emptyset$, is an elementary collapse of Z, then we have $Z = X \cup Y$, where Y and $X \cap Y$ are cones. Since cones are ramifications, and since the void complex is a ramification, we can again prove by induction that any collapsible complex is a ramification. Thus, we have $\mathbb{E} \subseteq \mathbb{R}$. See [7] and [10] (Sec. 8). See also [3] (Def. 3.14 and Prop. 5.17) where a slightly different definition of a collapsible complex is used.

The Bing's house [13] is a classical example of an object that is contractible but not collapsible, see Fig. 1(a). This two dimensional object is made of two rooms. Two tunnels allow to enter to the upper room by the lower face, and to the lower room by the upper face. Two small walls are attached to the tunnels in order to make this object acyclic.

In [7], it was noticed that the Bing's house B is a ramification. Let us consider the two complexes B_1 and B_2 of Fig. 1(b) and (c). We have $B = B_1 \cup B_2$. If B is correctly triangulated, then we can see that B_1, B_2, and $B_1 \cap B_2$ are all collapsible. Since $\mathbb{E} \subseteq \mathbb{R}$, these three complexes are ramifications. Thus, the Bing's house B is a ramification. But the Bing's house is not collapsible, in fact there is nowhere we can start a collapse sequence. In consequence, the inclusion $\mathbb{E} \subseteq \mathbb{R}$ is strict.

5.3   Ramifications and Contractible Complexes

Now, let us consider the collection $\mathbb{H}$ made of all contractible complexes. We have $\mathbb{H} \subseteq \mathbb{D}$, this inclusion is strict (see [10]).

It was shown (Prop. 5.17 of [3]) that any buildable complex (or any ramification) is contractible. The arguments given for the proof are based on the

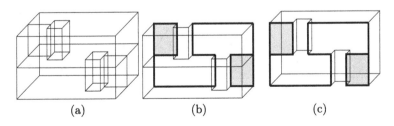

(a)   (b)   (c)

Fig. 1. (a): A Bing's house B with two rooms, (b): An object $B_1 \subseteq B$, (c): An object $B_2 \subseteq B$. We have $B = B_1 \cup B_2$, the object $B_1 \cap B_2$ is outlined in (b) and (c).

Hurewicz theorem (Th. 4.32 of [15]). It follows that these arguments do not allow to build an effective sequence of collapses and expansions that transform any ramification into the void complex. In fact, it is undecidable to determine whether a finite simplicial complex is contractible or not (for example see [16], Appendix). Thus, such an effective sequence cannot, in general, be given.

In the extension of this paper (Appendix A of [11]), we provide a direct proof that permits to build such a sequence. We illustrate an aspect of this proof with the decomposition $B = B_1 \cup B_2$ of the Bing's house given Fig. 1.

Let α be a vertex disjoint from B. Since B_1 is collapsible, B expands onto the complex $C = \alpha B_1 \cup B_2$ (Proposition 3, by replacing collapses by expansions). Now, we can collapse C onto the complex $D = \alpha(B_1 \cap B_2) \cup B_2$ (Proposition 2). Since $B_1 \cap B_2$ is collapsible, the complex D collapses onto B_2 (Proposition 3), which is collapsible.

Thus, the sequence $B \nearrow \alpha B_1 \cup B_2 \searrow \alpha(B_1 \cap B_2) \cup B_2 \searrow B_2 \searrow \emptyset$ gives an homotopic deformation between B and $\emptyset$; the symbol $\nearrow$ stands for expansions and the symbol $\searrow$ for collapses. Now, let us consider a complex $B' = B_1' \cup B_2'$ where B_1' and B_2' are two copies of B such that $B_1' \cap B_2'$ is a ramification. The complex B' is a ramification but, since B_1' and B_2' are not collapsible, the above sequence is no longer valid. Furthermore, this process may be iterated by considering two copies of B', and so on. In the extension [11], we handle this problem by proposing an inductive construction which, at each step, allows us to perform the above sequence.

Thus, we have $\mathbb{R} \subseteq \mathbb{H}$. Are there contractible complexes that are not ramification? This question corresponds to a conjecture formulated by J. Jonsson [3] (Problem 5.21). We give a positive answer to this question in the Appendix B of the extension of this paper [11]. The counter-example is given by the dunce hat [14] which is another classical example of an object that is contractible but not collapsible, see Fig. 2(a). See also Appendix A of this paper where the contractibility of this object is shown. Note that we only proved that a specific triangulation of the dunce hat is not a ramification. This leaves open the question for any triangulation of this complex.

The following proposition summarizes the facts given in this section.

Proposition 4. *We have* $\mathbb{E} \subsetneq \mathbb{R} \subsetneq \mathbb{H} \subsetneq \mathbb{D}$.

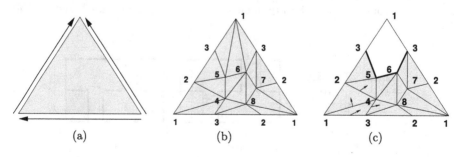

Fig. 2. (a): The dunce hat, the three edges of the triangle have to be identified with the arrows, (b): a triangulation D of the dunce hat, (c): a subcomplex of D.

6   Ramification Pairs

In the last section, we mentioned some previously published results related to ramifications. As far as we know these are the only ones that may be found in the literature. By providing counter-examples, and by giving an appropriate homotopic deformation, we also clarified the link between ramifications, collapsible and contractible complexes. Now, in order to achieve a better understanding of these objects, we will extend the collection $\mathbb{R}$ to a collection $\ddot{\mathbb{R}}$, which is composed of couples of complexes. It should be noted that the following completion $\langle \widetilde{\mathbb{R}} \rangle$ has already been introduced in a previous paper [8]. Nevertheless, it was always associated with another completion (its dual), so that all the following results are new. Note that all the proofs of the results given in this section may be found in the extension [11].

We set $\ddot{\mathbb{S}} = \{(X,Y) \mid X, Y \in \mathbb{S}, X \preceq Y\}$ and $\ddot{\mathbb{C}} = \{(A,B) \in \ddot{\mathbb{S}} \mid A, B \in \mathbb{C}\}$. The notation $\ddot{\mathcal{K}}$ stands for an arbitrary subcollection of $\ddot{\mathbb{S}}$.

We define the completion $\langle \widetilde{\mathbb{R}} \rangle$ on $\ddot{\mathbb{S}}$: For any (S,T), (S',T') in $\ddot{\mathbb{S}}$,

$\quad -> $ If (S,T), (S',T'), $(S \cap S', T \cap T') \in \ddot{\mathcal{K}}$, then $(S \cup S', T \cup T') \in \ddot{\mathcal{K}}$. $\langle \widetilde{\mathbb{R}} \rangle$

We set $\ddot{\mathbb{R}} = \langle \ddot{\mathbb{C}} \cup \ddot{\mathbb{I}}; \widetilde{\mathbb{R}} \rangle$, where $\ddot{\mathbb{I}} = \{(X,X) \mid X \in \mathbb{S}\}$.

Each couple of $\ddot{\mathbb{R}}$ is a *ramification pair*.

In fact, the collection $\ddot{\mathbb{R}}$ may be generated with a smaller starting collection. We have $\ddot{\mathbb{R}} = \langle \ddot{\mathbb{C}}^{\#}; \widetilde{\mathbb{R}} \rangle$, where $\ddot{\mathbb{C}}^{\#} = \ddot{\mathbb{C}} \cup \{(\{\emptyset\}, \{\emptyset\})\}$, see the extension [11].

The four couples given in Fig. 3 correspond to the four couples appearing in the definition of the completion $\langle \widetilde{\mathbb{R}} \rangle$. In this specific illustration, we observe that, if (X,Y) is one of these four couples, then Y collapses onto X (under an appropriate triangulation).

We introduce the notion of a Δ-form, the symbol Δ corresponds to a binary relation over $\ddot{\mathbb{S}}$ and $\mathbb{S}$.

Let $(X,Y) \in \ddot{\mathbb{S}}$ and $Z \in \mathbb{S}$. We write $\Delta(X,Y,Z)$ if there exists a vertex α, disjoint from Y, such that $Z = \alpha X \cup Y$. In this case, we write $\alpha(X,Y)$ for the complex Z, and we say that $\alpha(X,Y)$ is a Δ-form. We also say that $\alpha(X,Y)$ is a Δ-form of (X,Y) or a Δ-form of Z.

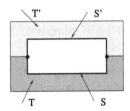

Fig. 3. Four couples (S,T), (S',T'), $(S \cap S', T \cap T')$, $(S \cup S', T \cup T')$, that are ramification pairs; S and S' are two simple open curves, $S \cap S'$ is made of two vertices.

If $Z \in \mathbb{S}$ and α is an arbitrary vertex, it may be seen that there exists a unique couple $(X, Y) \in \ddot{\mathbb{S}}$ such that $Z = \alpha(X, Y)$. We have:
$$X = \{x \in Z \mid x \cap \underline{\alpha} = \emptyset \text{ and } x \cup \underline{\alpha} \in Z\} \text{ and } Y = \{x \in Z \mid x \cap \underline{\alpha} = \emptyset\}.$$
The complex X is the so-called link of the face $\underline{\alpha}$ in Z, and Y is the so-called deletion of $\underline{\alpha}$ from Z, see [3]. Thus, we have:
$$\alpha(X, Y) = \alpha(X', Y') \text{ if and only if } (X, Y) = (X', Y').$$
Note that we have $X = \emptyset$ and $Y = Z$ whenever α is disjoint from Z.

We now clarify the correspondence between $\ddot{\mathbb{R}}$ and $\mathbb{R}$ induced by Δ-forms:

1) If $(X, Y) \in \ddot{\mathbb{S}}$, then, up to a renaming of the vertex α, the couple (X, Y) has a unique Δ-form $Z = \alpha(X, Y)$. Thus, up to this renaming, there is a unique complex in $\mathbb{S}$ which is the Δ-form of a couple in $\ddot{\mathbb{S}}$.
2) If $Z \in \mathbb{S}$, and for a given α, there is a unique couple (X, Y) such that $Z = \alpha(X, Y)$. Now, for all possible choices of α, we observe that there are precisely $k + 1$ different such couples, where k is the number of vertices included in Z (we have to consider the case where α is disjoint from Z). Thus, in general, there are several different couples in $\ddot{\mathbb{S}}$ which are the Δ-forms of a complex in $\mathbb{S}$.

Proposition 5. *Let $\alpha(X', Y')$ and $\alpha(X'', Y'')$ be two Δ-forms.*

1) *We have $\alpha(X' \cup X'', Y' \cup Y'') = \alpha(X', Y') \cup \alpha(X'', Y'')$.*
2) *We have $\alpha(X' \cap X'', Y' \cap Y'') = \alpha(X', Y') \cap \alpha(X'', Y'')$.*

By induction on $\mathbb{R}$ and $\ddot{\mathbb{R}}$, Proposition 5 leads to the following relation between these two collections.

Theorem 2. *Let $(X, Y) \in \ddot{\mathbb{S}}$ and let $Z \in \mathbb{S}$ such that $\Delta(X, Y, Z)$. We have $(X, Y) \in \ddot{\mathbb{R}}$ if and only if $Z \in \mathbb{R}$.*

Replacing X by $\emptyset$ in the previous theorem, we obtain the corollary:

Corollary 1. *We have $\mathbb{R} = \{X \in \mathbb{S} \mid (\emptyset, X) \in \ddot{\mathbb{R}}\}$.*

Let us consider an arbitrary collection $\ddot{\mathbb{K}} \subseteq \ddot{\mathbb{S}}$. We define the *kernel of* $\ddot{\mathbb{K}}$ as the collection $\mathbb{K} = \{X \in \mathbb{S} \mid (\emptyset, X) \in \ddot{\mathbb{K}}\}$. We consider the two properties:

- If $(X, Y) \in \ddot{\mathbb{K}}$, then we have $Z \in \mathbb{K}$ whenever $\Delta(X, Y, Z)$.   (∇)
- If $Z \in \mathbb{K}$, then we have $(X, Y) \in \ddot{\mathbb{K}}$ whenever $\Delta(X, Y, Z)$.   (Δ)

We say that $\ddot{\mathbb{K}}$ is a ∇-structure if $\ddot{\mathbb{K}}$ satisfies (∇).
We say that $\ddot{\mathbb{K}}$ is a Δ-structure if $\ddot{\mathbb{K}}$ satisfies both (∇) and (Δ).

Now, let us start from an arbitrary collection $\mathbb{K} \subseteq \mathbb{S}$.
We define $\ddot{\mathbb{K}}^+ = \{(X, Y) \in \ddot{\mathbb{S}} \mid \alpha(X, Y) \in \mathbb{K} \text{ for some vertex } \alpha\}$.
By construction, the kernel of $\ddot{\mathbb{K}}^+$ is precisely $\mathbb{K}$ and $\ddot{\mathbb{K}}^+$ is a Δ-structure.

If $\ddot{\mathbb{K}} \subseteq \ddot{\mathbb{S}}$ and the kernel of $\ddot{\mathbb{K}}$ is $\mathbb{K}$, then we have $\ddot{\mathbb{K}} = \ddot{\mathbb{K}}^+$ whenever $\ddot{\mathbb{K}}$ is a Δ-structure. Thus, a Δ-structure is uniquely determined by its kernel.

Returning to the case of ramifications pairs, Theorem 2 shows that we have $\ddot{\mathbb{R}} = \ddot{\mathbb{R}}^+$. Thus, the kernel of $\ddot{\mathbb{R}}$ is precisely the collection $\mathbb{R}$, and the collection $\ddot{\mathbb{R}}$ is a Δ-structure.

Informally, since $\ddot{\mathbb{R}}$ is a Δ-structure, we recover a property that is satisfied by the neighborhood (the link) of each vertex of an arbitrary ramification. If we pick any vertex α in a ramification Z (for example a Bing's house), then the couple (X, Y) such that $Z = \alpha(X, Y)$ may be recursively decomposed by $\langle \widetilde{\mathbb{R}} \rangle$, until an elementary couple.

7   Ramifications and the Five Completions

In previous works, we tried to build a framework, based on completions, for unifying certain notions of combinatorial topology. It turns out that five completions, acting on $\ddot{\mathbb{S}}$, appear to be particularly relevant for this purpose. In this section, we wish to relate ramifications to these completions.

We recall the five completions (the symbols $\widetilde{T}$, $\widetilde{U}$, $\widetilde{L}$ stand respectively for "transitivity", "upper confluence", and "lower confluence"):

For any $S, T \in \mathbb{S}$,

$\rightarrow$ If $(S \cap T, T) \in \ddot{\mathcal{K}}$, then $(S, S \cup T) \in \ddot{\mathcal{K}}$. $\langle \widetilde{X} \rangle$

$\rightarrow$ If $(S, S \cup T) \in \ddot{\mathcal{K}}$, then $(S \cap T, T) \in \ddot{\mathcal{K}}$. $\langle \widetilde{Y} \rangle$

For any $(R, S), (S, T), (R, T) \in \ddot{\mathbb{S}}$,

$\rightarrow$ If $(R, S) \in \ddot{\mathcal{K}}$ and $(S, T) \in \ddot{\mathcal{K}}$, then $(R, T) \in \ddot{\mathcal{K}}$. $\langle \widetilde{T} \rangle$

$\rightarrow$ If $(R, S) \in \ddot{\mathcal{K}}$ and $(R, T) \in \ddot{\mathcal{K}}$, then $(S, T) \in \ddot{\mathcal{K}}$. $\langle \widetilde{U} \rangle$

$\rightarrow$ If $(R, T) \in \ddot{\mathcal{K}}$ and $(S, T) \in \ddot{\mathcal{K}}$, then $(R, S) \in \ddot{\mathcal{K}}$. $\langle \widetilde{L} \rangle$

The largest collection is obtained by considering all the five completions. Let $\ddot{\mathbb{D}} = \langle \ddot{\mathbb{C}}; \widetilde{X}, \widetilde{Y}, \widetilde{T}, \widetilde{U}, \widetilde{L} \rangle$. Each couple of $\ddot{\mathbb{D}}$ is a *dyad* or an *acyclic pair*. In [8], we proved that $\ddot{\mathbb{D}}$ is a Δ-structure and that the kernel of $\ddot{\mathbb{D}}$ is precisely the collection $\mathbb{D}$ of dendrites.

In this section, we focus our attention on collections based on a subset of the five above completions. These completions are chosen because the kernel of $\ddot{\mathbb{D}}$ is $\mathbb{D}$, which corresponds to the remarkable collection made of all acyclic complexes. In particular, it includes all contractible complexes. We will use the following fact.

Proposition 6. *Let $\ddot{\mathbb{K}}$ be a ∇-structure and $\ddot{\mathbb{L}}$ be a Δ-structure. Let $\mathbb{K}$ and $\mathbb{L}$ be the kernels of $\ddot{\mathbb{K}}$ and $\ddot{\mathbb{L}}$, respectively. If $\mathbb{K} \subseteq \mathbb{L}$, then we have $\ddot{\mathbb{K}} \subseteq \ddot{\mathbb{L}}$. Furthermore, if $\mathbb{K} \subsetneq \mathbb{L}$, then $\ddot{\mathbb{K}} \subsetneq \ddot{\mathbb{L}}$.*

Proof. Suppose $\mathbb{K} \subseteq \mathbb{L}$. Let $(X, Y) \in \ddot{\mathbb{K}}$. Since $\ddot{\mathbb{K}}$ is a ∇-structure, $\alpha(X, Y) \in \mathbb{K}$. Thus $\alpha(X, Y) \in \mathbb{L}$. But since $\ddot{\mathbb{L}}$ be a Δ-structure, we have $(X, Y) \in \ddot{\mathbb{L}}$. By the very definition of a kernel, if $\ddot{\mathbb{K}} = \ddot{\mathbb{L}}$, then we must have $\mathbb{K} = \mathbb{L}$. $\square$

7.1   Ramification Pairs and Collapsibility

We denote by $\ddot{\mathbb{E}}$ the collection of all couples $(X, Y) \in \ddot{\mathbb{S}}$ such that the complex Y collapses onto X. Thus, the kernel of $\ddot{\mathbb{E}}$ is precisely the collection $\mathbb{E}$ made of all collapsible complexes. It has been shown [10] that $\ddot{\mathbb{E}}$ has an exact characterization with a subset of the five above completions. We have $\ddot{\mathbb{E}} = \langle \ddot{\triangle}; \tilde{X}, \tilde{T} \rangle$, where $\ddot{\triangle}$ is the collection composed of all all couples of cones $(\alpha X, \alpha Y)$, with $(X, Y) \in \ddot{\mathbb{S}}$.

Let $(X, Y) \in \ddot{\mathbb{E}}$. Let $Z = \alpha(X, Y)$. Since Y collapses onto X, the complex $Z = \alpha X \cup Y$ collapses onto the cone αX, which is collapsible. Thus Z is collapsible, i.e., $Z \in E$. It means that $\ddot{\mathbb{E}}$ is a ∇-structure. Since $\mathbb{E} \subsetneq \mathbb{R}$ (Proposition 4) then, by Proposition 6, we have $\ddot{\mathbb{E}} \subsetneq \ddot{\mathbb{R}}$.

Now, we consider the collection $\ddot{\mathbb{E}}^+$, which is a Δ-structure. By definition of a collection $\ddot{\mathbb{K}}^+$, a couple (X, Y) is in $\ddot{\mathbb{E}}^+$ if and only if $\alpha(X, Y) = \alpha X \cup Y$ is collapsible. Thus, the kernel of $\ddot{\mathbb{E}}^+$ is also the collection $\mathbb{E}$ and, by Proposition 6, we have $\ddot{\mathbb{E}} \subseteq \ddot{\mathbb{E}}^+$. One may ask whether we have $\ddot{\mathbb{E}} = \ddot{\mathbb{E}}^+$. A positive answer would imply that $\ddot{\mathbb{E}}$ is a Δ-structure. In fact this equality does not hold.

We have the following counter-example. Let Y be the complex depicted Fig. 2(c), and let X be the closed curve that is outlined. It may be seen that Y collapses onto X. The first steps of a possible sequence of elementary collapses are depicted by arrows. Let $X' = X \cup \{\{1\}, \{1, 3\}\}$, let α be a new vertex, and let $Z = \alpha X' \cup Y$. We observe that $(\{\alpha, 1\}, \{\alpha, 1, 3\})$ is a free pair for Z. Thus Z collapses onto $Z' = \alpha X \cup Y$. But, since Y collapses onto X, Z' collapses onto αX, thus Z collapses onto αX. Since the cone αX is collapsible, the complex Z is collapsible. Thus $\alpha(X', Y) \in \mathbb{E}$, it means that $(X', Y) \in \ddot{\mathbb{E}}^+$. But Y does not collapse onto X' since there is nowhere to start the collapse. Thus $(X', Y) \notin \ddot{\mathbb{E}}$.

In consequence the inclusion $\ddot{\mathbb{E}} \subseteq \ddot{\mathbb{E}}^+$ is strict. The collection $\ddot{\mathbb{E}}$ is a ∇-structure but not a Δ-structure.

Again, since the kernel of $\ddot{\mathbb{E}}^+$ is a proper subset of the kernel of $\ddot{\mathbb{R}}$, by Proposition 6, we may assert that $\ddot{\mathbb{E}}^+ \subseteq \ddot{\mathbb{R}}$ and that this inclusion is strict. Thus, starting from $\ddot{\mathbb{E}}$, we have build a new collection $\ddot{\mathbb{E}}^+$ which allows us to be closer to $\ddot{\mathbb{R}}$.

7.2 Ramification Pairs and Contractibility

We consider the collection $\ddot{\mathbb{W}}$ such that a couple (X, Y) is in $\ddot{\mathbb{W}}$ if and only if $\alpha(X, Y) = \alpha X \cup Y$ is contractible. By construction $\ddot{\mathbb{W}}$ is a Δ-structure, the kernel of $\ddot{\mathbb{W}}$ is precisely the collection $\mathbb{H}$ made of all contractible complexes. Again, it has been shown (Theorem 5 of [10]) that $\ddot{\mathbb{W}}$ admits an exact characterization with a subset of the five above completions. We have $\ddot{\mathbb{W}} = \langle \ddot{\mathbb{C}}; \tilde{X}, \tilde{Y}, \tilde{T}, \tilde{U} \rangle$.

We see that $\ddot{\mathbb{W}} \subseteq \ddot{\mathbb{D}}$, this inclusion is strict [10].

Since $\mathbb{R}$ is a proper subset of $\mathbb{H}$, by Proposition 6, we have $\ddot{\mathbb{R}} \subseteq \ddot{\mathbb{W}}$, and this inclusion is strict.

7.3 Properties Related to the Five Completions

The following theorem summarizes the results given above.

Theorem 3. *We have* $\ddot{\mathbb{E}} \subsetneq \ddot{\mathbb{E}}^+ \subsetneq \ddot{\mathbb{R}} \subsetneq \ddot{\mathbb{W}} \subsetneq \ddot{\mathbb{D}}$.

We emphasize that the collections $\ddot{\mathbb{E}}$, $\ddot{\mathbb{W}}$, $\ddot{\mathbb{D}}$, have an exact characterization based on the five completions. It means that these collections are fully described by global properties. The collections $\ddot{\mathbb{E}}$ and $\ddot{\mathbb{E}}^+$ are closely related since they have the same kernel $\mathbb{E}$ which is made of all collapsible complexes. Also it is worth pointing out that each couple in the collections $\ddot{\mathbb{E}}$, $\ddot{\mathbb{E}}^+$, $\ddot{\mathbb{W}}$, may be obtained by a sequence of local operations. Collapses/expansions and perforations/fillings (introduced in [10]) are sufficient for that purpose.

In the above sections, completions appeared as components of certain completion systems $\langle \mathbf{X}; K \rangle$. Here, we consider a completion as a property by itself.

Proposition 7. *The collection* $\ddot{\mathbb{R}}$ *satisfies the properties* $\langle \widetilde{X} \rangle$ *and* $\langle \widetilde{U} \rangle$.

Proof

1) Let $S, T \in \mathbb{S}$ such that $(S \cap T, T) \in \ddot{\mathbb{R}}$. Since $\ddot{\mathbb{I}} \subseteq \ddot{\mathbb{R}}$, we have $(S \cap T, T)$, (S, S), $(S \cap T, S \cap T) \in \ddot{\mathbb{R}}$. Thus, by $\langle \widetilde{R} \rangle$, we obtain $(S, S \cup T) \in \ddot{\mathbb{R}}$; the collection $\ddot{\mathbb{R}}$ satisfies $\langle \widetilde{X} \rangle$.
2) Let $(R, S), (S, T), (R, T) \in \ddot{\mathbb{S}}$ such that $(R, S) \in \ddot{\mathbb{R}}$ and $(R, T) \in \ddot{\mathbb{R}}$. Since $\ddot{\mathbb{I}} \subseteq \ddot{\mathbb{R}}$, we have (R, T), (S, S), $(R \cap S = R, T \cap S = S) \in \ddot{\mathbb{R}}$. Thus, by $\langle \widetilde{R} \rangle$, we obtain $(S, T) \in \ddot{\mathbb{R}}$; the collection $\ddot{\mathbb{R}}$ satisfies $\langle \widetilde{U} \rangle$.   □

Now we give two counter-examples which show that the collection $\ddot{\mathbb{R}}$ does not satisfy the properties $\langle \widetilde{L} \rangle$ and $\langle \widetilde{T} \rangle$.
The complex D represents the triangulation of the dunce hat given Fig. 2(b).

1) The complex D is contractible. By Theorem 5 of [1] (and also by Proposition 6 of [9]), there exists Y such that Y collapses onto D and Y is collapsible. It follows that $(D, Y) \in \ddot{\mathbb{R}}$ and $(\emptyset, Y) \in \ddot{\mathbb{R}}$. But, since D is not a ramification, we have $(\emptyset, D) \notin \ddot{\mathbb{R}}$. Thus $\ddot{\mathbb{R}}$ does not satisfy $\langle \widetilde{L} \rangle$.
2) Let Y be the complex depicted Fig. 2(c), and let X be the closed curve that is outlined. We have pointed out, in Sect. 7.1, that Y collapses onto X. Let $X' = X \cup \{\{1\}, \{1, 3\}\}$. We see that X' collapses onto X. Thus, $(X, Y) \in \ddot{\mathbb{R}}$, $(X, X') \in \ddot{\mathbb{R}}$. Since $\ddot{\mathbb{R}}$ satisfies $\langle \widetilde{U} \rangle$, we have $(X', Y) \in \ddot{\mathbb{R}}$. Let γ be the vertex corresponding to the label "1". We have $\gamma X \cap Y = X'$. Since $\ddot{\mathbb{R}}$ satisfies $\langle \widetilde{X} \rangle$, we obtain $(\gamma X, \gamma X \cup Y) \in \ddot{\mathbb{R}}$. But $D = \gamma X \cup Y$. We obtain $(\gamma X, D) \in \ddot{\mathbb{R}}$. Since γX is a cone, we have $(\emptyset, \gamma X) \in \ddot{\mathbb{R}}$. But we have not $(\emptyset, D) \in \ddot{\mathbb{R}}$. Thus $\ddot{\mathbb{R}}$ does not satisfy $\langle \widetilde{T} \rangle$.

Thus, we proved the following. Note that the question remains open for the property $\langle \widetilde{Y} \rangle$.

Proposition 8. *The collection* $\ddot{\mathbb{R}}$ *satisfies none of the properties* $\langle \widetilde{L} \rangle$ *and* $\langle \widetilde{T} \rangle$.

8   Conclusion

In this paper, we extended the collection $\mathbb{R}$ of ramifications to a collection $\ddot{\mathbb{R}}$ of ramification pairs. We followed an approach developed in earlier papers, where we make a relation between a collection $\ddot{\mathbb{K}}$ of couple of complexes and a collection $\mathbb{K}$ of complexes; $\mathbb{K}$ is the kernel of $\ddot{\mathbb{K}}$ and $\ddot{\mathbb{K}}$ is a structure on $\mathbb{K}$.

It's turn out that $\ddot{\mathbb{R}}$ has a noticeable property with respect to its kernel $\mathbb{R}$. In particular $\ddot{\mathbb{R}}$ is uniquely determined by $\mathbb{R}$.

We made a comparison between $\ddot{\mathbb{R}}$ and two others collection $\ddot{\mathbb{E}}$ and $\ddot{\mathbb{W}}$. The kernel $\mathbb{E}$ of $\ddot{\mathbb{E}}$ corresponds to collapsible complexes and the kernel $\mathbb{H}$ of $\ddot{\mathbb{W}}$ consists of all contractible complexes. We showed that $\ddot{\mathbb{E}} \subsetneq \ddot{\mathbb{R}} \subsetneq \ddot{\mathbb{W}}$.

The collection $\ddot{\mathbb{E}}$ is not uniquely determined by $\mathbb{E}$. Thus, from $\mathbb{E}$, we built an extension $\ddot{\mathbb{E}}^+$ of $\ddot{\mathbb{E}}$. We showed that $\ddot{\mathbb{E}} \subsetneq \ddot{\mathbb{E}}^+ \subsetneq \ddot{\mathbb{R}}$. Thus, starting from $\ddot{\mathbb{E}}$, we obtained a new collection $\ddot{\mathbb{E}}^+$ which allows us to be closer to $\ddot{\mathbb{R}}$.

A   Appendix

In this appendix, we present a sequence of expansions and collapses which shows the contractibility of the dunce hat. We give this sequence for the reader who wants to better understand this object which is used several times in this paper for crucial counter-examples.

Let D be the triangulation of the dunce hat of Fig. 2(b). Let X be the cell whose facet is the set $\{3, 5, 6\}$, thus $X \cap D$ is the closed curve that is highlighted in (c). Let γ be the vertex corresponding to the label "1", and let $E = \gamma X \cup D$. The pair $(\{3, 5, 6\}, \{\gamma, 3, 5, 6\})$ is a free pair for E, thus D is an elementary collapse of E. Let F be the complex given Fig. 2(c) and let $G = F \cup X$. It may be seen that E collapses onto G. First we remove the pair $(\{1, 3, 5\}, \{1, 3, 5, 6\})$, then the pair $(\{1, 5\}, \{1, 5, 6\})$, then the pair $(\{1, 6\}, \{1, 6, 3\}$. Now we observe that the complex G collapses onto the cell X, the first steps of a collapse sequence are represented Fig. 2(c). Since X is collapsible, the following sequence shows the contractibility of D:

$$D \nearrow E \searrow G \searrow X \searrow \emptyset$$

References

1. Whitehead, J.H.C.: Simplicial spaces, nuclei, and m-groups. Proc. Lond. Math. Soc. **2**(45), 243–327 (1939)
2. Björner, A.: Topological methods. In: Graham, R., Grötschel, M., Lovász, L. (eds.) Handbook of Combinatorics, pp. 1819–1872. North-Holland, Amsterdam (1995)
3. Jonsson, J.: Simplicial Complexes of Graphs. Springer, Heidelberg (2008)
4. Yung Kong, T.: Topology-preserving deletion of 1's from 2-, 3- and 4-dimensional binary images. In: Ahronovitz, E., Fiorio, C. (eds.) DGCI 1997. LNCS, vol. 1347, pp. 1–18. Springer, Heidelberg (1997). https://doi.org/10.1007/BFb0024826
5. Couprie, M., Bertrand, G.: New characterizations of simple points in 2D, 3D and 4D discrete spaces. IEEE Trans. PAMI **31**(4), 637–648 (2009)
6. Bertrand, G.: On critical kernels. Comptes Rendus de l'Académie des Sciences, Série Math. **345**, 363–367 (2007)
7. Bertrand, G.: Completions and simplicial complexes, hal-00761162 (2012)
8. Bertrand, G.: New structures based on completions. In: Gonzalez-Diaz, R., Jimenez, M.-J., Medrano, B. (eds.) DGCI 2013. LNCS, vol. 7749, pp. 83–94. Springer, Heidelberg (2013). https://doi.org/10.1007/978-3-642-37067-0_8

9. Bertrand, G.: Completions and simple homotopy. In: Barcucci, E., Frosini, A., Rinaldi, S. (eds.) DGCI 2014. LNCS, vol. 8668, pp. 63–74. Springer, Cham (2014). https://doi.org/10.1007/978-3-319-09955-2_6

10. Bertrand, G.: Completions, perforations and fillings. In: Lindblad, J., Malmberg, F., Sladoje, N. (eds.) DGMM 2021. LNCS, vol. 12708, pp. 137–151. Springer, Cham (2021). https://doi.org/10.1007/978-3-030-76657-3_9

11. Bertrand, G.: Some results on ramifications. Research report, hal-03638665 (2022)

12. Giblin, P.: Graphs, Surfaces and Homology. Chapman and Hall (1981)

13. Bing, R.H.: Some aspects of the topology of 3-manifolds related to the Poincaré conjecture. Lect. Mod. Math. **2**, 93–128 (1964)

14. Zeeman, E.C.: On the dunce hat. Topology **2**, 341–358 (1964)

15. Hatcher, A.: Algebraic Topology. Cambridge University Press, Cambridge (2001)

16. Tancer, M.: Recognition of collapsible complexes is NP-complete. arXiv:1211.6254 (2012)

Algorithms for Pixelwise Shape Deformations Preserving Digital Convexity

Lama Tarsissi[1,2], Yukiko Kenmochi[3(✉)], Hadjer Djerroumi[1],
David Coeurjolly[4], Pascal Romon[5], and Jean-Pierre Borel[6]

[1] Univ Gustave Eiffel, ESIEE Paris, CNRS, LIGM, Champs-sur-Marne, France
[2] Sorbonne University Abu Dhabi, Abu Dhabi, United Arab Emirates
[3] Normandie Univ, UNICAEN, ENSICAEN, CNRS, GREYC, Caen, France
yukiko.kenmochi@unicaen.fr
[4] Univ Lyon, CNRS, INSA Lyon, UCBL, LIRIS, UMR5205, Villeurbanne, France
[5] Univ Gustave Eiffel, CNRS, LAMA, Champs-sur-Marne, France
[6] Univ Limoges, CNRS, XLIM, Limoges, France

Abstract. In this article, we propose algorithms for pixelwise deformations of digital convex sets preserving their convexity using the combinatorics on words to identify digital convex sets via their boundary words, namely Lyndon and Christoffel words. The notion of removable and insertable points are used with a geometric strategy for choosing one of those pixels for each deformation step. The worst-case time complexity of each deflation and inflation step, which is the atomic deformation, is also analysed.

1   Introduction

Convexity is an elementary geometric property of digital sets in digital image processing. There are various applications which require deforming digital convex sets while preserving their convexity. Various definitions of digital convex sets exist, among which we choose the one based on the convex hull [15]. Indeed, Brleck et al. have characterized such digital convex sets via the boundary words, which are encoded by the Freeman chain code [14]; for short, a 4-connected digital set is digital convex if and only if the Lyndon factorization of its boundary word is made of Christoffel words [7]. Thanks to this approach based on combinatorics on words, we recently considered the following question: given a finite 4-connected, digital convex set C, how can one find a point x of C (resp. its complement $\overline{C}$) such that $C \setminus \{x\}$ (resp. $C \cup \{x\}$) is still 4-connected and digitally convex? In order to answer this question, we characterized the two types of points; they are called removable and insertable points [21,22].

In this article, following the approach based on combinatorics on words, we propose algorithms for pixelwise deformations of digital convex sets that preserve their convexity using the characterizations of removable and insertable points. The main contribution of this article is factorizing the inflation and deflation

© Springer Nature Switzerland AG 2022
E. Baudrier et al. (Eds.): DGMM 2022, LNCS 13493, pp. 84–97, 2022.
https://doi.org/10.1007/978-3-031-19897-7_8

Fig. 1. Digitally convex sets with and without 4-connectivity (left and center) and digitally non-convex set (right). The sequence of border points is also illustrated by a thick black polygonal line for each 4-connected set (left and right).

algorithms, whose time complexities are analysed, in order to propose the general deformation algorithm. A geometric strategy based on distance map is also used for choosing one of removal and insertable pixels for each deformation step. Given a pair of digital convex sets, we show that the proposed algorithms create a sequence of digital convex sets, which is such a deformation between them. Some experimental results are illustrated.

2 Basic Notions

2.1 Digital Convex Set

In $\mathbb{R}^2$, a subset S is convex if for any pair of points $x, y \in S$, every point on the straight line segment joining x and y is also within S. This notion, however, cannot be straightforwardly applied to subsets in $\mathbb{Z}^2$; various notions of convexity of a subset X of $\mathbb{Z}^2$ have been proposed. In this article, we focus on the following one [15] based on the convex hull, denoted by $conv(X)$ and also called H-convexity [13].

Definition 1 ([15]). *A subset X of $\mathbb{Z}^2$ is digitally convex if $X = conv(X) \cap \mathbb{Z}^2$.*

Figure 1 illustrates examples of digital convex and non-convex sets, based on this notion. The following remark warns us to pay attention to the connectivity separately from the convexity in $\mathbb{Z}^2$ (see Fig. 1 (center)).

Remark 1. Digital convexity does not imply connectivity in $\mathbb{Z}^2$.

Concerning the connectivity of a digital convex set, there exists a homeomorphism that makes the set almost 4-connected [10] while an alternative definition for digital convexity, called full convexity, that encompasses arithmetic lines and naturally entails connectivity has been proposed [17].

Let us call convex polygons with vertices in $\mathbb{Z}^2$, digital convex polygons. The following property [1,3] will help us to analyse the complexity of our deformation algorithms later.

Property 1 ([1,3]). Given a digital convex polygon of diameter N, the number of its vertices is bounded by $\mathcal{O}(N^{\frac{2}{3}})$.

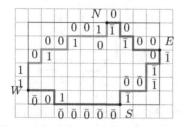
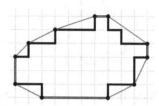

Fig. 2. The boundary word of a digital convex 4-connected set decomposed into four parts such that each part consists of a binary word (left), and the Lyndon points (black points) of the boundary word drawn as the black thick polygonal line (right). (Color figure online)

2.2 Boundary Words and Some Basic Notions of Words

Let $C \subset \mathbb{Z}^2$ be a finite, 4-connected digitally convex set. The border points of C can be tracked by classical border following algorithms (for example, see [2] for "left-hand-on-wall" border following), which generate a 4-connected sequence of the border points of C. Note that the sequence can include dead-ends and thus sometimes turnaround sub-sequences if the set contains thin parts of one pixel width. Here, we encode the sequence with Freeman code [14], called the boundary word of C, denoted by $Bd(C)$, in the clockwise order of border points. Boundary words are thus defined over an alphabet of four letters $0, 1, \bar{0}, \bar{1}$, which are associated to the right, up, left and down directions, respectively. The boundary word of a digital convex 4-connected set C is decomposed into four parts such that each part consists of two letters, as seen in Fig. 2 (left): WN, NE, ES and SW.

Let us present some basic notions of words (see [18] for more complete overview): a nonempty finite set of letters is called an alphabet A; in this article, we have the four letters $0, 1, \bar{0}, \bar{1}$ as mentioned above. A word w is a sequence of concatenated letters from A. The empty word ϵ is a sequence of zero letter. A^* denotes the set of all finite words over A. The length of w is denoted by $|w|$ while $|w|_a$ represents the number of occurrences of a in w. The n-times concatenation of w is written by w^n. A word is said *primitive* if it is not the power of a nonempty word. A word w is *conjugate* of a word w' if w' can be obtained from w by cyclically shifting the letters.

2.3 Lyndon Words and Lyndon Factorization

We give the definition of Lyndon words, which is a necessary notion for the sequel.

Definition 2 ([19]). *A word w over a totally ordered alphabet is a Lyndon word if it is the smallest among all its conjugates.*

For example, $w = 00101$ where $0 < 1$ is a Lyndon word as w is the smallest among all its conjugates. The following proposition will play a leading role in our algorithms.

Proposition 1 ([8]). *Every non-empty word w over a totally ordered alphabet can be written uniquely as $w = \ell_1^{n_1} \ell_2^{n_2} \ldots \ell_k^{n_k}$ such that every factor ℓ_i is a Lyndon word and $\{\ell_i\}_i$ is a lexicographically decreasing sequence.*

This decomposition of w into ℓ_i is called Lyndon factorization. Given a word w of length N, the Lyndon factorization of w is calculated in $\mathcal{O}(N)$ time with a constant space [12]. The points on a word w that separate different Lyndon factors are called Lyndon points. Note that the two extremities of w are also Lyndon points. Let us consider a finite, 4-connected digitally convex set $C \subset \mathbb{Z}^2$ and its boundary word w. Then, the Lyndon points of w correspond to the vertices of the convex hull of C geometrically (see Fig. 2 (right)).

2.4 Christoffel Words

Christoffel words are another important notion in this article. Their geometrical definition can be formulated as:

Definition 3 ([4]). *The lower Christoffel word of slope $\frac{b}{a}$ is determined by encoding with Freeman chain code the Christoffel path, which is the discrete path from the origin O to the point $P(a, b)$ such that:*

– *the path lies below the line segment OP;*
– *the integer points in the region enclosed by the path and the line segment OP are exactly those of the path.*

Any Christoffel word with $\gcd(a, b) = 1$ is called primitive. Some properties of Christoffel words are presented as follows:

– A Christoffel word describes a shortest discrete path, so that it is always composed from two letters.
– Let c_1, c_2 be two Christoffel words over the alphabet $\{0, 1\}$. Then lexicographically $c_1 < c_2$ iff $slope(c_1) < slope(c_2)$ [6].
– Every primitive Christoffel word is a Lyndon word [5].

The converse of the last one is not true; for example, 0011 is a Lyndon word but not a Christoffel word.

In this article, we need the following specific points, called furthest points of Christoffel words: Fig. 3 (left) illustrates an example of the lower Christoffel word of slope $\frac{4}{7}$ with its furthest point.

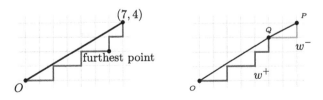

Fig. 3. The lower Christoffel word of slope $\frac{4}{7}$ with the furthest point (left) and its split (right).

Definition 4. *Given a primitive Christoffel word of slope $\frac{b}{a}$, the furthest point is uniquely defined on the path as the point whose vertical distance to the line segment joining $(0,0)$ and (a,b) is maximum.*

We will also need the diagonally opposite point of a furthest point, which is above the line segment OP (see Fig. 3), called the closest upper point.

2.5   Digital Convex Sets with Combinatorics on Words

By using Lyndon and Christoffel words, digital convex sets are characterized.

Proposition 2 ([7]). *A 4-connected set $C \subset \mathbb{Z}^2$ is digitally convex iff its boundary word is decomposed into four binary subwords and each subword has the unique Lyndon factorization $\ell_1^{n_1} \ell_2^{n_2} \ldots \ell_k^{n_k}$ such that all ℓ_i are primitive Christoffel words.*

The geometric interpretation of this proposition is that C is digitally convex iff the Lyndon factorization of $Bd(C)$ exactly corresponds to the segments of the convex hull of C (see Fig. 1 (left and right) for positive and negative examples). This characterization will be used in the rest of this article.

3   Removable and Insertable Points

In this section, we consider the following problem: given a finite 4-connected digitally convex set C, how can one find a point x of C (resp. the complement $\overline{C}$) such that $C \setminus \{x\}$ (resp. $C \cup \{x\}$) is still digitally convex and 4-connected? The former points are called removable points while the latter ones are called insertable points (see Fig. 4 for examples). Their characterizations have been studied previously [21,22]. We recall them in this section.

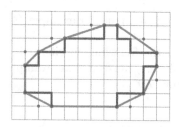

Fig. 4. Removable and insertable points, depicted in green and red respectively, for the boundary path of a digitally convex set, drawn as the blue polygonal line. (Color figure online)

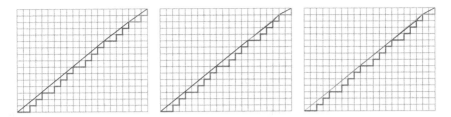

Fig. 5. Procedure of insertability verification on the left with propagation: the closest upper point $(19, 15)$ of ℓ_i (ℓ_i in brown) is inserted (left); as $\ell_{i-1} \leq L_0$ (ℓ_{i-1} in blue and $L_0 = \ell^+$ in red) is not satisfied but we have $\ell_{i-1} = \ell_i L_0$, we obtain $L_1 = \ell_{i-1} L_0$ (L_1 in green (=red+blue)) (center); as $\ell_{i-2} \leq L_1$ (ℓ_{i-2} in pink) is not satisfied but we have $\ell_{i-2} = \ell_{i-1} L_1^2$, we obtain $L_2 = \ell_{i-2} L_1$ (L_2 in green (=red+blue+pink)) (right).

3.1 Removable Points

Let us consider that the boundary word w of a digitally convex 4-connected set C and its Lyndon factorization $\mathcal{L}(w)$ are given. Then we have the following theorem.

Theorem 1 ([21]). *A point x of C is removable iff x is a Lyndon point of $\mathcal{L}(w)$ and a simple point with respect to C.*

As the digital convexity does not imply the connectivity, as mentioned above, we need to add the simpleness condition that is also locally characterized [16]. Thanks to this theorem, we can find a position k where we can apply the following switch operator, which corresponds to removing the point at k. The switch operator on a word $w = a_1 \ldots a_n$ at position $k < n$ is defined by $\mathrm{switch}_k(w) = a_1 \ldots a_{k-1} a_{k+1} a_k a_{k+2} \ldots a_n$ where each a_i is a letter. If $a_k a_{k+1}$ consists of consecutive reverse letters, namely $0\bar{0}$, $\bar{0}0$, $1\bar{1}$, $\bar{1}1$, this operator will simply remove both of them, instead of the substitution.

Once a chosen removable point is removed by the switch operator, the following proposition tells us that updating the Lyndon factorization, namely updating the list of Lyndon points, can be made locally.

Proposition 3 ([21]). *Let u and v be two consecutive Christoffel words of the Lyndon factorization of a boundary word such that $u > v$. After applying the switch operator at $|u|$ on the binary word uv, if we obtain its Lyndon factorization $\mathcal{L}(\mathrm{switch}_{|u|}(uv)) = \ell_1^{n_1} \ldots \ell_m^{n_m}$, then $u > \ell_1$ and $\ell_1 > \ldots > \ell_m$.*

3.2 Insertable Points

Let us consider that the boundary word w of a digitally convex 4-connected set C and its Lyndon factorization $\mathcal{L}(w)$ are given. Then we have the following proposition.

Proposition 4 ([22]). *If a point x of $\overline{C}$ is insertable, then x is a closest upper point of $\mathcal{L}(w)$.*

It should be mentioned that the converse is not always true. Indeed, an insertable point is geometrically a point such that its convex hull with C does not contain any other integer point, and the proposition indicates that the union of C and a closest upper point is not always digitally convex.

Instead of the switch operator for removable points, here we use the split operator that is defined for a primitive Christoffel word c, $|c| > 1$, such that $split(c) = switch_k(c)$ where k is the furthest point of c (see [22] for the definition in the case of $|c| = 1$). In order to insert a closest upper point of C, the following standard factorization is used.

Definition 5 ([5]). *Any Christoffel word c with $|c| > 1$ can be written in a unique way as a product $c = uv$ such that u and v are both primitive Christoffel words. The couple (u, v) is called the standard factorization of c.*

Note that the standard factorization of c can be computed in $\mathcal{O}(\log |c|)$ due to its geometric interpretation [20].

The proposition below implies that the standard factorization gives the result of the split operator without knowing the position of the furthest point (see Fig. 3 (right) for an example of application of this split operator).

Proposition 5 [11]. *Let c be a primitive Christoffel word, $|c| > 1$, such that its standard factorization is given by $c = c^- c^+$. Then, we have $split(c) = c^+ c^-$ with $c^+ > c^-$.*

The following is the characterization of the insertability of such a closest upper point $x \in \overline{C}$.

Proposition 6 ([22]). *Given the boundary word w of a digitally convex 4-connected set C and its Lyndon factorization $\mathcal{L}(w) = \ell_1^{n_1} \ldots \ell_m^{n_m}$, let x be the closest upper point in $\overline{C}$ of the j-th Lyndon factor of $\ell_i^{n_i}$ in $\mathcal{L}(w)$ such that $split(\ell_i) = \ell_i^+ \ell_i^-$ where $\ell_i = \ell_i^- \ell_i^+$. Let us say that:*

1. *x is insertable on the left if $\exists k \in \mathbb{Z}^*, \ell_{i-k-1} \geq L_k$ such that for every $h \leq k$, L_h is recursively defined by*

$$L_h = \begin{cases} \ell_i^{j-1} \ell_i^+ & \text{for } h = 0 \\ \ell_{i-h}^{n_{i-h}} L_{h-1} & \text{for } h \geq 1 \text{ if } \exists m_{h-1} \in \mathbb{Z}^+, \ell_{i-h} = \ell_{i-h-1} L_{h-1}^{m_{h-1}} \end{cases}$$

2. *similarly, x is insertable on the right if $\exists k \in \mathbb{Z}^*, \ell_{i+k+1} \leq R_k$ such that for every $h \leq k$, R_h is recursively defined by*

$$R_h = \begin{cases} \ell_i^- \ell_i^{n_i-j} & \text{for } h = 0 \\ R_{h-1} \ell_{i+h}^{n_{i+h}} & \text{for } h \geq 1 \text{ if } \exists m_{h-1} \in \mathbb{Z}^+, \ell_{i+h} = R_{h-1}^{m_{h-1}} \ell_{i+h-1} \end{cases}$$

Then, x is insertable if x is insertable on both sides.

This proposition indicates that the insertability cannot always be verified locally; see Fig. 5 for an example with propagation. On the other hand, Lyndon re-factorization is not necessarily applied after adding an insertable point as a simple concatenation of Christoffel words provides new Lyndon factors, multiplicities and points.

Algorithm 1: Deflation

---

input : digitally convex 4-connected sets $\mathcal{A}$, $\mathcal{B}$ such that $\mathcal{A} \supset \mathcal{B}$

output: a sequence $\mathcal{T}_\ominus$ of points to remove from $\mathcal{A}$ to obtain $\mathcal{B}$

1 $w \leftarrow$ the boundary word of $\mathcal{A}$, $\mathcal{L} \leftarrow$ Lyndon Factorization of w;

2 calculate $d_\mathcal{A}^\mathcal{B}(x)$ for all $x \in \mathcal{A} \setminus \mathcal{B}$;

3 set the current deflated set $\mathcal{C} \leftarrow \mathcal{A}$, $\mathcal{T}_\ominus \leftarrow \varnothing$;

4 $\mathcal{Q}_\ominus \leftarrow$ UpdateRemovable$(\mathcal{Q}_\ominus, \mathcal{L}, \varnothing, (1, \ldots, |\mathcal{L}|), d_\mathcal{A}^\mathcal{B}, \mathcal{C})$;

5 **while** $\mathcal{Q}_\ominus \neq \varnothing$ **do**

6     $i \leftarrow$ find_max$(\mathcal{Q}_\ominus)$;

7     $(\ell, n, p) \leftarrow \mathcal{L}[i]$, push p to $\mathcal{T}_\ominus$, $\mathcal{C} \leftarrow \mathcal{C} \setminus \{p\}$;

8     $I_{old} \leftarrow (i - 1, i)$;

9     $(\mathcal{L}, i') \leftarrow$ UpdateLyndonFactorizationDueToSwitch$(\mathcal{L}, i)$;

10     $I_{new} \leftarrow (i - 1, i, \ldots, i')$;

11     $\mathcal{Q}_\ominus \leftarrow$ UpdateRemovable$(\mathcal{Q}_\ominus, \mathcal{L}, I_{old}, I_{new}, d_\mathcal{A}^\mathcal{B}, \mathcal{C})$;

12 **end**

13 **return** $\mathcal{T}_\ominus$

Function 2: UpdateLyndonFactorizationDueToSwitch

---

input : Lyndon factorization $\mathcal{L}$ and switch operator position k

output: updated Lyndon factorization $\mathcal{L}$ and last new factor index h

1 $(\ell_1, n_1, p_1) \leftarrow \mathcal{L}[k - 1]$, $(\ell_2, n_2, p_2) \leftarrow \mathcal{L}[k]$;

2 $w \leftarrow$ switch$_{|\ell_1|} \ell_1 \ell_2$;

3 $\mathcal{L}_{new} \leftarrow$ the Lyndon factorization of w;

4 remove $\mathcal{L}[k - 1], \mathcal{L}[k]$;

5 $h = k - 1$;

6 **if** $n_1 > 1$ **then** insert $(\ell_1, n_1 - 1, p_1)$ at $\mathcal{L}[h]$, $h \leftarrow h + 1$;

7 insert $\mathcal{L}_{new}$ at $\mathcal{L}[h]$, $h \leftarrow h + |\mathcal{L}_{new}|$;

8 **if** $n_2 > 1$ **then** insert $(\ell_2, n_2 - 1, p_2 - |\ell_2|_0 e_1 - |\ell_2|_1 e_2)$ at $\mathcal{L}[h]$;

9 **else** $h \leftarrow h - 1$;

10 **return** $\mathcal{L}, h$

---

4 Deformation Preserving Digital Convexity

We now achieve our purpose of this article: given a pair of 4-connected digital convex sets, $\mathcal{A}, \mathcal{B} \subset \mathbb{Z}^2$, such that $\mathcal{A} \cap \mathcal{B} \neq \varnothing$, we would like to make a sequence of 4-connected digital convex sets, which represents a pixelwise deformation from $\mathcal{A}$ to $\mathcal{B}$. For each step, we remove or add a point of $\mathbb{Z}^2$ thanks to the notions of removable and insertable points. In order to choose a point among all the removable and insertable points, we use the following geometric information based on the distance map.

4.1 Priority Distance for Pixel Choices

Let $d(x, \mathcal{A})$ be the Euclidean distance between a point $x \in \mathbb{Z}^2$ and $\mathcal{A}$. Then we define the relative distance for $x \in \mathcal{A} \setminus \mathcal{B}$ from $\mathcal{A}$ to $\mathcal{B}$ such that $d_\mathcal{A}^\mathcal{B}(x) =$

Function 3: UpdateRemovable

---

input : removable point set $\mathcal{Q}_\ominus$, Lyndon factorization $\mathcal{L}$, old and new factor
lists I_{old} and I_{new}, priority d, digital set $\mathcal{C}$

output: updated removable point set $\mathcal{Q}_\ominus$

1   $i \leftarrow \text{pop}(I_{old})$, $j \leftarrow \text{pop}(I_{new})$;
2   replace the Lyndon point of $\mathcal{L}[i]$ by that of $\mathcal{L}[j]$ in $\mathcal{Q}_\ominus$;
3   **while** $I_{old} \neq \varnothing$ **do** $i \leftarrow \text{pop}(I_{old})$, remove the Lyndon point of $\mathcal{L}[i]$ from $\mathcal{Q}_\ominus$;
4   **while** $I_{new} \neq \varnothing$ **do**
5     $i \leftarrow \text{pop}(I_{new})$, $x \leftarrow$ the Lyndon point of $\mathcal{L}[i]$;
6   **if** x *is simple with respect to* $\mathcal{C}$ **then**
7   push x in $\mathcal{Q}_\ominus$;
8   **foreach** $x' \in \mathcal{N}_8(x) \setminus \{x\}$ **do**
9   **if** $\exists j \in \mathcal{Q}_\ominus, x'$ *is the Lyndon point of* $\mathcal{L}[j]$ *and not simple to* $\mathcal{C}$ **then**
10   remove x' from $\mathcal{Q}_\ominus$
11   **end**
12   **end**
13 **end**
14 **return** $\mathcal{Q}_\ominus$

$\frac{d(x,\mathcal{B})}{d(x,\mathcal{B})+d(x,\overline{\mathcal{A}})}$. We can observe that $d_{\mathcal{A}}^{\mathcal{B}}(x)$ is close to 0 when x is close to $\mathcal{B}$, $d_{\mathcal{A}}^{\mathcal{B}}(x)$ is close to 1 when x is close to $\overline{\mathcal{A}}$, and all the distances are between 0 and 1. Note that discrete points in $\mathcal{A} \setminus \mathcal{B}$ will be removed during the deformation while those in $\mathcal{B} \setminus \mathcal{A}$ will be added. For the points in $\mathcal{B} \setminus \mathcal{A}$, we use $d_{\mathcal{B}}^{\mathcal{A}}$.

4.2 Deflation Algorithm

Let us first consider the easiest case such that $\mathcal{A} \supset \mathcal{B}$. Let $\mathcal{L}$ be the Lyndon factorization of the boundary word of $\mathcal{A}$. During deflation, $\mathcal{L}$ is updated for each step of removing a point, which is chosen by the priority $d_{\mathcal{A}}^{\mathcal{B}}$. The priority $d_{\mathcal{A}}^{\mathcal{B}}$ is in descending order; the highest priority is given to pixels of largest $d_{\mathcal{A}}^{\mathcal{B}}$. The following data structures are used in the deflation algorithm:

- $\mathcal{L}$: Lyndon factorization of a boundary word whose i-th element is (ℓ_i, n_i, p_i); ℓ_i is the Lyndon factor, n_i is the multiplicity, and p_i is the (left) Lyndon point,
- $\mathcal{Q}_\ominus$: set of removable points represented by Lyndon factor indices i,
- $\mathcal{T}_\ominus$: sequence of removed points.

In the following, N represents the length of the boundary word of $\mathcal{A}$ (or $\mathcal{B}$) so that the number of the Lyndon points of the Lyndon factorization $\mathcal{L}$ is bounded by $\mathcal{O}(N^{\frac{2}{3}})$ according to Property 1.

    Algorithm 1 shows the procedure of deflation from $\mathcal{A}$ to $\mathcal{B}$, which call the two functions, UpdateLyndonFactorizationDueToSwitch (Function 2) and UpdateRemovable (Function 3). All the information of Lyndon factorization is stored in $\mathcal{L}$. The kernel of the algorithm is updating $\mathcal{L}$ efficiently for each removal step,

---

Algorithm 4: Inflation

input : digitally convex 4-connected sets $\mathcal{A}, \mathcal{B}$ such that $\mathcal{A} \subset \mathcal{B}$
output: a sequence $\mathcal{T}_\oplus$ of points to add

1 $w \leftarrow$ the boundary word of $\mathcal{A}$, $\mathcal{L} \leftarrow$ Lyndon Factorization of w;
2 calculate $d_{\mathcal{B}}^{\mathcal{A}}(x)$ for all $x \in \mathcal{B} \setminus \mathcal{A}$;
3 set the current inflated set $\mathcal{C} \leftarrow \mathcal{A}$, $\mathcal{T}_\oplus \leftarrow \varnothing$;
4 $\mathcal{Q}_\oplus \leftarrow$ AddInsertable$(\mathcal{Q}_\oplus, \mathcal{L}, (1, \ldots, |\mathcal{L}|), d_{\mathcal{B}}^{\mathcal{A}})$;
5 **while** $\mathcal{Q}_\oplus \neq \varnothing$ **do**
6 $\quad$ $(i, j, left, right) \leftarrow$ find_max$(\mathcal{Q}_\oplus)$;
7 $\quad$ $(\ell, n, p) \leftarrow \mathcal{L}[i]$;
8 $\quad$ $(\ell^+, \ell^-) \leftarrow split(\ell)$;
9 $\quad$ $x \leftarrow$ the closed upper point of the j-th ℓ, push x to $\mathcal{T}_\oplus$, $\mathcal{C} \leftarrow \mathcal{C} \cup \{x\}$;
10 $\quad$ $I_{old} \leftarrow (i - left, i - left + 1, \ldots, i + right)$;
11 $\quad$ remove $\mathcal{L}[i - left, \ldots, i + right]$;
12 $\quad$ insert $\ell_{i-left}^{n_{i-left}} \ldots \ell_{i-1}^{n_{i-1}} \ell^-$ at $\mathcal{L}[i - left]$;
13 $\quad$ insert $\ell^+ \ell_{i+1}^{n_{i+1}} \ldots \ell_{i+right}^{n_{i+right}}$ at $\mathcal{L}[i - left + 1]$;
14 $\quad$ $I_{new} \leftarrow (i - left, i - left + 1)$;
15 $\quad$ $\mathcal{Q}_\oplus \leftarrow$ DelInsertable$(\mathcal{Q}_\oplus, I_{old})$;
16 $\quad$ $\mathcal{Q}_\oplus \leftarrow$ AddInsertable$(\mathcal{Q}_\oplus, \mathcal{L}, I_{new}, d_{\mathcal{B}}^{\mathcal{A}})$
17 **end**
18 **return** $\mathcal{T}_\oplus$

---

which is described in Function 2: only the two Lyndon factors adjacent to a chosen removable point are modified by the Lyndon factorization after the switch operation. In other words, we can observe that the update is made locally. As the length of each Lyndon factor is $\mathcal{O}(N)$ in worst case, the time complexity of Function 2 is $\mathcal{O}(N)$. Finding the maximum element of $\mathcal{Q}_\ominus$ (Line 6) and its update (Line 11) need $\mathcal{O}(\log N)$ for each removal step, if we store the sorted removable points of $\mathcal{Q}_\ominus$ in a tree structure such as a heap [9], as the size of $\mathcal{Q}_\ominus$ is bounded by $\mathcal{O}(N^{\frac{2}{3}})$, which is the same size of $\mathcal{L}$. Note that simplicity can be verified efficiently by using its local characterization [16] (see Function 3). Thus, the overall complexity of each deflation step of Algorithm 1 is $\mathcal{O}(N)$.

4.3 Inflation Algorithm

Let us consider the case such that $\mathcal{A} \subset \mathcal{B}$. Here we add points one-by-one to $\mathcal{A}$ until obtaining $\mathcal{B}$ with the priority $d_{\mathcal{B}}^{\mathcal{A}}$. The inflation algorithm requires the following data structures with the Lyndon factorization $\mathcal{L}$ presented for the deflation algorithm.

- $\mathcal{Q}_\oplus$: set of insertable points, each of which is represented by a pair of a Lyndon factor and a multiplicity index (i, j), and their propagation ranges for left and right, $(left, right)$
- $\mathcal{T}_\oplus$: sequence of inserted points.

Function 5: AddInsertable

    input : insertable point set $\mathcal{Q}_\oplus$, Lyndon factorization $\mathcal{L}$, indext set I to verify
    output: updated $\mathcal{Q}_\oplus$

1 **while** $I \neq \varnothing$ **do**
2      $i \leftarrow \text{pop}(I)$, $(\ell, n, p) \leftarrow \mathcal{L}[i]$;
3   **foreach** $j = 1, \ldots, n$ **do**
4          $(insertable^-, k^-) \leftarrow InsertableLeft(i, j, \mathcal{L})$;
5          $(insertable^+, k^+) \leftarrow InsertableRight(i, j, \mathcal{L})$;
6   **if** $insertable^- \wedge insertable^+$ **then** push (i, j, k^-, k^+) to $\mathcal{Q}_\oplus$;
7   **end**
8 **end**
9 **return** $\mathcal{Q}_\oplus$

Function 6: DelInsertable

    input : insertable point set $\mathcal{Q}_\oplus$, Lyndon factorization $\mathcal{L}$, index set I to delete
    output: updated $\mathcal{Q}_\oplus$

1 **while** $I \neq \varnothing$ **do**
2      $i \leftarrow \text{pop}(I)$, $(\ell, n, p) \leftarrow \mathcal{L}[i]$;
3   **foreach** $j = 1, \ldots, n$ **do**
4   remove the element associated to the point index (i, j) from $\mathcal{Q}_\oplus$
5   **end**
6 **end**
7 **return** $\mathcal{Q}_\oplus$

Note that any insertable point is a closest upper point (Proposition 4), which exists uniquely for each j-th Lyndon factor ℓ_i. Thus, keeping $\mathcal{L}$ in the same way as the deflation is also enough for the inflation. When we add a point in the boundary of a digital convex set, the simplicity is obviously satisfied; no simplicity verification is necessary.

Algorithm 4 shows the inflation procedure from $\mathcal{A}$ to $\mathcal{B}$ guided by the priority $d_{\mathcal{B}}^{\mathcal{A}}$. Similarly to the deflation, the kernel of the inflation algorithm is also updating $\mathcal{L}$ efficiently for each point insertion. However, this update may affect *left* and *right* neighbors in the left and right propagations where *left*, *right* can be more than 1, contrary to the deflation case, in which *left* = *right* = 1. Instead, those affected neighboring Lyndon factors are always replaced by exactly two Lyndon factors (see Lines 12 and 13 in Algorithm 4). In other words, no Lyndon re-factorization is needed for the insertion. These left and right neighboring ranges, *left* and *right*, are respectively calculated in the functions, InsertableLeft and InsertableRight, both of which are called in Function 5 (see Function 7 for the InsertableLeft; InsertableRight is omitted here due to its similarity). In fact, those functions verify the instability of the point corresponding to the given factor and multiplicity indecencies, i and j, in left and right sides with the propagation verification. This part is based on Proposition 6. As this propagation cannot be theoretically bounded, the complexity of Function 5 is in $\mathcal{O}(N)$.

Function 7: InsertableLeft

    input : insertion factor index i, multiplicity index j, Lyndon factorization $\mathcal{L}$
    output: boolean $insertable$, left propagation range r

1   $(\ell, n, p) \leftarrow \mathcal{L}[i]$, $(\ell^-, \ell^+) \leftarrow$ standard factorization of ℓ;
2   $w \leftarrow \ell^{j-1}\ell^+$;
3   $propag \leftarrow true$, $insertable \leftarrow false$, $r \leftarrow 0$;
4   **while** $propag = true$ **do**
5 | $(\ell_{prev}, n_{prev}, p_{prev}) \leftarrow \mathcal{L}[i - r - 1]$;
6 | **if** $\ell > w$ **then** $insertable \leftarrow true$, $propag \leftarrow false$;
7 | **else if** $\ell_{prev} = w$ **then** $insertable \leftarrow true$, $propag \leftarrow false$, $r \leftarrow r + 1$;
8 | **else if** $\exists n \in \mathbb{Z}^+, \ell_{prev} = \ell w^n$ **then** $w \leftarrow \ell_{prev}^{n_{prev}} w$, $r \leftarrow r + 1$;
9 | **else** $propag \leftarrow false$;
10 **end**
11 **return** $(insertable, r)$

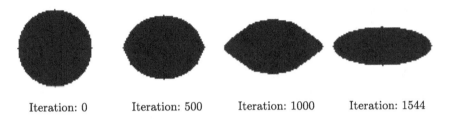

Iteration: 0   Iteration: 500   Iteration: 1000   Iteration: 1544

Fig. 6. Deformation from a digitized disk to a digitized ellipse, both of which are digitally convex.

In other words, if there is no propagation, this complexity can be reduced to $\mathcal{O}(\log N)$. This can be done if we strengthen the insertability condition such that $\ell_{i-1} \geq \ell_i^{j-1}\ell_i^+$ and $\ell_i^-\ell_i^{n_j-1} \geq \ell_{i+1}$ instead of those of Proposition 6.

Note that the size of $\mathcal{Q}_\oplus$ is almost equal to the number of furthest points, which can be given by $\sum_i m_i$ where m_i is the multiplicity for the i-th factor ℓ_i of the Lyndon factorization of the boundary word of the current deformed shape. If we set $M = \max_i m_i$, then we can also say that the size of $\mathcal{Q}_\oplus$ is in $\mathcal{O}(MN^{\frac{2}{3}})$. Thus the time complexity of updating $\mathcal{Q}_\oplus$ (Functions 6 and 5) are in $\mathcal{O}(\log N)$ as $M \leq N$. Then the overall complexity of each inflation step of Algorithm 4 is $\mathcal{O}(N)$ due to the propagation in the insertability verification. We remind that Lyndon re-factorization is not necessary for the inflation case, so that this $\mathcal{O}(N)$ comes only from the insertability verification propagation.

4.4 General Deformation Algorithm

Now let us consider more general case such that $\mathcal{A} \cap \mathcal{B} \neq \varnothing$. We start from an initial digitally convex set $\mathcal{A}$ and obtain $\mathcal{B}$ by adding points of $\mathcal{B} \setminus \mathcal{A}$ and removing points of $\mathcal{A} \setminus \mathcal{B}$. The algorithm is given by Algorithm 8, which is a simple fusion of Algorithms 1 and 4. Figure 6 shows an experimental result for a

---

Algorithm 8: Digital convexity preserving deformation

---

input : overlapped digitally convex 4-connected sets $\mathcal{A}$, $\mathcal{B}$
output: a sequence $\mathcal{T}$ of pixels to remove (with $-$) and to add (with $+$)

1 $w \leftarrow$ the boundary word of $\mathcal{A}$, $\mathcal{L} \leftarrow$ Lyndon Factorization of w;
2 calculate $d_{\mathcal{A}}^{\mathcal{B}}(x)$ for $x \in \mathcal{A} \setminus \mathcal{B}$ and $d_{\mathcal{B}}^{\mathcal{A}}(x)$ for all $x \in \mathcal{B} \setminus \mathcal{A}$;
3 set the current deflated set $\mathcal{C} \leftarrow \mathcal{A}$, $\mathcal{T} \leftarrow \varnothing$;
4 $\mathcal{Q}_{\ominus} \leftarrow$ UpdateRemovable($\mathcal{Q}_{\ominus}, \mathcal{L}, \varnothing, (1, \ldots, |\mathcal{L}|), d_{\mathcal{A}}^{\mathcal{B}}, \mathcal{C}$);
5 $\mathcal{Q}_{\oplus} \leftarrow$ AddInsertable($\mathcal{Q}_{\oplus}, \mathcal{L}, (1, \ldots, |\mathcal{L}|), d_{\mathcal{B}}^{\mathcal{A}}$);
6 **while** $\mathcal{Q}_{\ominus} \cup \mathcal{Q}_{\oplus} \neq \varnothing$ **do**
7     $x \leftarrow$ the Lyndon point corresponding to find_max($\mathcal{Q}_{\ominus}$);
8     $y \leftarrow$ the closest upper point corresponding to find_max($\mathcal{Q}_{\oplus}$);
9   **if** $d_{\mathcal{A}}^{\mathcal{B}}(x) \leq d_{\mathcal{B}}^{\mathcal{A}}(y)$ **then**
10   push $(x, -)$ to $\mathcal{T}$;
11   ... // `Deflation (Lines 6-10 of Algorithm 1)`
12   **else**
13   push $(y, +)$ to $\mathcal{T}$;
14   ... // `Inflation (Lines 6-14 of Algorithm 4)`
15   **end**
16     $\mathcal{Q}_{\ominus} \leftarrow$ UpdateRemovable($\mathcal{Q}_{\ominus}, \mathcal{L}, I_{old}, I_{new}, d_{\mathcal{A}}^{\mathcal{B}}, \mathcal{C}$);
17     $\mathcal{Q}_{\oplus} \leftarrow$ DelInsertable($\mathcal{Q}_{\oplus}, I_{old}$), $\mathcal{Q}_{\oplus} \leftarrow$ AddInsertable($\mathcal{Q}_{\oplus}, \mathcal{L}, I_{new}, d_{\mathcal{B}}^{\mathcal{A}}$);
18 **end**
19 **return** $\mathcal{T}$

---

deformation from a digitized disk of 2821 points (most left) to a digitized ellipse (most right).

5 Conclusion

In this article, using the combinatorics on words to identify digital convex sets via their boundary words, we proposed algorithms for pixelwise inflation, deflation and more general deformation of digital convex sets preserving their convexity. Given a pair of digital convex sets, we showed that each proposed algorithm creates a sequence of digital convex sets, namely a deformation between them. The worst-case time complexity for each inflation and deflation iteration step was analyzed: $\mathcal{O}(N)$ for both where N is the length of the boundary word of a given digital convex.

References

1. Acketa, D.M., Žunić, J.: On the maximal number of edges of convex digital polygons included into an m x m-grid. J. Comb. Theor. Ser. A **69**, 358–368 (1995)
2. Alexander, J.C., Thaler, A.I.: The boundary count of digital pictures. J. ACM **18**(1), 105–112 (1971)

3. Andrews, G.: A lower bound for the volume of strictly convex bodies with many boundary lattice points. Trans. Am. Math. Soc. **106**, 270–279 (1963)
4. Berstel, J.: Tracé de droites, fractions continues et morphismes itérés. In: Mots, pp. 298–309. Hermès (1990)
5. Borel, J.P., Laubie, F.: Quelques mots sur la droite projective réelle. J. de Théorie des Nombres de Bordeaux **5**(1), 23–51 (1993)
6. Borel, J.P., Laubie, F.: Construction de mots de Christoffel. Comptes Rendus de l'Académie des Sciences. Série I. Mathématique **313**(8), 483–485 (1991)
7. Brlek, S., Lachaud, J.O., Provençal, X., Reutenauer, C.: Lyndon + christoffel = digitally convex. Pattern Recogn. **42**(10), 2239–2246 (2009)
8. Chen, K.T., Fox, R.H., Lyndon, R.C.: Free differential calculus, iv. the quotient groups of the lower central series. Ann. Math. **68**(1), 81–95 (1958)
9. Cormen, T.H., Leiserson, C.E., Rivest, R.L., Stein, C.: Introduction to algorithms, 2nd edn. MIT Press, Cambridge (2001)
10. Crombez, L.: Digital convex + unimodular mapping = 8-connected (all points but one 4-connected). In: Lindblad, J., Malmberg, F., Sladoje, N. (eds.) DGMM 2021. LNCS, vol. 12708, pp. 164–176. Springer, Cham (2021). https://doi.org/10.1007/978-3-030-76657-3_11
11. Dulio, P., Frosini, A., Rinaldi, S., Tarsissi, L., Vuillon, L.: First steps in the algorithmic reconstruction of digital convex sets. In: Brlek, S., Dolce, F., Reutenauer, C., Vandomme, É. (eds.) WORDS 2017. LNCS, vol. 10432, pp. 164–176. Springer, Cham (2017). https://doi.org/10.1007/978-3-319-66396-8_16
12. Duval, J.P.: Factorizing words over an ordered alphabet. J. Algorithms **4**(4), 363–381 (1983)
13. Eckhardt, U.: Digital lines and digital convexity. In: Bertrand, G., Imiya, A., Klette, R. (eds.) Digital and Image Geometry. LNCS, vol. 2243, pp. 209–228. Springer, Heidelberg (2001). https://doi.org/10.1007/3-540-45576-0_13
14. Freeman, H.: On the Encoding of Arbitrary Geometric Configurations. IRE Trans. Electron. Comput. EC-**10**(2), 260–268 (1961)
15. Kim, C.E.: On the Cellular Convexity of Complexes. IEEE Trans. Pattern Anal. Mach. Intell. PAMI-**3**(6), 617–625 (1981)
16. Kong, T.Y., Rosenfeld, A.: Digital topology: introduction and survey. Comput. Vision Graph. Image Process. **48**(3), 357–393 (1989)
17. Lachaud, J.-O.: An alternative definition for digital convexity. J. Math. Imaging Vision , 1–18 (2022). https://doi.org/10.1007/s10851-022-01076-0
18. Lothaire, M.: Algebraic combinatorics on words, Encyclopedia of Mathematics and its Applications, vol. 90. Cambridge University Press, Cambridge (2002)
19. Lyndon, R.C.: On burnside's problem. Trans. Am. Math. Soc. **77**(2), 202–215 (1954)
20. Roussillon, T.: An arithmetical characterization of the convex hull of digital straight segments. In: Barcucci, E., Frosini, A., Rinaldi, S. (eds.) DGCI 2014. LNCS, vol. 8668, pp. 150–161. Springer, Cham (2014). https://doi.org/10.1007/978-3-319-09955-2_13
21. Tarsissi, L., Coeurjolly, D., Kenmochi, Y., Romon, P.: Convexity preserving contraction of digital sets. In: Palaiahnakote, S., Sanniti di Baja, G., Wang, L., Yan, W.Q. (eds.) ACPR 2019. LNCS, vol. 12047, pp. 611–624. Springer, Cham (2020). https://doi.org/10.1007/978-3-030-41299-9_48
22. Tarsissi, L., Kenmochi, Y., Romon, P., Coeurjolly, D., Borel, J.P.: Convexity preserving deformations of digital sets: characterization of removable and insertable points. Technical report, LIGM (2022)

Full Convexity for Polyhedral Models
in Digital Spaces

Fabien Feschet[1]($\boxtimes$) (iD) and Jacques-Olivier Lachaud[2]($\boxtimes$) (iD)

[1] Université Clermont Auvergne, CNRS, ENSMSE, LIMOS,
63000 Clermont-Ferrand, France
`fabien.feschet@u-auvergne.fr`
[2] Université Savoie Mont Blanc, CNRS, LAMA, 73000 Chambéry, France
`jacques-olivier.lachaud@univ-smb.fr`

Abstract. In a recent work, *full convexity* has been proposed as an alternative definition of digital convexity. It solves many problems related to its usual definitions, for instance: fully convex sets are digitally convex in the usual sense, but are also connected and simply connected. However, full convexity is not a monotone property hence intersections of fully convex sets may be neither fully convex nor connected. This defect might forbid digital polyhedral models with fully convex faces and edges. This can be detrimental since classical standard and naive planes are fully convex. We propose in this paper an *envelope operator* which solves in arbitrary dimension the problem of extending a digital set into a fully convex set. This extension naturally leads to digital polyhedra whose cells are fully convex. We present first a generic envelope operator which add points in required directions in parallel and prove that it builds a fully convex set. Then a relative envelope operator is proposed, which can be used to force digital planarity of fully convex sets. We provide experiments showing that our method produces coherent polyhedral models for any polyhedron in arbitrary dimension.

Keywords: Digital geometry · Digital convexity · Polyhedral model

1 Introduction

Convexity is a classical property in various domains of mathematics and computer science. It allows for instance guarantees for optimization, containment property via its separability with hyperplanes, and many convergence results in real or discrete analysis need convexity assumptions. While it has been primarily developed in $\mathbb{R}^d$, several extensions have been proposed in the past. Two main paths are possible for extending convexity: either going more abstract to adapt convexity to generic spaces or building more specialized versions for dedicated spaces like the digital space $\mathbb{Z}^d$ for instance. Most general extensions of convexity

This work has been partly funded by CoMeDiC ANR-15-CE40-0006 research grant.

É. Baudrier et al. (Eds.): DGMM 2022, LNCS 13493, pp. 98–109, 2022.
https://doi.org/10.1007/978-3-031-19897-7_9

rely on hull systems [Lau06], K-convexity and simplicial convexity [Lli02] or closure (hull) operators [And06]. Those general extensions do not necessarily embed a geometric vision of convexity, so convex sets do not have a geometric structure in the same veins as in $\mathbb{R}^d$. More resembling extensions rely on anti-matroids notably with the anti-exchange property [RS03] or cellular extensions based on discrete hyperplanes [Web01, RS03]. They induce spaces of convex sets with more geometric interpretations, but also fail to be connected in some situations. Several extensions have also been proposed in the optimization community using convexity and digital convexity as certificates of optimality [MS01]. For digital spaces $\mathbb{Z}^d$, digital convexity was first defined as the intersection of real convex sets of $\mathbb{R}^d$ with $\mathbb{Z}^d$ (e.g. see survey [Ron89]). Many works have then tried to enforce the connectedness of such sets, for instance by relying on digital lines [KR82b, Eck01] or extensions of digital functions [Kis04]. Most works are limited to 2D, and 3D extensions do not solve all geometric issues [KR82a].

This paper considers the recently introduced notion of *full convexity* [Lac21, Lac22]. It extends digital convex sets while enforcing connectedness of fully convex sets. This notion is also computational in the sense that verifying full convexity is an easy task. Furthermore classical standard and naive planes are fully convex, so this convexity is appealing for building polyhedral models in any dimensions. However, since intersections of fully convex sets are not always fully convex, full convexity cannot be used directly for building faces and edges of polyhedra. Indeed the full convexity does not verify the monotonicity property of classical hull operators and thus fully convex hull is not a properly defined hull operator. This is a problem if we wish to build digital polyhedra in arbitrary dimension. In 3D, graceful lines and planes have been proposed in [BB02] to define edges consistent with triangular faces. It permits to fix varying arithmetical thickness between interior and boundary of digital triangles by construction but it is limited to 3D.

Our objective is to define polyhedral models in digital space $\mathbb{Z}^d$ which are based on full convexity. Our proposal lets us freely choose the thickness of digital faces, is canonic in arbitrary dimension, and benefits from the nice properties of fully convex sets. Indeed, naive, standard or even thicker pieces of arithmetical planes can be reconstructed in the proposed unified framework.

We start by defining the *fully convex envelope*, that is a pre-hull operator without the monotonicity property, which builds a fully convex set containing the input digital set. Our process is iterative, fully parallel at each iteration and ends after a finite number of iterations. It uses solely classical operators in the cubical complex $\mathscr{C}^d$ associated to $\mathbb{Z}^d$. We then adapt it to define a fully convex enveloppe *relative to another fully convex set*. Since thick enough digital planes are known to be fully convex, we can define fully convex subsets of digital planes in arbitrary dimension. The simultaneous use of those two operators builds edges and faces for meshes with planar faces or meshes with non planar faces. Experiments show that the induced *polyhedral models* are visually appealing and preserve the connectivity graphs between faces and edges of original models.

2   Full Convexity and Fully Convex Envelope

2.1   Definitions

Cubical Cell Complex. We consider the (cubical) cell complex $\mathscr{C}^d$ induced by the lattice $\mathbb{Z}^d$, such that its 0-cells are the points of $\mathbb{Z}^d$, its 1-cells are the open unit segments joining two 0-cells at distance 1, its 2-cells are the open unit squares formed by these segments, ..., and its d-cells are the d-dimensional unit hypercubes with vertices in $\mathbb{Z}^d$. We denote $\mathscr{C}^d_k$ the set of its k-cells. We call *complex/subcomplex* any subset of cells of $\mathscr{C}^d$, e.g. any single cell is a subcomplex. A *digital set* is a subset of $\mathbb{Z}^d$.

The *(topological) boundary* ∂Y of a subset Y of $\mathbb{R}^d$ is the set of points in its closure but not in its interior. The *star* of a cell σ in $\mathscr{C}^d$, denoted by $\mathrm{Star}\,(\sigma)$, is the set of cells of $\mathscr{C}^d$ whose boundary contains σ and it contains the cell σ itself. The *closure* $\mathrm{Cl}\,(\sigma)$ of σ contains σ and all the cells in its boundary. We extend these definitions to any subcomplex K of $\mathscr{C}^d$ by taking unions:

$$\mathrm{Star}\,(K) := \bigcup_{\sigma \in K} \{\mathrm{Star}\,(\sigma)\},$$

$$\mathrm{Cl}\,(K) := \bigcup_{\sigma \in K} \{\mathrm{Cl}\,(\sigma)\}.$$

In combinatorial topology, a subcomplex K with $\mathrm{Star}\,(K) = K$ is *open*, while being *closed* when $\mathrm{Cl}\,(K) = K$. The *body* of a subcomplex K, i.e. the union of its cells in $\mathbb{R}^d$, is written $\|K\|$. We denote by $\mathrm{Extr}\,(K) = \mathrm{Cl}\,(K) \cap \mathbb{Z}^d$.

Intersection Complex. If Y is any subset of the Euclidean space $\mathbb{R}^d$, we denote by $\bar{\mathscr{C}}^d_k[Y]$ the set of k-cells whose topological closure intersects Y, i.e.

$$\bar{\mathscr{C}}^d_k[Y] := \{c \in \mathscr{C}^d_k, \bar{c} \cap Y \neq \emptyset\}. \tag{1}$$

The complex that is the union of all, $\bar{\mathscr{C}}^d_k[Y], 0 \leqslant k \leqslant d$, is called the *intersection (cubical) complex of* Y and is denoted by $\bar{\mathscr{C}}^d[Y]$.

It is worth to note that, for any complex K, $\mathrm{Star}\,(K) = \bar{\mathscr{C}}^d[\|K\|]$. Hence, for any subset $Y \subset \mathbb{R}^d$, it is natural to define $\mathrm{Star}\,(Y) := \bar{\mathscr{C}}^d[Y]$, which coincides with the standard definition of star on subsets of $\mathscr{C}^d$ or $\mathbb{Z}^d$.

Skeleton. We define a kind of converse operation to the star. For any complex $K \subset \mathscr{C}^d$, the *skeleton of* K is (with K' any subset of K)

$$\mathrm{Skel}\,(K) := \bigcap_{K' \subset K \subset \mathrm{Star}(K')} K'. \tag{2}$$

Lemma 1. *For any complex* K, $K \subset \mathrm{Star}\,(\mathrm{Skel}\,(K))$.

Lemma 2. *For any digital set* X *we have* $\mathrm{Skel}\,(\mathrm{Star}\,(X)) = X$ *using lemma (1).*

Lemma 3. *For any open complex K, $\mathrm{Star}(\mathrm{Skel}(K)) = K$.*

Proof. $(\supset)$ $K \subset \mathrm{Star}(\mathrm{Skel}(K))$ by lemma (1).
$(\subset)$ $\mathrm{Skel}(K) \subset K$ because $\mathrm{Skel}(K)$ is the intersection of subsets of K. $\mathrm{Star}()$ being increasing, $\mathrm{Star}(\mathrm{Skel}(K)) \subset \mathrm{Star}(K) = K$ since K is open.

2.2   Full Convexity

For a set $A \subset \mathbb{R}^d$, its *convex hull* $\mathrm{CvxH}(A)$ is the intersection of all convex sets that contains A.

Definition 1 (Full convexity). *A digital set $X \subset \mathbb{Z}^d$ is* digitally k-convex *for $0 \leqslant k \leqslant d$ whenever*

$$\bar{\mathscr{C}}_k^d[X] = \bar{\mathscr{C}}_k^d[\mathrm{CvxH}(X)]. \tag{3}$$

Subset X is fully (digitally) convex *if it is digitally k-convex for all $k, 0 \leqslant k \leqslant d$.*

The following characterization will be useful:

Lemma 4 *A digital set X is fully convex iff $\mathrm{Star}(X) = \mathrm{Star}(\mathrm{CvxH}(X))$.*

2.3   Fully Convex Envelope

Convex hull is one of the most fundamental tool in continuous geometry. We wish to design a digital analogue to convex hull. The question is then how to build a fully convex set from an arbitrary digital subset of $\mathbb{Z}^d$. For instance can we build this fully convex envelope with intersections of fully convex set ? We do have this rather straightforward property:

Lemma 5 *If A and B are digitally 0-convex, then $A \cap B$ is digitally 0-convex.*

Proof

$$\mathrm{CvxH}(A \cap B) \cap \mathbb{Z}^d \subset \mathrm{CvxH}(A) \cap \mathrm{CvxH}(B) \cap \mathbb{Z}^d \quad (\mathrm{CvxH}(\cdot) \text{ is increasing})$$
$$= A \cap B \quad\quad\quad\quad (A \text{ and } B \text{ are digitally 0-convex})$$

$\square$

However, intersections of fully convex sets are generally not fully convex. As a very simple example, just pick $A = \{(0,0),(1,1),(2,1)\}$ and $B = \{(0,0),(1,0),(2,1)\}$, which are both fully convex. Then the set $A \cap B = \{(0,0),(2,1)\}$ is not fully convex, not even connected.

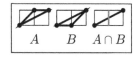

Therefore, we propose another way to build a fully convex set from an arbitrary digital set, which uses the cells intersected by the convex hull of this set, and which is defined through an iterative process.

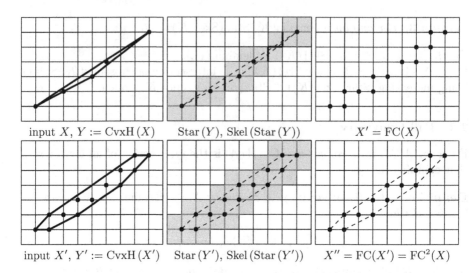

input X, $Y := \mathrm{CvxH}(X)$	Star (Y), Skel $(\mathrm{Star}(Y))$	$X' = \mathrm{FC}(X)$
input X', $Y' := \mathrm{CvxH}(X')$	Star (Y'), Skel $(\mathrm{Star}(Y'))$	$X'' = \mathrm{FC}(X') = \mathrm{FC}^2(X)$

Fig. 1. Illustration of FC operation and fully convex envelope construction. Left: input digital set X and its convex hull, middle: Star $(\mathrm{CvxH}(X))$ (gray and thick black) and its skeleton (thick black), right: extremal points of the skeleton, i.e. $\mathrm{FC}(X)$. Here X is digitally 0-convex but not fully convex. $\mathrm{FC}(X)$ is not even digitally 0-convex, while $\mathrm{FC}(\mathrm{FC}(X))$ is fully convex and is therefore the fully convex envelope to X

Each iteration composes these operations, for $X \subset \mathbb{R}^d$:

$$\mathrm{FC}(X) := \mathrm{Extr}(\mathrm{Skel}(\mathrm{Star}(\mathrm{CvxH}(X))))$$

First the Euclidean convex hull of the set is computed, letting $Y = \mathrm{CvxH}(X)$, then its covering Star (Y) by cells of the cellular grid is determined. The skeleton of these cells is their smallest subset such that Star $(\mathrm{Skel}(\mathrm{Star}(Y))) = Y$. Finally $\mathrm{FC}(X)$ is composed of the grid vertices of the skeleton cells. The last operation implies that $\mathrm{FC}(X) \subset \mathbb{Z}^d$. Refer to Fig. 1 for an illustration of FC operation and fully convex envelope computation.

Definition 2 (Fully convex envelope). *For any integer $n \geqslant 0$, the n-th convex envelope of $X \subset \mathbb{R}^d$ is the n times composition of operation FC.*

$$FC^n(X) := \underbrace{FC \circ \cdots \circ FC}_{n\ times}(X).$$

The fully convex envelope *of X is the limit of $FC^n(X)$ when $n \to \infty$:*

$$FC^*(X) := \lim_{n \to \infty} FC^n(X).$$

We have to show that this process has a limit for every subset X.

Theorem 1. *For any finite digital set $X \subset \mathbb{Z}^d$, there exists a finite n such that $FC^n(X) = FC^{n+1}(X)$, which implies that $FC^*(X)$ exists and is equal to $FC^n(X)$.*

It is the immediate consequence of Lemma 6 and Lemma 7 below: the first one tells that FC is increasing, the second that X and $FC(X)$ have the same bounding box.

Lemma 6. *For any $X \subset \mathbb{Z}^d, X \subset FC(X)$.*

Proof. Let $x \in X \subset \mathbb{Z}^d = \mathscr{C}_0^d$. Obviously $x \in \text{CvxH}(X)$. It follows that $x \in \text{Star}(\text{CvxH}(X))$ and, since $\text{Star}(\cdot)$ is idempotent, $\text{Star}(x) \subset \text{Star}(\text{CvxH}(X))$. The whole star of x belonging to the subcomplex $K := \text{Star}(\text{CvxH}(X))$, the 0-cell x belongs to the skeleton of K. Since all 0-cells of a subcomplex are extremal points, it is an extremal point of $\text{Skel}(K)$, which concludes.   □

Lemma 7. *For any finite $X \subset \mathbb{Z}^d$, X and $FC(X)$ have the same bounding box.*

Proof. Let $p \in \mathbb{Z}^d$ be the lowest point of the axis-aligned bounding box of X, i.e. $\forall i, 1 \leqslant i \leqslant d, p^i = \min_{z \in X} z^i$. Obviously, it is also the lowest point of the bounding box of $\text{CvxH}(X)$. Let $K := \text{Star}(\text{CvxH}(X))$. Since $\forall x \in \text{CvxH}(X), p^i \leqslant x^i$, any cell c of K that lie below point q along some coordinate axis j has a twin cell $e \in K$ in its boundary, such that e is closed along coordinate j and $e^j = p^j$. Continuing the argument along every coordinate axis k where e is below point p, we know that there is a digital point $z \in K$ in the boundary of c, such that z is not below p. Point z being a 0-cell it follows that $z \in \text{Skel}(K)$ while all m-cells incident to z, $m > 0$, are not in $\text{Skel}(K)$. We have just shown that no cells of $\text{Skel}(K)$ can be lower than p. The reasoning is the same for the uppermost point.   □

A first observation is that operation FC does not modify fully convex sets, so the fully convex envelope of a fully convex set X is X itself.

Lemma 8. *If $X \subset \mathbb{Z}^d$ is fully convex, then $FC(X) - X$. So $FC^*(X) - X$.*

Proof. Indeed we have

$$
\begin{aligned}
FC(X) &= \text{Extr}(\text{Skel}(\text{Star}(\text{CvxH}(X)))) \\
&= \text{Extr}(\text{Skel}(\text{Star}(X))) & \text{(Lemma 4)} \\
&= \text{Extr}(X) & \text{(Lemma 2)} \\
&= X & (X \subset \mathbb{Z}^d)
\end{aligned}
$$

□

Reciprocally, non fully convex sets are modified through operation FC.

Lemma 9. *If $X \subset \mathbb{Z}^d$ is not fully convex, then $X \subsetneq FC(X)$*

Proof. By Lemma 6 we already know that $X \subset FC(X)$. Let us show that there is a digital point $z \in FC(X)$ that is not in X. Since X is not fully convex, there exists some cell $c \in \text{Star}(\text{CvxH}(X))$ such that $c \notin \text{Star}(X)$. It is possible that there are other cells c' in $\bar{c}$ such that $c' \in \text{Star}(\text{CvxH}(X))$ and $c' \notin \text{Star}(X)$. In this case we pick one, say b, with lowest dimension.

Let $z \in \bar{b} \cap \mathbb{Z}^d$ be a grid vertex of this cell (which may be b itself). Then $z \notin X$. Otherwise, Star $(z) \subset$ Star (X), hence the cell b, which belongs to Star (z) (through the equivalence $z \subset \bar{b} \Leftrightarrow b \in$ Star (z)), would thus belong to Star (X), a contradiction with the hypothesis.

Let us show now that $z \in FC(X)$. Recall that

$$FC(X) = \text{Extr}\left(\text{Skel}\left(\text{Star}\left(\text{CvxH}\left(X\right)\right)\right)\right).$$

We have $b \in$ Star $(\text{CvxH}(X))$. Furthermore b belongs to the skeleton of Star $(\text{CvxH}(X))$, since it is a cell of Star $(\text{CvxH}(X))$ with lowest dimension in the closure of c. Finally grid vertex z is an extremal point of b, so belongs to $FC(X)$. We conclude since $z \notin X$ holds.   □

Note that the Lemma also indicates *where* operation FC add digital points. Indeed, they are the vertices of the cells touched by the convex hull but not by the digital set itself. Lemmas 8 and 9 lead immediately to a characterization of fully convex sets:

Theorem 2. $X \subset \mathbb{Z}^d$ *is fully convex iff* $X = FC(X)$.

It also induces the most important property of the fully convex envelope operation: it always outputs fully convex sets.

Theorem 3. *For any finite* $X \subset \mathbb{Z}^d$, $FC^*(X)$ *is fully convex.*

Proof. By Theorem 1, $FC^*(X)$ exists and there exists some n such that $FC^*(X) = FC^n(X)$. Hence, $FC(FC^n(X)) = FC^n(X)$. By Theorem 2, $FC^n(X)$ is fully convex, and so is $FC^*(X)$.   □

The operator $FC^*(.)$ is thus increasing and idempotent. It however fails to be monotone because Skel $(.)$ is not a monotone operator with respect to inclusion. So, it is not a hull operator [And06]. Nevertheless, it induces a preorder relation $\mathcal{R}_{FC^*}$ on digital sets using

$$X \mathcal{R}_{FC^*} Y \iff FC^*(X) = FC^*(Y).$$

It induces equivalent classes among the set of digital sets. It has its own topology through its associated Alexandrov topology.

2.4   Algorithmic Aspects

We now look at the algorithmic aspects of computing FC^*. Since the computation of FC^* is done in a loop, we compute the complexity for each iteration. At the beginning of iteration k the points set is $FC^{k-1}(X)$. Using Quickhull, the convex hull can be computed in $O(nf_r/r)$ [BDH96] with n the number of input points, r the number of processed points and f_r the maximum number of facets of r vertices ($f_r = O(r^{\lfloor d/2 \rfloor}/\lfloor d/2 \rfloor!)$). Obviously $r \leqslant n$, such that the complexity is bounded by $O(f_n)$ with $f_n = O(n^{\lfloor d/2 \rfloor}/\lfloor d/2 \rfloor!)$. Here, n is the number of points in $FC^{k-1}(X)$. As described in [Lac21], Star $(\text{CvxH}(.))$ can be computed using 2^d

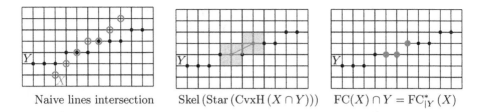

| Naive lines intersection | Skel (Star (CvxH ($X \cap Y$))) | FC(X) $\cap$ Y = FC$^*_{|Y}$ (X) |

Fig. 2. Relative fully convex envelope for naive lines having disconnected intersection.

Quickhull calls with the morphological characterizations of full convexity. It is the most intensive part of the computation. Then, Skel and Extr are extracted by simple traversal over the volume of Star (CvxH (.)). It is thus linear in the volume of Star (CvxH (.)) which is bounded above by the volume of the bounding box of FC^{k-1}(X). Hence the complexity of one iteration is bounded by $O(n^{\lfloor d/2 \rfloor})$. A precise bound on the number of iterations is still under study. In practice 1–4 iterations are generally observed in 3D, but we have come along examples with depth about ten.

3   Relative Fully Convex Envelope

We now specialize operator FC in order to stay into a given fully convex set. This creates fully convex sets relative to a given fully convex set. Given $Y \subset \mathbb{Z}^d$ a fully convex set and $X \subset Y$, the FC operator relative to Y is defined as

$$FC_{|Y} (X) := FC(X) \cap Y.$$

As previously, $FC^n_{|Y} (X) := FC_{|Y} \circ \cdots \circ FC_{|Y}(X)$, composed n times. The *fully convex envelope of X relative to Y* is obtained at the limit:

$$FC^*_{|Y} (X) := \lim_{n \to \infty} FC^n_{|Y} (X).$$

We thus have $FC^*(X) = FC^*_{|\mathbb{Z}^d} (X)$. In practice, for X not included in Y, we compute $FC^*_{|Y} (X \cap Y)$ to get the fully convex envelope of $X \cap Y$.

As seen on Figure (2), the relative fully convex envelope extends sets only using points of the fully convex set Y. So when considering two naive lines X and Y having disconnected intersection, both subsets $FC^*_{|Y} (X \cap Y)$ and $FC^*_{|X} (X \cap Y)$ are fully convex, hence are connected intersections.

Theorem 4. *For any finite $X \subset \mathbb{Z}^d$ and any fully convex set $Y \subset \mathbb{Z}^d$, the digital set $FC^*_{|Y} (X \cap Y)$ is fully convex and is included in Y.*

Proof. Let $X' = X \cap Y$. To see that $FC^*_{|Y} (X')$ is well defined, we rely on previous properties of $FC^*()$. By construction, since $FC()$ is increasing, so is $FC_{|Y}()$. Moreover Lemma (7) readily extends to say that X' and $FC_{|Y} (X')$

have the same bounding box. It is also true that if X' is fully convex then $FC_{|Y}(X') = X' \cap Y$ and so $FC^*_{|Y}(X') = X'$. Let us now see why Lemma (9) also extends to this situation. We hence suppose that X' is not fully convex. Let us then consider any cell b such that $b \in \mathrm{Star}(\mathrm{CvxH}(X'))$ but $b \notin \mathrm{Star}(X')$. Since $\mathrm{CvxH}(X') \subset \mathrm{CvxH}(Y)$, we deduce that $b \in \mathrm{Star}(\mathrm{CvxH}(Y)) = \mathrm{Star}(Y)$ since Y is fully convex. Moreover as in Lemma (9), we have $\bar{b} \cap \mathbb{Z}^d \cap X' = \emptyset$. But since $Y \subset \mathbb{Z}^d$ and $b \in \mathrm{Star}(Y)$, we deduce that $\bar{b} \cap \mathbb{Z}^d \cap Y \neq \emptyset$. Hence at least one point in Y is added by $FC_{|Y}()$. This implies that $X' \subsetneq FC_{|Y}(X')$. We can thus mimic Theorem 1 and Theorem 2 to get that $FC^*_{|Y}(X')$ exists and is fully convex. It is included in Y by construction.   □

Arithmetical planes with thickness at least as thick as naive planes are fully convex [Lac21, Theorem 7]. Hence the set Y can be chosen to be either a naive or a standard plane. Then the fully convex hull of X relative to Y is a fully convex subset of Y containing $X \cap Y$. Hence, $FC^*_{|X \cap Y}(X)$ is a simply connected piece of the arithmetical plane Y. To compute $FC^*_{|Y}(.)$, we only have to incorporate the intersection with Y at each iteration. This is directly linked to the complexity of deciding if a point p is in Y. If Y is a digital plane then this complexity is constant but in general it can be up to the order of $O(\log(\sharp Y))$.

4   Digital Polyhedron

We now present digital models for Euclidean polyhedra based on envelopes. A *polyhedron* $\mathscr{P}$ is a collection of finite convex sets called *cells*, such that each cell σ is characterized by a finite number of points $V(\sigma)$ called vertices. Cell σ is a *face* of cell σ' if $V(\sigma) \subset V(\sigma')$. The vertices V of the polyhedron are the union of the vertices of all cells. Generally an abstract dimension is attached to cells, 0 for vertices, 1 for edges, 2 for faces, etc., and must be consistent with the face relation. We take an interest here in polyhedra with maximal dimension $d-1$, i.e. surfaces, whose $(d-1)$-cells are called *facets*. Figure 3, left, shows two polyhedra in 3D space: a quadrangulated surface $\mathscr{Q}$ with non planar facets and a triangulated surface $\mathscr{T}$ with planar facets.

Assuming each vertex of $\mathscr{P}$ is a point of $\mathbb{Z}^d$, the *(generic) digital polyhedron* $\mathscr{P}^*$ associated to $\mathscr{P}$ is the collection of digital cells that are subsets of $\mathbb{Z}^d$, such that: if σ is a cell of $\mathscr{P}$, then σ^* is a cell of $\mathscr{P}^*$ with $\sigma^* := FC^*(V(\sigma))$. Such a digital polyhedron is illustrated on Fig. 3, top row.

When vertices of facets are coplanar, we can build a digital polyhedron whose facets are pieces of arithmetic planes. Pure simplicial complexes of dimension $d-1$ are important examples of such polyhedron. For $T \subset \mathbb{Z}^d$ made of coplanar points, let us denote by $P_1(T)$ the median standard plane (resp. $P_\infty(T)$ the median naive plane) defined by T.

The *standard (resp. naive) digital polyhedron* $\mathscr{P}^*_1$ (resp. $\mathscr{P}^*_\infty$) is the collection of digital cells subsets of $\mathbb{Z}^d$, defined as follows. For $p \in \{1, \infty\}$, if σ is a facet of $\mathscr{P}$, then σ^*_p is a cell of $\mathscr{P}^*_p$ with $\sigma^*_p := FC^*_{|P_p(V(\sigma))}(V(\sigma))$. For any cell τ that is not a facet, then it has as many geometric realizations as incident facets σ

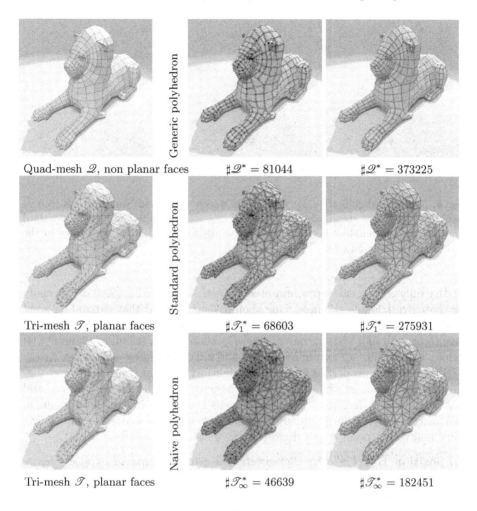

Fig. 3. Discretization of Euclidean polyhedral models without or with planar facets (left), at gridstep $h = 1$ (middle) and $h = 0.5$ (right).

and each pair (τ, σ) is digitized as $(\tau, \sigma)_p^* := \mathrm{FC}^*_{|\sigma_p^*}(V(\tau))$. Cell pairs have the same role as *half-edges* in winged-edge data structures and more generally *darts* in combinatorial maps. Note that other thicknesses could be chosen for digital polyhedron but naive and standard are the most common ones. A standard (resp. naive) digital polyhedron associated to a triangulated mesh is illustrated on Fig. 3, middle row (resp. bottom row). They require less digital points than the generic digital points, while keeping their separation properties.

To better understand the three defined polyhedra, let us consider a single triangle and its edges and vertices: its three digital models are displayed on Fig. 4. All induced cells are fully convex, but we notice that standard cells are thinner while naive cells are even thinner. What might be surprising is that

generic faces $\sharp\mathscr{T}^* = 1193$ standard faces $\sharp\mathscr{T}_1^* = 985$ naive faces $\sharp\mathscr{T}_\infty^* = 567$

Fig. 4. A generic digital triangle $\mathscr{T}^*$ with its darker edges and black vertices (p, q, r) (left); corresponding standard digital triangle $\mathscr{T}_1^*$ which lies in the median standard plane $P_1(p, q, r)$ (middle); corresponding naive digital triangle $\mathscr{T}_\infty^*$ which lies in the median naive plane $P_\infty(p, q, r)$ (right).

relative fully convex enveloppe may create larger subset than expected, especially for the naive triangle example. One should keep in mind that expanding a set inside a naive plane to become fully convex is a very restrictive transform: edges have to expand more within naive plane P_∞ than within standard plane P_1. Of course, this is quite an extreme example and edges are narrower in most cases.

The following properties are quite straightforward, but show that every digital polyhedron covers well the cells of its associated Euclidean polyhedron, and that the inclusion/face property between cells is satisfied in the digital domain. Digitizing a polyhedron at different gridstep h is just a matter of embedding every real vertex point q as a digital vertex $q^* = \text{round}(q/h)$ (see Fig. 3).

Proposition 1. *Let σ^* be a digital cell of a generic, standard or naive digital polyhedron. Then it is fully convex, hence digitally connected and simply connected. We have* $\text{Star}(\text{CvxH}(V(\sigma))) \subset \text{Star}(\sigma^*)$. *For any cell τ such that σ is a face of τ,* $\text{Star}(\tau^*)$ *cover* $\text{Star}(\text{CvxH}(V(\sigma)))$.

5   Conclusion and Perspectives

We provide in this paper an envelope operator for full convexity $\text{FC}^*(.)$. For any digital set X, $\text{FC}^*(X)$ is proved to be fully convex and $X \subset \text{FC}^*(X)$. Furthermore this operator leaves fully convex sets unchanged. Moreover, the operator is well defined in arbitrary dimension as well as computable. This operator can be restricted to stay within a fully convex set Y, leading to the relative enveloppe operator $\text{FC}^*_{|Y}(X)$. It builds fully convex sets within Y. Since classical naive and standard planes are fully convex, this leads to a straightforward computation of digital analogues to polyhedral models of $\mathbb{R}^d$. The obtained results are quite appealing: we can control the incidence relationship between cells, while their full convexity guarantees their topological and geometrical properties. These digital polyhedral models embrace both meshes with planar or non planar faces.

In future works, we would like to study more precisely the iterative process of FC*(.), in order to localize where full convexity defects reside. This could accelerate the operator and give more practical bounds on the number of iterations. Incremental quickhull should also be considered. A more general goal is to extend the enveloppe process to a true convex hull operator. The difficulty is to ensure the monotone property, but if we succeed, the full convexity would then be a digital analogue to convexity for digital spaces.

References

[And06] Ando, K.: Extreme points axioms for closure spaces. Discret. Math. **306**, 3181–3188 (2006)

[BB02] Brimkov, V.E., Barneva, R.P.: Graceful planes and lines. Theoret. Comput. Sci. **283**(1), 151–170 (2002)

[BDH96] Barber, C.B., Dobkin, D.P., Huhdanpaa, H.: The quickhull algorithm for convex hulls. ACM Trans. Math. Softw. **22**(4), 469–483 (1996)

[Eck01] Eckhardt, U.: Digital lines and digital convexity. In: Bertrand, G., Imiya, A., Klette, R. (eds.) Digital and Image Geometry. LNCS, vol. 2243, pp. 209–228. Springer, Heidelberg (2001). https://doi.org/10.1007/3-540-45576-0_13

[Kis04] Kiselman, C.O.: Convex functions on discrete sets. In: Klette, R., Žunić, J. (eds.) IWCIA 2004. LNCS, vol. 3322, pp. 443–457. Springer, Heidelberg (2004). https://doi.org/10.1007/978-3-540-30503-3_32

[KR82a] Kim, C.E., Rosenfeld, A.: Convex digital solids. IEEE Trans. Pattern Anal. Mach. Intel. **6**, 612–618 (1982)

[KR82b] Kim, C.E., Rosenfeld, A.: Digital straight lines and convexity of digital regions. IEEE Trans. Pattern Anal. Mach. Intel. **2**, 149–153 (1982)

[Lac21] Lachaud, J.-O.: An alternative definition for digital convexity. In: Lindblad, J., Malmberg, F., Sladoje, N. (eds.) DGMM 2021. LNCS, vol. 12708, pp. 269–282. Springer, Cham (2021). https://doi.org/10.1007/978-3-030-76657-3_19

[Lac22] Lachaud, J.-O.: An alternative definition for digital convexity. J. Math. Imaging Vis. (2022). (To appear)

[Lau06] Lau, D.: Function Algebras on Finite Sets. Springer-Verlag, Chm (2006). https://doi.org/10.1007/3-540-36023-9

[Lli02] Llinares, J.-V.: Abstract convexity, some relations and applications. Optimization **51**(6), 797–818 (2002)

[MS01] Murota, K., Shioura, A.: Relationship of m/l-convex functions with discrete convex functions by Miller and Favati-Tardella. Discrete Appl. Math. **115**, 151–176 (2001)

[Ron89] Ronse, C.: A bibliography on digital and computational convexity (1961–1988). IEEE Trans. Pattern Anal. Mach. Intell. **11**(2), 181–190 (1989)

[RS03] Roy, A.J., Stell, J.G.: Convexity in discrete space. In: Kuhn, W., Worboys, M.F., Timpf, S. (eds.) COSIT 2003. LNCS, vol. 2825, pp. 253–269. Springer, Heidelberg (2003). https://doi.org/10.1007/978-3-540-39923-0_17

[Web01] Webster, J.: Cell complexes and digital convexity. In: Bertrand, G., Imiya, A., Klette, R. (eds.) Digital and Image Geometry. LNCS, vol. 2243, pp. 272–282. Springer, Heidelberg (2001). https://doi.org/10.1007/3-540-45576-0_16

Implicit Encoding and Simplification/Reduction of nGmaps

Florian Bogner[✉], Jiří Hladůvka, and Walter Kropatsch

Pattern Recognition and Image Processing Group, Vienna University of Technology, Vienna, Austria
florian.bogner@tuwien.ac.at, {jiri,krw}@prip.tuwien.ac.at
https://www.prip.tuwien.ac.at

Abstract. This paper aims to present a new method of translating labeled 3D scans of biological tissues into Generalized Maps (nGmaps). Creating such nGmaps from labeled images is a solved problem in 2D and 3D using incremental algorithms. We present a new approach that works in arbitrary dimensions. To achieve this in an effective manner, we perform the necessary operations implicitly using theory rather than explicitly in memory. First we define implicit nGmaps. We then present a scheme to construct said nGmap representing an nD pixel/voxel-grid implicitly. Thirdly we give a description of the process needed to reduce such implicit nGmap. We demonstrate that our implicit approach is able to reduce nGmaps in a fraction of otherwise necessary memory.

Keywords: Generalized Maps · nGmaps · Implicit representation · Memory savings

1 Introduction

For analysing CT scans of biological tissues, methods are needed to process the images. Assume that we have a microscopic 3D raster image of tissue and want to run a simulation of physiological processes within, for example leaf tissue and its inherent osmotic movements, respiration and further aspects of biological interest. Assume furthermore that the image is already segmented, meaning each pixel[1] is labeled. This means we know the specific cell or air-pocket a pixel belongs to.

1.1 Problem Statement

For such a simulation we need a data structure where cells and the connections between them are the primary objects. A data structure that meets these requirements and we therefore choose to use, is the n-dimensional Generalized Map (or nGmap for short) [3]. Thus we are faced with the problem of converting the labeled image into an nGmap.

---

[1] In this paper we use pixel as generic term for any dimension, i.e. including voxels in 3D and hypervoxels in 4D.

© Springer Nature Switzerland AG 2022
É. Baudrier et al. (Eds.): DGMM 2022, LNCS 13493, pp. 110–122, 2022.
https://doi.org/10.1007/978-3-031-19897-7_10

1.2   Prior Work

For 2D-images an algorithm already exists [3]. For 3D-images there is an algorithm for Combinatorial Maps [1]. While this algorithm could be adapted for 3Gmaps, we present a new method that generalizes to arbitrary dimensions.

1.3   Content

To coherently present our approach we first need to explain nGmaps and their specifics. In the following section we give a recap on nGmaps as well as new definitions.

In later sections we will present a new algorithm to translate labeled images into nGmaps. This includes two steps:

1. The implicit construction of the pixel-grid.
2. The contraction of the pixel-grid to adequately represent the labeled regions.

2   Basic Definitions

2.1   nGmap - The Intuitive Definition

An nGmap is a data structure similar to a graph or a mesh. It encodes topological information of a subdivision of an n-dimensional manifold. It consists of so-called i-cells for i from 0 to n. The number i describes the dimension of the i-cell. A 0-cell is a point, a 1-cell is a line bound by two points, i.e. two 0-cells. A 2-cell is a surface patch bound by 1-cells and so on. In general, a $(i+1)$-cell is bounded by i-cells.

For $i \neq j$ we call an i-cell A *incident* to a j-cell B, if A is in the boundary of B or vice-versa.

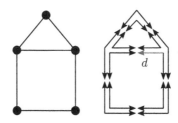

Fig. 1. Left: A 2Gmap consisting of five 0-cells, six 1-cells and three 2-cells (including the outside one). Right: The same 2Gmap depicted via its darts, which are drawn as arrows.

These i-cells, however, are not the primary elements used to encode the nGmap, instead so called *darts* are. A dart can be thought of as the intersection of incident i-cells, one for each dimension, i.e. for $i \in \{0, \ldots, n\}$. For example, the marked dart d in Fig. 1 corresponds to the middle-right 0-cell, the top-horizontal 1-cell and the square 2-cell. This interpretation of a dart as an intersection of i-cells is quite important for the intuitive understanding of nGmaps.

For a given i if two darts share the same j-cells for $j \neq i$, but have a different i-cell, we call them i-*linked*. It turns out that a dart is only ever i-linked to a maximum of one other dart. Therefore we can define the involution α_i as the function that maps a dart to its i-linked partner or itself if it has none.

It turns out that the set of darts and the involutions $\alpha_0, \ldots, \alpha_n$ fully describe the structure of the nGmap. Therefore we define an nGmap in the formal definition by its darts and subsequently also define i-cells in terms of darts.

2.2   The Formal Definition

Definition 1. Involution: *A function $f : X \to X$ is called an involution if*

$$\forall x \in X : f(f(x)) = x$$

Definition 2. nGmap: *For $n \in \mathbb{N}_0$ an nGmap or n-dimensional Generalized Map is a tuple $(D, \alpha_0, \ldots, \alpha_n)$, where:*

- *D is a finite set of darts.*
- *For $i \in \{0, \ldots, n\}$ the function $\alpha_i : D \to D$ is an involution.*
- *For $i, j \in \{0, \ldots, n\}, |i - j| \geq 2$ the composition $\alpha_i \circ \alpha_j$ is an involution.*

To define i-cells we first need to define the term orbit:

Definition 3. Orbit: *Let A be a set, $B \subset A$ and $f_1, \ldots, f_n : A \to A$ be functions. Then the orbit of B under $f_1, \ldots, f_n$:*

$$\langle f_1, \ldots, f_n \rangle (B)$$

is the smallest super-set of B closed under $f_1, \ldots, f_n$, i.e.:

- *$B \subset \langle f_1, \ldots, f_n \rangle (B) \subset A$*
- *$\forall i \in \{1, \ldots, n\}, \forall x \in \langle f_1, \ldots, f_n \rangle (B) : f_i(x) \in \langle f_1, \ldots, f_n \rangle (B)$*
- *These are all.*

For $x \in A$ the orbit is defined as the orbit of the singleton set $\{x\}$.

For example, the orbit of a single element under a single function is

$$\langle f \rangle (x) = \{x, f(x), f(f(x)), f(f(f(x))), \ldots\}$$

In formal definition i-cells are sets of darts. To a given dart we can find the corresponding i-cell for given i as follows:

Definition 4. i-cell: *Let $(D, \alpha_0, \ldots, \alpha_n)$ be an nGmap, $d \in D$ a dart and $i \in \{0, \ldots, n\}$. The i-cell containing d is defined as the set of darts:*

$$c_i(d) := \langle \alpha_0, \ldots, \alpha_{i-1}, \alpha_{i+1}, \ldots, \alpha_n \rangle (d)$$

This definition is motivated by the intuitive understanding that α_i changes i-cell and in turn α_j (for $j \neq i$) remains with the same i-cell. Therefore by traversing the orbit of d under α_j for $j \neq i$ we never leave the i-cell and because we consider all α_j, we find every dart of the i-cell.

3   Motivation

3.1   The Naive Algorithm

Let us consider a naive algorithm for the problem:

- Generate a 3Gmap with one cubic 3-cell for every pixel.
- To merge all adjacent pixels with the same label, remove in-between 2-cells.
- Form membranes by merging adjacent 2-cells that border the same 3-cells.
- Form membrane edges by removing every 0-cell that has less than three incident 1-cells.

We now have created a 3Gmap from a labeled image, effectively solving the problem in theory. However, let us estimate the memory requirements: Assuming that a dart is a class consisting of four pointers, one for each involution. A pointer takes up eight bytes in a 64-bit system. A cube in a 3Gmap consists of 48 darts. The 3D-images of the plant scans that motivated this work have a resolution of about 2000^3 pixels. So in total we have

$$2000^3 \times 48 \times 4 \times 8 \text{ bytes} \cong 12 \text{ terabytes}$$

Clearly the memory requirements for creating the pixel-grid mentioned in step one render the naive algorithm infeasible for such a scan. Our solution to circumvent the huge memory requirements is to represent the pixel-grid implicitly, instead of explicitly representing it in memory. Furthermore the reduction as in step 2 onward of the naive algorithm, can be represented implicitly. These two processes will be topic of Sects. 4 and 5 respectively. However one more tool needs to be defined as groundwork before.

3.2   Implicit nGmaps

Definition 5. Implicit nGmap: *For $n \in \mathbb{N}_0$ an implicit nGmap is a tuple $(D, D', \alpha_0, \ldots, \alpha_n)$, where:*

- *D is a (not necessarily finite) set of darts.*
- *$D' \subset D$ is a finite set of seed-darts.*
- *$\forall i : \alpha_i : D \to D$ is a function. (Not necessarily an involution.)*
- *$(\langle \alpha_0, \ldots, \alpha_n \rangle(D'), \alpha_0, \ldots, \alpha_n)$ is an nGmap, which is called the* Construction.

The idea here is that not all elements of D are darts in the nGmap we want to define. The darts in D' are called seed-darts because from them the orbits grow.

To distinguish, we will also call nGmaps as of Definition 2 *explicit nGmaps*. One major difference between the two types is less of theoretical nature and more related to actual implementations in code:

- Explicit nGmaps can be thought of as being stored in memory, with the α-involutions being implemented via lookup-table or memory pointers. They are mutable.
- Implicit nGmaps however can be thought of as being computed on the fly. Their α-involutions are procedures without state. Therefore they do not occupy much memory, but as a downside they are immutable.

4   Implicit Encoding of the Pixel-Grid

In this section we define an nGmap representing an infinite nD grid. By defining an infinite rather than a finite grid corresponding to the size of the image, we can avoid special cases related to the boundary.

4.1   Darts

As the set of darts we use[2]:

$$D := \mathbb{Z}^n \times \mathbb{N}_{<2^n \cdot n!}$$

A dart is a tuple $d = (p, s) \in D$. The first component $p \in \mathbb{Z}^n$ is called the pixel-position. The second component $s \in \mathbb{N}_{<2^n \cdot n!}$ is called the subpixel-position. Note that there are $2^n \cdot n!$ darts in an nGmap representing a bounded nD-hypercube.

The following is a scheme to enumerate all darts in the interior of an nD cube. Recall that a dart represents the intersection of one i-cell for each i from 0 to n. Thus we describe a dart first by its position via those i-cells and then transform that description into an integer.

4.2   Positional Dart Descriptions

We construct our Positional Dart Description by answering a series of questions.

First: In which n-cell is the dart? We only have one n-cell, so the answer is trivial.

---

[2] Because the grid is infinite, the construction technically is not an nGmap. One can modify $D := \mathbb{Z}_k^n \times \mathbb{N}_{<2^n \cdot n!}$ using the cyclic group $\mathbb{Z}_k$ for some sufficiently large number k. The nGmap then represents a grid on a large torus and D is finite. When implementing D in code using for example 32-bit ints, this automatically happens with $k = 2^{32}$.

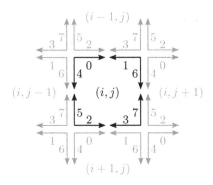

Fig. 2. Representation of one pixel and its neighbors in 2D. (s is attached to each dart, while $p = (i, j)$ is written in the center of each pixel instead of duplicated 8 times.)

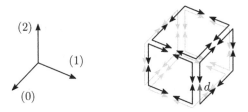

Fig. 3. 3Gmap of a cube, with one dart d marked as an example. Note the orientation and direction of the coordinate axes.

Next: In which $(n − 1)$-cell is the dart? There are two $(n − 1)$-cells for each coordinate axis, so we can describe the $(n − 1)$-cell by the coordinate axis it is perpendicular to and whether it is facing the positive or the negative direction, here called the "top" or the "bottom" respectively. The example dart d in Fig. 3 is in a 2-cell perpendicular to the (1) coordinate axis and on the "top". The descriptions thus begins with:

$$1 \uparrow \ldots$$

The identified $(n−1)$-cell now itself consists of $(n−2)$-cells, so our description continues recursively. In which one does it lie?

The example dart d is on the 1-cell perpendicular to the (0) axis and on the "top". Finally, d is on the "bottom" side of the (2) axis. The full description therefore is:

$$1 \uparrow 0 \uparrow 2 \downarrow$$

In general, a description is a list of length $2n$, a permutation of $\{0, \ldots, n − 1\}$ interleaved with arrows $\downarrow$ or $\uparrow$. As a sanity check, lets calculate the total possibilities: each of the n arrows can be up or down, so we have 2^n possibilities here. The axis numbers can be permuted in $n!$ ways. These are independent, so in total we have $2^n \cdot n!$ possibilities. This exactly matches the number of darts in the nD hypercube.

4.3   Mixed Radix Numbers

To transform a Positional Dart Description into a number, we use a mixed radix numbering system with the signature $(\ldots 5, 2, 4, 2, 3, 2, 2, 2, 1, 2)$ as explained below.

Table 1. Mixed radix system for our dart numbering system

Radix	...	n	2	...	5	2	4	2	3	2	2	2	1	2
Digit worth	...	$2^n n!$	$2^{n-1} n!$	...	768	384	96	48	16	8	4	2	2	1

In a usual numbering system with base b, each digit is worth b times the one on the right. In a mixed radix system, the relative worth of a digit is different for each digit according to its signature. The most common mixed radix system is used to measure time with the signature $(7, 24, 60, 60)$. Each week has 7 d, each day has 24 h, each hour has 60 min, each minute has 60 s.

Table 2. Mixed radix system for time

Name	Weeks	Days	Hours	Minutes	Seconds
Radix	–	7	24	60	60
Digit worth	604800	86400	3600	60	1

To transform a dart description, we translate each part into a digit of the mixed radix. For the arrows we can simply put 0 for ↓ and 1 for ↑.

The axis numbers are not translated directly, i.e. are not the translated digits themselves. Instead, the translation of an axis is its index on the list of not-yet-used axes. This is best explained by example. Let us translate the 4D dart description 0 ↑ 3 ↓ 1 ↑ 2 ↑. At first, no axes were used, so the list is $[0, 1, 2, 3]$. 0 has index 0. Our number thus starts as

$$(01??????)_b$$

The remaining list is now $[1, 2, 3]$ and 3 has index 2 on that list, so the number continues as

$$(0120????)_b$$

The remaining list is now $[1,2]$. 1 has index 0, and afterward 2 has index 0 so the complete number is

$$(01200101)_b$$

Finally lets translate the number into the decimal system using the 'Digit worth' entries from Table 1.

$$(01200101)_b = 0 \cdot 96 + 1 \cdot 48 + 2 \cdot 16 + 0 \cdot 8 + 0 \cdot 4 + 1 \cdot 2 + 0 \cdot 2 + 1 \cdot 1 = 83$$

Notice how we have for every digit exactly as many choices as is the radix for this digit. Therefore the dart numbers lie flush without gaps.

4.4 Involutions

We define the involutions by lookup-tables (LUT). The LUT maps a subpixel-position to another subpixel-position as well as an offset to the pixel-position. Because only α_n leaves the n-cell i.e. the pixel, the offset is actually only required for this single involution.

We thus define:

$$\alpha_i((p, s)) := \begin{cases} (p, \alpha_i^*(s)) & i < n \\ (p + \Delta p(s), \alpha_i^*(s)) & i = n \end{cases}$$

where α_i^* and Δp are called lookup-tables.

Table 3. Lookup-tables for the 2D case corresponding to Fig. 2 as well as an implementation using bit-flipping magic on the binary representation. (The hat means bit negation.)

s	$\alpha_0^*(s)$	$\alpha_1^*(s)$	$\alpha_2^*(s)$	$\Delta p(s)$
0	1	4	2	$(-1, 0)$
1	0	6	3	$(-1, 0)$
2	3	5	0	$(1, 0)$
3	2	7	1	$(1, 0)$
4	5	0	6	$(0, -1)$
5	4	2	7	$(0, -1)$
6	7	1	4	$(0, 1)$
7	6	3	5	$(0, 1)$
abc	$ab\hat{c}$	$\hat{a}cb$	$a\hat{b}c$	N/A

But how can we define these lookup tables? Let us again turn to the Positional Dart Description. First, notation: For $i \in \{0, \ldots, n-1\}$, let $x_i \in \{0, \ldots, n-1\}$ be the axis that the i-cell is perpendicular to and $I_i \in \{\downarrow, \uparrow\}$ be the bottom-top-indicator. $\hat{I}_i$ shall denote the opposite arrow of I_i itself. A general description then looks like this:

$$x_{n-1}I_{n-1} \ldots x_1 I_1 x_0 I_0$$

- α_0^*: The involution α_0 changes 0-cell while staying in the same i-cell for $i > 0$. Thus the start of the description stays the same and only in the last part we swap which side we are on. Thus:

$$\alpha_0^*(x_{n-1}I_{n-1} \ldots x_1 I_1 x_0 I_0) = x_{n-1}I_{n-1} \ldots x_1 I_1 x_0 \hat{I}_0$$

- α_i^* for $0 < i < n$: The involution α_i changes i-cell while staying in the same j-cell for $j \neq i$. Therefore the description before $x_i I_i$ stays the same. The i-cell changes, therefore x_i must change. The original i-cell and the image i-cell intersect in an $(i-1)$-cell. This $(i-1)$-cell is perpendicular to both axes x_i and x_{i-1}. Therefore the image i cell is perpendicular to x_{i-1}. This intersecting $(i-1)$-cell is now on the x_i side of the image i-cell. x_i and x_{i-1} therefore swap places in the description. The arrows swap with them. Afterwards, we are in the same $(i-2)$-cell and so on, so the suffix of the description does not change as well.

$$\alpha_i^*(\ldots x_i I_i x_{i-1} I_{i-1} \ldots) = \ldots x_{i-1} I_{i-1} x_i I_i \ldots$$

- α_n^* and Δp: The involution α_n moves us from one n-cell to another, in particular the one that shares the same $(n-1)$-cell. The orientation in regard to the other axes does not change. The direction we move is dependent on I_{n-1}. Therefore we find that:

$$\alpha_n^*(x_{n-1} I_{n-1} \ldots x_1 I_1 x_0 I_0) = x_{n-1} \hat{I}_{n-1} \ldots x_1 I_1 x_0 I_0$$

$$\Delta p(x_{n-1} I_{n-1} \ldots x_1 I_1 x_0 I_0) = \begin{cases} e_{x_{n-1}} & I_{n-1} = \uparrow \\ -e_{x_{n-1}} & I_{n-1} = \downarrow \end{cases}$$

where e_k is the k-th unit vector.

Note that these definitions elegantly fulfil condition 2 and 3 of Definition 2. Table 3 is generated with these definitions.

4.5   Labels

With the structure of the grid fully defined, we finally need to associate every dart with a label. For a dart $d = (p, s)$ we associate:

- If the pixel-position p is within the image, we associate the label from that pixel in the image.
- Otherwise we associate an additionally created label not occurring in the image called the Out-Of-Bounds-Label. By treating the OOBL as just another label, we can avoid having to consider special cases on the boundary of the image.

Going forward, we denote the set of labels including the OOBL as $\mathbb{L}$ and the association between darts and labels as the function $L : D \to \mathbb{L}$.

5   Implicit Reductions and Contractions

Given an nGmap $(D, \alpha_0, \ldots, \alpha_n)$ and a label function $L : D \to \mathbb{L}$ we want to define new involutions β_i and the set D' such that $(D, D', \beta_0, \ldots, \beta_n)$ is an implicit nGmap. Note that the original nGmap doesn't have to be the pixel-grid from the previous section. All that is required is that the label function L is *consistent*, meaning all darts from a n-cell map to the same label. We define the β-functions iteratively from the highest dimension to the lowest and then discuss finding an appropriate set of seed-darts D'.

5.1   Defining β_n

Recall the intuitive understanding of α_n. It changes n-cell while staying in the same i-cell for $i < n$. Since we don't want to remove n-cells, but only merge them later on, we can just define $\beta_n := \alpha_n$.

5.2   Defining β_{n-1}

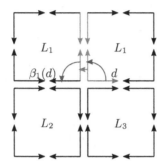

Fig. 4. 2D Example: The grey 1-cell ought to be removed, as it has the same label on both sides. The relevant involutions α_1 and β_2 are illustrated in blue and pink respectively. (Color figure online)

Intuitively the involution α_{n-1} changes to a different $(n-1)$-cell, but such an $(n-1)$-cell may ought to be removed, if the labels on both sides are equal. In this case, illustrated in Fig. 4, we would have to move past that $(n-1)$-cell onto the next one via $\alpha_{n-1} \circ \beta_n \circ \alpha_{n-1}$. Of course, it has to be checked if this next $(n-1)$-cell ought to be removed as well. We define the condition that describes if the $(n-1)$-cell $c_{n-1}(d)$ of a given dart d is removable:

$$R_{n-1}(d) :\Leftrightarrow L(d) = L(\beta_n(d))$$

With this, we can define the new involution:

$$\beta_{n-1}(d) := \begin{cases} \alpha_{n-1}(d) \\ \qquad \text{if } \neg R_{n-1}(\alpha_{n-1}(d)) \\ \alpha_{n-1} \circ \beta_n \circ \alpha_{n-1}(d) \\ \qquad \text{elif } \neg R_{n-1}(\alpha_{n-1} \circ \beta_n \circ \alpha_{n-1}(d)) \\ \alpha_{n-1} \circ \beta_n \circ \alpha_{n-1} \circ \beta_n \circ \alpha_{n-1}(d) \\ \qquad \text{elif } \neg R_{n-1}(\alpha_{n-1} \circ \beta_n \circ \alpha_{n-1} \circ \beta_n \circ \alpha_{n-1}(d)) \\ \dots \\ (\alpha_{n-1} \circ \beta_n)^k \circ \alpha_{n-1}(d) \\ \qquad \text{elif } \neg R_{n-1}\left((\alpha_{n-1} \circ \beta_n)^k \circ \alpha_{n-1}(d)\right) \end{cases}$$

The implicit nGmap $(D, D', \alpha_0, \ldots, \alpha_{n-2}, \beta_{n-1}, \beta_n)$ represents an nGmap where the pixels are merged into bigger n-cells according to their labels, but the $(n-1)$-cells are still the sides of a pixel. They can be simplified further.

5.3   Defining β_i for $i \leq n - 2$

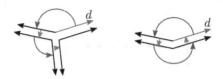

Fig. 5. 2D Example: The left 0-cell is not removable, while the right 0-cell is removable. On the right $\beta_1 \circ \beta_2 \circ \beta_1 \circ \beta_2 (d)$ loops back, which is not the case on the left.

For the next step, we need to simplify the $(n-2)$-cells, which is accomplished by an appropriate definition of β_{n-2}. Further we also need to simplify the $(n-3)$-cells via β_{n-3} and so on. All of these steps are alike and follow the same definition. For this we need to define when an i-cell is removable. This is only fulfilled if the i-cell has two incident $(i+1)$- cells. [4]

$$R_i(d) :\Leftrightarrow \forall d' \in \langle \beta_{i+2}, \ldots \beta_n \rangle (d) : d' = \beta_{i+1} \circ \beta_{i+2} \circ \beta_{i+1} \circ \beta_{i+2} (d')$$

With this removability criterion illustrated in Fig. 5 we can define:

$$\beta_i(d) := \begin{cases} \alpha_i(d) & \text{if } \neg R_i(\alpha_i(d)) \\ \alpha_i \circ \beta_{i+1} \circ \alpha_i(d) & \text{elif } \neg R_i(\alpha_i \circ \beta_{i+1} \circ \alpha_i(d)) \\ \alpha_i \circ \beta_{i+1} \circ \alpha_i \circ \beta_{i+1} \circ \alpha_i(d) & \text{elif } \neg R_i(\alpha_i \circ \beta_{i+1} \circ \alpha_i \circ \beta_{i+1} \circ \alpha_i(d)) \\ \ldots \\ (\alpha_i \circ \beta_{i+1})^k \circ \alpha_i(d) & \text{elif } \neg R_i\left((\alpha_i \circ \beta_{i+1})^k \circ \alpha_i(d)\right) \end{cases}$$

The implicit nGmap $(D, D', \alpha_0, \ldots, \alpha_{n-3}, \beta_{n-2}, \beta_{n-1}, \beta_n)$ has simplified n-cells and $(n-1)$-cells, but still has non-simplified i-cells for $i < n - 1$. This pattern continues until finally the implicit nGmap $(D, D', \beta_0, \ldots, \beta_n)$ represents the complete reduction of the image.

5.4   Finding Seed-Darts

To build the *Construction* of $(D, D', \beta_0, \ldots, \beta_n)$ we need one seed-dart for each connected component, since the orbit $\langle \beta_0, \ldots, \beta_n \rangle$ of one dart in a connected component cannot reach another connected component. If we know, like in our example-case, the image shows only one connected component, a single seed-dart suffices to create the nGmap for the whole image.

To find a suitable seed-dart we look for a dart d that is not removable at all:

$$\forall i \in \{0, \ldots, n - 1\} : \neg R_i(d)$$

5.5   Construction

During the *Construction*, i.e. the traversal of the orbit of the seed darts, we create an explicit nGmap. Every encountered implicit dart is associated with an explicit dart. As explicit nGmap we used nGmaps from the CGAL [2]. This explicit nGmap can now be used for further processing, simulations etc.

Optionally, if a bounded nGmap is demanded, the n-cell of the OOBL can be removed. Notice how thanks to the OOBL there were no special cases dealing with the border of the image.

5.6   Limitations

In certain cases the algorithm fails to detect every i-cell. In 3D we found two such cases:

– If a region of one label is completely surrounded by another region, then the 2-cell separating them will erroneously get removed fully. However, this does not occur in plant tissue, as biological cells neither float nor contain each other.
– If two regions are touching and the 2-cell between them is surrounded by a ring-shaped third region, the 2-cell does not get recognized and the first two regions appear disconnected in the resulting 3Gmap. Sadly, this configuration is common in our CT scans.

6   Results

Since the reduction happens implicitly, only the final nGmap needs to be explicitly processed. This means memory is only used for the *Construction* of the final nGmap. This minimizes the necessary memory.

Furthermore, processing only the required minimum of darts allows the algorithm to be fast. For example, a 512^3 labeled image of a leaf cross-section takes about 5 min to be processed on VSC4 (without multi-threading). A 400^3 synthetic image (20^3 checkerboard pattern) takes about 4 min.

In future work, the algorithm needs further refinement to mitigate limitations mentioned in Sect. 5.6. Furthermore, the process could be parallelized to further speed up computations. Finally, in-depth performance profiling and comparison to other approaches should be conducted.

Acknowledgments. This project was supported by the Vienna Science and Technology Fund (WWTF), project LS19-013. The computational results presented have been achieved in part using the Vienna Scientific Cluster (VSC).

References

1. Damiand, G.: Topological model for 3d image representation: definition and incremental extraction algorithm. Comput. Vis. Image Underst. **109**, 260–289 (2008). https://doi.org/10.1016/j.cviu.2007.09.007

2. Damiand, G.: Generalized maps. In: CGAL User and Reference Manual. CGAL Editorial Board, 5.4 (edn.) (2022). https://doc.cgal.org/5.4/Manual/packages.html#PkgGeneralizedMaps
3. Damiand, G., Lienhardt, P.: Combinatorial Maps Efficient Data Structures for Computer Graphics and Image Processing. A K Peters/Crc Press (2014)
4. Illetschko, T.: Minimal combinatorial maps for analyzing 3d data. Technical Report PRIP-TR-110, PRIP, TU Wien (2006). https://www.prip.tuwien.ac.at/pripfiles/trs/tr110.pdf

Topological Analysis of Simple Segmentation Maps

Maria-Jose Jimenez$^{(\boxtimes)}$ and Belen Medrano

Universidad de Sevilla, Av. Reina Mercedes s/n, 41012 Seville, Spain
{majiro,belenmg}@us.es

Abstract. In this paper, we propose a geometry-aware topological analysis of a segmentation of an image into regions which might correspond, for example, to a geographical map or to segmented cells in a microscopic image of a biological packed tissue. The regions must satisfy that the centroid of each one lies inside the region itself. We propose a novel simplicial complex modeling such data, for persistent homology computation, that better respects the geometry of the regions than existing techniques. More specifically, our approach joins benefits from previous models by encoding both neighbouring relations between the regions, as well as spatial distribution of the set of centroids. In addition, we introduce geometric information regarding distances between centroids and boundaries delimiting each region.

Keywords: Persistent homology · Segmentation map · Shape descriptor of regions · Topological organization of regions

1 Introduction

A segmentation map is a partition of an image into different regions, each one representing an object or a specific area on the image. In this paper, we are inspired by the problem of analysing the organization of cells in a biological packed tissue. The process of segmentation of a microscopic image of a packed tissue into cells is usually based on edge detection that partitions the plane into different regions. A well-founded idea in cellular biology is using Voronoi diagrams built on the set of centroids of cells for estimating their morphology. In [10], the authors evaluated the suitability of approximating the cells by such a mathematical construction. More generally, the approximation of regions on a map by Voronoi regions is a classical problem. In [13], a measure of the error of the approximation was given.

There have been several works in the literature [1,2,9,14] analysing the organization of the cells using Topological Data Analysis (TDA) tools. In [9,14], the

This research was funded by Ministerio de Ciencia e Innovación - Agencia Estatal de Investigación/10.13039/501100011033, grant PID2019-107339GB-I00 and Agencia Andaluza del Conocimiento, grant PAIDI-2020 P20-01145. Authors listed in alphabetical order.

É. Baudrier et al. (Eds.): DGMM 2022, LNCS 13493, pp. 123–135, 2022.
https://doi.org/10.1007/978-3-031-19897-7_11

authors used a contact graph representing neighbouring relations between the cells as ground for the analysis. In [1], they used, instead, alpha complexes, which are constructed out of the set of centroids of the regions, to model the inner structure of the cells. By doing this, they were implicitly approximating the regions of the cells by Voronoi regions of their centroids (see Fig. 1). In a more recent work, [2], they use two different approaches: (1) topological analysis of the contact graph constructed from neighbouring regions; (2) topological analysis of the point cloud of centroids of the regions. Once persistent homology is computed out of the generated complexes in each case, different topological features summarising the corresponding outputs can be computed. We are concerned with the design of a simplicial complex that models the regions in a segmentation of the plane, capturing both neighbouring information and spatial distribution of the centroids, while at the same time, saving geometric information related to the shape of the regions. An effective construction will be described over a labelled image representing a segmentation into regions of a subset of the plane $\Omega = [0, M - 1] \times [0, N - 1]$, with some constraints.

The paper is organized as follows: in Sect. 2, we introduce some fundamentals of TDA; we define a new simplicial complex associated to a specific type of segmentation into regions of a subset of the plane in Sect. 3; Sect. 4 will describe the specific input labelled images to be processed, as well as the algorithm to get the new complex out of them; some application examples will be shown in Sect. 5; finally, some conclusions and future research lines are drawn in Sect. 6.

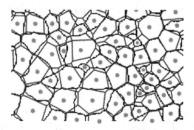

Fig. 1. Image of a cell tissue: cells are delimited in black; centroids of each region are depicted in red; Voronoi region corresponding to the set of centrois in blue. (Color figure online)

2   Tools from TDA

In this section, we recall the tools from Topological Data Analysis (TDA) used in this paper. Main theoretical concepts may be consulted in [5].

Simplicial Complex and Filtration. An (abstract) *simplicial complex* $\mathcal{K}$ is formed by a finite set V (whose elements are called vertices) together with a collection of subsets of V, called simplices of $\mathcal{K}$, such that for all $\sigma \in \mathcal{K}$ and for all non-empty $\sigma' \subset \sigma$, $\sigma' \in \mathcal{K}$.

$\sigma = \{v_0, \ldots, v_k\} \in \mathcal{K}$ is called a k-*simplex* or simplex of dimension k. Each ℓ-simplex contained in σ with $\ell < k$ is called a *face* of σ. The *dimension* of a simplicial complex is the maximum of the dimensions of its simplices.

A *filtration* over a simplicial complex $\mathcal{K}$ is a finite increasing sequence of simplicial complexes

$$\mathcal{K}_1 \subset \mathcal{K}_2 \subset \ldots \subset \mathcal{K}_n = \mathcal{K}.$$

This way, a partial ordering of all the simplices in a simplicial complex $\mathcal{K}$ produces a filtration over $\mathcal{K}$ whenever $\sigma \subset \mu$ implies that the order index of σ is less or equal than the one of μ.

Sometimes, a filtration is defined using a monotonic function $f : \mathcal{K} \to \mathbb{R}$ such that, for any two simplices $\sigma, \mu \in \mathcal{K}$, if σ is a face of μ, then $f(\sigma) \leq f(\mu)$. That way, if $a_1 \leq \ldots \leq a_n$ are the function values of all the simplices in $\mathcal{K}$, then the subcomplexes $\mathcal{K}_i = f^{-1}(-\infty, a_i]$, for $i = 1 \ldots n$ define a filtration over $\mathcal{K}$.

Clique Complex. Given an undirected graph G with set of vertices V, the clique complex $X(G)$ is an abstract simplicial complex formed by the sets of vertices in the cliques (complete subgraphs) of G. This way, each clique of k vertices of G corresponds to a simplex of dimension $k - 1$ of $X(G)$.

Vietoris-Rips Complex. Given a metric space (X, d), the *Vietoris-Rips complex* for X associated to a parameter $\epsilon \in (0, +\infty)$, $VR(X, d, \epsilon)$, is the simplicial complex with vertex set X and k-simplices spanned by $\{x_0, \ldots, x_k\}$ whenever $d(x_i, x_j) \leq \epsilon$ for all $0 \leq i, j \leq k$. Then, the *Vietoris-Rips filtration* is defined by increasing the filter value ϵ.

Alpha Complex. Given a finite collection $\mathcal{F}$ of sets, the *nerve* of $\mathcal{F}$ [5], consists of all non-empty subcollections whose sets have a non-empty common intersection, that is,

$$Nrv\ \mathcal{F} = \{X \subseteq \mathcal{F} : \bigcap X \neq \emptyset\}.$$

Notice that $Nrv\ \mathcal{F}$ is always an abstract simplicial complex.

The Nerve Theorem [3] states that if $\mathcal{F}$ is a finite collection of closed, convex sets in Euclidean space, then $Nrv\ \mathcal{F}$ has the same homotopy type than $\bigcup \mathcal{F}$. Such a theoretical result provides the ground for the construction of the alpha complex (see [5]).

A *Voronoi diagram* is a partitioning of the plane depending on a set of vertices $V = \{v_1, \ldots, v_n\}$: for each vertex v_i, the Voronoi region $\mathcal{V}_i$ associated to v_i is given by

$$\mathcal{V}_i = \{x \mid d(v_i, x) \leq d(v_j, x), \quad \forall j = 1, \cdots n, \ j \neq i\}.$$

That is, each region $\mathcal{V}_i$ is formed by points of the plane for which that vertex is the closest point (see the first row of images in Fig. 3).

Fixed a value α, consider B_α^i as the ball of center v_i and radius α and consider the region $U_\alpha^i = B_\alpha^i \cap V_i$. An *alpha complex* $\mathcal{K}_\alpha$ is a simplicial complex such that a k-simplex $\{v_0 \ldots v_k\}$ belongs to $\mathcal{K}_\alpha$ when the intersection of the regions $U_\alpha^i = B_\alpha^i \cap V_i$, $\forall i = 0, \cdots k$, is not empty. If the vertices are in *general position* in the plane, no k-simplex will arise with k greater than 2 and the final simplicial complex is known as the *Delaunay triangulation* [5, p. 63]. See Fig. 3, first row. Varying α, the simplicial complexes $\mathcal{K}_\alpha$ induce a filtration over the Delaunay triangulation called *alpha filtration*.

Persistent Homology and Barcodes. The concept of *homology class*, from algebraic topology, defines n-dimensional holes rigorously and computes them using linear algebra (see [8] for rigorous definitions). Intuitively, 0-dimensional holes are connected components, 1-dimensional holes are tunnels and 2-dimensional holes are cavities. *Persistent homology* [6,15] is the main tool in TDA defined for tracking the persistence or the lifetime (encoding birth and death moments) of holes along a filtration. For example, in the case of an alpha complex constructed from a point cloud on the plane, 0-dimensional and 1-dimensional persistent homology describe, respectively, the evolution of connected components and holes of the collection of regions (intersection of Voronoi diagrams and balls) as the balls radii increase. The persistence of each n-dimensional hole can be represented using an interval of the form $[b, d]$, where b is the birth time, that is, the index for which K_b is the simplicial complex (in the filtration) where the hole first appears, and d is the death time, if K_d is the first simplicial complex (in the filtration) where the hole disappears. If the hole remains until the final simplicial complex, we write $d = \infty$. This codification in terms of intervals (or bars) $[b, d]$ is called the n-dimensional *persistence barcode*, n-barcode for short, (see Fig. 2). A formal definition of homology and persistent homology together with algorithms for computing it can be found in [5].

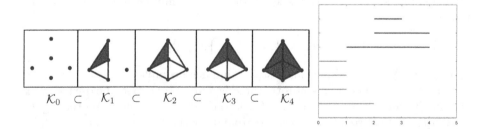

$$\mathcal{K}_0 \subset \quad \mathcal{K}_1 \subset \quad \mathcal{K}_2 \subset \quad \mathcal{K}_3 \subset \quad \mathcal{K}_4$$

Fig. 2. Left: example of a filtration over a complex $\mathcal{K} = \mathcal{K}_4$. Rigth: 0 and 1-barcodes representing connected components and holes (in red and blue, respectively). (Color figure online)

Notice that small perturbations on the boundary between regions would lead to small perturbations on the positions of the centroids and the stability results for persistent homology (see [4]) would imply small changes in the barcodes.

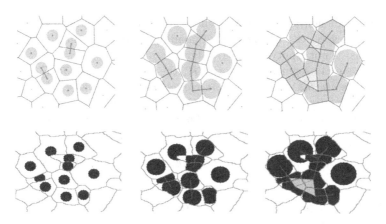

Fig. 3. An example of simplicial complexes (in red) associated to a partition into regions, each one, from left to right, corresponding to a higher value of the radius α. Top row: alpha complexes (forgetting the real boundaries of the regions). Bottom row: complexes associated to the regions. (Color figure online)

3   Simple Segmentation Complex

We set the definition of simple segmentation map on the plane and a filtration that is inspired by the one of alpha complex. However, in this case, we cannot guarantee that the Nerve Theorem is satisfied, so we have to prove that the filtration is correctly built.

Definition 1. *Consider $\Omega = [0, N-1] \times [0, M-1] \subset \mathbb{R}^2$. We say that $\mathcal{R} = \{R_j\}_{j=1,\cdots n}$ is a* simple segmentation map *of Ω if $\Omega = \bigcup_{j=1,\cdots n} R_j$ such that:*

- *each region R_j is homeomorphic to a disk, delimited by a boundary curve, which is a Jordan curve, whose points are considered to be part of that region as well;*
- *the centroids $\{c_1, \ldots, c_n\}$ lie inside each region;*
- *two regions have either empty intersection or they share a set of curve segments on the boundary of both regions (and we say then that they are* neighbour regions*).*

Notice that the regions may not be convex and that the intersection between two regions may be formed by several disjoint curve segments.

Filtration Associated to a Simple Segmentation Map. We define now a filtered simplicial complex associated to a simple segmentation map $\mathcal{R}$.

Definition 2. *Given a simple segmentation map, $\mathcal{R}$, whose set of centroids is $\{c_1, \ldots, c_n\}$, consider $B_\alpha(c_i)$ as the ball of center c_i and radius α. For each α, consider the region $U_\alpha^{\mathcal{R}}(c_i) = B_\alpha(c_i) \cap R_i$ and define the simplicial complex $\mathcal{K}_\alpha^{\mathcal{R}}$ with simplices $\{c_1 \ldots c_k\} \in \mathcal{K}_\alpha^{\mathcal{R}}$ when $\bigcap_{i=1\ldots k} U_\alpha^{\mathcal{R}}(c_i) \neq \emptyset$.*

In other words, a simplex lies in $\mathcal{K}_\alpha^\mathcal{R}$ when the intersection of balls with radius α and centers its vertices with their corresponding regions (of which those vertices are centroids) is not empty (see Fig. 3, bottom row).

Given a simple segmentation map $\mathcal{R}$, the *contact graph of $\mathcal{R}$ is a graph whose vertices are the centroids of the regions of $\mathcal{R}$ and whose edges represent neighbouring relations between regions. No multiple edges are allowed, even if the intersection of two regions are two (or more) disjoint curve segments.*

Proposition 1. *Given a simple segmentation map, $\mathcal{R}$, $\{\mathcal{K}_\alpha^\mathcal{R}\}_\alpha$ is a filtration over the simplicial complex generated as the clique complex of the contact graph of the regions.*

Proof. In order to prove that the filtration is well constructed, we only must check that the minimum value α for which a simplex σ belongs to $\mathcal{K}_\alpha^\mathcal{R}$ is always greater than the minimum value β for which any of its proper faces μ belongs to $\mathcal{K}_\beta^\mathcal{R}$. Notice first, that all the vertices of the complex (the centroids) lie in $\mathcal{K}_0^\mathcal{R}$. Assume that σ is a 2-simplex (associated to three regions R_1, R_2, R_3) and μ a 1-simplex. Then α is the maximum distance from the three centroids c_1, c_2, c_3 to their intersection point P_{123}. If μ represents, the neighbouring relation between R_1 and R_2, then β is the minimum of the maxima values of the pairs of distances of each point of the boundary curve to both centroids c_1 and c_2. Then, if $\alpha \leq \beta$, since $\alpha > d(c_1, P_{123})$ and $\alpha > d(c_2, P_{123})$, that would mean that $\beta > d(c_1, P_{123})$ and $\beta > d(c_2, P_{123})$, what is a contradiction (since P_{123} also belongs to the boundary curve between R_1 and R_2). Since all the neighbouring relations of more than 3 regions also occur due to one intersection point and the value of the filter is given by the maximum distance to each centroid, then the statement is true in all the cases.   □

4   Labelled Images Representing Simple Segmentation Maps and Associated Simplicial Complex

In order to put into practice the model that we propose, we constrain to the setting of labelled images representing a simple segmentation map. In this section we introduce first, the input images that we are going to process and later, we set the algorithm that produces the filtered complex that will be the object of study. For basic concepts of digital topology, see [12].

By *labelled image* we refer to a digital image $L : D \to \{0, 1, 2, \dots, m\}$, where $D = ([0, M-1] \times [0, N-1]) \cap \mathbb{Z}^2$, for given M and N, and the labels $\ell_i \in \{0, 1, 2, \dots, m\}$, have a specific meaning, such as the result of a segmentation process.

Our input images are labelled images, as in Fig. 4 representing regions that are bounded by boundary pixels. Notice that the regions "touching" first/last row/column might correspond to regions that are not fully represented, so we will consider them as *non-valid regions*. We are interested in computing a simplicial complex representing the structure of valid regions.

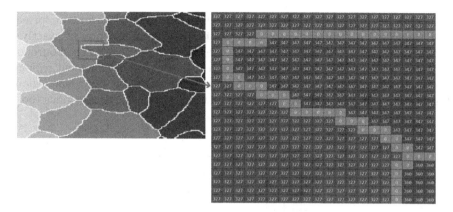

Fig. 4. Left: a labelled image. Each label has been assigned a gray level. Boundary pixels have been drawn in white. Right: a portion of the labelled image containing, partially, three different regions. Color is used only for visualization purposes. (Color figure online)

We will assume that each region is bounded by a 4−connected curve with no *simple point* [11]. Recall that a 4−connected curve is a closed path of pixels that are 4−adjacent; a pixel is a simple point if the change of its value does not change the topology of the image, that is, removing it from the boundary curve (by changing its value to the label of the neighbour region), the boundary is still a 4-connected curve. Hence, the deletion of a simple point of a 4−connected curve still produces a 4−connected curve bounding the same region. If the set of boundary pixels has simple points, then, a morphological thinning can be accomplished to remove them (see, for example, [7], Ch. 9, p.660). We assume that the regions are big enough so that their boundary pixels do not disconnect them (there is no Jordan paradox).

Sometimes, a segmentation process provides a labelled image that partitions the image in 4-connected regions with no holes, whose centroids lie inside each region, but with no boundary pixels. Then, the set of boundary pixels could be defined by: 1) taking those pixels having a 4-adjacent pixel with a different label; 2) thinning that set of pixels, so that it has no simple points for 4−connectivity. We put all the constraints together to define our simplesegmentation image.

Definition 3. *Let* $L : ([0, M − 1] \times [0, N − 1]) \cap \mathbb{Z}^2 \to \{0, 1, 2, \ldots, m\}$ *be an* $M \times N$ *labelled image. We will say that* L *is a* simple segmentation image *if it is a union of sets of 4-connected pixels with the same label such that:*

- *labels* $\{1, 2, \ldots, m\}$ *are assigned, each one, to a 4-connected component with no holes, whose centroid lies inside the region;*
- *label 0 is assigned to a 4-connected component of pixels with no simple points, called* boundary pixels, *with as many holes as regions with labels from 1 to* m *and such that each region is bounded by a 4−connected curve of boundary pixels.*

Taking into account that the boundary pixels delimiting each region are formed by 4–connected paths, it is easy to check that in the intersection of exactly 3 regions (and not 4), there will be only one pixel whose 3×3 neighbourhood contains pixels from the 3 regions, as in Fig. 4. In the case of 4 incident regions, two cases may arise: (1) a 3×3 cross, where there would be 4 incident regions (Fig. 5, left); (2) the configuration in Fig. 5, right, with 4 incident regions too.

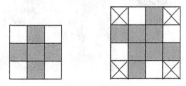

Fig. 5. Left: A 3×3 cross configuration of boundary pixels (in grey) delimiting 4 different regions. Right: Configuration of pixels on the boundary (grey) containing a 2×2 block of boundary pixels. Pixels with a cross can lie either on the boundary or not. (Color figure online)

Lemma 1. *If there is a configuration of 2×2 pixels on the boundary, then, the 4×4 neighbourhood around those 2×2 pixels, will have the configuration of pixels given in Fig. 5, on the right, and, hence, will contain 4 incident regions.*

Proof. A thorough check of all cases shows that, first, any other configuration in the 4×4 neighbourhood would contain a simple point in the set of boundary pixels and also that if there were three (or fewer) incident regions, then, there would be a simple point.   □

See Fig. 5. The 3×3 neighbourhood of each pixel in the 2×2 configuration will contain pixels with labels of, at most, two different regions. Thus, the 4 incident regions will be adjacent two by two. However, in this case, it seems natural to consider that the four regions appearing in the 4×4 neighbourhood, should be mutually adjacent. That way, we provide the following definition for adjacency of regions (or neighbour regions):

Definition 4. *Let L be a simple segmentation image. Then, we will say that two regions with labels ℓ_1 and ℓ_2, are neighbour regions if either there is a 3×3 neighbourhood centered on a boundary pixel that contains pixels with labels ℓ_1 and ℓ_2 or there is a 2×2 configuration of boundary pixels such that the 4×4 neighbourhood centered on it contains pixels with labels ℓ_1 and ℓ_2.*

Simplicial Complex Associated to a Simple Segmentation Image. We will use the concept of neighbour regions defined above to construct a simplicial complex out of the input labelled image, together with a filtration. Hence, each region is represented by its centroid (as a vertex), pairs of adjacent regions will produce

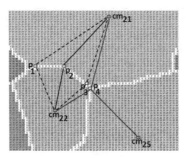

Fig. 6. Illustration of Algorithm 1: $\{cm_i\}_{i=21,22,25}$ are centroids of regions of labels $i = 21, 22, 25$ and $\{p_i\}_{i=1,...,4}$ sample boundary pixels. All the pixels $\{p_i\}$ belong to the boundary between regions with labels 21 and 22. Their distances to the corresponding centroids are: $d(p_1, cm_{21}) = 31.18$ and $d(p_1, cm_{22}) = 18.39$; $d(p_2, cm_{21}) = 25.29$ and $d(p_2, cm_{22}) = 17.87$; $d(p_3, cm_{21}) = 29.19$ and $d(p_3, cm_{22}) = 14.15$; $d(p_4, cm_{21}) = 30.67$ and $d(p_4, cm_{22}) = 14.28$. Then, the filter value for the edge $\{21, 22\}$ is 25.29, whenever there is no other pixel on the boundary whose greater distance to one of both centroids is lower than that amount. Finally, the pixel p_4 lies on the boundary of regions 21, 22 and 25 and $d(p_4, cm_{25}) = 24.01$, so the triangle $\{21, 22, 25\}$ is assigned filter value 30.67.

edges between them, triples of neighbour regions, triangles and configurations of Fig. 5 will lead to tetrahedra in the simplicial complex. For that aim, we follow the steps in Algorithm 1 to build a complex as well as a filtration (stored in F): roughly, run over all the boundary pixels and consider the centroids of labels arising in the 3×3 neighbourhood $E3(i, j)$ of each boundary pixel (i, j). When there are two labels, consider the pair of distances to both centroids and take the minimum between the greater distance of each pair of distances; when there are 3 or 4 labels, simply take the maximum distance for the filter value. See Fig. 6 for an illustration of the process. The 4×4 neighbourhood $E4(i, j)$ is considered only in exceptional cases of configuration as in Fig. 5 right.

Proposition 2. *The list F of Algorithm 1 produces a filtration over the simplicial complex representing all the neighbouring relations as in Definition 4.*

Proof. The proof follows the same argument as in Proposition 1. Notice that now, the maximum number of incident regions is 4 by construction.

5 Discussion

Our method can be considered as an improvement of the alpha complex in the case where the points in a point cloud on the plane have an associated region to which the increasing balls have to be restricted.

In Fig. 7, top row, we can see that the 1-simplices of the simple segmentation complex coincides with the contact graph, while in the case of the alpha complex

Algorithm 1. Computation of the Simple Segmentation complex

Input: A simple segmentation image L;

$\{\ell_1, \ell_2, \ldots, \ell_m\} \leftarrow$ labels of valid regions;

$\{c_1, c_2, \ldots, c_m\} \leftarrow$ centroids of valid regions;

Output: $F \leftarrow$ a filtration over the simplicial complex representing neighbour regions in L.

```
 1: F ← { (c_i, ℓ_i, 0) }_{i=1,...n}
 2: L ← { }
 3: for each pixel (i, j) on the boundary do
 4:     L(i, j) ← {ℓ_1, ℓ_2, ...} labels of regions in E3(i, j)
 5:     for each pair of labels {ℓ_{k1}, ℓ_{k2}} in L(i, j) do
 6:         d_{k1} ← d((i, j), c_{k1})
 7:         d_{k2} ← d((i, j), c_{k2})
 8:         if there is no other element in F containing {ℓ_{k1}, ℓ_{k2}} then
 9:             F ← F ∪ ((i, j), {ℓ_{k1}, ℓ_{k2}}, max{d_{k1}, d_{k2}})
10:         else
11:             if ((i', j'), {ℓ_{k1}, ℓ_{k2}}, d') ∈ F then
12:                 if d' > max{d_{k1}, d_{k2}} then
13:                     Replace (i', j') by (i, j)
14:                     Replace d' by max{d_{k1}, d_{k2}}
15:                 end if
16:             end if
17:         end if
18:     end for
19:     for each subset of three labels {ℓ_{k1}, ℓ_{k2}, ℓ_{k3}} in L(i, j) do
20:         d_{kl} ← d((i, j), c_{kl}), for l = 1, 2, 3
21:         if there is no other element in F containing {ℓ_{k1}, ℓ_{k2}, ℓ_{k3}} then
22:             F ← F ∪ ((i, j), {ℓ_{k1}, ℓ_{k2}, ℓ_{k3}}, max_l{d_{kl}})
23:         else
24:             if ((i', j'), {ℓ_{k1}, ℓ_{k2}, ℓ_{k3}}, d') ∈ F then
25:                 if d' > max_l{d_{kl}} then
26:                     Replace (i', j') by (i, j)
27:                     Replace d' by max_l{d_{kl}}
28:                 end if
29:             end if
30:         end if
31:     end for
32:     for each subset of four labels {ℓ_{k1}, ℓ_{k2}, ℓ_{k3}, ℓ_{k4}} in L(i, j) do
33:         d_{kl} ← d((i, j), c_{kl})
34:         F ← F ∪ ((i, j), {ℓ_{k1}, ℓ_{k2}, ℓ_{k3}, ℓ_{k4}}, max_l{d_{kl}})
35:     end for
36:     if the pixels (i + 1, j), (i, j + 1) and (i + 1, j + 1) are all boundary pixels too,
37:     none of them from the zero padding, then
38:         L(i, j) ← {ℓ_{k1}, ℓ_{k2}, ℓ_{k3}, ℓ_{k4}} labels of regions in E4(i, j)
39:         d_{kl} ← d((i + ½, j + ½), c_{kl})
40:         F ← F ∪ ((i, j), {ℓ_{k1}, ℓ_{k2}, ℓ_{k3}, ℓ_{k4}}, max_l{d_{kl}})
41:         for all the combinations of three labels of L(i, j) (denoted by L_3(i, j)) do
42:             F ← F ∪ ((i, j), {ℓ_{kl}}_{kl ∈ L_3(i,j)}, max_{kl ∈ L_3(i,j)}{d_{kl}})
43:         end for
44:     end if
45: end for
```

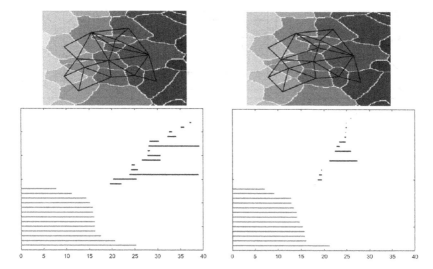

Fig. 7. Top left: contact graph representing neighbour regions. Top right: Delaunay triangulation of the corresponding set of centroids. Bottom: 0 and 1-barcodes (in red and blue, respectively) corresponding to the simple segmentation complex (left) and alpha complex (right), built upon the graphs above. (Color figure online)

can include edges relating no neighbouring regions. Besides, the two holes that are formed due to the presence of the biggest cells, which correspond to the two longest blue bars in both barcodes, are more obvious in the barcode of our complex (left).

Besides, in order to illustrate the fact that our method captures more information about the geometry of the regions, we have applied it to different segmentation images provided by some geometric tessellations (see Fig. 8). They satisfy that: (1) each valid region has always the same number of neighbours; (2) the spatial distribution of their centroids are also the same. Therefore, first, the barcodes computed from the contact graph with a filtration given by the number of neighbour regions will be the same; second, the three barcodes computed from the Vietoris-Rips complexes upon the set of centroids, as well as the ones obtained by the alpha complexes, are also exactly the same (middle row of Fig. 8). However, as shown in Fig. 8, bottom row, our method provides barcodes that can be considered as geometric and topological signatures that characterise the tessellations. Some observations on Fig. 8: in the barcode from Vietoris-Rips filtration, there are no holes since the moment at which the balls touch each other two by two, the triangles are also added; in the case of the alpha complex, small holes (blue bars) are born when the balls on the centroids touch each other two by two and die when each three of them meet; when considering the intersection of balls with the regions themselves (bottom row), the way in which they intersect are different in each of the three tessellations, what is reflected in both 0 and 1-barcodes. Supplementary material can be found in https://github.com/belenmg/Simple-segmentation-maps/tree/main/Illustrations to illustrate the evolution of the complexes.

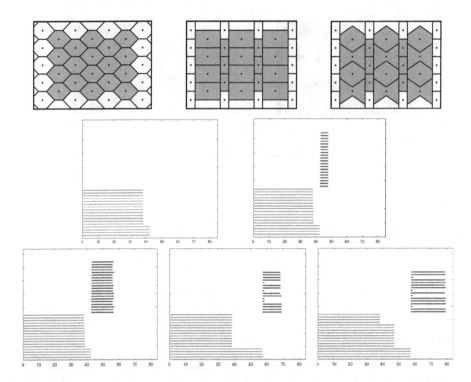

Fig. 8. Top row: three different tessellations of the plane with 18 valid regions. The set of centroids of the 18 regions are identical in the three pictures. Middle row: 0 and 1-dimensional barcodes (in red and blue, respectively) computed from Vietoris-Rips filtration (left) and alpha complex (right) of the set of centroids. Bottom row: from left to right, 0 and 1-dimensional barcodes (in red and blue, respectively) computed from the simple segmentation complex of each tessellation (from left to right). (Color figure online)

6   Conclusions and Future Work

In this paper, we have introduced a novel filtered simplicial complex to model the organization and shapes of regions in a simple segmentation image. This complex is more geometric-aware than the ones used previously in the topological study of cells organization in packed tissues and its topological analysis reveals a great potential to capture geometric properties of the regions. More specifically, the persistence barcode that produces can detect regularity patterns that are not captured by classical simplicial complexes. The authors intend to explore applications in different fields like cellular biology, materials science or crystallographic image processing.

References

1. Atienza, N., Escudero, L.M., Jimenez, M.J., Soriano-Trigueros, M.: Characterising epithelial tissues using persistent entropy. In: Marfil, R., Calderón, M., Díaz del Río, F., Real, P., Bandera, A. (eds.) CTIC 2019. LNCS, vol. 11382, pp. 179–190. Springer, Cham (2019). https://doi.org/10.1007/978-3-030-10828-1_14
2. Atienza, N., Jimenez, M.J., Soriano-Trigueros, M.: Stable topological summaries for analyzing the organization of cells in a packed tissue. Mathematics **9**(15) (2021). https://doi.org/10.3390/math9151723. https://www.mdpi.com/2227-7390/9/15/1723
3. Borsuk, K.: On the imbedding of systems of compacta in simplicial complexes. Fund. Math. **35**, 217–234 (1948)
4. Cohen-Steiner, D., Edelsbrunner, H., Harer, J., Mileyko, Y.: Lipschitz functions have L p-stable persistence. Found. Comput. Math. **10**(2), 127–139 (2010)
5. Edelsbrunner, H., Harer, J.: Computational Topology: An Introduction. American Mathematical Society (2010)
6. Edelsbrunner, H., Letscher, D., Zomorodian, A.: Topological persistence and simplification. Discret. Comput. Geom. **28**(4), 511–533 (2002). https://doi.org/10.1007/s00454-002-2885-2
7. Gonzalez, R.C., Woods, R.E.: Digital Image Processing. Pearson International Content (2018)
8. Hatcher, A.: Algebraic Topology. Cambridge University Press (2002)
9. Jimenez, M.J., Rucco, M., Vicente-Munuera, P., Gómez-Gálvez, P., Escudero, L.M.: Topological data analysis for self-organization of biological tissues. In: Brimkov, V.E., Barneva, R.P. (eds.) IWCIA 2017. LNCS, vol. 10256, pp. 229–242. Springer, Cham (2017). https://doi.org/10.1007/978-3-319-59108-7_18
10. Kaliman, S., Jayachandran, C., Rehfeldt, F., Smith, A.S.: Limits of applicability of the Voronoi tessellation determined by centers of cell nuclei to epithelium morphology. Front. Physiol. **7**(551) (2016). https://doi.org/10.3389/fphys.2016.00551
11. Klette, G.: Simple points in 2D and 3D binary images. In: Petkov, N., Westenberg, M.A. (eds.) CAIP 2003. LNCS, vol. 2756, pp. 57–64. Springer, Heidelberg (2003). https://doi.org/10.1007/978-3-540-45179-2_8
12. Kong, T., Rosenfeld, A.: Digital topology: introduction and survey. Comput. Vis. Graph. Image Process. **48**(3), 357–393 (1989). https://doi.org/10.1016/0734-189X(89)90147-3. https://www.sciencedirect.com/science/article/pii/0734189X89901473
13. Suzuki, A., Iri, M.: Approximation of a tessellation of the plane by a Voronoi diagram. J. Oper. Res. Soc. Jpn. **29**(I) (1986)
14. Villoutreix, P.: Randomness and variability in animal embryogenesis, a multi-scale approach. Ph.D. thesis, Université Sorbonne Paris Cité (2015)
15. Zomorodian, A., Carlsson, G.: Computing persistent homology. Discret. Comput. Geom. **33**(2), 249–274 (2004). https://doi.org/10.1007/s00454-004-1146-y

Discrete Tomography and Inverse Problems

On the Decomposability of Homogeneous Binary Planar Configurations with Respect to a Given Exact Polyomino

Michela Ascolese$^{(\boxtimes)}$ and Andrea Frosini

Dipartimento di Matematica e Informatica, Università di Firenze, Firenze, Italy
{michela.ascolese,andrea.frosini}@unifi.it

Abstract. A binary planar configuration A associates to each point in $\mathbb{Z}^2$ an element in $\{0,1\}$. Provided a finite window probe P, we locally inspect A by moving P in all its possible positions and counting the 1s elements that fit inside it. In case all the computed values have the same value k, then we say that A is k-homogeneous w.r.t. P. A recent conjecture states that a binary planar configuration is k-homogeneous with respect to an exact polyomino P, i.e., a polyomino that tiles the plane by translation, if and only if it can be decomposed into k configurations that are 1-homogeneous with respect to P. In this paper we define a class of exact polyominoes called perfect pseudo-squares ($\mathcal{PPS}$) and we investigate the periodicity behaviors of the homogeneous configurations that are related to them. Then, we show that some elements in $\mathcal{PPS}$ allow 2-homogeneous or 3-homogeneous non-decomposable planar configurations, so providing evidence that the conjecture does not hold for the whole class of exact polyominoes.

Keywords: Discrete tomography · Discrete geometry · Tiling · Exact tile

AMS Classification: 05B45 · 52C20

1 Introduction

The study of structural properties of binary planar configurations of $\mathbb{Z}^2$ by inspecting them through a finite size window is a longstanding problem both in Combinatorics and Discrete Geometry. In particular, it is strictly related with the notion of planar periodicity, as an example in terms of different configurations that appear in the window, as stated by the Morse-Hedlund theorem in [10] and in [12] and successive studies, and with the notion of patterns detection and reconstruction [7,8], so intersecting the main topics of the wide area of Computerized and Discrete Tomography and with the notion of planar tilings.

Our researches focus on this last connection and base on the studies by M. Nivat and co-authors in [2,7,12] about homogeneous configurations that may be revealed inspecting a binary planar configuration according to a chosen window probe, later deepened and generalized in [1,3,4].

© Springer Nature Switzerland AG 2022
É. Baudrier et al. (Eds.): DGMM 2022, LNCS 13493, pp. 139–152, 2022.
https://doi.org/10.1007/978-3-031-19897-7_12

More specifically, provided a binary planar configuration A of $\mathbb{Z}^2$ and a window probe P, usually a finite 4-connected set of points called polyomino, we move P in all possible positions of A and count the sum of the visible points, i.e., the number of elements 1s that fit inside the probe. If the obtained values have a constant value k, with k ranging from 0 to the area of P, then the configuration A is said to be k-homogeneous w.r.t. P, and its elements show interesting periodical behaviors. In [2], it is also observed that only exact polyominoes, i.e., polyominoes that tile the plane by translation, allow 1-homogeneous configurations. Details on the characterization and geometrical properties of exact sets are in [2].

Furthermore, in case of specific exact polyominoes, called *decomposable*, a decomposition theorem also holds, which allows us to split each k-homogeneous configuration A into k sub-configurations that are 1-homogeneous. In particular, in [7,11], the authors obtain the result for rectangular polyominoes. Later, in [1], the class is extended including diamonds and all those exact polyominoes that are balls in a generalized norm L^1 of $\mathbb{Z}^2$. The authors of [7] finally conjectured that a binary planar configuration is k-homogeneous with respect to an exact polyomino P if and only if it can be decomposed into k disjoint configurations that are 1-homogeneous. Moving from that, we define the class of perfect pseudo-squares polyominoes, $\mathcal{PPS}$, and we investigate the periodical behavior of the homogeneous planar configurations induced by their tilings. Then, we detect some elements of $\mathcal{PPS}$ that allow 2-homogeneous or 3-homogeneous non-decomposable planar configurations. So, we provide evidence that the conjecture in [7] does not hold, in general, for the whole class of exact polyominoes. A remarkable fact is that the computed non-decomposable configurations still show a periodical behaviour, which is different from that one characteristic of the tilings of the related exact polyominoes. Finally, some new research lines that come out from our investigation are pointed out.

2   Definitions and Previous Results

In this section we provide basic notions concerning exact polyominoes and tilings, and recall some previous results useful in the sequel.

A *polyomino* is a finite subset of the square-lattice $\mathbb{Z}^2$ whose points are 4-connected. Furthermore, we consider polyominoes that have no holes, i.e., whose boundary is a single (closed) non intersecting path. The length n of the boundary path defines the *perimeter* of the polyomino, and is always even since the boundary is closed. The boundary of a polyomino P can be coded through the Freeman chain code [5,6] as a word on the 4-letters alphabet $\Sigma = \{N, S, E, W\}$, so that each letter represents a step in the directions North, South, East and West, respectively. We decide to consider the boundary word w_P obtained by travelling the polyomino P clockwise, and up to circular shifts of its letters. To avoid ambiguities, we consider the word w_P to start in the lower-left point of the polyomino boundary. Figure 1 shows three polyominoes with the standard representation as a set of cells on a squared surface, and their Freeman coding (the starting point of the coding is also highlighted).

Σ^* is the set of all the finite words defined on the alphabet Σ, with $\varepsilon \in \Sigma^*$ the empty word, and let us indicate the opposite directions N and S (E and W, respectively) as *conjugate* letters.

We define three operators on a word $w = w_1 w_2 \ldots w_n \in \Sigma^*$:

1. the *conjugate of w, $\bar{w}$,* is the word obtained by replacing each letter of w with its conjugate;
2. the *reversal of w, $Rev(w)$,* defined as $Rev(w) = w_n w_{n-1} \ldots w_1$;
3. the composition of the previous operations, $\hat{w} = Rev(\bar{w})$.

We further introduce the notation $|w|_x$ to point out the occurrences of the letter x in the word w. We finally remind that a word $w \in \Sigma^*$ is *periodic* if there exist a non-empty word $u \in \Sigma^*$ and $k \geq 2$ such that $w = u^k$, where u^k stays for the concatenation of k copies of the word itself, $u^k = uu \ldots u$.

We will focus on the so called *exact* polyominoes, or *tiles*, defined in [2] as those polyominoes that tile the plane by translation. A *tiling* of the plane by a polyomino P is defined as a set of non-overlapping translated copies of P that covers all the plane; we indicate a tiling of the plane with P by T_P. An example of tiling is provided in Fig. 2. We highlight the following

Property 1. *A polyomino P tiles the plane by translation if and only if it can be surrounded with copies of itself.*

Beauquier and Nivat characterized exact polyominoes in relation to their boundary word, providing the following result.

Theorem 1 ([2]). *A polyomino P is exact if and only if there exist $x_1, x_2, x_3 \in \Sigma^*$ such that*

$$w_P = x_1 x_2 x_3 \hat{x}_1 \hat{x}_2 \hat{x}_3,$$

where at most one of the words is empty. This factorization may be not unique.

We will refer to this decomposition as a *BN-factorization*. The terms x_i and $\hat{x}_i$, for $i = 1, 2, 3$, define the translations of the polyomino P in $\mathbb{Z}^2$. These translations completely define the tiling we obtain.

Starting from their BN-factorization, exact polyominoes can be further divided in two classes: *pseudo-hexagons*, if x_1, x_2 and x_3 are all non-empty words, and *pseudo-squares*, if one of the words is empty. The name is due to the fact that the polyomino can be surrounded with six (respectively, four) copies of itself, in reference to the respective regular polygons.

Remark 1. *Since the BN-factorization of w_P is not unique, then a polyomino can tile the plane both as a pseudo-square and as a pseudo-hexagon. For example, the polyomino depicted in Fig. 1c, $w_P = $ NWNWNEESESESWW, is both a pseudo-square and a pseudo-hexagon, w.r.t. the BN-factorizations*

$$w_P = (\text{NWNWN})(\text{EE})(\text{SESES})(\text{WW}),$$
$$w_P = (\text{NWNWN})(\text{E})(\text{E})(\text{SESES})(\text{W})(\text{W}).$$

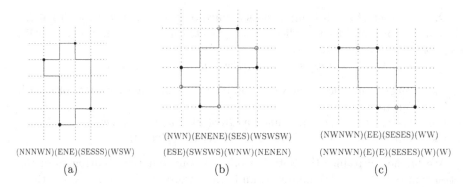

(NNNWN)(ENE)(SESSS)(WSW)

(NWN)(ENENE)(SES)(WSWSW)
(ESE)(SWSWS)(WNW)(NENEN)

(NWNWN)(EE)(SESES)(WW)
(NWNWN)(E)(E)(SESES)(W)(W)

(a)   (b)   (c)

Fig. 1. Three examples of exact polyominoes, with the corresponding BN-factorizations. The factors x_i are highlighted on the boundary of the polyomino with small dots. Different colors of dots refer to different factorizations (if any).

By Property 1, we can observe that the configuration T_P is *periodic*, i.e., there exist one or two linearly independent vectors v_i such that the tiling is invariant by translation along their directions (see [2])[1]. We define the set $D_{T_P} = \{v_i\}$ as the set of the *directions of periodicity* of the tiling T_P. The directions of periodicity of T_P can be deduced from the BN-factorization of the polyomino that induces that tiling (see Proposition 4). We remind again that different factorizations of the word w_P induce different tilings, and consequently different sets of directions of periodicity. In particular,

Theorem 2 ([3]). *An exact polyomino tiles the plane as pseudo-square in at most two distinct ways.*

3   The Class of Perfect Pseudo-squares and Their Properties

In this section we will focus our attention on the different classes of exact polyominoes.

Exact polyominoes can be characterized in subclasses, w.r.t. some properties of their boundary word. These properties reflect in some geometrical features and in the tilings induced by the polyomino. We define at first the class of the *perfect pseudo-squares*, $\mathcal{PPS}$, as the set of the exact polyominoes that are pseudo-squares but are not pseudo-hexagons. As arises from their BN-factorizations, the polyominoes depicted in Fig. 1(a) and Fig. 1(b) are examples of perfect pseudo-squares.

Proposition 1. *Given a pseudo-square* $w_P = x_1 x_2 \hat{x}_1 \hat{x}_2$, *if* x_1 *or* x_2 *is periodic, then* $P \notin \mathcal{PPS}$.

---

[1] In this study, we include in the class of *periodic tilings* also those called *half-periodic* in [2].

Proof. By contradiction. Let us suppose without loss of generality that x_1 is periodic. Then $x_1 = u^k$ for some non-empty word u and $k \geq 2$. Consequently, $\hat{x}_1 = \hat{u}^k$ by definition of the operator. We can decompose x_1 as $x_1 = uu^{k-1}$, getting the following BN-factorization of the boundary word of P,

$$w_P = uu^{k-1}x_2\hat{u}\hat{u}^{k-1}\hat{x}_2.$$

We obtained a BN-factorization of w_P in three non-empty factors, i.e., P is also a pseudo-hexagon, in contradiction with the definition of perfect pseudo-square. □

The condition stated in Proposition 1 does not allow to characterize the class of perfect pseudo-squares. Indeed, if we consider the BN-factorization of a pseudo-square P, the periodicity of one of its factors is sufficient to state that P is not perfect, but is not necessary as shown below.

Example 1. *The pseudo-square* $w_P = (NNWNN)(ENENE)(SSESS)(WSWSW)$ *is not perfect, since it admits a second BN-factorization as a pseudo-hexagon,*

$$w_P = (N)(NENEN)(ESSE)(S)(SWSWS)(WNNW).$$

Nevertheless, in its BN-factorization as a pseudo-square, none of the factors is periodic.

We will focus on the properties of the tilings obtained with perfect pseudo-squares. A first property directly follows from Theorem 2.

Proposition 2. *A perfect pseudo-square tiles the plane in at most two distinct ways.*

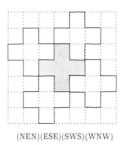
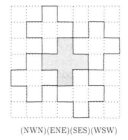

(NEN)(ESE)(SWS)(WNW)   (NWN)(ENE)(SES)(WSW)

Fig. 2. A double square with its possible tilings. The directions of periodicity are $\{(1, 2), (2, -1)\}$ (on the left) and $\{(-1, 2), (2, 1)\}$ (on the right).

Figure 2 shows an example of double tiling of the plane induced by two different BN-factorizations of the same perfect pseudo-square.

As a consequence, the boundary word of a perfect pseudo-square admits at most two distinct BN-factorizations. We indicate the tiles that admit a double factorization as *double squares*. This definition is motivated by the following result, stating that the BN-factorizations of a double square must alternate.

Lemma 1 ([4]). *If the boundary word of an exact polyomino satisfies $x_1 x_2 \hat{x}_1 \hat{x}_2 \equiv_d y_1 y_2 \hat{y}_1 \hat{y}_2$, with $0 \le d \le |x_1|$ and $\{x_1, x_2, \hat{x}_1, \hat{x}_2\} \neq \{y_1, y_2, \hat{y}_1, \hat{y}_2\}$, then the factorization must alternate, i.e., $0 < d < |x_1| < d + |y_1|$.*

The notation $x \equiv_d y$ means that the word x equals the circular shift of the word y of d letters, while $|\cdot|$ stands for the length of the word. The above result leads to the following.

Proposition 3. *Let P be a perfect pseudo-square. Each possible tiling T_P has two directions of periodicity.*

Proof. By contradiction. Let us suppose that there exists a tile T_P with one only direction of periodicity. Since a tiling obtained by the same BN-factorization of all its tiles has two directions of periodicity (indicated as *regular* tiling in [2]), then there exists at least one tile in T_P whose border is factorized with factors from both its two (according to Theorem 2) possible pseudo-square BN-factorizations. Lemma 1 states that the two BN-factorizations of a perfect pseudo-square are shifted and arranged so that there do not exist common starting or ending points in their factors, so they can not occur in the same tile factorization, reaching a contradiction.   □

So, in case of perfect pseudo-squares, we will use the term *couple(s) of directions of periodicity*.

Example 2. *The polyomino depicted in Fig. 1(b) is a double square. Its couples of directions of periodicity are $\{(-1,2),(3,2)\}$ and $\{(2,-1),(-2,3)\}$. On the other hand, the perfect pseudo-square in Fig. 1(a) is not double, and admits only one couple of directions of periodicity, $\{(-1,4),(2,1)\}$.*

The couple(s) of directions of periodicity of P can be computed starting from w_P and its BN-factorization(s). Indeed, we remind that the factors x_1 and x_2 uniquely identify the translations of the polyomino that define the tiling, thus allowing to determine the couple of directions of periodicity of the tiling itself.

Proposition 4. *Let us consider P a perfect pseudo-square and $w_P = x_1 x_2 \hat{x}_1 \hat{x}_2$ a BN-factorization. Each factor x_i, with $i = 1, 2$, defines a direction of periodicity in the related tiling, $v_i = (|x_i|_E - |x_i|_W, |x_i|_N - |x_i|_S)$.*

Proof. Let us consider the factor x_2. We remind that it defines a side of the boundary of P that matches with the corresponding side $\hat{x}_2$ for one copy of P in the tiling, i.e., it identifies a translation of the polyomino in $\mathbb{Z}^2$. The direction of translation is provided by the steps along the side x_1 of the polyomino, i.e. the steps of one path leading from x_2 to $\hat{x}_2$ in the BN-decomposition of the tile boundary. So, the horizontal component of the direction of periodicity can be computed as the difference between E and W steps in x_1, i.e. $|x_1|_E - |x_1|_W$. Analogously, the vertical component of the direction of periodicity can be computed as the difference between N and S steps in x_1, i.e. $|x_1|_N - |x_1|_S$. The same reasoning can be used to obtain the direction v_2 from the factor x_2.   □

The result can be extended to compute the directions of periodicity of each tiling, regardless the class of the exact polyomino that induces it.

BN-factorizations of tiles that are both pseudo-square and pseudo-hexagon

For sake of completeness we provide the mutual position of BN-factorizations of those tiles that are both pseudo-square and pseudo-hexagon. The following propositions hold.

Proposition 5 [9]. *Let x and z be two non-empty words and y be a word such that $xy = yz$. Then there exist unique words u and v and a unique integer $i \geq 0$ such that $x = uv$, $y = (uv)^i u$ and $z = vu$.*

Proposition 6. *Let P be an exact polyomino that admits two different BN-factorizations, as pseudo-square, say $w_P = x_1 x_2 \hat{x}_1 \hat{x}_2$, and as pseudo-hexagon, say $w'_P = y_1 y_2 y_3 \hat{y}_1 \hat{y}_2 \hat{y}_3$, such that $w_P \equiv_d w'_P$, with $0 \leq d \leq |x_1|$. The following cases only arise:*

i) *the two factorizations are such that*

$$x_1 x_2 \hat{x}_1 \hat{x}_2 = y_1 y_2 x_2 \hat{y}_1 \hat{y}_2 \hat{x}_2,$$

and so $d = 0$. In this case the word x_1 is periodic.

ii) *the two factorizations are such that*

$$x_1 x_2 \hat{x}_1 \hat{x}_2 \equiv_d y_1 y_2 y_3 \hat{y}_1 \hat{y}_2 \hat{y}_3$$

with $d \neq 0$, and no two factors in w_P and w'_P start in the same point.

Proof. Regarding case i), an example is the polyomino in Remark 1. The periodicity of x_1 directly follows from Proposition 5.

An example of the configuration of case ii) is the polyomino provided in Example 1.

One further possible arrangement of the two factorizations is left, i.e., $x_1 x_2 \hat{x}_1 \hat{x}_2 = y_1 y_2 y_3 \hat{y}_1 \hat{y}_2 \hat{y}_3$, again with $d = 0$, and such that $x_1 \neq y_1 y_2$. From Proposition 5, we get the periodicity of $x_1 x_2$. So the boundary of the polyomino degenerates into a single line, and the case never occurs.   □

Cases i) and ii) of Proposition 6 also differ in the number of directions of periodicity generated by the tilings. As a matter of fact, if the tiling is obtained as a pseudo-square, there always exist two directions of periodicity (see Proposition 3). On the other hand, when the tiling is obtained as a pseudo-hexagon, it holds that

1. in case i) of Proposition 6 there exists only one direction of periodicity, i.e., the direction computed from x_2. A second direction of periodicity is prevented by the possible non-regularity of the tiling in the sense of [2];
2. in case ii) of Proposition 6 there exists two directions of periodicity, that can be computed from the factors y_i of the BN-factorization. This follows after observing that the generated tiling is always regular (see [2]).

We conclude this section with the following

Theorem 3. *The class of pseudo-hexagons, $\mathcal{PH}$, and the class of double squares, $\mathcal{DS}$, are such that $\mathcal{PH} \cap \mathcal{DS} = \emptyset$, i.e. $\mathcal{DS} \subset \mathcal{PPS}$.*

The method used in Theorem 1 of [3] can be adapted to obtain the proof, as observed by the authors.

4   Homogeneous Non-decomposable Configurations

In this section we introduce the notion of *configuration of the plane* and a related conjecture expressed in [7]. We further recall some previous results that support this conjecture in some specific cases, and finally provide counterexamples to prove that it does not hold in general.

We define a *configuration of the plane* as a subset A of points of the discrete plane $\mathbb{Z}^2$. We will consider the plane as a binary configuration in which the element 1 indicates the presence of a point of the subset A, while 0 indicates its absence. We adopt, for the binary planar configurations, a definition of periodicity similar to that on tilings: a configuration A is called *periodic* w.r.t. the vector v if it holds that $a \in A$ if and only if $a + v \in A$, where $a + v$ stands for a translation of the point a along the direction v. We underline that if a configuration is periodic, then the subset A is infinite. We want to study the planar configurations by progressively inspecting them through a polyomino W, called *window*, that, moving by translation, allows to reveal from place to place a finite number of points of the binary configuration, in accordance to its dimension.

We point out that the scan of the plane with a window is a generalization of the concept of linear projection on a finite set of points: in this case, we do not collect quantitative data of the set along a direction, but we scan the configuration through a two-dimensional object. We highlight that the value of these scans can only vary from 0 to w, with w the area of the window, i.e. the number of discrete points that constitute the polyomino W.

Given a configuration A and a window W, we say that A is *k-homogeneous* w.r.t. W if each scan has value k, in other words, the window W reveals exactly k elements of the configuration A for each possible position in the plane. In [11] it was proved that there exists a 1-homogeneous configuration w.r.t. W if and only if W is an exact polyomino. Moreover, in this case the configuration turns out to be periodic w.r.t. the directions of periodicity of W.

The k-homogeneous configurations raise interest when $k \geq 2$. In [11] it is proved that if the window W is a rectangle, then each k-homogeneous configuration can be split into k disjoint 1-homogeneous (periodic) ones. In [7], the authors conjectured that this decomposition result can be extended to each window W that is an exact polyomino.

Conjecture 1: if W is an exact polyomino and the configuration A is k-homogeneous w.r.t. the window W, then there exist k disjoint configurations $A_1, \ldots, A_k$ such that A_i is 1-homogeneous for each i, and $A = \bigcup_{i=1}^{k} A_i$.

Finally, in [1], the authors extended the decomposition to the *diamond* polyominoes, and then to all the exact polyominoes that are open sets in a generalized norm L^1 of $\mathbb{Z}^2$. With the aim of providing experimental evidence to the conjecture, we define the algorithm *Exh-Dec* that takes as input a perfect pseudo-square polyomino W and an integer k, and checks the possibility of decomposing all the possible k-homogeneous configurations w.r.t. W.

Exh-Dec is a simple brute force algorithm that acts in two phases: first, it recursively generates all the possible k-homogeneous configurations w.r.t. W, and then it proceeds in the decomposition of each of them into k disjoint 1-homogeneous configurations. Since W is a perfect pseudo-square polyomino, then, by [2] and Proposition 4, each 1-homogeneous configuration can be identified by means of a couple of directions of periodicity computed from its BN-factorizations.

We sketch below the main steps of the two phases of *Exh-Dec*.

Generation of k-homogeneous configurations

Input: a perfect pseudo-square W of area w and a homogeneity value $k \in \{1, 2, \ldots, w\}$.

Step 1: Create a void matrix A, whose dimensions include the input tile and four surrounding copies of itself, that provide the pattern of the planar tiling. This matrix will simulate the configuration in the discrete plane $\mathbb{Z}^2$;

Step 2: place the window W in the center of A and insert in W all the possible binary configurations having k elements 1. For each configuration start a different line of computation;

Step 3: move W in A of a single (discrete) step at a time, reaching all the non already visited positions. For each new position, change the non already assigned elements of W in all possible ways in order to obtain a configuration with k elements 1. If no such assignment is allowed, then prune the line of computation. For each remaining different configuration start a new line of computation;

Step 4: for each line of computation, repeat Step 3 until reaching the border of A;

Output: provide as output the remaining k-homogeneous configurations A.

Some remarks:

1. Usually most of the lines of computation are pruned. The reason lies in the fact that, after the first few steps, each k-homogeneous configuration has a well defined placement of the elements in order to keep the required homogeneity;

2. Since A is finite, the k-homogeneity may be lost while approaching to its borders. This will not impact the following phase, since A can be chosen huge enough to ignore border phenomena.

Decomposition

In this second phase, we use the periodicity of a tiling by W to decompose each computed k-homogeneous configuration. Since W is perfect pseudo-square, then Proposition 3 assures that its tilings by translation have one or two couples of directions of periodicity.

Input: a perfect pseudo-square W and a configuration A that is k-homogeneous w.r.t. W.

Step 1: compute the couple(s) of directions of periodicity u and v of W;

Step 2: starting from a randomly chosen element $x = 1$ of A, detect the elements of A that form with x a periodical configuration either w.r.t. u or to v. Update A by removing the obtained configuration;

Step 3: repeat Step 2 for k times and check that all the elements are removed from A. If so, then A has been decomposed into k disjoint 1-homogeneous configurations. Otherwise, the configuration is non-decomposable;

Output: flag for the *success* or *failure* of the decomposition process.

If *Exh-Dec* successfully performs the decomposition of all the computed k-homogeneous configurations for each possible $k = 1, \ldots, w$, then we have the experimental evidence that the decomposition theorem holds for the chosen tile W of area w.

On the other hand, if we find a k-homogeneous configuration that is not periodic w.r.t. W, then the decomposition fails. In this case, we say that the tile W is k-*non-decomposable* or simply *non-decomposable*.

We underline that the procedure *Exh-Dec* deeply relies on the fact that W is a perfect pseudo-square in the detection of the periodical 1-homogeneous configurations that decompose A. A generalized version of the algorithm can be defined keeping in mind that the couples of directions of periodicity may reduce to one single direction in case of non-regular tilings, as observed in [2].

Experimental Results

Hereafter we briefly report the results obtained after the exhaustive generation of all the perfect pseudo-squares with perimeter n ranging from 10 to 22, and the performances of *Exh-Dec* on them. Then, we highlight four polyominoes that are non-decomposable, and we provide one of the related homogeneous configurations.

The results of the exhaustive computation up to perimeter $n = 16$ are shown in Table 1, together with their BN-factorization as a square and the related couple of directions of periodicity. In case of double square, both the BN-factorizations and the two couples of directions of periodicity are provided.

We point out that there exists one single double square of perimeter $n = 12$, i.e., the polyomino in Fig. 2, and one single of perimeter $n = 16$, while no double squares exist of perimeter $n = 14$.

Moving to greater perimeter, the following statistics hold (up to rotations and symmetry):

$n = 18$: there exist 90 perfect pseudo-squares and one double square;
$n = 20$: there exist 273 perfect pseudo-squares and 3 double squares;
$n = 22$: there exist 836 perfect pseudo-squares, none of which is a double square.

Concerning the decomposability of the computed perfect pseudo-square polyominoes, *Exh-Dec* decomposes all of them up to perimeter $n = 18$ and homogeneity value $k = 2$. This result partially supports Conjecture 1.

Table 1. List of all the perfect pseudo-squares with perimeter $12 \leq n \leq 16$ (exhaustively generated), and their couples of directions of periodicity. The two BN-factorizations of the double square polyominoes are grouped.

Boundary word	Perimeter	Directions of periodicity	
NWN|ENE|SES|WSW	12	$\{(-1,2),(2,1)\}$	$\{(1,2),(2,-1)\}$
NEN|ESE|SWS|WNW			
NNWN|ENE|SESS|WSW	14	$\{(-1,3),(2,1)\}$	$//$
NNNWN|ENE|SESSS|WSW	16	$\{(-1,4),(2,1)\}$	$//$
NNWN|ENNE|SESS|WSSW	16	$\{(-1,3),(2,2)\}$	$//$
NNEN|ENEE|SWSS|WWSW	16	$\{(1,3),(3,1)\}$	$//$
NNWN|ENEE|SESS|WWSW	16	$\{(-1,3),(3,1)\}$	$//$
NWN|ENNEE|SES|WWSSW	16	$\{(-1,2),(3,2)\}$	$//$
NWN|ENENE|SES|WSWSW	16	$\{(-1,2),(3,2)\}$	$\{(2,-1),(-2,-3)\}$
ESE|SWSWS|WNW|NENEN			
NWWN|ENNE|SEES|WSSW	16	$\{(-2,2),(2,2)\}$	$//$

Increasing the perimeter of perfect pseudo-squares to $n = 20$, we unexpectedly found a counterexample to Conjecture 1. More precisely, we found a perfect pseudo-square W', and a specific configuration of the plane A', that is 2-homogeneous w.r.t. W' and it does not allow a decomposition into two 1-homogeneous configurations (see Fig. 3). We also observe that A' shows a periodical pattern different from that one obtained from the couples of periodicity related to W'. Figure 3 shows the polyomino W' and the configuration A'.

It is easy to check that the configuration depicted in Fig. 3 is 2-homogeneous w.r.t. W'. To check the failure of the decomposition procedure, let us consider two disjoint configurations, A_1 (circled elements) and A_2 (squared elements). We start by labelling the elements in the scan in Fig. 3(a). Then we move the polyomino step by step as shown in the scans of Fig. 3 from (b) to (d), labelling the internal elements so that no two elements of the same configuration lie in the scan. We point out that each scan from (b) to (d) has one single possible label for the inside elements. Finally, in Fig. 3(e), we obtain a scan with two elements belonging to the same configuration A_2, so preventing its 1-homogeneity.

As already pointed out, A' is also periodic w.r.t. a couple of directions that does not correspond to the directions related to W'.

As a matter of fact, there exist three other perfect pseudo-squares of perimeter $n = 20$ that are 2-non-decomposable. These polyominoes are depicted in Fig. 4 (for brevity sake we omit the related configurations). It is interesting to note that the four perfect pseudo-square polyominoes have different shapes, suggesting that there could be several different classes of exact polyominoes where the decomposition theorem does not hold. This remark opens a new, interesting research line.

We end the report of our experimental results by showing another surprising example of the exact polyomino W'', of perimeter $n = 24$, that is 2-decomposable

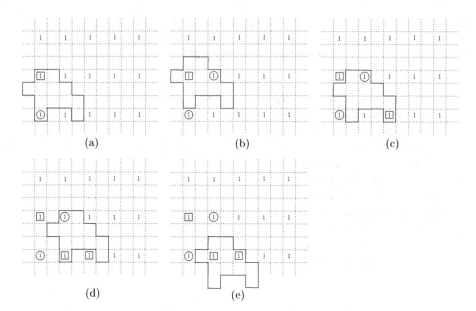

Fig. 3. The steps of the decomposition procedure of the configuration A' that is 2-homogeneous w.r.t. the polyomino W'. In (e) the scan that provides the non-decomposability is shown.

and 3-non-decomposable (see Fig. 5). This last example rises interest in finding a connection between the shape of an exact polyomino and the allowed values of k for its decomposability.

5   Discussion and Open Problems

In this work we consider exact polyominoes and the tilings they allow on the discrete plane $\mathbb{Z}^2$, with the aim of investigating a related conjecture, here Conjecture 1, stated in [7] and concerning the existence of a decomposition theorem for homogeneous configurations related to exact polyominoes.

In the first part of the paper, we recall the main definitions about exact polyominoes, we define the class $\mathcal{PPS}$ of perfect pseudo-squares and we prove some useful properties. We underline that the boundary word identifies the characteristics of an exact polyomino and the properties of the related tilings. So, we restrict the study of Conjecture 1 to the class $\mathcal{PPS}$, and we perform experimental tests on them.

We found unexpected results that give evidence that Conjecture 1 does not hold in general. In particular, we stress that a decomposition theorem can not be obtained for some elements of $\mathcal{PPS}$ from perimeter $n = 20$ on, since for the first value $k = 2$ of homogeneity. As a matter of fact, $k = 2$ can be regarded as the simplest instance of the conjectured decomposition theorem.

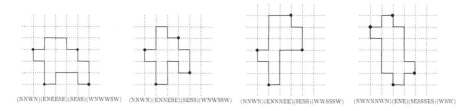

(NNWN)(ENEESE)(SESS)(WNWWSW)   (NNWN)(ENNESE)(SESS)(WNWSSW)   (NNWN)(ENNNEE)(SESS)(WWSSSW)   (NWNNNWN)(ENE)(SESSSES)(WSW)

Fig. 4. The smallest (w.r.t. the perimeter) four perfect pseudo-squares that allow a 2-homogeneous non-decomposable configuration.

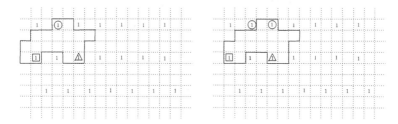

Fig. 5. A 3-homogeneous configuration w.r.t. the perfect pseudo-square $W'' = $ (NNEN)(EENEESEE)(SWSS)(WWNWWSWW), of perimeter $n = 24$. The decomposition procedure fails in the circled elements on the right scan.

Going forward, we found the exact polyomino W'' of greater perimeter that shows a different behavior: it is decomposable w.r.t. all its 2-homogeneous configurations, while there exists a 3-homogeneous configuration that turns out to be non-decomposable. This result outcomes to be relevant in defining some sub-classes of $\mathcal{PPS}$ where the decomposability is related to the index k of the k-homogeneous configurations.

Several new problems arise from our researches. Among them:

  i) it could be interesting to extend the study of the non-decomposability of homogeneous configurations to a generic exact polyomino. Obviously, this investigation has to consider that a tile can generate non-regular tilings, preventing, in that cases, the presence of a couple of directions of periodicity in the homogeneous configurations;

 ii) the characterization of the perfect pseudo-square polyominoes that are non-decomposable is also left as open problem. At a first sight, the few small ones detected seem to have no common characteristics both in their shapes, in their BN-factorizations, and in the index k of k-homogeneity of the non-decomposable configurations. This fact may imply that more than one class of non-decomposable polyominoes could be defined;

iii) it could be suitable to deepen the study of the periodicity of the homogeneous configurations related to non-decomposable perfect pseudo-squares. In fact, it is observed that those configurations show a couple of directions of periodicity that are different from those ones that can be computed from the considered tile. It is also curious that quite often one among the horizontal and vertical directions (sometimes both) is present.

References

1. Battaglino, D., Frosini, A., Rinaldi, S.: A decomposition theorem for homogeneous sets with respect to diamond probes. Comput. Vis. Image Underst. **17**, 319–325 (2013)
2. Beauquier, D., Nivat, M.: On translating one polyomino to tile the plane. Discret. Comput. Geom. **6**, 575–592 (1991)
3. Blondin Massé, A., Brlek, S., Labbé, S.: A parallelogram tile fills the plane by translation in at most two distinct ways. Discret. Appl. Math. **170**(7–8), 1011–1018 (2012)
4. Blondin Massé, A., Garon, A., Labbé, S.: Combinatorial properties of double square tiles. Theor. Comput. Sci. **502**, 98–117 (2013)
5. Freeman, H.: Boundary encoding and processing. In: Lipkin, B., Rosenfeld, A. (eds.) Picture Processing and Psychopictorics, pp. 241–266. Academic Press, New York (1970)
6. Freeman, H.: On the encoding of arbitrary geometric configurations. IRE Trans. Electron. Comput. **10**, 260–268 (1961)
7. Frosini, A., Nivat, M.: On a tomographic equivalence between $(0, 1)$-matrices. Pure Math. Appl. **16**(3), 1–15 (2005)
8. Frosini, A., Nivat, M.: Binary matrices under the microscope: a tomographical problem. Theor. Comput. Sci. **370**, 201–217 (2007)
9. Lothaire, M.: Combinatorics on Words. Cambridge University Press, Cambridge (1997)
10. Morse, M., Hedlund, G.A.: Symbolic dynamics. Am. J. Math. **60**(4), 815–866 (1938)
11. Nivat, M., Sous-ensembles homogènes de $\mathbb{Z}^2$ et pavages du plan. C.R. Acad. Sci. Paris. Ser. I **335**, 83–86 (2002)
12. Nivat, M.: Invited talk at ICALP97

Properties of SAT Formulas Characterizing Convex Sets with Given Projections

Niccolò Di Marco$^{(\boxtimes)}$ and Andrea Frosini

Dipartimento di Matematica e Informatica, Università di Firenze, Firenze, Italy
{niccolo.dimarco,andrea.frosini}@unifi.it

Abstract. One of the most interesting and challenging problems in Discrete Tomography concerns the faithful reconstruction of an unknown finite discrete set from its horizontal and vertical projections. The computational complexity of this problem has been considered and solved in case of horizontal and vertical convex polyominoes, by coding the possible solutions through a 2-SAT formula. On the other hand, the problem is still open in case of (full) convex polyominoes. As a matter of fact, the previous polynomial-time reconstruction strategy does not naturally generalize to them. In particular, it has been observed that the convexity constraint on polyominoes involves, in general, a k-SAT formula φ, preventing, up to now, the polynomiality of the entire process, assuming that $P \neq NP$. Our studies focus on the clauses of φ. We show that they can be reduced to 2-SAT or 3-SAT only and that a subset of the variables involved in the reconstruction may appear in the 3-SAT clauses of φ, thus detecting some situations that lead to a polynomial time reconstruction. Some examples of situations where 3-SAT formulas arise are also provided.

Keywords: Discrete tomography · Reconstruction · SAT formulas

1 Introduction

Discrete Tomography concerns the retrieval of geometrical information about the internal (and so sometime inaccessible) structure of combinatorial objects from quantitative measurements of their primary constituents along linear (or multidimensional, in general) subspaces. These measurements are usually addressed as *projections*. The involved combinatorial objects may vary from generic discrete sets to constrained ones as graphs or hypergraphs [12,14]. Among the studied problems in discrete tomography, we find the retrieval of necessary and sufficient conditions for a pair of vectors to be the horizontal and vertical projections of an $m \times n$ binary matrix. Since, in general, the number of matrices sharing the same projections grows exponentially with their dimension, in most applications some further information are needed to obtain a solution as close as possible to the original object. Research tackle the algorithmic challenges of limiting the class of

© Springer Nature Switzerland AG 2022
É. Baudrier et al. (Eds.): DGMM 2022, LNCS 13493, pp. 153–166, 2022.
https://doi.org/10.1007/978-3-031-19897-7_13

possible solutions in different ways, e.g. increasing the number of projections or adding geometrical information. Among connected (finite) sets a dominant role deserved *polyominoes*, that are 4−connected sets of points of the integer lattice, considered up to translation.

In [1] the authors present an algorithm (called, from now on, $HVRec$) to reconstruct hv-convex set in polynomial time starting from their horizontal and vertical projections. Note that the term polynomial is here referred to as a deterministic polynomial computation. This algorithm consists of two separate parts: it first reconstructs an internal hv−convex kernel of points which is common to all the convex polyominoes having the input projections. Then, it expands the kernel to reach the desired projections maintaining the hv−convexity by means of a $2-SAT$ formula φ, one of whose valuations can be computed in polynomial time.

On the other hand, the computational complexity of the reconstruction of (full) convex polyominoes is still an open problem, but it may benefit from the strategy described above. Note that different definitions of convexity and different approaches are studied from the community [13,15].

In particular, in our study, we define a generalization of $HVRec$, called $CRec$, that leads in this direction: starting from the couple of horizontal and vertical projections our algorithm executes the kernel reconstruction performed by $HVRec$ with the further inclusion of the points in its *convex hull*, so obtaining a convex kernel.

The last part of the algorithm expands the kernel reaching the desired projections by using, in general, a formula whose valuations represent all the possible solutions to the reconstruction problem. Unfortunately, shifting to full convexity, this formula, say φ_{Conv}, becomes a generic SAT one, sliding its computational complexity to the non polynomiality, assuming that $P \neq NP$. Our studies deepen the characteristics of φ_{Conv} and offer a perspective to decrease its computational complexity.

In [8,11], the authors stressed that, differently from the hv-convex polyominoes reconstruction, the formula φ_{Conv} imposes convexity on a specific region of the border of the kernel without, in general, providing the global convexity of the whole border. This observation turns out to be the main reason that prevents the extension of $HVRec$ to the case of convex polyominoes and the motivation of our study.

In Sect. 2, we provide the basic notions of Discrete Tomography and combinatorics on words that lead to the formulation of the reconstruction problem. Some useful known results are also highlighted. Section 3 is devoted to the definition of the strategy for the reconstruction of convex polyominoes from horizontal and vertical projections. Some properties that simplify the SAT formulas φ characterizing the convex constraint are shown, according to the positions of the elements that have to be included in the kernel to reach the desired projections. Since, from experimental computations, it's not easy to detect cases in which general SAT formula are needed, we provide an example where a $3-SAT$ formula is required. The last Sect. 4 contains final comments and hints for future researches.

2   Definitions and Preliminary Results

A planar discrete set S is a finite subspace of points of the integer lattice $\mathbb{Z}^2$, considered up to translation. The dimension of S are those of its minimal bounding rectangle. The set S can be naturally represented as a set of unitary cells centered on the points of S (see Fig. 1, (a)).

A *row* (resp. *column*) of S is its intersection with an infinite strip of cells whose centers lie on horizontal (resp. vertical) lines.

To each discrete set of dimension $m \times n$ we can associate two integer vectors $H = (h_1, \ldots, h_m)$ and $V = (v_1, \ldots, v_n)$ such that for each $1 \le i \le m, 1 \le j \le n$, h_i and v_j are the number of cells of S which lie on row i and column j, respectively. We call H and V *horizontal* and *vertical projections* of S, respectively.

A wide literature links the geometrical and topological characteristics of discrete sets with their projections. A special focus is on the properties of connectedness and convexity, providing several results on the related sub-classes.

So, we define *polyomino P* a $4-$connected (i.e. connected on horizontal and vertical directions) planar discrete set. A P is $h-$convex (resp. $v-$convex) if each rows (resp. columns) is connected (see Fig 1, (b)). If P is both h-convex and $v-$convex then we say that P is $hv-$convex.

Concerning the convexity along all the possible (discrete) directions, several definitions take care of pathological situations that may arise when continuous convex shapes are discretized into convex discrete sets. In case of *polyominoes*, the notion of convexity turns in the natural simple form of the equivalence between the polyomino and its *convex hull* (see Fig 1, (c) and (d)).

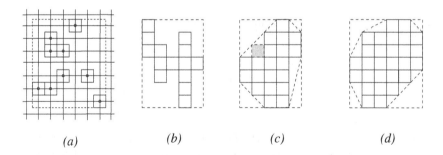

(a)	*(b)*	*(c)*	*(d)*

Fig. 1. (a) a generic discrete set of points in $\mathbb{Z}^2$ and its representation as a set of cells on a squared surface; (b) a vertically convex polyomino; (c) a hv-convex polyomino. Its convex hull and the cell that prevents it to be full convex are dashed and highlighted; (d) a (full) convex polyomino and its convex hull.

Coding the boundary of a convex polyomino
We recall that the *Freeman code* [9] associates to each polyomino its *boundary word*, i.e., the word on four letter alphabet $A' = \{0, \bar{0}, 1, \bar{1}\}$ obtained by coding the path that clockwise follows the boundary of the cell representation of the

polyomino starting from a specific point. The letters $\bar{0}$ and $\bar{1}$ represent the horizontal step and the vertical step when travelled in the opposite directions with respect to 0 and 1, respectively. If the polyomino is hv-convex, we can identify four points W, N, E and S as the points where the polyomino's boundary first touches the west, north, east and south sides of its minimal bounding rectangle, respectively, when moving clockwise along it.

These four points determine four paths, according to the starting and ending points, i.e., WN, NE, ES and SW-paths, as depicted in Fig. 2. A path is WN-convex (resp. NE, ES and SW-convex) if it is the WN-path (resp. NE, ES and SW-path) of a convex polyomino. Each of the four paths is *monotone*, i.e., it uses only two of the four Freeman coding steps.

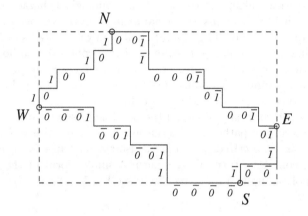

Fig. 2. The boundary of an hv-convex polyomino and its decomposition into four monotone paths.

In the following, we consider the WN-paths only, reminding that all the results can be extended to the other three paths.

Christoffel words

Let a, b be two co-prime numbers, the *lower* Christoffel path of slope $\frac{a}{b}$ is defined as the connected path in the discrete plane joining the origin $O(0,0)$ to the point (b, a) such that it is the nearest path strictly below the Euclidean line segment joining these two points. Analogously, the upper Christoffel path is defined as the nearest path that lies above the line segment.

To both Christoffel path one can associate so-called *Christoffel word* on the binary alphabet $A = \{0, 1\}$, such that the letter 0 represents a horizontal step and the letter 1 a vertical step. The Christoffel word commonly indicates the word w related to its lower Christoffel path and whose slope $\rho(w)$ equals $\frac{a}{b}$. Since the upper and the lower Christoffel paths of the same slope are the mirror images of each other, the two related words $\widetilde{w}$ and w are the mirror images of each other, i.e., $w(i) = \widetilde{w}(|w| - i + 1)$, where $s(i)$ denotes the ith character of the string s. By definition, $|w| = |\widetilde{w}| = a + b$. We indicate by $|w|_\alpha$ the number of occurrences of the letter α in w. Figure 3 represents the situation.

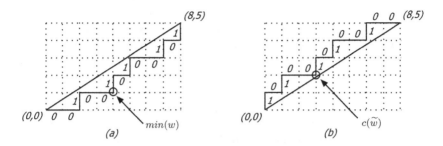

Fig. 3. (a) the lower Christoffel path of the segment with slope $\frac{5}{8}$ and the related Christoffel word w; (b) the upper Christoffel path of the same segment and the word $\widetilde{w}$ associated to it. The points $min(w)$ and $c(\widetilde{w})$ are also highlighted.

It is well known that each Christoffel word w different from 0 or 1 can be uniquely split as concatenation of two smaller Christoffel words w_1 and w_2, providing the so called *standard factorization* introduced in [3]. Such factorization involves the closest point $c(w)$ from the line segment of slope $\rho(w) = \frac{a}{b}$. So, it holds that $w = w_1 w_2$, where w_1 is the word leading from $(0,0)$ to $c(w)$, and w_2 is the word leading from $c(w)$ to (a,b). By abuse of notation we indicate with $c(w)$ also its index position in w. Moreover, we indicate as $min(w)$ the (index of the) furthest point from the same line (see Fig. 3). Note that $min(w)$ is also (the index of) the closest point in the upper Christoffel word $\widetilde{w}$, i.e., $min(w) = c(\widetilde{w})$.

It is worthwhile noticing the following property of Christoffel words.

Proposition 1. *The WN-path of a convex polyomino can be decomposed into a sequence of Christoffel words having decreasing slope.*

In [4] it has been proved that such a decomposition is unique and it can be obtained by the Lyndon factorization of the WN-path.

3   A Strategy to Reconstruct Convex Polyominoes

The retrieval of geometrical or structural properties of an unknown discrete object, modeled as a discrete set of points at a certain resolution, from projections, and, at best, its full reconstruction, is one of the main topic in Discrete Tomography. In general, there exist point sets sharing the same projections, so it is important to consider some a priori information, if available, about the unknown object that may lead to its faithful reconstruction.

3.1   The Reconstruction of hv-convex Polyominoes from Horizontal and Vertical Projections

In [1] the authors defined the algorithm $HV\,Rec$ that reconstructs (in polynomial time w.r.t. its size) a hv-convex polyomino compatible with an input couple of horizontal and vertical projections H and V, if it exists. In [5], it has been proved that hv-convex polyominoes does not guarantee the uniqueness of the reconstruction. This is due to the presence of specific configurations of points called *switching components* (see [12] for a survey in the topic and [6,7,11] for the characterization and related results). The strategy of the algorithm concerns the detection of two hv-convex subsets of cells: the *kernel* whose cells belong to all the solutions, if any, and the *shell* whose cells are outside from all the solutions. Finally, the cells not yet assigned will be included in the final solutions according to the valuation of a 2-SAT formula.

In more details, $HV\,Rec$ gets as input two vectors H and V and performs the two following steps:

Step 1 *(kernel and shell computation)*: according to each possible positions of the elements that lie in first and last rows and columns (rows 1 and m and columns 1 and n) of the minimal bounding rectangle including the solution, it detects the cells that are common to all the $hv-$convex polyominoes having H and V as horizontal and vertical projections, say the *kernel*. At the same time, it also detects the cells that are external to the polyomino and that constitute the *shell* of the polyomino. Both the detection tasks are iteratively performed by using a sequence of *filling operations* that take advantage from the convexity constraints and from the knowledge of the vectors H and V.

So, $HV\,Rec$ converges to the final kernel by approximating it both from inside and from outside. The process ends when the filling operations either fail, meaning that no solution exists, or leave the kernel and the shell unchanged. In the latter case, the cells that remain unidentified, so not belonging to the kernel or to the shell, are grouped in an ambiguous set $\mathcal{X}$. If $\mathcal{X} = \emptyset$, then the polyomino has been successfully reconstructed. Otherwise, it contains the elements of the switching components of the polyomino and Step 2 is required.

Step 2 *(2-SAT formula definition and valuation)*: each switching component in $\mathcal{X}$ is detected and its cells alternatively labelled associating them to a new boolean variable x or $\overline{x}$. Recall that $\overline{x}$ stands for not x. Finally, a $2 - SAT$ formula involving all the variables is defined. It imposes on the cells of $\mathcal{X}$ both the hv-convexity of the kernel and of the shell, preserving the coherence with the vectors H and V. Each valuation of the 2-SAT formula (obtained in polynomial time) leads to a feasible solution of the reconstruction problem in the sense that a cell of $\mathcal{X}$ belongs to the polyomino if and only it is labelled with a variable whose valuation is true. The computed polyomino is then provided as output.

Example 1. Figure 4, (*a*) depicts a part of the *NW* path of a polyomino reconstructed after performing Step 1 of *HVRec*. Assume that the cells labelled $x_1, \ldots, x_7$ belong to the set $\mathcal{X}$. Step 2 requires to define a sequence of 2-SAT formulas coding the *hv*-constraint on them both related to the kernel and to the shell. More precisely, the *h*-convexity of the kernel imposes the formulas $x_2 \to x_3$, and $x_6 \to x_7$, while the *v*-convexity imposes $x_6 \to x_5$. Note that the *h*-convexity of the shell imposes their equivalent clauses. These clauses (by abuse of notation we indicate the implication $(x \to y)$ as a clause due to its logical equivalence to the clause $(\neg x \wedge y)$) that guarantee a valuation that preserves the *hv*-convex of the final solution, as desired.

3.2   A Strategy to Reconstruct Convex Polyominoes from Horizontal and Vertical Projections

The authors of [8,11], underline that there is no natural generalization of *HVRec* when *hv*-convexity constraint changes into full convexity. In particular it is proved that the 2-SAT formula imposing the *hv*-convexity defined in Step 2 may change into a *k*-SAT formula, with $k \geq 2$.

Relying on *HVRec*, we approach a reconstruction strategy, say *CRec* (Convex Reconstruction), by defining a slightly modified version of Step 1 indicated with StepConv 1, where the computation of the kernel includes the constraint of its convexity. This action is performed by including in the kernel, after each iteration of the filling operations, the cells of its convex hull. The shell computation is left unchanged.

Concerning the modifications required in Step 2, indicated with StepConv 2, our aim is to provide a set of clauses that impose the convexity of the set $\mathcal{X}$. Some properties of these clauses are hereafter provided, leaving open the computational complexity of their valuation. As an example, a similar way forward is used in [10] where the reconstruction of a sub-class of *hv*-convex polyominoes is performed by means 3-SAT Horn clauses whose valuation requires polynomial time.

So, let us consider the points of $\mathcal{X}$ that lie above the *WN*-path of the convex kernel identified in StepConv 1 (for the points related to the three remaining *NE*, *ES* and *SW* convex paths we proceed analogously). We recall that the membership of these points to one of the convex polyominoes consistent with the input horizontal and vertical projections *H* and *V*, if any, has to be determined.

The clauses of the formula we are going to define in StepConv 2 consider how the inclusion of each point in $\mathcal{X}$ reflects on the others, in order to preserve the convexity of the structure.

Convexity preserved by 2-SAT clauses
As a first result, we show that a single point inclusion in the kernel can be performed if the point is minimal w.r.t. the Christoffel word where it lies on. In particular, from [3], Theorem 1 and its corollaries, it follows

Proposition 2. *Let w be the Christoffel word of slope $\frac{a}{b}$ and denote by $w(i : j)$ the substring of w between indeces i and j. Then, it exists only one index $1 < i < a + b$ in which $w(i, i + 1) = 01$, such that $w_1 = w(1 : i - 1) 1$ and*

$w_2 = 0 \, w(i+2:a+b)$ *(we consider the substring* $w(i+2:a+b) = w(a+b)$ *if* $i = a+b-1$*) are both Christoffel words. Furthermore, it holds that* $i = min(w)$*.*

Relying on Proposition 2, we obtain the following result stating that the inclusion in a NW Christoffel word w of a point different from $min(w)$ does not preserve the convexity of w and so of the whole NW path.

Proposition 3. *Let* w *be the Christoffel word of slope* $\frac{a}{b}$ *and* $min(w) < i < a + b$*. The Christoffel path of slope* $\frac{|w(1:i-1)|_1+1}{|w(1:i-1)|_0}$ *includes the point* $c(\widetilde{w})$*. On the other hand, if* $1 < i < min(w)$*, then the Christoffel path of slope* $\frac{|w(i+1:a+b)|_1}{|w(i+1:a+b)|_0+1}$ *includes the point* $c(\widetilde{w})$*.*

Proof. Recall that the two indeces $c(\widetilde{w})$ and $min(w)$ are equal in the upper and lower Christoffel paths, respectively. Let us proceed by contradiction assuming that the Christoffel path of slope $\frac{|w(1:i-1)|_1+1}{|w(1:i-1)|_0}$ does not include the point $c(\widetilde{w})$, when $min(w) < i < a + b$. We consider the integer points $O = (0,0)$, $B = (b,a)$ and $C = (|w(1:i-1)|_0, |w(1:i-1)|_1 + 1)$. By the proof of Theorem 3.3 in [2], it holds that the triangle OCB contains at least one integer point. Let D be the point of OBC closest to the line segment of slope $\frac{a}{b}$. It follows that the triangle ODB contains no integer points and furthermore, by assumption, D is different from $c(\widetilde{w})$, reaching a contradiction. A similar argument holds if $0 < i < min(w)$. □

We can rephrase this proposition in a more algorithmic fashion.

Corollary 1. *Let* w *be a Christoffel word in the* WN*-path of the kernel obtained after StepConv 1. If the point* $c(\widetilde{w})$ *belongs to the shell, then all the points of* $\mathcal{X}$ *that lie on* w *also belong to the shell.*

Corollary 2. *Let* w *be a Christoffel word in the* WN*-path of the kernel obtained after StepConv 1. If we include in the final convex solution a point of* $\mathcal{X}$ *lying above its* WN*-path, then also the point* $c(\widetilde{w})$ *has to be included in order to preserve the convexity.*

Example 2. Let us assume that StepConv 1 provided the WN-path depicted Fig. 4 and that the cells $x_1, \ldots x_7$ belong to $\mathcal{X}$. The $WN-$path is the Christoffel word of slope $m = \frac{6}{11}$, $w = 0010010010010\,\mathbf{01}\,01$, where the boldface entries indicate its minimal point. If we require to add the point x_3, then the computation of the related convex hull (straight line in Fig. 4, (b)) imposes the inclusion of the points x_4 and x_5. So the two new clauses $x_3 \to x_4$ and $x_3 \to x_5$ has to be added to the clauses already defined in Example 1. Note that x_5 is $c(\widetilde{w})$ as expected by Proposition 3. Note that the inclusion of the points x_3, x_4 and x_5 in the kernel produces the Christoffel words $w_1 = 00100101$ and $w_2 = (001)^3$ (that is not primitive) whose slopes preserve the decreasing order.

Furthermore, we underline that, if the word w of Example 2 is followed by a Christoffel word of slope m, with $\frac{1}{2} < m < \frac{6}{11}$, then the slope of the

(a)

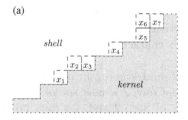

(b)

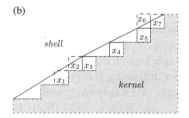

Fig. 4. The WN-path of the kernel computed by StepConv 1 and the points belonging to $\mathcal{X}$. In (a) the situation is depicted, while in (b) it is shown the convex hull computed after the inclusion in the kernel of x_3. To preserve convexity, it is also required the inclusion of x_4 and x_5.

WN-path is not preserved after the splitting, and some more points may need to be included. This situation may also arise if we include in the WN-path of the kernel in Fig. 4 the point $c(\widetilde{w}) = x_5$ only. The following example shows the situation for a different WN-path.

Example 3. Let w_1 and w_2 be two Christoffel words of a WN-path of a convex kernel, with $\rho(w_1) = \frac{3}{5} > \rho(w_2) = \frac{11}{20}$ as in Fig. 5. Including in the kernel the point $x = c(\widetilde{w}_1)$ changes w_1 into two new Christoffel words $u_1 v_1$, with $\rho(u_1) = \frac{2}{3}$ and $\rho(v_1) = \frac{1}{2}$. Now, the sequence of slopes $\rho(u_1), \rho(v_1)$ and $\rho(w_2)$ is not decreasing. Furthermore, $(v_1\, w_2) = 001\,00100100100100101000100100100101$ is not a Christoffel word, so the corresponding path is not $WN-$convex. To get back convexity, we need to include a second point in the polyomino, i.e., the point $y = c(\widetilde{w}_2)$ that belongs to the convex hull H_x depicted in Fig. 5, on the right, obtaining the word $w_3 = 0010010010010010\,\mathbf{10}\,0100100100100101$, where the included point is in boldface (see Fig. 5 for a visual representation).

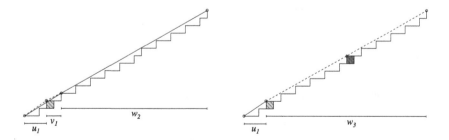

Fig. 5. The inclusion of one single point (on the left) prevent the WN-convexity of the path. A second point (on the right) has to be added to keep it back.

The $WN-$convexity is so imposed by the clause $x \to y$ that, in terms of Christoffel words, produces $v_1\, w_2$ changes into w_3.

In [8], the authors consider all possible cases that arise when a single point is included in a WN-path, and determine when further points need to be included to preserve global convexity. In the sequel, we characterize these situations through logical implications.

So, following the strategy of $HVRec$, Step 2, and keeping in mind the situations in Examples 2 and 3, we define in StepConv 2 a new set of clauses $\varphi_{Conv2SAT}$ that includes both those clauses φ_{hv} of Step 2 and, for each point $x \in \mathcal{X}$, the clauses $\varphi_x = (x \to x_1) \wedge \cdots \wedge (x \to x_n)$, with $x_1, \ldots, x_n$ being the points belonging to the convex hull H_x computed after the inclusion of x in the kernel.

We underline that, if the convex hull H_x related to a point $x \in \mathcal{X}$ includes points of the shell, since their values can be considered 0, then the same value is transferred to x.

So, a new reconstruction algorithm for a subclass of convex polyominoes, say $CRec_{2SAT}$, can be defined considering the modifications of Step 1 and Step 2 above described and involving 2-SAT clauses only.

Example 4. Consider the Christoffel word $w = 00101001010010101$ of slope $\rho = \frac{7}{10}$ that is in the $NW-$path of a kernel. Let $x_1, \ldots x_6$ be points of $\mathcal{X}$ (see Fig. 6 (a)). Let us include x_2 to the kernel, i.e. $x_2 = 1$. Then, we must also include x_1, x_3 and x_5 obtaining $\varphi_{x_2} = (x_2 \to x_1) \wedge (x_2 \to x_3) \wedge (x_2 \to x_5)$ (see Fig. 6 (b)). On the other hand, if x_6 is included, then x_3 and x_5 must be also included in the kernel obtaining $\varphi_{x_6} = (x_6 \to x_3) \wedge (x_6 \to x_5)$ (see Fig. 6 (c)). Finally, if we include both x_2 and x_6 and we compute again the convex hull, then we realize that that also the points x_1, x_3, x_4, and x_5 have to be included, with x_4 being a new one (see Fig. 6 (d)). So, a new clause has to be added to $\varphi_{x_2} \wedge \varphi_{x_6}$, i.e. $(x_2 \wedge x_6) \to x_4$. This provides an example of a situation where a $3 - SAT$ formula is required.

Relying on the above example, it may happen, in general, that an inclusion of a subset of k points of $\mathcal{X}$ leads to a $(k+1)$-SAT formula. In the sequel, we define some properties that allow to simplify the SAT clauses leading to a normal form involving only 3-SAT clauses. Up to now, no polynomial time valuation is known for this class of formulas.

3.3   Properties of the k-SAT Formulas to Impose Global Convexity

In [11], the author proved that in case of a specific class of convex polyominoes, adding points of $\mathcal{X}$ to the convex kernel can be performed in polynomial time. This result is related to some specific switching components of the elements of the class. For an extensive analysis of hv-convex switching components see [6,7]. Note that the class studied in [11] has non-empty intersection with that characterized by the formula $\varphi_{Conv2SAT}$ already know, however being distinct classes. So, the general versions of the algorithm to reconstruct convex polyominoes from projections may benefit from the following results about the SAT formula φ_{Conv} that has to be defined in StepConv 2 to express the global convexity. Given two

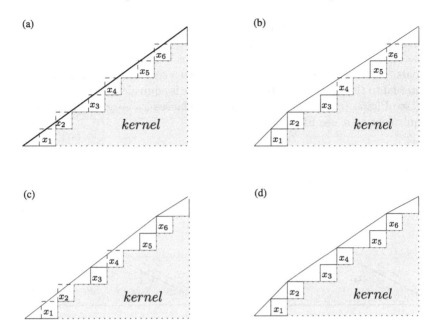

Fig. 6. An example of $3 - SAT$ in the convex reconstruction problem.

points x and y in $\mathcal{X}$, we define the partial order $x <_o y$ if and only if x precedes y while moving clockwise along the WN-path of the kernel.

Proposition 4. *Let $c = (x_1 \wedge x_2) \rightarrow x$ be a clause included φ_{Conv}. Then $x_1 <_o x <_o x_2$.*

Proof. Let us assume, w.l.o.g., that $x <_o x_1 <_o x_2$. This implies that either the segment $[x_1; x_2]$ is one of the sides of the convex hull computed after the kernel inclusion of x_1 and x_2 or it lies behind it. In the first case it holds $x_1 \rightarrow x$, while in the second case either $x_1 \rightarrow x$ or $x_2 \rightarrow x$ according to which among x_1 or x_2 is a vertex of the convex hull. So, c is equivalent to one of the two above clauses and it has not to be included in φ_{Conv}. □

Proposition 5. *Let $c = (x_1 \wedge x_2 \wedge x_3) \rightarrow x$ be a clause included in φ_{Conv}. There exists a 3-SAT clause that is equivalent to c.*

Proof. Let us assume w.l.o.g. that $x_1 < x < x_2 < x_3$, where $x_i < x_j$ means that the point x_i precedes x_j while moving clockwise along the WN-path of the kernel. Then that the convex hull computed after the kernel inclusion of the three points x_1, x_2, and x_3 includes also x. Two cases arise:

i) the segment $[x_2; x_3]$ lies inside the convex hull. In this case either the segment $[x_1; x_3]$ or $[x_1; x_2]$ is a side of the convex hull. In the first case the point x_2 lies below it, see Fig. 7, (a). Then $(x_1 \wedge x_3) \rightarrow x_2$, and c turns out to be equivalent to $c_1 = (x_1 \wedge x_3) \rightarrow x$. Analogously, in the second case x_3 lies inside the convex hull, see Fig. 7, (b). Then $x_2 \to x_3$, and c turns out to be equivalent to $c_1 = (x_1 \wedge x_2) \to x$.

$ii)$ the segment $[x_2; x_3]$ is a side of the convex hull. A reasoning similar to $i)$ holds. Two cases arise: either $[x_1; x_2]$ is also a side of the convex hull or it is internal to the convex hull. In the first case c is equivalent to $c_1 = (x_1 \wedge x_2) \to x$, see Fig. 7, (c). In the latter case, the clauses $x_2 \to x_1$ and $x_2 \to x$ are equivalent to c, see Fig. 7, (d).

$\square$

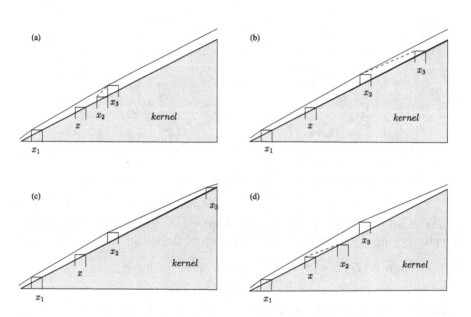

Fig. 7. The different configurations of points related to the clause c in the proof of Proposition 5. Internal segments are dashed while a full line indicates the convex hull.

A similar reasoning on a generic SAT clause leads to

Corollary 3. *Let* $c = (x_1 \wedge \cdots \wedge x_k) \to x$ *be a clause in* φ_{Conv}, *with* $k \geq 2$. *There exists a set of clauses in 3-SAT that are equivalent to* c.

Proposition 6. *If a clause* c *of* φ_{Conv} *includes both* x *and* $\bar{x}$, *then* c *can be reduced to* $2-SAT$.

Proof. The proof directly follows from the definitions of the logical operators. Two cases arise: if $c = (x_1 \wedge \bar{x}_1) \to x$, then the implication is a tautology. Otherwise, if $c = (x_1 \wedge x) \to \bar{x}_1$, then only three valuations of x_1 and x are admissible to have $c = 1$: either $x_1 = 1$ and $x = 0$ (so being $\bar{x}_1 = 0$) or $x_1 = 0$ and x is either 0 or 1. These valuations are equivalent to the 2-SAT clause $x_1 \to \bar{x}$.

$\square$

Experimental evidence allows us to conjecture the following statement:

Conjecture 1. Consider two consecutive Christoffel words w_1 and w_2 such that adding $c(\widetilde{w}_1)$ or $c(\widetilde{w}_2)$ separately does not cause the addition of anything else in the convex hull. Then, if $(x_i \wedge x_j) \to x_k$ is a clause of φ_{Conv}, then x_i and x_j cannot be both minimum of the two consecutive Christoffel words.

From the above conjecture it follows that, in case StepConv 1 detects a set $\mathcal{X}$ whose elements are minimum of the Christoffel words of the kernel border, then the reconstruction procedure can be performed in polynomial time.

4 Conclusions and Final Remarks

In this paper we propose a generalization of the reconstruction algorithm for hv-convex polyominoes from the horizontal and vertical projections in [1] to the class of full convex polyominoes. The algorithm that we define, called $CRec$, uses in its last step a boolean formula φ_{Conv} to encode both the convexity constraint of the solution and its consistence with the given projections. Unfortunately, the formula turns out to be in SAT, leaving open the computational complexity of the whole process.

We show some examples where a simplest 2-SAT formula is required to achieve the final solution. On the other hand, some cases are provided where a 3-SAT formula is needed to encode the possibility of inclusion in the solution of some points in specific positions. Furthermore, we provide some properties of the SAT formula φ_{Conv} involved in the reconstruction in order to simplify its valuation.

A deeper investigation could be addressed to define the characteristics related to the switching components that may be present in a convex polyominoes in order to find a possible physical distance between groups of positive and negative variables. Such a result may lead to a further simplification of φ_{Conv} with the final aim of defining a polynomial time reconstruction strategy for the whole class of convex polyominoes or for some sub-classes of interest.

References

1. Barcucci, E., Del Lungo, A., Nivat, M., Pinzani, R.: Reconstructing convex polyominoes from horizontal and vertical projections. Theor. Comput. Sci. **155**(2), 321–347 (1996)
2. Berstel, J., Lauve, A., Reutenauer, C., Saliola, F.V.: Combinatorics on Words: Christoffel Words and Repetitions in Words. Mathematical Society, Amer (2009)
3. Borel, J.-P., Laubie, F.: Quelques mots sur la droite projective réelle. J. Théor. Nombres Bordeaux **5**(1), 23–51 (1993)
4. Brlek, S., Lachaud, J.O., Provençal, X., Reutenauer, C.: Lyndon + Christoffel= digitally convex. Pattern Recogn. **42**(10), 2239–2246 (2009)
5. Del Lungo, A., Nivat, M., Pinzani, R.: The number of convex polyominoes reconstructible from their orthogonal projections. Discrete Math. **157**(1–3), 65–78 (1996)

6. Dulio, P., Frosini, A.: On some geometric aspects of the class of hv-convex switching components. In: Lindblad, J., Malmberg, F., Sladoje, N. (eds.) Discrete Geometry and Mathematical Morphology. DGMM 2021 Lecture Notes in Computer Science, vol. 12708, pp. 1–13. Springer, Cham (2021). https://doi.org/10.1007/978-3-030-76657-3_21

7. Dulio, P., Frosini, A.: Characterization of hv-convex sequences. J. Math. Imaging Vis. **64**, 771–785 (2022)

8. Dulio, P., Frosini, A., Rinaldi, S., Tarsissi, L., Vuillon, L.: Further steps on the reconstruction of convex polyominoes from orthogonal projections. J. Comb. Optim. 1–20 (2021)

9. Freeman, H.: On the encoding of arbitrary geometric configurations. IRE Trans. Electron. Comput. $EC - 10$, **2**, 260–268 (1961)

10. Frosini, A., Vuillon, L.: Tomographic reconstruction of 2-convex polyominoes using dual horn clauses. Theoret. Comput. Sc. **777**(2), 329–337 (2019)

11. Gerard, Y.: Regular switching components. Theoret. Comput. Sci. **777**, 338–355 (2019)

12. Herman, G.T., Kuba, A. (eds.): Discrete Tomography: Foundations Algorithms and Applications. Birkhauser, Boston (1999)

13. Kadu, A., van Leeuwen, T.: A convex formulation for binary tomography. IEEE Trans. Comput. Imaging **6**, 1–11 (2020)

14. Kocay, W., Li, P.C.: On 3-hypergraphs with equal degree sequences. Ars Combin. **82**, 145–157 (2006)

15. Tasi, T.S., Nyúl, L.G., Balázs, P.: Directional convexity measure for binary tomography. In: Ruiz-Shulcloper, J., Sanniti di Baja, G. (eds.) Progress in Pattern Recognition, Image Analysis, Computer Vision, and Applications. CIARP 2013. Lecture Notes in Computer Science, vol. 8259, pp. 1–13 (2013). https://doi.org/10.1007/978-3-642-41827-3_2

Multivariate and PDE-Based Mathematical Morphology, Morphological Filtering

Morphological Counterpart of Ornstein–Uhlenbeck Semigroups and PDEs

Jesús Angulo[(✉)]

MINES ParisTech, PSL-Research University, CMM-Centre de Morphologie
Mathématique, Fontainebleau, France
`jesus.angulo@mines-paristech.fr`

Abstract. Morphological semigroups and corresponding Partial Differential Equations are equivalent respectively to Hopf-Lax semigroups and the Cauchy problem of a family of first-order Hamilton–Jacobi equations. They are related to Maslov idempotent analysis too.

The Ornstein–Uhlenbeck operator and Ornstein–Uhlenbeck semigroup play the role of the Laplacian and the heat kernel semigroup if the Lebesgue measure is replaced by the standard Gaussian measure.

In this paper we revisit some contributions on the idempotent analogue of the semigroups associated with the Ornstein–Uhlenbeck semigroup, which are based on a Maslov measure, as well as the associated first-order Hamilton–Jacobi equation. We study the relevance of the corresponding semigroups in the context of morphological erosions and dilations.

Keywords: Ornstein–Uhlenbeck operator · Hamilton–Jacobi pde · Mathematical morphology · Morphological semigroups

1 Introduction

Morphological semigroups and corresponding Partial Differential Equations (PDEs) are equivalent respectively to Hopf-Lax semigroups and the Cauchy problem of a family of first-order Hamilton-Jacobi equations. More precisely, the following canonic morphological PDE plays a central role in continuous mathematical morphology [2,7,13,18]:

$$\begin{cases} \frac{\partial u}{\partial t} = \pm\frac{1}{2}\|\nabla u\|^2, \ x \in \mathbb{R}^n, \ t > 0 \\ u(x,0) = f(x), \quad x \in \mathbb{R}^n \end{cases} \tag{1}$$

such that the corresponding viscosity solutions are given by

$$u(x,t) = \sup_{y \in \mathbb{R}^n} \left\{ f(y) - \frac{\|x-y\|^2}{2t} \right\} \quad (\text{for } + \text{ sign}), \tag{2}$$

$$u(x,t) = \inf_{y \in \mathbb{R}^n} \left\{ f(y) + \frac{\|x-y\|^2}{2t} \right\} \quad (\text{for } - \text{ sign}), \tag{3}$$

E. Baudrier et al. (Eds.): DGMM 2022, LNCS 13493, pp. 169–181, 2022.
https://doi.org/10.1007/978-3-031-19897-7_14

which correspond to a dilation $(f \oplus p_t)$ and an erosion $(f \ominus p_t)$ of function $f(x)$ defined as

$$(f \oplus p_t)(x) = \sup_{y \in \mathbb{R}^n} \{f(y) + p_t(y - x)\}, \tag{4}$$

$$(f \ominus p_t)(x) = \inf_{y \in \mathbb{R}^n} \{f(y) - p_t(y - x)\}, \tag{5}$$

using as structuring function $p_t(x)$ the so-called multiscale quadratic (or parabolic) structuring function:

$$p_t(x) = -\frac{\|x\|^2}{2t}. \tag{6}$$

Due to its properties of semigroup, dimension separability and invariance to transform domain [12, 16, 17], the structuring function $p_t(x)$ can be considered as the canonic one in morphology, playing a similar role to the Gaussian kernel in linear filtering [14]. An alternative interpretation of the quadratic structuring function as the equivalent of the Gaussian kernel is based on Maslov's idempotent analysis [21, 22].

Other variations of that family of Hamilton–Jacobi models cover the flat morphology by disks [18]; i.e., $u_t = \pm \|\nabla u\|$, as well as operators with more general P-power concave structuring functions, i.e., $u_t = \pm \|\nabla u\|^P$, $P > 1$. For the application of the latter model to adaptive morphology, see [15]. In the most general case, this family of morphological PDEs and semigroups are formulated in the framework of length spaces [5] and a similar counterpart in ultrametric spaces [6]. A general theory of morphological counterparts of linear shift-invariant scale-spaces based on the Cramér-Fourier transform has been proposed [24]. We note that the case considered here is not shift-invariant and therefore it is outside of the scope of that theory.

The Ornstein–Uhlenbeck operator and Ornstein–Uhlenbeck semigroup play the role of the Laplacian and the heat kernel semigroup if the Lebesgue measure is replaced by the standard Gaussian measure. In this paper we revisit a series of contributions by Avantaggiati and Loreti [8–11], where the idempotent analogue of the semigroups associated with the Ornstein–Uhlenbeck semigroup has a simple Maslov measure equivalent of the Gaussian measure, and they are the viscosity solution of a first-order Hamilton–Jacobi equation.

Aim and Organisation of the Paper. In the paper we study and formalise the relevance of the corresponding semigroups in the context of morphological multiscale erosions and dilations for nonlinear signal and image processing. The rest of the paper is organized as follows. Section 2 provides a background on Ornstein–Uhlenbeck semigroups and its stochastic interpretation. In Sect. 3, we introduce Ornstein–Uhlenbeck morphological operators and study their properties, in particular the corresponding semigroups. The Ornstein–Uhlenbeck morphological PDE is considered in Sect. 4. Section 5 presents some preliminary experiment on the application of these semigroups. Conclusions and perspectives close the paper in Sect. 6.

2   Background: Ornstein–Uhlenbeck Operator and Semigroups

Ornstein–Uhlenbeck (O–U) semigroups can be introduced from either a functional analysis viewpoint or a stochastic differential equation viewpoint. We briefly revisit the main elements of both theories, see classical references [19,20].

Working on $X = L^2(\mathbb{R}^n)$, the Laplacian of a function $f \in X$, $\Delta f(x) = (\text{div} \circ \nabla f)(x)$, is the infinitesimal generator of the heat semigroup P_t. Namely, given any function f, there exists the limit $\lim\limits_{t \to 0^+} \frac{P_t f - f}{t} = \Delta f$. We recall that the heat semigroup is just given by the convolution of f with a Gaussian kernel, i.e., $P_t(f)(x) = \frac{1}{(4\pi t)^{n/2}} \int_{\mathbb{R}} f(y) e^{-\|x-y\|^2/4t} dy$, $t > 0$.

The O–U operator and semigroup play the role of the Laplacian and the heat kernel semigroup if the Lebesgue measure $d\mu$ is replaced by the standard Gaussian measure $d\gamma^\alpha$, with parameter $\alpha > 0$:

$$d\gamma^\alpha(y) = (2\pi)^{-n/2} \exp(-(\alpha\|y\|^2)/2)dy.$$

Let us consider that we are working on the space of the bounded and continuous functions $C_b(X)$. The O–U semigroup is defined by the family of scale operators

$$N_t^\alpha f(x) = \int_X f\left(e^{-\alpha t}x + \sqrt{1 - e^{-2\alpha t}}y\right) d\gamma^\alpha(y), \quad t \geq 0, \ \alpha > 0.$$

The O–U semigroup $\{N_t^\alpha\}_{t\geq 0}$ is a linear operator satisfying the following properties: For any function $f \in C_b(X)$ and any $\alpha > 0$, $t, s \geq 0$,

1. (Preservation of positivity) $N_0^\alpha = \text{Id}$.
2. (Conservative) $N_t^\alpha 1 - 1$ and $N_t^\alpha f \geq 0$ if $f \geq 0$.
3. (Contractive) $\|N_t^\alpha f\|_\infty \leq \|f\|_\infty$.
4. (Additive semigroup) $N_t^\alpha \circ N_s^\alpha = N_{t+s}^\alpha$.
5. (Continuity and convergence) The map $t \to N_t^\alpha f$ is continuous from $\mathbb{R}^+$ to $L^2(\mathbb{R}^n, d\gamma^\alpha)$ and $\lim\limits_{t \to 0^+} N_t^\alpha f(x) = f(x), \forall x \in X$.
6. (Invariant measure) The Gaussian measure is the unique invariant probability measure, i.e., $\int_X N_t^\alpha f d\gamma^\alpha = \int_X f d\gamma^\alpha$. More generally, for $f, g \in C_b(X)$, one has $\int_X g(x) N_t^\alpha f(x) d\gamma^\alpha(x) = \int_X f(x) N_t^\alpha g(x) d\gamma^\alpha(x)$.

The O–U differential operator is a generalization of the Laplace operator Δ:

$$\mathcal{L}_\alpha f(x) := \Delta f(x) - \alpha x \cdot \nabla f(x), \quad \alpha > 0.$$

This operator is the infinitesimal generator of the O–U semigroup. Namely, $\frac{\partial}{\partial t} N_t^\alpha f(x) = \mathcal{L}_\alpha (N_t^\alpha f)(x) = N_t^\alpha (\mathcal{L}_\alpha f)(x) = \Delta N_t^\alpha f(x) - \alpha x \cdot \nabla N_t^\alpha f(x)$, with $\frac{\partial}{\partial t} N_t^\alpha f(x)|_{t=0} = \Delta f(x) - \alpha x \cdot \nabla f(x)$.

Let us now consider the stochastic viewpoint. The O–U process is a stochastic Markov process viewed as a modification of the random walk which tends to drift back towards its long-term mean (mean reverting), with a greater attraction when the process is further from the central location. It can be physically viewed as the model of the velocity of a massive Brownian particle under the influence of friction. More precisely, an O–U process X_t satisfies the following SDE:

$$dX_t = \theta \left(\mu - X_t \right) dt + \sigma dB_t,$$

where the parameters are $\theta > 0$, μ and $\sigma > 0$ and B_t is a Brownian process.

There is also a relationship with the Fokker–Planck equation representation, which provides the linear parabolic PDE for the probability density function $p(x, t)$ of the random variable described by a SDE. In the particular case of the O–U SDE, the Fokker–Planck equation is

$$\frac{\partial p}{\partial t} = \theta \frac{\partial p}{\partial x} \left[(x - \mu)p \right] + \frac{\sigma^2}{2} \frac{\partial^2 p}{\partial x^2}.$$

Its Green function for an initial condition consisting of a unit point mass at location x_0 is given by a Gaussian distribution with mean: $\mu + (x_0 - \mu)e^{-\theta t} = x_0 e^{-\theta t} + \left(1 - e^{-\theta t} \right) \mu$ and variance: $\frac{\sigma^2}{2\theta} \left(1 - e^{-2\theta t} \right)$.

The stationary solution of this equation is the limit for $t \to +\infty$, which is the Gaussian distribution with mean μ and variance $\sigma^2/(2\theta)$.

3   Ornstein–Uhlenbeck Erosion and Dilation Semigroups

Theoretical foundations of Maslov idempotent measure theory [1,23] are based on replacing in the structural axioms of probability theory the role of the classical semiring $S_{(+,\times)} = (\mathbb{R}_+, +, \times, 0, 1, \leq)$ of positive real numbers by the idempotent semiring: $S_{(\max,+)} = (\bar{\mathbb{R}}, \max, +, -\infty, 0, \leq)$. In this context, a change of the measure involves a consistent counterpart to the standard probability theory.

Indeed, we can start by considering that the counterpart of the Gaussian measure $d\gamma^\alpha(x)$ in standard $(+, \times)$-analysis is the quadratic one in $(\max, +)$-analysis:

$$d\gamma^\alpha(x) = (2\pi)^{-n/2} \exp(-(\alpha \|x\|^2)/2) dx \xrightarrow[\text{idempotent}]{\text{analogy}} dm^\alpha(x) = \alpha \|x\|^2, \quad \alpha > 0$$

3.1   O–U Erosion, Adjoint Dilation and Complement Dilation

It seems natural to conjecture that the multiscale O–U erosion for any $f : X \to \bar{\mathbb{R}}$ is given by

$$\begin{aligned}
E_t^\alpha f(x) &= \inf_{z \in X} \left\{ f \left(e^{-\alpha t} x + \sqrt{1 - e^{-2\alpha t}} z \right) + dm^\alpha(z) \right\} \\
&= \inf_{z \in X} \left\{ f \left(e^{-\alpha t} x + \sqrt{1 - e^{-2\alpha t}} z \right) + \alpha \|z\|^2 \right\} \\
&= \inf_{y \in X} \left\{ f(y) + \frac{\alpha}{1 - e^{-2\alpha t}} \|e^{-\alpha t} x - y\|^2 \right\}.
\end{aligned} \tag{7}$$

The last step is just based on the change of variable

$$y = e^{-\alpha t} x + \sqrt{1 - e^{-2\alpha t}} z \iff z = \left(1 - e^{-2\alpha t} \right)^{-1/2} \left(y - e^{-\alpha t} x \right).$$

The erosion E_t^α can be seen as a generalization of the quadratic canonic operators when $\alpha \to 0$, i.e., $e^{-2\alpha t} \approx 1 - 2\alpha t$, then $\frac{\alpha}{1-e^{-2\alpha t}} \to \frac{1}{2t}$.

From an algebraic viewpoint, one can say that E_t^α, $t \geq 0$, $\alpha > 0$, is an erosion since the following two properties hold:

1. (Increaseness) If $f(x) \leq g(x)$, $\forall x \in X$, then

$$E_t^\alpha f(x) \leq E_t^\alpha g(x), \quad \forall x \in X.$$

2. (Commutation with infimum) For any $f, g : X \to \bar{\mathbb{R}}$,

$$E_t^\alpha \left(f \wedge g \right)(x) = E_t^\alpha f(x) \wedge E_t^\alpha g(x), \quad \forall x \in X.$$

The proof is obvious from the property of the infimum. The second one is a particular case of the linearity in the sense of the (min, +) of semiring of $\mathbb{R}_{\min} = \mathbb{R} \cup \{+\infty\}$: $\forall \lambda \in \mathbb{R}_{\min}$, $E_t^\alpha \left(\lambda + (f \wedge g) \right) = \lambda + [E_t^\alpha f \wedge E_t^\alpha g]$.

We can now introduce the corresponding multi-scale dilation by means of the adjunction property.

Proposition 1. *For every function* $f : X \to \bar{\mathbb{R}}$, *and for any* $t \geq 0$ *and* $\alpha > 0$, *the* adjoint O–U dilation *to* E_t^α *is given by*

$$D_t^\alpha f(x) = \sup_{y \in X} \left\{ f(y) - \frac{\alpha}{1 - e^{-2\alpha t}} \|e^{-\alpha t} y - x\|^2 \right\}, \tag{8}$$

Therefore, the duality by adjunction is satisfied, i.e., for any two functions f *and* g, *the pair* (D_t^α, E_t^α) *provides the relationship*

$$D_t^\alpha f(x) \leq g(x) \iff f(x) \leq E_t^\alpha g(x), \quad \forall x \in X. \tag{9}$$

It is easy to prove that the operator D_t^α is increasing; i.e., if $f(x) \leq g(x)$, $\forall x \in X$, then $D_t^\alpha f(x) \leq D_t^\alpha g(x)$ and commutates with the supremum; i.e., for any f and g, $D_t^\alpha (f \vee g)(x) = D_t^\alpha f(x) \vee D_t^\alpha g(x)$. Therefore we state that it is dilation.

The composition of the adjoint pair (D_t^α, E_t^α) provides morphological multi-scale opening and closing. The study of the corresponding O–U morphological filters is out of the scope of this paper.

There is another dilation associated to E_t^α which can be introduced by the duality associated to the complement (involution by negative, i.e., $f \mapsto -f$), named here *O–U complement dilation* $\bar{D}_t^\alpha$ and given by

$$\bar{D}_t^\alpha f(x) = -E_t^\alpha (-f)(x) = -\inf_{y \in X} \left\{ -f(y) + \frac{\alpha}{1 - e^{-2\alpha t}} \|e^{-\alpha t} x - y\|^2 \right\}$$

$$= \sup_{y \in X} \left\{ f(y) - \frac{\alpha}{1 - e^{-2\alpha t}} \|e^{-\alpha t} x - y\|^2 \right\}. \tag{10}$$

3.2 Semigroup Property

The property of additive semigroup of the O–U erosion and the complement O–U dilation gives us the basics to use these operators in the context of scale-space signal and image processing.

Proposition 2. *For any function $f : X \to \bar{\mathbb{R}}$, and for any pair of scale parameters $t, s \geq 0$, we have a semigroup for the O–U erosion and the O–U complement dilation:*

$$E_t^\alpha E_s^\alpha f(x) = E_{t+s}^\alpha f(x) \tag{11}$$
$$\bar{D}_t^\alpha \bar{D}_s^\alpha f(x) = \bar{D}_{t+s}^\alpha f(x) \tag{12}$$

A proof for the semigroups associated to more general Hamiltonians is provided in [8]. For the sake of understanding of their behaviour, we provide here a proof for the one-dimensional case.

Proof. Let's consider $X \subseteq \mathbb{R}$. For any t and s and given α, starting from

$$E_s^\alpha f(x) = \inf_{w \in X} \left\{ f\left(e^{-\alpha s} x + \sqrt{1 - e^{-2\alpha s}} w \right) + \alpha w^2 \right\},$$

one has that $E_t^\alpha E_s^\alpha f(x) =$

$$\inf_{z \in X} \left\{ \inf_{w \in X} \left[f\left(e^{-\alpha s} \left(x e^{-\alpha t} + \sqrt{1 - e^{-2\alpha t}} z \right) + \sqrt{1 - e^{-2\alpha s}} w \right) + \alpha w^2 \right] + \alpha z^2 \right\} =$$

$$\inf_{z \in X} \left\{ \inf_{w \in X} \left[f\left(e^{-\alpha(s+t)} x + e^{-\alpha s} \sqrt{1 - e^{-2\alpha t}} z + \sqrt{1 - e^{-2\alpha s}} w \right) + \alpha w^2 \right] + \alpha z^2 \right\}.$$

The following change of variable $(z, w) \to (u, v)$ is considered

$$\begin{cases} e^{-\alpha s} \sqrt{1 - e^{-2\alpha t}} z + \sqrt{1 - e^{-2\alpha s}} w = \sqrt{1 - e^{-2\alpha(s+t)}} u \\ -\sqrt{1 - e^{-2\alpha s}} z + e^{-\alpha s} \sqrt{1 - e^{-2\alpha t}} w = v \end{cases}$$

Squaring and adding gives

$$\left(1 - e^{-2\alpha(t+s)} \right) (z^2 + w^2) = (1 - e^{-2\alpha(t+s)}) u^2 + v^2,$$

and thus

$$z^2 + w^2 = u^2 + \frac{1}{1 - e^{-2\alpha(t+s)}} v^2.$$

Introducing the new variables, one has

$$\inf_{z \in X} \left\{ \inf_{w \in X} \left[f\left(e^{-\alpha(s+t)} x + e^{-\alpha s} \sqrt{1 - e^{-2\alpha t}} z + \sqrt{1 - e^{-2\alpha s}} w \right) + \alpha w^2 \right] + \alpha z^2 \right\} =$$

$$\inf_{u \in X} \left\{ \inf_{v \in X} \left[f\left(e^{-\alpha(s+t)} x + \sqrt{1 - e^{-2\alpha(s+t)}} u \right) + \alpha u^2 \right] + \frac{\alpha}{1 - e^{-2\alpha(t+s)}} v^2 \right\}.$$

We note that the sum of the first two terms does not depend on v and the minimum with respect to v will correspond to $v = 0$. So in conclusion we obtain:

$$\inf_{u \in X} \left\{ f\left(e^{-\alpha(t+s)} x + \sqrt{1 - e^{-2\alpha(t+s)}} u \right) + \alpha u^2 \right\} = E_{t+s}^\alpha f(u).$$

Similarly for $\bar{D}_t^\alpha$ just using its definition by complement.   □

3.3 Other Properties

Coming back to the link with Maslov's measures, we have the following result.

Proposition 3. (Avantaggiati and Loreti, 2009). *The Maslov measure* $dm^\alpha(x) = \alpha\|x\|^2$ *is* $(\min, +)$ *idempotent invariant with respect to the O–U erosion in the sense that for any* $f, g : X \to \mathbb{R}_{\min}$ *we have*

$$\inf_{x \in X} \{g(x) + E_t^\alpha f(x) + dm^\alpha(x)\} = \inf_{x \in X} \{f(x) + E_t^\alpha g(x) + dm^\alpha(x)\}, \quad (13)$$

with the particular case $g(x) = +\infty$, $\forall x \in X$, *which yields*

$$\inf_{x \in X} \{E_t^\alpha f(x) + \alpha\|x\|^2\} = \inf_{x \in X} \{f(x) + \alpha\|x\|^2\}. \quad (14)$$

The invariance of the idempotent measure with respect to the erosion semigroup consistently provides an additional feature of the major role played by $dm^\alpha(x) = \alpha\|x\|^2$ as counterpart of the Gaussian measure in morphological operators.

Finally, let us consider a property of regularization for Lipschitz functions which is another fundamental aspect of morphological semigroups [4].

Proposition 4. *Let* $f : \mathcal{K} \to \bar{\mathbb{R}}$ *be a Lipschitz function of constant* L *defined on a compact set* $\mathcal{K}$. *Then there exists a constant* K *such that for any* $t > 0$ *and* $\alpha > 0$ *one has*

$$|E_t^\alpha f(x) - f(x)| \le K \left(1 - e^{-\alpha t}\right) \quad (15)$$

where K *depends on* L, α, t *and the diameter of the set* $\mathcal{K}$. *A similar result is obtained for the O–U dilation semigroup.*

The proof is provided in [8] as part of other results.

4 Ornstein–Uhlenbeck Morphological PDE

The Ornstein–Uhlenbeck morphological PDE with parameter $\alpha > 0$ as a Cauchy problem is given by

$$\begin{cases} \frac{\partial u}{\partial t} = -\frac{1}{2}\|\nabla u\|^2 - \alpha x \cdot \nabla u, & x \in \mathbb{R}^n, \ t > 0 \\ u(x, 0) = f(x), & x \in \mathbb{R}^n \end{cases} \quad (16)$$

such that the corresponding viscosity solutions are given by

$$u(x, t) = \inf_{y \in X} \left\{ f(y) + \frac{\alpha}{1 - e^{-2\alpha t}}\|e^{-\alpha t}x - y\|^2 \right\},$$

which thus corresponds to the O–U erosion E_t^α; i.e., $u(x, t) = E_t^\alpha f(x)$. We note that $\alpha = 0$ provides the canonic morphological PDE (1).

4.1   Heuristic Derivation Using the Semigroup Property

Without providing a rigorous proof of the viscosity solution, let us sketch a heuristic derivation. First, we note that

$$u(x, t+s) = \inf_{y \in X} \left\{ u(y, t) + \frac{\alpha}{1 - e^{-2\alpha s}} \| e^{-\alpha s} x - y \|^2 \right\}.$$

Thus, using the same semigroup property, one has for $0 \le s < t$ and for all $y \in \mathbb{R}^n$

$$u(x, t) \le u(y, s) + \frac{\alpha}{1 - e^{-2\alpha(t-s)}} \| y - e^{-\alpha(t-s)} x \|^2$$

$$u(x, t) - u(y, s) \le \frac{\alpha}{1 - e^{-2\alpha(t-s)}} \| y - e^{-\alpha(t-s)} x \|^2.$$

We set now

$$h = 1 - e^{-\alpha(t-s)}; \quad y - e^{-\alpha(t-s)} x = -hz$$

which implies

$$s = t - \frac{1}{\alpha} \log \frac{1}{1 - h}; \quad y = x - h(x + z).$$

Using that change of variable, we have

$$u(x, t) - u\left(x - h(x + z), t - \frac{1}{\alpha} \log \frac{1}{1 - h} \right) \le \frac{\alpha}{1 - (1 - h)^2} \| hz \|^2$$

$$\frac{u(x, t) - u\left(x - h(x + z), t - \frac{1}{\alpha} \log \frac{1}{1 - h} \right)}{h} \le \frac{\alpha h}{1 - (1 - h)^2} \| z \|^2$$

Noticing that

$$\lim_{h \to 0+} \frac{h}{1 - (1 - h)^2} = \frac{1}{2}$$

we can consider the limit when $h \to 0$ and to introduce the gradient and get

$$\nabla u(x, t)(x + z) + \frac{1}{\alpha} \frac{\partial u}{\partial t}(x, t) \le \frac{\alpha}{2} \| z \|^2.$$

For any $z \in \mathbb{R}^n$,

$$z \cdot \nabla u(x, t) - \alpha \frac{1}{2} \| z \|^2 = \frac{1}{\alpha} \left(\alpha z \cdot \nabla u(x, t) - \frac{1}{2} \| \alpha z \|^2 \right).$$

Finally, using the Legendre–Fenchel transform of the quadratic norm $\frac{1}{2} \| \cdot \|^2$, it is obtained

$$\frac{\partial u}{\partial t}(x, t) + \alpha x \cdot \nabla u(x, t) + \frac{1}{2} \| \nabla u \|^2 \le 0.$$

4.2 General Hamiltonian

The previous O–U morphological PDE is just a particular case of a family of Cauchy problems for that class of Hamilton–Jacobi equations, with a Hamiltonian which will depend on a parameter $p > 1$, such that $p = 2$ corresponds to (16). This PDE has been formulated and studied for initial data fulfilling the Lipschitz condition in [8] and for initial data being lower semicontinuous in [10]. We provide their main original result.

Theorem 1. (Avantaggiati and Loreti, 2008). *Let us assume a Lipschitz continuous function $f : \mathbb{R}^n \to \mathbb{R}$, which will be the initial condition for the following initial-value Hamilton–Jacobi first-order partial differential equation*

$$\begin{cases} u_t(x,t) + H\left(Du(x,t)\right) + \alpha x \cdot Du(x,t) = 0, & in \ \mathbb{R}^n \times (0,+\infty), \\ u(x,0) = f(x), & in \ \mathbb{R}^n, \end{cases} \tag{17}$$

Let us assume that the Hamiltonian $H : \mathbb{R}^n \to \mathbb{R}$ is an even, non-negative, convex function and positively homogenous of degree p, with $p > 1$. Then the viscosity solution of the Cauchy problem (17) is given by the following Hopf–Lax–Oleinik semigroup:

$$u(x,t) = \inf_{z \in \mathbb{R}^n} \left[f\left(e^{-\alpha t} x + \left(\frac{1 - e^{-\alpha p t}}{\alpha p} \right)^{1/p} z \right) + L(z) \right] \tag{18}$$

$$= \inf_{y \in \mathbb{R}^n} \left[f(y) + \left(\frac{\alpha p}{1 - e^{-\alpha p t}} \right)^{q-1} L\left(y - e^{-\alpha t} x \right) \right], \tag{19}$$

where the Lagrangian $L(q)$ is the one-dimensional Legendre–Fenchel transform of the function $H(p)$, i.e.,

$$L(q) = H^*(q) = \sup_{p \in \mathbb{R}_+} \{ p\,q - H(p) \}, \quad q \in \mathbb{R}_+. \tag{20}$$

We note that, by standard results on the Legendre–Fenchel transform, L is also an even, non-negative, convex function and positively homogenous of degree q, with $\frac{1}{p} + \frac{1}{q} = 1$.

The corresponding semigroups have therefore a more general form which depends on the shape parameter p.

5 Preliminary Experiments

For the preliminary experiments of this paper, we are illustrating the effects of the morphological semigroups on 1D signals; i.e., $X \subset \mathbb{R}$. Let us first consider the shape of the multiscale structuring functions of the O–U erosion and associated dilations. We define the *O–U structuring function* as

$$q_t^\alpha(x,y) = -\frac{\alpha}{1 - e^{-2\alpha s}} \| e^{-\alpha s} x - y \|^2, \quad x, y \in X, \tag{21}$$

such that $E_t^\alpha f(x) = \inf_{y \in X} \{f(y) - q_t^\alpha(x, y)\}$ and $\bar{D}_t^\alpha f(x) = \sup_{y \in X} \{f(y) + q_t^\alpha(x, y)\}$. We remind from (6) that $\lim_{\alpha \to 0} q_t^\alpha(x, y) = p_t(x - y)$.

In Fig. 1, the structuring functions $q_t^\alpha(x, y)$ and $p_t(x - y)$ are compared with respect to variations on α, with fixed scale $t = 1$. In the plots, we fix the point x where the structuring function is centered. As expected for α close to zero, Fig. 1(a), one has just parabolic structuring functions which are translated. When α increases, Fig. 1(b)–(c), the structuring function is attracted to the origin which involves that the maximum of the concave function is not located at x and therefore the O–U erosion is not anti-extensive. For large α, Fig. 1(d), structuring functions are deformed and introduced a significantly higher effect of penalization for the same t.

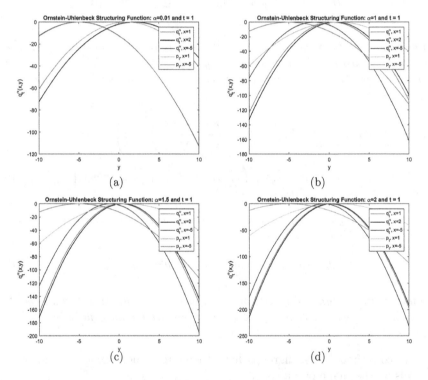

Fig. 1. O–U structuring function $q_t^\alpha(x, y)$ compared with respect to variations on α, with fixed scale $t = 1$: (a) $\alpha = 0.01$, (b) $\alpha = 1$, (c) $\alpha = 1.5$, (d) $\alpha = 2$. We fix the point x where the structuring function is centered.

A comparison of the effect of O–U erosions E_t^α and complement dilations $\bar{D}_t^\alpha$ on two signals is provided in Fig. 2. For the periodic $f(x)$, Fig. 2(a) depicts the erosions at scale $t = 0.1$ and three values of α. We note that the effect of drift back towards the origin by increasing α involves a delay of the location of the minima with respect to their position in the signal f. The case Fig. 2(b) of

$\alpha = 1$ and three values of t shows the typical multiscale effect of semigroups. In Fig. 2(c) and (d) provides multiscale operators for $\alpha = 0.01$ (equivalent to shift invariant parabolic ones) and $\alpha = 5$.

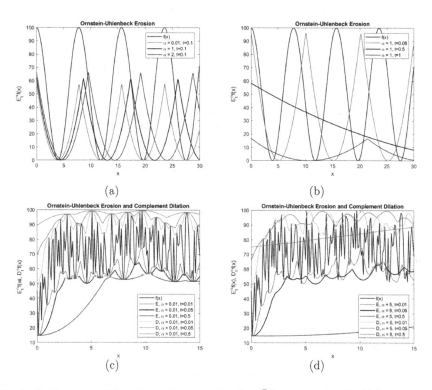

Fig. 2. O–U erosion E_t^α and complement dilation $\bar{D}_t^\alpha$ on two signals $f(x)$, compared with respect to variations on α and t: (a) $t = 0.1$ and three values of α, (b) $\alpha = 1$ and three values of t, (c) $\alpha = 0.01$ and three values of t, (d) $\alpha = 5$ and three values of t.

6   Conclusions and Perspectives

The O–U morphological semigroups satisfy the dimensionality separability property on the space, with potentially different parameters $\alpha_1, \cdots, \alpha_n$ for each dimension of X. Using this property, we will explore the formulation of spatio-temporal morphological semigroups with appropiate α for space and time.

We will consider the stochastic viewpoint of O–U semigroups to formalize the morphological counterpart. Our starting viewpoint will be similar to that of Bellman–Maslov random walks that was proposed in [3].

Finally, the evolution of shape of the generalized O–U structuring functions including a power parameter p with respect to α should be studied.

References

1. Akian, M.: Densities of idempotent measures and large deviations. Trans. Am. Math. Soc. **351**(11), 4515–4543 (1999)
2. Alvarez, L., Guichard, F., Lions, P.-L., Morel, J.-M.: Axioms and fundamental equations of image processing. Arch. Ration. Mech. **123**(3), 199–257 (1993). https://doi.org/10.1007/BF00375127
3. Angulo, J., Velasco-Forero, S.: Stochastic morphological filtering and bellman-maslov chains. In: Hendriks, C.L.L., Borgefors, G., Strand, R. (eds.) ISMM 2013. LNCS, vol. 7883, pp. 171–182. Springer, Heidelberg (2013). https://doi.org/10.1007/978-3-642-38294-9_15
4. Angulo, J.: Lipschitz Regularization of Images supported on Surfaces using Riemannian Morphological Operators. HAL hal-01108130v2 (2014)
5. Angulo, J.: Morphological PDE and dilation/erosion semigroups on length spaces. In: Benediktsson, J.A., Chanussot, J., Najman, L., Talbot, H. (eds.) ISMM 2015. LNCS, vol. 9082, pp. 509–521. Springer, Cham (2015). https://doi.org/10.1007/978-3-319-18720-4_43
6. Angulo, J., Velasco-Forero, S.: Morphological semigroups and scale-spaces on ultrametric spaces. In: Angulo, J., Velasco-Forero, S., Meyer, F. (eds.) ISMM 2017. LNCS, vol. 10225, pp. 28–39. Springer, Cham (2017). https://doi.org/10.1007/978-3-319-57240-6_3
7. Arehart, A.B., Vincent, L., Kimia, B.B.: Mathematical morphology: the Hamilton-Jacobi connection. In: Proceedings of IEEE 4th International Conference on Computer Vision (ICCV 1993), pp. 215–219 (1993)
8. Avantaggiati, A., Loreti, P.: Hopf-Lax type formulas and hypercontractivity. Ricerche mat. **57**, 171–202 (2008)
9. Avantaggiati, A., Loreti, P.: Idempotent aspects of Hopf-Lax type formulas. Tropical and idempotent mathematics. Contemp. Math. **495**, 103–114 (2009)
10. Avantaggiati, A., Loreti, P.: Lax type formulas with lower semicontinuous initial data and hypercontractivity results. Nonlinear Diff. Equat. Appl. NoDEA **20**(3), 385–411 (2012). https://doi.org/10.1007/s00030-012-0157-2
11. Avantaggiati, A., Loreti, P.: An approximation of Hopf-Lax type formula via idempotent analysis. Contemp. Math. **616**, 47–59 (2014)
12. van den Boomgaard, R., Dorst, L.: The morphological equivalent of Gaussian scale-space. In: Sporring, J., Nielsen, M., Florack, L., Johansen, P. (eds.) Gaussian Scale-Space Theory, vol. 8, pp. 203–220. Springer, Dordrecht (1997). https://doi.org/10.1007/978-94-015-8802-7_15
13. Brockett, R.W., Maragos, P.: Evolution equations for continuous-scale morphology. IEEE Trans. Signal Process. **42**(12), 3377–3386 (1994)
14. Burgeth, B., Weickert, J.: An explanation for the logarithmic connection between linear and morphological system theory. Int. J. Comput. Vision **64**, 157–169 (2005)
15. Diop, E.H.S., Angulo, J.: Multiscale image analysis based on robust and adaptive morphological scale-spaces. Image Anal. Stereol. **34**(1), 39–50 (2014)
16. Jackway, P.T., Deriche, M.: Scale-space properties of the multiscale morphological dilation-erosion. IEEE Trans. Pattern Anal. Mach. Intell. **18**(1), 38–51 (1996)
17. Maragos, P.: Slope transforms: theory and application to nonlinear signal processing. IEEE Trans. Signal Process. **43**(4), 864–877 (1995)
18. Maragos, P.: Differential morphology and image processing. IEEE Trans. Image Process. **5**(1), 922–937 (1996)

19. Bogachev, V.I.: Gaussian Measures. American Mathematical Society, Providence (1998)
20. Davies, B.: Heat Kernels and Spectral Theory. Cambridge University Press, Cambridge (1989)
21. Litvinov, G.L., Maslov, V.P., Shpiz, G.B.: Idempotent functional analysis: an algebraic approach. Math. Notes **69**(5–6), 696–729 (2001)
22. Maslov, V.: Méthodes Opératorielles. Editions Mir, Moscow (1987)
23. Del Moral, P.: Maslov Optimization Theory: Optimality versus Randomness. In: Idempotency Analysis and its Applications, Kluwer (1997)
24. Schmidt, M., Weickert, J.: Morphological counterparts of linear shift-invariant scale-spaces. J. Math. Imaging Vis. **56**, 352–366 (2016)

A Novel Approach for Computation of Morphological Operations Using the Number Theoretic Transform

Vivek Sridhar[(✉)] and Michael Breuß

Institute for Mathematics, Brandenburg Technical University,
Platz der Deutschen Einheit 1, 03046 Cottbus, Germany
{sridhviv,breuss}@b-tu.de

Abstract. The fundamental operations of mathematical morphology are dilation and erosion. They are often implemented using a sliding window with the purpose to compute maximum respectively minimum of pixel values within the corresponding mask.

We reformulate the problem of morphological dilation respectively erosion of an image with a non-flat filter as a convolution of their umbras. To this end, we propose to make use of the number theoretic transform to compute the convolution in this setting. In contrast to other possible schemes, this transform represents a completely discrete computational approach. It allows exact convolution of sequences made up of integers. Therefore we propose by the combination of umbra framework and number theoretic transform a well-engineered combination.

There is no restriction on size or shape of the structuring element, and also flat and non-flat filters can be realised.

Keywords: Morphological dilation · Morphological erosion · Number theoretic transform

1 Introduction

Mathematical morphology is a highly successful field in image processing that is concerned with the analysis of shapes and structures in images, see for instance [7–9] for an account of theory and applications. The basic building blocks of many of its processes are dilation and erosion. Since these operations are dual, it is convenient to focus on dilation for the construction of algorithms, as we do in the following. In dilation, a pixel value is set to the maximum of the grey values within a filter mask centred upon it. This mask is called structuring

The current work was supported by the European Regional Development Fund (EFRE 85037495). Furthermore, the authors acknowledge the support by BTU Graduate Research School (STIBET short-term scholarship for international PhD Students sponsored by the German Academic Exchange Service (DAAD) with funds of the German Federal Foreign Office).

E. Baudrier et al. (Eds.): DGMM 2022, LNCS 13493, pp. 182–192, 2022.
https://doi.org/10.1007/978-3-031-19897-7_15

element (SE), and it can be either flat or non-flat [15]. A flat SE describes the shape of the mask over which the dilation is performed, whereas a non-flat SE also contains additive offsets.

A remarkable property of morphological filters is the high efficiency that can be gained in their implementations. Many of these methods may be classified as follows, cf. [18]. A first family of schemes aims to reduce the size of the SE or to decompose it, thus reducing the number of comparison operations that needs to be performed over the SE. In a second family of methods a given image is analysed so that redundant operations that may arise in some image parts could be reduced. However, most of these methods are limited in terms of shape, size or flatness of SE, or specific hardware that is needed, cf. [9, 11–14, 16].

There are just few fast methods that allow SE of arbitrary shape and size. A very popular example is the classic scheme from [17] that relies on histogram updates. However, as also for [17], the algorithmic complexity of most methods relies inherently on size (and often also shape) of the SE. Since the SE is moved over an image in implementations relying on sliding window technique, the computational effort also relates to image size.

An alternative construction of fast algorithms relies on the possibility to formulate operations over an SE as convolutions, which may be realized via a fast transform. In a first work [3], binary dilation respectively erosion are represented by convolution of characteristic functions of underlying sets. In [10] this approach was extended in a straightforward way to grey scale images. This was done by decomposing an image into its level sets, and each level set was processed like a binary image. By construction, the method is limited to flat filters of particular shape. A different extension of [3] has been proposed in [4], making use of an analytical approximation of morphological operations. The resulting method is suitable for flat and non-flat SE, without restriction on shape or size. However, as analyzed in [4,5], this comes at the expense of a shift and smoothing effect in the tonal histogram. In order to address this issue, it has recently been proposed [21] to consider the umbra of image and filter as the computational setting for the convolution. However, in all these works the Fast Fourier Transform (FFT) has been used, while it is evident that a computational method should be considered where the integer based nature of the tonal domain is taken into account.

The *main contribution of this work* is to propose the Number Theoretic Transform (NTT) as the conceptually adequate method for computing discrete convolutions in the umbra setting. In doing this, we extend and refine results of previous work as sketched above, especially of [21]. Let us also note, that by this work we also identify a well-engineered combination of domain and proper tool in terms of umbra setting and NTT, with no limitation with respect to size or shape of structuring elements.

The NTT is a specialized version of the discrete Fourier transform with similar computational scaling properties as the FFT, with the specific property that it enables exact convolution of integer sequences. While standard discrete Fourier transform can also be used to perform such a convolution, it is susceptible to round-off errors due to use of float arithmetic. In contrast, the NTT deals purely with integers that can be exactly represented and processed. As another point of potential interest that could be explored, FFT makes use of 32 bit float data type, whereas NTT may employ just 16 bit integer arithmetic.

2   Basic Definitions

In this section we recall some basic notions of mathematical morphology and discrete convolutions. Upon the latter we also state some observations that help in constructing the proposed method.

Morphological Operations. We consider an N-dimensional grey-value image as a function $f : F \to L$. In accordance, $F \subseteq \mathbb{Z}^N$ is the set of (N-dimensional) indices in the image, which is the **domain** of the image. By L we denote the tonal range of the image, so that in case of 8-bit grey-value imagery, L is the set of integers in the range $[0, 255]$. A grey-value SE, flat or non-flat, can be defined as $b : B \to L$, $B \subseteq \mathbb{Z}^N$. Then the dilation of image f by filter b is denoted by $f \oplus b$ and is computed for each $x \in F$ as

$$(f \oplus b)(x) = \begin{cases} 0 \text{ if } \not\exists y \in B : x - y \in F \\ \max\{f(x - y) + b(y) | x - y \in F \text{ and } y \in B\}, \text{ otherwise} \end{cases} \quad (1)$$

See Fig. 1 for a visual account of dilation and its dual operation erosion.

Discrete Linear Convolution. We begin by describing discrete linear convolution in one dimension (1D), as the multidimensional case is a straightforward extension. Consider two 1D discrete signals $f : F \to \mathbb{R}$ and $g : G \to \mathbb{R}$, where $F, G \subseteq \mathbb{Z}$. Their $f \circledast g$ results in a 1D discrete signal $h : \mathbb{Z} \to R$ as by:

$$h[k] = (f \circledast g)[k] = (g \circledast f)[k] = \sum_{i=-\infty}^{\infty} f[i]g[k - i] , \quad \forall k \in \mathbb{Z} \quad (2)$$

Thereby, f and g are sufficiently *padded* with 0s, i.e., $f[i] = 0$, if $i \notin F$, and $g[i] = 0$, if $i \notin G$.

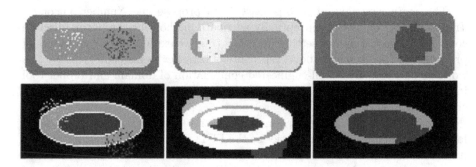

Fig. 1. Left: One *sample image* of size 101×101. **Centre:** Dilation. **Right:** Erosion. Dilation and erosion are realised each with a 5×5 flat filter.

If F and G are finite subsets of $\mathbb{Z}$, we might be interested in $h = (f \circledast g)$ only over a finite subset $H \subseteq \mathbb{Z}$. This subset is determined by the *mode* of convolution [6]. We say that in *full mode* of convolution, the output $h_{full} = (f \circledast_{full} g)$ omits all the elements whose computation only involves *padded* parts of the inputs:

$$h_{full}[k] = (f \circledast_{full} g)[k] = \sum_{i\,:\,i \in F \wedge (k-i) \in G} f[i]g[k-i] \tag{3}$$

Furthermore, in *same mode* of convolution, $h_{same} = (f \circledast_{same} g)$ is the same size as f and is centred with respect to the output of the *full mode* of convolution.

For describing the multidimensional linear discrete convolution, let $\tilde{f} : \tilde{F} \to R$ and $\tilde{g} : \tilde{G} \to R$ be now N-dimensional discrete signals, i.e. $\tilde{F}, \tilde{G} \subseteq \mathbb{Z}^N$. The convolution $(\tilde{f} \circledast \tilde{g}) = \tilde{h} : \mathbb{Z}^N \to \mathbb{R}$ is defined as:

$$\tilde{h}[k_1, k_2, \ldots, k_N] = (\tilde{f} \circledast \tilde{g})[k_1, k_2, \ldots, k_N]$$
$$= \sum_{i_1=-\infty}^{\infty} \sum_{i_2=-\infty}^{\infty} \cdots \sum_{i_N=-\infty}^{\infty} \tilde{f}[i_1, i_2, \ldots, i_N]\tilde{g}[k_1 - i_1, k_2 - i_2, \ldots, k_N - i_N] \tag{4}$$

valid for all $k_1, k_2, \ldots k_N \in \mathbb{Z}$, and where $\tilde{f}$ and $\tilde{g}$ are again sufficiently *padded*.

If $\tilde{F}$ and $\tilde{G}$ are finite and N-dimensional rectangles, i.e. $\tilde{F} = F_1 \times F_2 \times \ldots F_N$, $\tilde{G} = G_1 \times G_2 \times \ldots G_N$, where all the $F_1, \ldots, F_N, G_1, \ldots, G_N \subseteq \mathbb{Z}^N$, the finite domain of interest of output $\tilde{H} \subseteq \mathbb{Z}^N$ is specified by the *mode* of convolution along each dimension, similar to 1-dimensional case.

Important Observations. We now recall some observations on discrete convolution that are important for computations in the umbra setting, see [21] for some more details.

1. Let f and g be 1-dimensional signals defined on $[0, n]$. Then,

   highest non-zero index of $(f \circledast_{full} g)$ =highest non-zero index of $f+$

   highest non-zero index of g

2. Let $f_0, f_1 \ldots, f_m$ be non-negative 1-dimensional signals defined on $[0, n]$, i.e. $f_j[k] \geq 0 \,\forall k \in [0, n], \forall j \in \{0, 1, \ldots, m\}$. Then

   highest non-zero index of $\displaystyle\sum_{j=0}^{m} f_j = \max_{i \in \{0,1,\ldots,m\}} \{\text{highest non-zero index of } f_i\}$

3. Consider $\tilde{f} : \tilde{F} \to R$ and $\tilde{g} : \tilde{G} \to R$, with $\tilde{F}, \tilde{G} \subseteq \mathbb{Z}^N$. Let F and G be *finite*. Then

   $$\tilde{h}[k_1, k_2, \ldots, k_N] = (\tilde{f} \circledast \tilde{g})[k_1, k_2, \ldots, k_N]$$
   $$= \sum_{i_1=-\infty}^{\infty} \left\{ \tilde{f}[i_1, :, :, \ldots :] \circledast \tilde{g}[k_1 - i_1, :, :, \ldots :] \right\}$$

Here, $\tilde{f}[i_1,:,:,\dots\,:]$ and $\tilde{g}[k_1 - i_1,:,:,\dots\,:]$ are $N - 1$ dimensional signals, obtained by fixing the index of the first dimension. Similarly, one can derive

$$\tilde{h}[k_1, k_2, \dots, k_N] = (\tilde{f} \circledast \tilde{g})[k_1, k_2, \dots, k_N]$$

$$= \sum_{i_1=-\infty}^{\infty} \dots \sum_{i_{N-1}=-\infty}^{\infty} \{\tilde{f}[i_1, \dots, i_{N_1},:] \circledast \tilde{g}[k_1 - i_1, \dots, k_{N-1} - i_{N-1},:]\}$$

Here, $\tilde{f}[i_1, \dots, i_{N_1},:]$ and $\tilde{g}[k_1 - i_1, \dots, k_{N-1} - i_{N-1},:]$ are 1-dimensional signals, with index of every dimension except the last fixed.

The above observations allows us to relate dilation of scalars to convolution of polynomials. Products of polynomials are essentially full convolutions of coefficient vectors. Thus, we have reduced N-dimensional dilations to $(N + 1)$-dimensional convolutions. We can use one of the many available transform with convolution property to speed up the computations.

2.1   Number Theoretic Transform

Transforms on a finite integer ring with interesting computational properties for realizing convolutions were first constructed in [22]. Especially, finite transforms in rings of integers modulo Fermat numbers, may be of interest for application in digital signal processing, see e.g. [24]. We briefly recall the Fermat Number Transform [23], which is the version of NTT we consider in this work, in a single dimension. Similar to FFT, multi-dimensional NTT can be obtained by successively taking NTT along each dimension.

Fermat Numbers are prime numbers of form $p_f = 2^{2^m} + 1$, for some $m \in \mathbb{N}$. There are only five known Fermat primes, i.e., 3, 5, 17, 257 and 65537. We can see that 2 is a primitive root of integer rings modulo a Fermat prime. Thus, we have, $2^{(p_f-1)} \equiv 1 \pmod{p_f}$ and moreover, $p_f - 1 = 2^{2^m}$. This allows defining the Radix-2 algorithm for Fermat Number Transform (FNT), similar to Cooley-Tukey FFT.

Let $x[n]$, $n = 0, 1, 2, \dots n_0 - 1$ be a 1-dimensional discrete signal of length n_0 ($n_0 < p_f$). Then the FNT of $x[n]$, in integer ring modulo Fermat prime p_f and the inverse of FNT is given by Algorithm 1.

Clearly, like any Radix-2 algorithm, we require the signal length be a power of 2. This is in practice achieved by padding. We also require the overall length of the signal to be less than p_f, see [22].

In FNT, the input is a discrete signal x of length n_0. As indicated, the signal is padded with 0s so that $n_0 = 2^q$, for some $q \in \mathbb{N}$. Since we are operating in integer ring modulo p_f, the number n_0 must be less than p_f and $x[n] < p_f$, $x[n] \in \mathbb{N}$, $\forall n \in \{0, 1, \dots, n_0 - 1\}$.

The Fermat number transform of x is obtained by calling the recursive function $\mathrm{FNT}(x, n_0, 2, p_f)$. The base case, for $n_0 = 1$, and the recursive calls are similar to Radix-2 FFT. The difference arises in how the two sub-parts, y_0 and y_1, are combined to give the result, i.e. the transformed signal y.

Algorithm 1. Radix-2 Fermat Number Transform

```
   /* Inputs are the signal x[.], the signal length n0, the primitive
      root ω and the Fermat prime pf                                    */
   /* To obtain FNT of x, we call, FNT (x,x.length, 2, pf).             */
 1 Function FNT(x, n0, ω , pf):
 2 │   if n0 == 1 then
 3 │   └─  return x
 4 │   x0 = {x[0], x[2], ... x[n − 2]}
 5 │   x1 = {x[1], x[3], ... x[n − 1]}
 6 │   y0 = FNT(x0, n0/2, ω², pf)
 7 │   y1 = FNT(x1, n0/2, ω², pf)
 8 │   α = 1
 9 │   for k = 0 to n0/2 − 1 do
   │   │   /* all calculations are in integer ring modulo pf            */
10 │   │   y[k] = y0[k] + y1[k]
11 │   │   y[k + n/2] = y0[k] − αy1[k]
12 │   └─  α = αω
13 │   return y
   │                                          // y is a signal of length n0
   └─
   /* To obtain inverse FNT of x, we call,
      inv_FNT (x,x.length, 2, pf).                                      */
14 Function inv_FNT(x, n0, ω , pf):
15 │   inv_ω ← multiplicative inverse of ω modulo pf
16 │   inv_n0 ← multiplicative inverse of n0 modulo pf
17 │   y = FNT(x0, n0, inv_ω, pf)
18 │   return inv_n0 y
   │       // multiplying, modulo pf, each element of signal y by inv_n0
   └─
```

In line 9 to 12 of Algorithm 1, all the computations are performed in the integer ring. The computations are exact, since there is no use of floating point data types or approximations as in FFT. We only require bit-shifts, addition and subtractions modulo p_f for the computations. We do not require multiplication [20]. The number 2 is a primitive root of integer ring modulo p_f. Therefore, for any element γ in the ring, we can write $\gamma \equiv 2^m \pmod{p_f}$, for some $m \in \{0, 1, \ldots p_f - 1\}$. Therefore, products of two elements of integer ring are essentially products of two powers of 2, modulo p_f. This can be achieved with the bit-shifts and modulo operations.

The computation of inverse of Fermat Number Transform is then supposed to be self-explanatory, as given in Algorithm 1.

For experiments in this paper, we do all the computation, for FNT, in integer ring modulo 65537. For a discrete signal of length n_0, the time complexity of FNT, similar to FFT, is $\mathcal{O}(n_0 \log n_0)$.

Let us note that, an extension of FNT without limitation on the length of the signals is proposed in [20]. But, we do not require it here.

3 Proposed Method

Let us first briefly outline the proposed method. The method takes into account the finite nature of tonal range of image and filter. The novel approach may be sketched as follows. First the N-dimensional image ($N = 2$ for standard imagery) is cast to $(N+1)$ dimensions adding the tonal range, exploring thus the umbra of the image; see [15] for more details on the latter notion. There we employ the NTT [1] to compute discrete linear convolution, and afterwards we take back the projection to N-D. This procedure induces no restrictions on flatness, shape or size of the filter. By the properties of the NTT as a dedicated variant of the discrete Fourier transform, it is also asymptotically faster than the classic fast method from [17].

Let us introduce appropriate notations. For this we largely follow the notation employed in [21]. We consider an N-dimensional grey-value image $f : F \rightarrow L$ and SE (flat or non-flat) $b : B \rightarrow L$, $F, B \subseteq \mathbb{Z}^N$. We construct $(N + 1)$-dimensional arrays, f_{Um} and b_{Um}, for f and b respectively, in a way that we have a 1-dimensional vector corresponding to each pixel, and the highest non-zero index of the vector equals the pixel value. Then as pointed out, convolution in the umbra setting follows, and the dilated image $(f \oplus b)$ can be obtained by appropriately projecting the $(N + 1)$-dimensional array $(f \oplus b)_{Um}$ on N-dimensions. A description of the steps one needs to perform in our method is given below, see [21] for presentation of a detailed example for an analogous construction of the umbra domain.

Step 1. Let $l_R = \max_{x \in F}\{f(x)\} + \max_{x \in B}\{b(x)\}$. First construct two $(N+1)$-dimensional arrays f_{Um} and b_{Um}. The first N dimensions, referred to as the *domain dimensions*, of f_{Um} and b_{Um} consists of all (N-dimensional) indices of F and B respectively. The last dimension, referred to as the *range dimension*, consists of indices $\{0, 1, \ldots l_R\}$. The arrays f_{Um} and b_{Um} are determined by the following two equations:

$$f_{Um}(x, y) = \begin{cases} 1 & \text{if } x \in F \text{ and } f(x) = y \\ 0 & \text{otherwise.} \end{cases} \tag{5}$$

$$b_{Um}(x, y) = \begin{cases} 1 & \text{if } x \in B \text{ and } b(x) = y \\ 0 & \text{otherwise.} \end{cases} \tag{6}$$

Note that the above construction makes it possible to have image and filter of any shape in the domain. The constructed arrays f_{Um} and b_{Um} will always be in the format of an $(N+1)$-dimensional hyper-rectangle, regardless the shape of image domain F and filter domain B.

Step 2. We compute $(f \oplus b)_{Um}$ by taking the linear convolution of f_{Um} and b_{Um} by using *same mode* on the *domain dimensions* and *full mode* on the *range dimension*. This step is realised by the NTT algorithm.

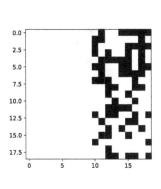
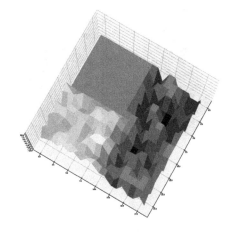

Fig. 2. Left: Base of the non-flat filter. **Right:** Umbra of the non-flat filter.

Step 3. $(f \oplus b)$ is determined from $(f \oplus b)_{Um}$ for each $x \in F$ by making use of:

$$(f \odot b)(x) = \begin{cases} \max\{y|(f \oplus b)_{Um}(x,y) \geq 1\} & \text{if } x \in F \oplus B \\ 0 & \text{otherwise.} \end{cases} \quad (7)$$

where $x \notin F \oplus B = \{x|x - x_b \in F \text{ for some } x_b \in B\} \Leftrightarrow \nexists y : (f \oplus b)_{Um}(x,y) \geq 1$. Thus $(f \oplus b)$ computed by (7) is the same as in (1).

4 Experiments

To demonstrate the viability of the proposed method, we perform dilation on the *Cameraman* image of size 490×487, by a non-flat filtering SE of size 19×19. The filter is generated using random.randint() of NumPy package [2], with range of values from 0 to 255. The base of the SE and its umbra are given in Fig. 2. We assume that all the pixels outside the image boundary are at $-\infty$, so that they do not contribute to the max evaluation.

We compare classical dilation with the method proposed in this paper. The classical dilation is computed pixel by pixel by using the Formula (1).

The workings of the proposed method is as explained in Sect. 3. We implement 1-dimensional FNT and inv_FNT, modulo the Fermat prime 65537, as described in algorithm 1. To compute the 3-dimensional FNT, we consecutively take 1-dimensional FNT along each dimension. The 3-dimensional inv_FNT is implemented by taking 3-dimensional inv_FNT along each dimension in the reverse order.

The proposed method gives exact result in our experiment and this can be verified from comparing the Figs. 3 **Centre** and **Right** in top row. We also see that the histogram of classical dilation exactly coincides with that of the proposed method, see Fig. 4, as expected.

Fig. 3. Top row. Left: *Cameraman* of size 490 × 487. **Centre:** Classical Dilation with non-flat filter. **Right:** Proposed method using NTT. **Second row.** Umbras of original image and dilated image.

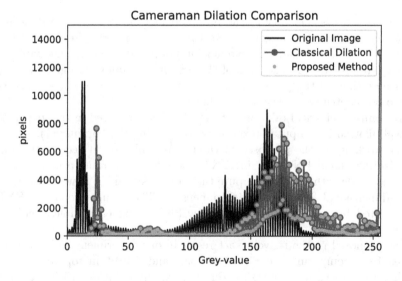

Fig. 4. Histogram of images.

5   Conclusion

We have proposed by combining umbra domain and number theoretic transform NTT a well-engineered framework and method to compute morphological dilation and erosion. By the algorithmic properties of the NTT, we have thus proposed a method to compute these elementary morphological processes over an image of size n_i, with any arbitrary non-flat filter of size $n_f \leq n_i$, in $\mathcal{O}(n_i \log n_i)$ operations.

This article represents from our point of view an important step in our work on fast transform methods for mathematical morphology. The general framework of number theoretic transforms, which appears to match the natural requirements on fast transforms in this field, may be explored with further benefit for defining dedicated implementations in future work. Let us note in this context, that number theoretic transforms represent by themselves a highly developing field, with many advances still possible.

Acknowledgements. The current work was supported by the European Regional Development Fund (EFRE 85037495). Furthermore, the authors acknowledge the support by BTU Graduate Research School (STIBET short-term scholarship for international PhD Students sponsored by the German Academic Exchange Service (DAAD) with funds of the German Federal Foreign Office).

References

1. Agarwal, R.C., Burrus, C.S.: Number theoretic transforms to implement fast digital convolution Proc. IEEE **63**(4), 550–560 (1975)
2. Harris, C.R., et al.: Array programming with NumPy. Nature **585**(7825), 357–362 (2020)
3. Tuzikov, A.V., Margolin, G.L., Grenov, A.I.: Convex set symmetry measurement via Minkowski addition. J. Math. Imaging Vis. **7**(1), 53–68 (1997)
4. Kahra, M., Sridhar, V., Breuß, M.: Fast morphological dilation and erosion for grey scale images using the Fourier transform. In: Elmoataz, A., Fadili, J., Quéau, Y., Rabin, J., Simon, L. (eds.) SSVM 2021. LNCS, vol. 12679, pp. 65–77. Springer, Cham (2021). https://doi.org/10.1007/978-3-030-75549-2_6
5. Sridhar, V., Breuss, M., Kahra, M.: Fast approximation of color morphology. In: Bebis, G., et al. (eds.) ISVC 2021. LNCS, vol. 13018, pp. 488–499. Springer, Cham (2021). https://doi.org/10.1007/978-3-030-90436-4_39
6. SciPy Documentation. https://docs.scipy.org/doc/scipy/reference/generated/scipy.signal.convolve.html. Accessed 2 Feb 2021
7. Serra, J., Soille, P. (eds.): Mathematical Morphology and its Applications to Image Processing, vol. 2. Springer, Cham (2012). https://doi.org/10.1007/978-94-011-1040-2
8. Najman, L., Talbot, H. (eds.): Mathematical Morphology: From Theory to Applications. Wiley, Hoboken (2013)
9. Roerdink, J.B.T.M.: Mathematical morphology in computer graphics, scientific visualization and visual exploration. In: Soille, P., Pesaresi, M., Ouzounis, G.K. (eds.) ISMM 2011. LNCS, vol. 6671, pp. 367–380. Springer, Heidelberg (2011). https://doi.org/10.1007/978-3-642-21569-8_32

10. Kukal, J., Majerová, D. Procházka, A.: Dilation and erosion of gray images with spherical masks. In: Proceedings of the 15th Annual Conference Technical Computing (2007)
11. Déforges, O., Normand, N., Babel, M.: Fast recursive grayscale morphology operators: from the algorithm to the pipeline architecture. J Real-Time Image Process. 8(2), 143–152 (2013)
12. Moreaud, M. Itthirad, F.: Fast algorithm for dilation and erosion using arbitrary flat structuring element: improvement of Urbach and Wilkinson's algorithm to GPU computing. In: 2014 International Conference on Multimedia Computing and Systems (ICMCS), pp. 289–294. IEEE (2014)
13. Lin, X. Xu, Z., A fast algorithm for erosion and dilation in mathematical morphology. In: 2009 WRI World Congress on Software Engineering, vol. 2, pp. 185–188. IEEE (2009)
14. Van Herk, M.: A fast algorithm for local minimum and maximum filters on rectangular and octagonal kernels. Pattern Recogn. Lett. 13(7), 517–521 (1992)
15. Haralick, R.M., Sternberg, S.R., Zhuang, X.: Image analysis using mathematical morphology. IEEE Trans. Pattern Anal. Mach. Intell. 4, 532–550 (1987)
16. Thurley, M.J., Danell, V.: Fast morphological image processing open-source extensions for GPU processing with CUDA. IEEE J. Sel. Top. Sig. Process. 6(7), 849–855 (2012)
17. Van Droogenbroeck, M., Talbot, H.: Fast computation of morphological operations with arbitrary structuring elements. Pattern Recogn. Lett. 17(14), 1451–1460 (1996)
18. Van Droogenbroeck, M., Buckley, M.J.: Morphological erosions and openings: fast algorithms based on anchors. J. Math. Imaging Vis. 22(2), 121–142 (2005)
19. Jones, R.: Connected filtering and segmentation using component trees. Comput. Vis. Image Underst. 75(3), 215–228 (1999)
20. Agarwal, R.C., Burrus, C.: Fast convolution using Fermat number transforms with applications to digital filtering. IEEE Trans. Acoust. Speech Sig. Process. 22(2), 87–97 (1974)
21. Sridhar, V. Breuß, M.: An exact fast Fourier method for morphological dilation and erosion using the umbra technique. In: 2022 19th Conference on Robots and Vision (CRV) (2022)
22. Pollard, J.M.: The fast Fourier transform in a finite field. Math. Comput. 25(114), 365–374 (1971)
23. Rader, C.M., The number theoretic DFT and exact discrete convolution. In: IEEE Arden House Workshop on Digital Signal Processing. Harriman NY (1972)
24. Krizek, M., Luca, F., Somer, L.: 17 Lectures on Fermat Numbers: From Number Theory To Geometry. Springer, Cham (2002). https://doi.org/10.1007/978-0-387-21850-2

Equivariance-Based Analysis of PDE Evolutions Related to Multivariate Medians

Martin Welk[(✉)]

Institute of Biomedical Image Analysis, UMIT TIROL – Private University
of Health Sciences, Medical Informatics and Technology,
Eduard-Wallnöfer-Zentrum 1, 6060 Hall/Tyrol, Austria
martin.welk@umit-tirol.at

Abstract. For multivariate data there exist several concepts generalising the median, which differ by their equivariance properties w.r.t. transformations of the data space (e.g. Euclidean, affine). In earlier work on the asymptotic analysis of multivariate image filters built upon these concepts, it was observed that several affine equivariant median filters approximate the same system of partial differential equations (PDEs). In this paper we discuss the equivariance properties of multivariate medians and their associated PDEs in more detail. We discuss what equivariance concept is the preferable generalisation of the very strong equivariance of the scalar-valued median (sometimes also denoted as morphological equivariance) w.r.t. arbitrary monotone transformations. Moreover, we derive multivariate PDE evolutions systematically from equivariance properties. It turns out that the approximation of the same PDE system by different affine equivariant medians is no coincidence but a necessary implication of their equivariance properties. As a by-product, a more general class of multivariate PDE evolutions with favourable equivariance properties arises.

Keywords: Multivariate images · Partial differential equations · Multivariate median · Affine equivariance · Morphological filters

1 Introduction

Curvature motion can be described in numerous ways. First of all, it is a curve evolution that can be used for contours or shapes, and which can be stated e.g. by a partial differential equation (PDE) for parametrised curves, and which is a gradient descent for the curve length functional [1]. In a grey-value image, its simultaneous application to all level lines gives rise to an image evolution which can be described by a PDE acting directly on the intensities [9]. As such, it can be used for structure-preserving image simplification. Moreover, it is closely related to median filtering: As proven in [4], space-continuous median filtering with a disc-shaped structuring element of radius ϱ asymptotically approximates

© Springer Nature Switzerland AG 2022
É. Baudrier et al. (Eds.): DGMM 2022, LNCS 13493, pp. 193–205, 2022.
https://doi.org/10.1007/978-3-031-19897-7_16

curvature motion up to evolution time $\varrho^2/6$. Remarkably, both curvature motion and the median filter are equivariant under arbitrary monotone intensity rescalings, i.e. their application commutes with such rescalings. As they share this strong property with a larger set of fundamental morphological operations, this property is also often called morphological equivariance (or invariance).

In earlier work, several steps have been undertaken to generalise this framework to multivariate (such as colour) images. Regarding median filtering, this requires a generalisation of the median concept to multivariate data for which different proposals have been made in literature since the beginnings of the 20th century, see e.g. [3,7,8,10,13] and the overview in [11]. Among the differences between these definitions, equivariance properties with regard to transformations of the data space play an important role as they are decisive for the applicability of the concepts to particular categories of data. Applications for the median filtering of multivariate images can be found e.g. in [6,12,19–21].

Asymptotic PDE approximation results for bivariate and multivariate median filtering have been presented in [14,16,17], see also extensions to adaptive median filtering with morphological amoebas as structuring elements [15]. A remarkable observation was that affine equivariant multivariate medians, despite not coinciding as such, consistently led to the same PDE evolutions, which suggests that common underlying principles of the PDE evolutions themselves related to equivariance can be worth considering. This is the purpose of the present contribution.

Our Contribution. We start by discussing the equivariance properties of multivariate medians and their associated PDEs. Referring to the morphological equivariance of the scalar-valued median and curvature motion PDE, we also address the question what is the best multivariate generalisation of that concept.

We then turn to derive multivariate PDE evolutions in a principled way from equivariance properties modelled after multivariate median concepts. In fact, the asymptotic approximation of the same PDE evolution by an entire class of affine equivariant multivariate median filters turns out to be necessary rather than just coincidential. For Euclidean equivariance, partial results are obtained. Considering slightly relaxed requirements, we find a more general class of multivariate PDE evolutions which deserve further study.

Structure of the Paper. The remainder of the paper is organised as follows. In Sect. 2 we recall multivariate median concepts from literature. We collect known facts about their equivariance properties. At the end of the section, we discuss what is the proper counterpart of morphological equivariance in the case of multivariate data. Section 3 lists existing results on the asymptotic approximation of PDEs by multivariate median filters, emphasising the role of equivariance properties in their derivation. In Sect. 4 we present the systematic direct derivation of bivariate image filtering PDEs from equivariance properties, culminating in a re-derivation of the PDE system associated with affine equivariant multivariate medians. Section 5 illustrates the theoretical findings by numerical examples of PDE evolutions. A short summary in Sect. 6 ends the paper.

2 Medians and Equivariance

In the following we shortly recall some definitions of multivariate medians and discuss their equivariance properties. Throughout this section we assume that $\mathcal{X}$ is a finite multiset of values $\boldsymbol{x} \in \mathbb{R}^d$, $d \geq 2$.

2.1 Multivariate Medians

The $\boldsymbol{L^1}$ **median** of $\mathcal{X}$ is defined as the point $\boldsymbol{\mu} \in \mathbb{R}^d$ that minimises the sum of Euclidean distances $|\boldsymbol{\mu} - \boldsymbol{x}|$ to all given points $\boldsymbol{x} \in \mathcal{X}$. Having been introduced in 1909 [13], this is the most widespread concept of multivariate median which has been intensively studied since and has also been used in image processing [6, 12, 19–21]. We remark that in exceptional situations (namely, if all data points are collinear, and $\mathcal{X}$ has even cardinality), the L^1 median is non-unique (set-valued) but do not detail this further as it is generally not relevant for our further investigation. Also the following multivariate medians can be set-valued in certain configurations which we will not detail further.

Oja's simplex median [7] instead defines the median as the $\boldsymbol{\mu} \in \mathbb{R}^d$ that minimises the sum of simplex volumes $|[\boldsymbol{\mu}, \boldsymbol{x}_1, \dots, \boldsymbol{x}_d]|$ for all d-tuples of data points $\boldsymbol{x}_1, \dots, \boldsymbol{x}_d \in \mathcal{X}$. Especially in the bivariate case $d = 2$ this means to minimise a sum of triangle areas. Note that we denote by $[\dots]$ the oriented simplex volume.

To avoid the high computational expense of the Oja median caused by the combinatorial complexity of its definition, [8] proposed the **transformation-retransformation $\boldsymbol{L^1}$ median (TR-$\boldsymbol{L^1}$ median)**, see also [5]. This median is computed by first applying an affine transform T to $\mathcal{X}$ to normalise the data points such that their covariance matrix becomes the $d \times d$ identity matrix, then applying the L^1 median and then using the inverse transform T^{-1} to yield the final median $\boldsymbol{\mu} \in \mathbb{R}^d$. If all $\boldsymbol{x} \in \mathcal{X}$ lie in a common affine subspace of $\mathbb{R}^d$, special consideration is needed such as applying the procedure in the subspace only.

The **half-space median** [10] is the point $\boldsymbol{\mu} \in \mathcal{X}$ of maximal half-space depth w.r.t. $\mathcal{X}$. Here, the half-space depth is the minimum over all hyperplanes $H \ni \boldsymbol{\mu}$ of the number of points $\boldsymbol{x} \in \mathcal{X}$ that lie on one side of H. Parametrising hyperplanes with unit normal vectors $\boldsymbol{n} \perp H$ this can be expressed as $\boldsymbol{\mu} = \underset{\boldsymbol{\mu} \in \mathbb{R}^d}{\operatorname{argmax}} \underset{\boldsymbol{n} \in \mathbb{R}^d, |\boldsymbol{n}|=1}{\min} \sum_{\boldsymbol{x} \in \mathcal{X}} \operatorname{sgn}\langle \boldsymbol{x} - \boldsymbol{\mu}, \boldsymbol{n} \rangle$. Clearly, for a given multiset $\mathcal{X}$ the half-space depth cannot exceed $\lfloor (\#\mathcal{X} - 1)/2 \rfloor$ where $\#$ symbolises cardinality but this value is not always realised.

As the last multivariate median concept, we mention the **convex-hull-stripping median** [3]. It is obtained by an iterative process: Starting with $\mathcal{X}_0 := \mathcal{X}$, one obtains $\mathcal{X}_{i+1}$ from $\mathcal{X}_i$, $i = 0, 1, 2, \dots$, by removing all points $\boldsymbol{x}$ that lie on the boundary of the convex hull of $\mathcal{X}_i$. This is repeated until one finds i with $\mathcal{X}_i \neq \varnothing = \mathcal{X}_{i+1}$. Each point $\boldsymbol{\mu}$ in the convex hull of $\mathcal{X}_i$ then is a convex-hull-stripping median of the initial $\mathcal{X}$.

2.2   Equivariance

Equivariance essentially describes commutativity between some operator acting in a suitable space and transformations of this space. In the following we will distinguish whether the set of admissible transformations is independent of the actual data set or not, and speak of absolute or relative equivariance, respectively.

Basic Definitions. Denote by S a suitable data space (e.g. $\mathbb{R}^d$). An operator φ mapping multisets $\mathcal{X}$ of values $x \in S$ to single values $\varphi(\mathcal{X}) \in S$ is called **absolutely equivariant** w.r.t. a set $\mathcal{T}$ of transformations $T : S \to S$ if for any multiset $\mathcal{X} \subset S$ and any transformation $T \in \mathcal{T}$ one has $\varphi(T\mathcal{X}) = T\varphi(\mathcal{X})$. Note that $T\mathcal{X}$ here denotes simultaneous application of T to all elements of $\mathcal{X}$.

An operator φ as stated before is called **relatively equivariant** w.r.t. $\mathcal{T}$ if $\mathcal{T}$ is a set-valued operator that assigns to each multiset $\mathcal{X} \subset S$ a set $\mathcal{T}(\mathcal{X})$ of transformations $T : S' \to S$ where $\mathcal{X} \cup \{\varphi(\mathcal{X})\} \subseteq S' \subseteq S$ such that for any $T \in \mathcal{T}(\mathcal{X})$ one has $\varphi(T\mathcal{X}) = T\varphi(\mathcal{X})$.

Equivariance is in fact a decisive feature when it comes to the application of filtering operators to given data. For example, application of an operator that possesses only Euclidean equivariance to data which do not have a meaningful Euclidean structure is dangerous as it implicitly imposes a random Euclidean structure on these data, and uses it to draw conclusions. This difficulty has in fact been a driving force behind the development of different multivariate median concepts in statistical literature.

The following equivariance properties of univariate and multivariate medians are largely known from the literature, see in particular [11] and the references therein.

Univariate Median. The classic median possesses two strong equivariance properties that together form the essence of its outstanding role as a robust central position measure. First, it is equivariant under (the set of all) strictly monotonically increasing functions $T : \mathbb{R} \to \mathbb{R}$. This is the **morphological equivariance** mentioned in Sect. 1, which is obviously an absolute equivariance property. Second, there is the **radial scaling equivariance:** Given a finite multiset $\mathcal{X} \subset \mathbb{R}$ with median μ, the median is unchanged if each $x \in \mathcal{X}$ is replaced with some $\mu + c(x - \mu)$ where the factors $c > 0$ can even be chosen independently for each x. As the set of admissible transformations obviously depends on $\mathcal{X}$, namely, of its median, this is a relative equivariance. Finally, the univariate median is equivariant under reflections. Formally, this is also an absolute equivariance property which we will shortly refer to as **centrality.**

Centrality, understood as equivariance under reflections at arbitrary hyperplanes, is shared by all multivariate medians under discussion (intuitively, it is crucial for calling an operator a median, or more generally a mean). Regarding other equivariances, the multivariate medians vary, so we will shortly discuss each of them.

L^1 **Median.** This median is much more restrictive in terms of absolute equivariance. It is equivariant under similarity transforms, i.e. under Euclidean transformations and global rescalings. On the other hand, it fully implements radial scaling equivariance as a relative equivariance property.

Oja and TR-L^1 medians. Both medians are absolutely equivariant under arbitrary affine transformations of the data space. Radial scaling equivariance holds for the Oja median in configurations where it is uniquely defined. Unfortunately, radial scaling equivariance is not preserved for the TR-L^1 median. We remark instead the following asymptotic property: Let a data multiset $\mathcal{X} = \{x_1, \ldots, x_N\}$ be given. If radial rescaling weights c_i for the data x_i are chosen as $c_i = 1 + \varepsilon C_i$ with fixed C_i for the individual points and a global variable parameter ε, then for $\varepsilon \to 0$ the TR-L^1 median of the multiset $\mathcal{X}' = \{c_1 x_1, \ldots, c_N x_N\}$ deviates from that of $\mathcal{X}$ by $\mathcal{O}(\varepsilon)$.

Half-Space Median. The half-space median shares with the previously mentioned two concepts the absolute equivariance under affine transformations. Moreover, as it depends only on the situation of points relative to straight lines, i.e. whether some point is located on the one or other side of that straight line, one can establish equivariance w.r.t. a somewhat larger set of global transformations, namely all projective transforms of the projective space $\mathrm{P}^d \supset \mathbb{R}^d$ that do not take any point from the convex hull of the data multiset $\mathcal{X}$ to infinity. As the set of admissible transformations depends on $\mathcal{X}$, this is a relative equivariance; we will refer to it as **restricted projective equivariance.** Note that for sequences $(\mathcal{X}_1, \mathcal{X}_2, \ldots)$ increasing beyond limits, i.e. with convex hulls $\mathrm{conv}(\mathcal{X}_i)$ that fulfil $\bigcup_{i=1}^{\infty} \mathrm{conv}(\mathcal{X}_i) = \mathbb{R}^d$, the corresponding sets $\mathcal{T}_i$ of admissible projective transforms converge to the set $\mathcal{T}^*$ of affine transforms, $\mathcal{T}_1 \supset \mathcal{T}_2 \supset \ldots$ with $\bigcap_{i=1}^{\infty} \mathcal{T}_i = \mathcal{T}^*$ because affine transforms are the only projective transforms that take no finite point to infinity.

Radial scaling equivariance does in general not hold for the half-space median; however, it is valid for those data multisets $\mathcal{X}$ for which the half-space median attains the maximum possible half-space depth $\lfloor (\#\mathcal{X} - 1)/2 \rfloor$.

Convex-Hull-Stripping Median. The equivariance properties of the convex-hull-stripping median resemble those of the half-space median as it possesses the same absolute affine equivariance and relative restricted projective equivariance. Radial scaling equivariance does not hold.

Generalisation of Morphological Equivariance. Looking back at the equivariance of the univariate median (and many morphological operators) under arbitrary monotone transformations of $\mathbb{R}$, the question arises what is the best counterpart one can establish for this in the multivariate case. For a tentative answer to this question, one can interpret increasing monotone transformations of $\mathbb{R}$ as orientation-preserving maps: they do not change the orientation of intervals, i.e. one-dimensional simplices $[x, y]$. Generalising this to the multivariate case, one is naturally led to consider transformations T of $\mathbb{R}^d$ that preserve the orientation of d-dimensional simplices $[x_0, \ldots, x_d]$. This boils down to requiring that the situation of any point in $\mathbb{R}^d$ relative to any hyperplane must not change.

Postulating this for all points in $\mathbb{R}^d$, one obtains affine equivariance. Alternatively, restricting the requirement to the convex hull of a given data multiset, one obtains again the restricted projective equivariance.

We suggest therefore to consider restricted projective equivariance as multivariate morphological equivariance.

3   Space-Continuous Analysis

All definitions from Subsect. 2.1 can be directly applied within median filtering procedures for discrete multivariate images. To study PDE limits, however, they need to be transferred to space-continuous multivariate images represented by smooth functions $u : \mathbb{R}^2 \to \mathbb{R}^d$. The selection of values around a given location $x \in \mathbb{R}^2$ is then accomplished using a compact neighbourhood of x as structuring element, and yields a density of intensities $\gamma : \mathbb{R}^d \to \mathbb{R}_0^+$ where $\mathbb{R}_0^+$ denotes the set of nonnegative real numbers. Given the smoothness of u, γ has compact support and is absolutely integrable; it may be normalised to total weight 1.

Medians of Multivariate Densities. With the exception of the convex-hull-stripping median, the multivariate median concepts can easily be transferred to the case of compactly supported absolutely integrable densities γ, essentially by replacing sums with integrals, see [17]. The convex-hull-stripping median is more difficult to transfer; as shown in [18] the iterative process turns into a shape evolution process similar to the affine morphological scale space [2]. As an asymptotic analysis of the final point of this shape evolution has not been accomplished so far, we leave this median concept aside for the further discussion in this subsection.

Limiting Process. Modelled after [4], we consider disc-shaped structuring elements $D_\varrho(x)$ of radius ϱ centered at x for the filtering of multivariate images u as specified above. The multivariate median of the density γ of image values within $D_\varrho(x)$ then is the value of the median-filtered image $M_\varrho u$. Similar to [4], one obtains results of the type $\lim_{\varrho \to 0} \frac{M_\varrho u(x) - u(x)}{\varrho^2/6} = Lu$ with some (spatial) differential operator L which justify to consider the time evolution PDE $u_t = Lu$ as the asymptotic evolution for the respective multivariate median filter.

Equivariant Normalisation. In [14,16] asymptotic evolutions of multivariate median filters were derived. In doing so, it was helpful to exploit the Euclidean and affine equivariance, respectively, of the underlying median operators in the data space as well as the Euclidean equivariance in the image plane contributed by the structuring element D_ϱ to normalise the function u around the location x.

In the bivariate case ($d = 2$) this is done as follows: First, translations in the image plane and data space are used to shift x and $u(x)$ to 0. Next, rotations around 0 in the image plane and data space are applied to make the Jacobian $Du(0)$ diagonal and positive semidefinite (at generic locations: positive definite). We call the normalisation up to this step **Euclidean normalisation**. Furthermore, if the median under consideration admits affine equivariance, an affine transform in the data space can be used to rescale the data (at non-degenerate locations) such that the Jacobian becomes the unit matrix. We refer to this as **affine normalisation**. Note that this is the continuous counterpart of the normalisation by the covariance matrix in the definition of the TR-L^1 median.

In the case $d > 2$, essentially the same kind of normalisation can be applied; however, the Jacobian is now a $d \times 2$-matrix and will be transformed in a way that its third and further rows are zero, and the 2×2-submatrix consisting of the first two rows satisfies the requirements (diagonal, positive semidefinite, positive definite, unit matrix) as specified before.

In the following we state the approximation results from [14,16] for the normalised cases; the general equations are obtained from these by applying the respective inverse transforms to the PDE $\boldsymbol{u}_t = L\boldsymbol{u}$. For brevity we focus on the bivariate case ($d = 2$).

Normalised PDE Approximations of Multivariate Median Filtering. In [14] it was shown that bivariate L^1 median filtering of an image $\boldsymbol{u} : \mathbb{R}^2 \to \mathbb{R}^2$, $(x, y)^\mathrm{T} \mapsto (u, v)^\mathrm{T}$ in Euclidean normalisation approximates the PDE system

$$u_t = Q(u_x/v_y)u_{xx} + (1 - Q(u_x/v_y))u_{yy} - 2(u_x/v_y)Q(u_x/v_y)v_{xy}$$
$$v_t = (1 - Q(v_y/u_x))v_{xx} + Q(v_y/u_x)v_{yy} - 2(v_y/u_x)Q(v_y/u_x)u_{xy} \qquad (1)$$

with a coefficient function Q that can be stated in terms of elliptic integrals. Specialising to the affine normalised situation, one has $Q(1) = 1/4$, thus

$$u_t = \tfrac{1}{4}u_{xx} + \tfrac{3}{4}u_{yy} - \tfrac{1}{2}v_{xy} , \qquad\qquad v_t = \tfrac{3}{4}v_{xx} + \tfrac{1}{4}v_{yy} - \tfrac{1}{2}u_{xy} , \qquad (2)$$

which is the PDE system for the Oja and TR-L^1 median filters in affine normalisation. In [16] it was proven that bivariate half-space median filtering approximates the same PDE system. For the L^1, Oja and TR-L^1 median filters also trivariate versions of these PDE systems are found in [14].

Equivariance. The definitions of absolute and relative equivariance translate straightforward to the case of PDEs evolutions. As can be expected, the PDE evolutions for L^1 median filtering such as (1) are equivariant under similarity transformations of the data space; the PDE evolutions such as (2) for the other medians are affine equivariant. Moreover, the fact that (2) also corresponds to the half-space median lets expect restricted projective equivariance which indeed holds. Remarkably, the just affine equivariance of the Oja and TR-L^1 medians is upgraded to restricted projective equivariance in the asymptotic limit.

4   Derivation of PDE Evolutions by Equivariance

In this section we turn around to derive bivariate image filtering PDEs from equivariance properties modelled after median filters. We start by assuming that $\boldsymbol{u}$ is a smooth bivariate image evolution which is described by some PDE system $\boldsymbol{u}_t = L\boldsymbol{u}$. To restrict the PDE system, we impose conditions one by one, modelled after the properties of median operators.

(I) Translation Equivariance. This allows us to shift the location of interest to $\mathbf{0}$ with $\boldsymbol{u}(\mathbf{0}) = \mathbf{0}$. We write down the spatial Taylor expansion up to second order at a non-degenerate location $\mathbf{0}$. Suppressing for the moment the time parameter, and considering only $\boldsymbol{x} = (x,y)^{\mathrm{T}} \in D_\varrho(\mathbf{0})$, we have

$$
\begin{aligned}
u(x,y) &= \alpha_1 x + \alpha_2 y + \tfrac{1}{2}\beta x^2 + \gamma xy + \tfrac{1}{2}\delta y^2 + \mathcal{O}(\varrho^3) \,, \\
v(x,y) &= \alpha_1' x + \alpha_2' y + \tfrac{1}{2}\beta' x^2 + \gamma' xy + \tfrac{1}{2}\delta' y^2 + \mathcal{O}(\varrho^3)
\end{aligned}
\tag{3}
$$

where we have replaced first and second order derivatives of $\boldsymbol{u}$ at $\mathbf{0}$ with variables.

Noting that medians of $\boldsymbol{u}$ within D_ϱ are $\mathcal{O}(\varrho^2)$, we seek a PDE evolution that is approximated by some filtering process in the limit $\varrho \to 0$ with step size $\mathcal{O}(\varrho^2)$. This implies that the PDE evolution is described by a bivariate function $\boldsymbol{p} = (p,q)^{\mathrm{T}}$ of the first and second order derivatives of $\boldsymbol{u}$ as

$$
\boldsymbol{u}_t = \boldsymbol{p}(\alpha_1, \alpha_2, \beta, \gamma, \delta, \alpha_1', \alpha_2', \beta', \gamma', \delta') \,.
\tag{4}
$$

(II) Centrality. We impose first centrality in its weakest form, w.r.t. the reflection on the origin, which implies

$$
\boldsymbol{p}(\alpha_1, \alpha_2, 0, 0, 0, \alpha_1', \alpha_2', 0, 0, 0) = \mathbf{0} \,.
\tag{5}
$$

(III) Scaling Equivariance. With this requirement it follows that $\boldsymbol{p}$ is homogeneous of degree 0 in α_1, α_2, α_1' and α_2' and of degree 1 in the remaining parameters,

$$
\boldsymbol{p}(\lambda\alpha_1, \lambda\alpha_2, \mu\beta, \mu\gamma, \mu\delta, \lambda\alpha_1', \lambda\alpha_2', \mu\beta', \mu\gamma', \mu\delta') = \mu\boldsymbol{p}(\alpha, \beta, \gamma, \delta, \alpha', \beta', \gamma', \delta')
\tag{6}
$$

for $\lambda > 0$, $\mu > 0$.

(IV) Euclidean Equivariance. Now we can apply Euclidean normalisation. In the normalised setting, we have $\alpha_2 = \alpha_1' = 0$. By homogeneity, $\boldsymbol{p}$ in fact only depends on the ratio α_1/α_2' instead of the two individual variables.

(V) Affine Equivariance. By affine normalisation we achieve $\alpha_1' = \alpha_2 = 1$, thus only the second order derivatives are left as parameters for $\boldsymbol{p} = \boldsymbol{p}(\beta, \gamma, \delta, \beta', \gamma', \delta')$.

We notice that in the affine normalised setting, there is a further degree of freedom: Simultaneous rotations and reflections in the image (x-y) and data (u-v) plane leave $\alpha_1 = \alpha_2' = 1$ untouched but transform the second order derivatives. Thus, $\boldsymbol{p}$ must be equivariant under these operations.

In particular, reflections on the y and v axes and similarly on the x and u axes imply

$$
p(0, \gamma, 0, \beta', 0, \delta') = 0 \,, \qquad\qquad q(\beta, 0, \delta, 0, \gamma', 0) = 0
\tag{7}
$$

as well as $\boldsymbol{p}(-\beta, -\gamma, -\delta, -\beta', -\gamma', -\delta') = -\boldsymbol{p}(\beta, \gamma, \delta, \beta', \gamma', \delta')$. By reflection on the diagonal $x = y$, we find $q(\beta, \gamma, \delta, \beta', \gamma', \delta') = p(\delta', \gamma', \beta', \delta, \gamma, \beta)$, thus reducing the problem to finding a single univariate function p. Using general rotations with rotation matrix $\boldsymbol{R} = \begin{pmatrix} \cos\varphi & \sin\varphi \\ -\sin\varphi & \cos\varphi \end{pmatrix}$ simultaneously in the x-y and u-v planes and taking first-order derivatives w.r.t. the rotation angle φ yields the differential equations

$$p_\beta - p_\delta - p_{\gamma'} = 0 \,, \qquad\qquad p_{\beta'} - p_{\delta'} - p_\gamma = 0 \qquad (8)$$

for p. Furthermore, second derivatives w.r.t. φ yield the additional conditions $p_\gamma = p_{\beta'} = p_{\delta'} = 0$, from which together with (7) we see that p is a 1-homogeneous function of only β, δ and γ'. According to Euler's Homogeneous Function Theorem p can be represented in the form

$$p(\beta, \delta, \gamma') = \beta p_\beta + \delta p_\delta + \gamma' p_{\gamma'} \,. \qquad (9)$$

Simplifying (9) with (8) we obtain the following intermediate result.

Proposition 1. *A bivariate PDE evolution in affine normalisation which is associated to a local filtering operator with centrality property and affine equivariance can only be of the form*

$$u_t = p(u_{xx}, u_{yy}, v_{xy}) \,, \qquad\qquad v_t = p(v_{yy}, v_{xx}, u_{xy}) \qquad (10)$$

with a 1-homogeneous function p that satisfies

$$p(\beta, \delta, \gamma') = (\beta + \gamma')p_\beta + (\delta - \gamma')p_\delta \,. \qquad (11)$$

(VI) Relative Equivariances. For further specification we need an additional requirement that can be derived from several relative equivariances. If we assume radial scaling equivariance, we can in particular replace (in the normalised setting under consideration) $\boldsymbol{u}(\boldsymbol{x})$ for each $\boldsymbol{x} \in D_\varrho$ with the scalar multiple $(1+\varepsilon\boldsymbol{x})\boldsymbol{u}(\boldsymbol{x})$ for some small ε and require that $\boldsymbol{u}_t$ remains unchanged. This implies

$$p(\beta + 2\varepsilon, \delta, \gamma' + \varepsilon) = p(\beta, \delta, \gamma') \,. \qquad (12)$$

Unfortunately, as discussed earlier, radial scaling equivariance does not hold for all multivariate median concepts in our investigation. Among the affine equivariant medians, it holds only for the Oja median (where the restriction to unique cases is no problem in the space-continuous case at generic locations). However, (12) can be derived alternatively from restricted projective equivariance (as it holds for the half-space median). Moreover, it can be shown that also the asymptotic radial scaling equivariance which holds for the TR-L^1 median is sufficient to ensure (12) since the effect of the radial rescaling with $(1 + \varepsilon\boldsymbol{x})$ for $\boldsymbol{x} \in D_\varrho$ on $\boldsymbol{p}$ is of order $\mathcal{O}(\varrho\varepsilon)$ and thereby vanishes in the limit $\varrho \to 0$.

Inserting (11) into (12) we obtain $(\beta + \gamma' + 3\varepsilon)p_\beta + (\delta - \gamma' - \varepsilon)p_\delta = (\beta + \gamma')p_\beta + (\delta - \gamma')p_\delta$ and finally

$$p_\delta = 3p_\beta \,. \qquad (13)$$

Together with (11) this yields $p(\beta, \delta, \gamma') = (\beta + 3\delta - 2\gamma')p_\beta$ which implies that p_β is constant and p a linear function. The single degree of freedom is the choice of p_β which amounts to a time rescaling. The consequence is our second result, summarised in the following proposition.

Proposition 2. *A bivariate PDE evolution in affine normalisation as in Proposition 1 which additionally satisfies asymptotic radial scaling equivariance or restricted projective equivariance is necessarily of the form (2).*

We remark that Proposition 2 explains the coincidence of the PDE asymptotics of Oja, TR-L^1 and half-space median in the bivariate case. Generalisations on one hand to trivariate and generally multivariate evolutions, and on the other hand to Euclidean equivariance are part of ongoing work.

The result of Proposition 1 states a more general class of affine equivariant PDE evolutions that are in a sense close to median-associated ones but without the last requirement of relative equivariances. Functions p that satisfy (11) for real β, δ, γ' with $\beta + \gamma' > 0$ and $\delta - \gamma' > 0$ are e.g. given by

$$p(\beta, \delta, \gamma') = \left((\beta + \gamma')^s + \vartheta(\delta - \gamma')^s\right)^{1/s} \tag{14}$$

with arbitrary parameters $\vartheta > 0$, $s > 0$, which includes the linear case for $s = 1$. To be usable for the PDE image filter, however, p needs to be defined on the entire parameter space $(\beta, \delta, \gamma') \in \mathbb{R}^3$. Such an extension is obviously possible for some values of s, particularly $s = m$ or $s = 1/m$ for odd natural numbers m. We believe that this larger class of PDE evolution deserves further study, and include some numerical examples in the next section.

5   Experiments

While the emphasis of this paper is largely on theory, we want to give an impression of the effect of the filters under consideration by a numerical example. Although practical relevance is expected rather for multivariate images with at least three channels such as RGB colour images or diffusion tensor images, it appears appropriate to stay in the bivariate setting in accordance with the analysis presented. As an example of a bivariate image we therefore present a colour image where the RGB colour space has been reduced to a yellow-blue (YB) colour space by averaging the red and green channels. All algorithms were implemented in C++.

Numerical Aspects. Whereas the PDE system (10), (14) is stated in affine normalisation, practical computation by a finite-difference scheme is best done by applying only Euclidean normalisation to a 3×3 patch and evaluating the PDE system therefore in the form

$$u_t = p(u_{xx}, u_{yy}, v_{xy}/v_y), \qquad\qquad v_t = p(v_{yy}, v_{xx}, u_{xy}/u_x). \tag{15}$$

Still, as already noted in [14], a straightforward discretisation by central differences is unstable. In [14, App. 6] a stable numerical scheme was devised that uses in particular min-mod stabilised upwind discretisations for the terms involving u_{xy}/u_x, v_{xy}/v_y. We use this scheme with minor adaptations to suit the more general function p from (14).

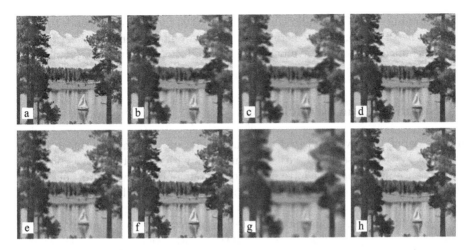

Fig. 1. Filtering of a bivariate colour image (512×512 pixels). **a** Colour image *sailboat* reduced to yellow-blue colour space. **b** Result of half-space median filtering with a discrete disc of radius 2 as structuring element, 15 iterations. **c** Corresponding evolution by the PDE system (10), (14), $s = 1$, $\vartheta = 3$, up to time $T = 2.5$ (60 time steps of size 0.041665); same evolution as (2) except for speed-up by a factor 4. **d** Same as c but with heuristic anti-diffusion to reduce numerical dissipation; see text. **e** PDE evolution (10), (14), $s = 3$, $\vartheta = 3$, $T = 2.5$ (same time steps as c, d). **f** Same as e but with heuristic anti-diffusion. **g** PDE evolution (10), (14), $s = 1/3$, $\vartheta = 3$, $T = 0.8333$ (1000 time steps of size 0.0008333). **h** Same as g but with heuristic anti-diffusion. (Colour figure online)

As finite difference discretisations of PDEs tend to add undesired blur to the results, it was proposed in [14] to modify the coefficients of the PDE evolutions by an anti-diffusion term which can be safely done just by reducing the coefficients of u_{xx}, u_{yy}, v_{xx}, v_{yy} by a uniform amount. We follow this recommendation and include exemplary results with this compensation.

Exemplary Results. Based on the original image shown in Fig. 1a, we show first a result of iterated half-space median filtering, Fig. 1b, using the implementation from [16]. Frame c shows the result of the corresponding PDE system (10), (14) with $s = 1$ at the appropriate evolution time, whereas frame d represents the same with the anti-diffusion compensation. Frames e–f show visually similar results obtained with $s = 3$ instead of $s = 1$. In frames g–h we used $s = 1/3$. To achieve a visually comparable degree of image smoothing, adjustments were required both for the total evolution time (reduced by a factor of 3) but also for the time step size (reduced by a factor of 50, necessitating dramatically more iterations).

Still, the computation time for all of the PDE evolutions is much less than that for the half-space median computation. With our C++ implementations that were in no way optimised for performance, computation times for single-core computation on an AMD Phenom(tm) II X6 1100T processor (manufactured around

2011) running at 3.2 GHz under Ubuntu Linux 20.04 ranged from seconds to minutes for the PDE evolutions whereas more than two hours were necessary for the half-space median filtering. We believe that significant speedups are possible by more efficient implementations.

6   Summary and Conclusions

In this paper we have re-visited previous results on multivariate image filtering PDE systems associated with median filters with special emphasis on their equivariance properties. We have presented a systematic derivation of such PDE systems on the basis of equivariance properties, for the time being in the bivariate case. As a result, we showed that the approximation of the same PDE system by several affine equivariant median filtering processes is no coincidence, but a necessity. As a by-product we have identified a more general class of PDE evolutions with homogeneous functions of second derivatives as right-hand sides that appear to be worth further study.

Ongoing work is directed at generalising the result of this paper to the general multivariate situation including practically meaningful cases like RGB colour images and diffusion tensor images. We also aim at extending the analysis to the case of Euclidean equivariance where part of our present line of argument cannot be transferred straightforwardly. As mentioned before, the larger class of homogeneous evolutions described above is of interest for further investigation, too.

Another direction for future research is the adequate interpretation of the PDE systems in question. The clear geometric intuition of curvature flow in the univariate case is so far not reflected in an appropriate understanding of the multivariate PDE evolution. A geometric interpretation will definitely strengthen the theoretical framework and promote applicability.

References

1. Alvarez, L., Guichard, F., Lions, P.L., Morel, J.M.: Axioms and fundamental equations in image processing. Arch. Ration. Mech. Anal. **123**, 199–257 (1993). https://doi.org/10.1007/BF00375127
2. Alvarez, L., Morales, F.: Affine morphological multiscale analysis of corners and multiple junctions. Int. J. Comput. Vis. **25**, 95–107 (1994)
3. Green, P.J.: Peeling bivariate data. In: Barnett, V. (ed.) Interpreting Multivariate Data, pp. 3–20. Wiley, Chichester (1981)
4. Guichard, F., Morel, J.M.: Partial differential equations and image iterative filtering. In: Duff, I.S., Watson, G.A. (eds.) The State of the Art in Numerical Analysis, pp. 525–562, No. 63 in IMA Conference Series (New Series), Clarendon Press, Oxford (1997)
5. Hettmansperger, T.P., Randles, R.H.: A practical affine equivariant multivariate median. Biometrika **89**(4), 851–860 (2002)

6. Kleefeld, A., Breuß, M., Welk, M., Burgeth, B.: Adaptive filters for color images: median filtering and its extensions. In: Trémeau, A., Schettini, R., Tominaga, S. (eds.) Computational Color Imaging. Lecture Notes in Computer Science, vol. 9016, pp. 149–158. Springer, Cham (2015). https://doi.org/10.1007/978-3-319-15979-9_15

7. Oja, H.: Descriptive statistics for multivariate distributions. Statist. Probab. Lett. **1**, 327–332 (1983)

8. Rao, C.R.: Methodology based on the L_1-norm in statistical inference. Sankhyā A **50**, 289–313 (1988)

9. Sethian, J.A.: Level Set Methods: Evolving Interfaces in Geometry, Fluid Mechanics. Computer Vision and Material Sciences. Cambridge University Press, Cambridge, UK (1996)

10. Small, C.G.: Measures of centrality for multivariate and directional distributions. Can. J. Stat. **15**(1), 31–39 (1987)

11. Small, C.G.: A survey of multidimensional medians. Int. Stat. Rev. **58**(3), 263–277 (1990)

12. Spence, C., Fancourt, C.: An iterative method for vector median filtering. In: Proceeding 2007 IEEE International Conference on Image Processing, vol. 5, pp. 265–268 (2007)

13. Weber, A.: Über den Standort der Industrien. Mohr, Tübingen (1909)

14. Welk, M.: Multivariate median filters and partial differential equations. J. Math. Imaging Vis. **56**, 320–351 (2016). https://doi.org/10.1007/s10851-016-0645-9

15. Welk, M.: PDE for bivariate amoeba median filtering. In: Angulo, J., Velasco-Forero, S., Meyer, F. (eds.) Mathematical Morphology and its Applications to Signal and Image Processing. Lecture Notes in Computer Science, vol. 10225, pp. 271–283. Springer, Cham (2017). https://doi.org/10.1007/978-3-319-57240-6_22

16. Welk, M.: Asymptotic analysis of bivariate half-space median filtering. In: Roth, P.M., Steinbauer, G., Fraundorfer, F., Brandstötter, M., Perko, R. (eds.) Proceedings of the Joint Austrian Computer Vision and Robotics Workshop, pp. 151–156. Verlag der Technischen Universität Graz, Graz (2020). https://doi.org/10.3217/978-3-85125-752-6-34

17. Welk, M.: Multivariate medians for image and shape analysis. Technical Report eess.IV:1911.00143v2, arXiv.org (2021), arXiv:1911.00143

18. Welk, M., Breuß, M.: The convex-hull-stripping median approximates affine curvature motion. In: Burger, M., Lellmann, J., Modersitzki, J. (eds.) Scale Space and Variational Methods in Computer Vision. Lecture Notes in Computer Science, vol. 11603, pp. 199–210. Springer, Cham (2019). https://doi.org/10.1007/978-3-030-22368-7_16

19. Welk, M., Feddern, C., Burgeth, B., Weickert, J.: Median filtering of tensor-valued images. In: Michaelis, B., Krell, G. (eds.) Pattern Recognition. Lecture Notes in Computer Science, vol. 2781, pp. 17–24. Springer, Berlin (2003). https://doi.org/10.1007/978-3-540-45243-0_3

20. Welk, M., Kleefeld, A., Breuß, M.: Non-adaptive and amoeba quantile filters for colour images. In: Benediktsson, J.A., Chanussot, J., Najman, L., Talbot, H. (eds.) Mathematical Morphology and its Applications to Signal and Image Processing. Lecture Notes in Computer Science, vol. 9082, pp. 398–409. Springer, Cham (2015). https://doi.org/10.1007/978-3-319-18720-4_34

21. Welk, M., Weickert, J., Becker, F., Schnörr, C., Feddern, C., Burgeth, B.: Median and related local filters for tensor-valued images. Signal Process. **87**, 291–308 (2007)

Morphological Adjunctions Represented by Matrices in Max-Plus Algebra for Signal and Image Processing

Samy Blusseau[1]([✉]), Santiago Velasco-Forero[1], Jesús Angulo[1], and Isabelle Bloch[2]

[1] Mines Paris, PSL University, Centre for Mathematical Morphology, 77300 Fontainebleau, France
samy.blusseau@minesparis.psl.eu
[2] Sorbonne Université, CNRS, LIP6, 75005 Paris, France

Abstract. In discrete signal and image processing, many dilations and erosions can be written as the max-plus and min-plus product of a matrix on a vector. Previous studies considered operators on symmetrical, unbounded complete lattices, such as Cartesian powers of the completed real line. This paper focuses on adjunctions on closed hypercubes, which are the complete lattices used in practice to represent digital signals and images. We show that this constrains the representing matrices to be doubly-0-astic and we characterise the adjunctions that can be represented by them. A graph interpretation of the defined operators naturally arises from the adjacency relationship encoded by the matrices, as well as a max-plus spectral interpretation.

Keywords: Morphological operators · Max-plus algebra · Graph theory

1 Introduction

Like linear filters can be represented by matrices in discrete image and signal processing, many morphological dilations and erosions can be seen as applying a matrix product to a vector, but in the minimax algebra. This is in particular the case for those defined with structuring functions, either flat or not, local or non-local [13,14], translation invariant or spatially variant [4,7,10,16]. They are commonly known to be the vertical-shift-invariant dilations and erosions [9,11]. While the matrix point of view is not the most appropriate for the implementation of these operators, especially translation-invariant ones, it is a valuable insight for their theoretical understanding. In particular, it can help predict and control complex behaviours such as those of iterated operators based on adjunctions with non-flat, spatially variant and input-adapted structuring functions [2,3]. Indeed, it is a flexible and general framework which embraces a very broad part of morphological literature, and it is supported by the rich theory of Minimax algebra [1,6].

© Springer Nature Switzerland AG 2022
É. Baudrier et al. (Eds.): DGMM 2022, LNCS 13493, pp. 206–218, 2022.
https://doi.org/10.1007/978-3-031-19897-7_17

In the abundant literature on spatially-variant morphological image processing, only a few approaches explicitly used the matrix formulation [2,3,13,14], whereas most contributions were limited to flat structuring elements and focused on the local effects of the adaptive strategy. On the theoretical side, the representation of morphological adjunctions by matrices was studied in a setting that does not directly apply to digital signal and image processing, as the co-domain is usually an unbounded lattice, stable under any vertical translation [11]. Although a method was proposed to convert these adjunctions to new ones on bounded lattices [12], it is not practical and does not allow for the interpretations that are exposed here.

In the present paper, we focus on complete lattices of the type $[a, b]^n$, where a and b represent the minimal and maximal possible signal values (typically $a = 0$ and $b = 255$ for 8-bits images), and n is an integer representing the size of the signal (typically, the number of pixels of an image, reshaped as a column vector). This is a theoretical contribution that can be viewed as a companion paper to previous studies where this framework has been successfully applied to adaptive anisotropic filtering [2,3][1]. In Sect. 2 we introduce the matrix-based morphological setting and prove simple but fundamental results: in particular, we characterise the adjunctions that can be represented by matrices and show that these matrices need to be doubly-0-astic. By viewing matrices as encoding adjacency, we provide in Sect. 3 a graph interpretation of iterated operators and their associated granulometries. In Sect. 4 we draw a link between these operators and some results on the spectrum of matrices in the max-plus algebra, before concluding in Sect. 5.

2   Matrix-Based Morphological Adjunctions

2.1   Notations

In this paper matrices will be denoted by capital letters, such as W, and their i-th row and j-th column coefficients by corresponding indexed lowercase letters w_{ij}. Similarly, vectors are written as boldface lowercase letters, such as $\mathbf{x}$, and their i-th component as x_i. Let $0 \leq a < b \in \mathbb{R}^+$ be two non-negative real numbers, $n \in \mathbb{N}^*$ a positive integer. The set $\{1, \ldots, n\}$ will be denoted by $[\![1, n]\!]$. Let $\mathcal{L} = ([a, b]^n, \leq)$ be the complete lattice equipped with the usual product partial ordering (Pareto ordering): $\mathbf{x} \leq \mathbf{y} \iff x_i \leq y_i, \ \forall i \in [\![1, n]\!]$. The supremum and infimum on $\mathcal{L}$ are induced by the Pareto ordering: for a family $(\mathbf{x}^{(k)})_{k \in K}$ of $\mathcal{L}$, $\bigvee_{k \in K} \mathbf{x}^{(k)}$ is the vector $\mathbf{y}$ defined by $y_i = \bigvee_{k \in K} x_i^{(k)}$, where K is any index set. Therefore $\mathbf{a} = (a, \ldots, a)^T$ and $\mathbf{b} = (b, \ldots, b)^T$ are respectively the smallest and largest elements in $\mathcal{L}$. For $\mathbf{x} \in \mathcal{L}$, we note $\mathbf{x}^c \doteq \mathbf{b} - \mathbf{x} + \mathbf{a}$, and for any $i \in \{1, \ldots, n\}$, $\mathbf{e}^{(i)}$ is the "impulse" vector in $\mathcal{L}$ such that $e_i^{(i)} = b$ and $e_j^{(i)} = a$ for $j \neq i$.

We note $\mathbb{R}_{\max} \doteq \mathbb{R} \cup \{-\infty\}$, $\mathbb{R}_{\min} \doteq \mathbb{R} \cup \{+\infty\}$ and $\mathcal{M}_n$ the set of $n \times n$ square matrices with coefficients in $\mathbb{R}_{\max}$. Like $(\mathbb{R}_{\max}, \vee, +)$, $(\mathcal{M}_n, \vee, \otimes)$ is an

---

[1] An online demo for [3] is available: https://bit.ly/anisop_demo.

idempotent semiring, with the addition $\vee$ and product $\otimes$ defined as follows. For $A, B \in \mathcal{M}_n$, $A \otimes B$ and $A \vee B$ are the $n \times n$ matrices defined respectively by $(A \otimes B)_{ij} = \bigvee_{k=1}^n a_{ik} + b_{kj}$ and $(A \vee B)_{ij} = a_{ij} \vee b_{ij} = \max(a_{ij}, b_{ij})$, for $1 \le i, j \le n$. Similarly, for $\mathbf{x} \in \mathbb{R}_{\max}^n$, $A \otimes \mathbf{x}$ is the vector such that $(A \otimes \mathbf{x})_i = \bigvee_{j=1}^n a_{ij} + x_j$. Note that $\vee$ and $\otimes$ are associative and $\otimes$ is distributive over $\vee$. Finally, the product of a scalar $\lambda \in \mathbb{R}_{\max}$ by a vector $\mathbf{x} \in \mathbb{R}_{\max}^n$ is $\lambda \otimes \mathbf{x} \doteq \lambda + \mathbf{x}$, the vector in $\mathbb{R}_{\max}$ such that $(\lambda \otimes \mathbf{x})_i = \lambda + x_i$. In [6] and [14], special subsets of $\mathcal{M}_n$ are introduced, that we will show to be essential to represent morphological adjunctions on $\mathcal{L}$.

Definition 1 (0-asticity [6]). *A matrix $W \in \mathcal{M}_n$ is said* **row-0-astic** *if for any $1 \le i \le n$, $\bigvee_{j=1}^n w_{ij} = 0$. Similarly, it is said* **column-0-astic** *if the supremum of each column is 0, and* **doubly-0-astic** *if the matrix is both row-0-astic and column-0-astic. Finally, W is simply said* 0-**astic** *if $\bigvee_{1 \le i,j \le n} w_{ij} = 0$.*

A special kind of doubly-0-astic matrices are those with zeros on the diagonal and non-positive coefficients elsewhere.

Definition 2 (CMW matrices [14]). *A matrix $W \in \mathcal{M}_n$ is called a Conservative Morphological Weights (CMW) matrix if $\forall i, j \in [\![1, n]\!]$, $w_{ij} \le 0$ and $w_{ii} = 0$.*

We now introduce the morphological framework on $\mathcal{L}$, based on the max-plus algebra product between matrices and vectors.

2.2   Dilations

For $W \in \mathcal{M}_n$, we consider the function δ_W from $\mathcal{L}$ to $\mathbb{R}_{\max}^n$ such that

$$\forall \mathbf{x} \in \mathcal{L}, \quad \delta_W(\mathbf{x}) = W \otimes \mathbf{x} = \left(\bigvee_{1 \le j \le n} \{w_{ij} + x_j\} \right)_{1 \le i \le n}. \tag{1}$$

In the processing of digital data such as images we usually want the input to be comparable with the output. Hence, we will constrain W such that $\delta_W(\mathcal{L}) \subseteq \mathcal{L}$. This has the following consequences:

$$\delta_W(\mathbf{b}) \le \mathbf{b} \Rightarrow \forall i \in [\![1, n]\!], \ b + (\bigvee_{j=1}^n w_{ij}) \le b \Rightarrow \forall i \in [\![1, n]\!], \ \bigvee_{j=1}^n w_{ij} \le 0$$

since $b > -\infty$. Similarly, $\delta_W(\mathbf{a}) \ge \mathbf{a} \Rightarrow \forall i \in [\![1, n]\!], \ \bigvee_{j=1}^n w_{ij} \ge 0$. Hence a necessary condition to have $\delta_W(\mathcal{L}) \subseteq \mathcal{L}$ is that W be row-0-astic (Definition 1). Conversely, the row-0-asticity for W implies that $\delta_W(\mathbf{a}) = \mathbf{a}$ and $\delta_W(\mathbf{b}) = \mathbf{b}$, and therefore that $\delta_W(\mathcal{L}) \subseteq \mathcal{L}$ by increasingness of δ_W. This leads to the following result.

Proposition 1. *Let $W \in \mathcal{M}_n$ and δ_W be the function defined by (1). Then δ_W is a dilation mapping $\mathcal{L}$ to $\mathcal{L}$ if and only if W is row-0-astic.*

Proof. If δ_W is a dilation mapping $\mathcal{L}$ to $\mathcal{L}$, then $\delta_W(\mathcal{L}) \subseteq \mathcal{L}$ which, as we showed, implies that W is row-0-astic. Conversely, we saw that a row-0-astic W implies $\delta_W(\mathcal{L}) \subseteq \mathcal{L}$. Therefore, we only have to verify that δ_W is a dilation, or equivalently that it commutes with the supremum. This is straightforward from the definition of $W \otimes \mathbf{x}$. □

2.3   Erosions and Adjunctions

Now we suppose that $W \in \mathcal{M}_n$ is row-0-astic, hence δ_W is a dilation from $\mathcal{L}$ to $\mathcal{L}$, and we are interested in its adjoint erosion α_W defined for any $\mathbf{y} \in \mathcal{L}$ by $\alpha_W(\mathbf{y}) = \bigvee E_\mathbf{y}$ where $E_\mathbf{y} = \{\mathbf{x} \in \mathcal{L}, \delta_W(\mathbf{x}) \leq \mathbf{y}\}$. Let us denote by ε_W the function from $\mathcal{L}$ to $\mathbb{R}_{\min}^n$ such that for any $\mathbf{y} \in \mathcal{L}$

$$\varepsilon_W(\mathbf{y}) = \left(\delta_{W^T}(\mathbf{y}^c)\right)^c = \left(W^T \otimes \mathbf{y}^c\right)^c = \left(\bigwedge_{1\leq j\leq n} \{y_j - w_{ji}\}\right)_{1\leq i\leq n}. \tag{2}$$

Then we can check that $\forall \mathbf{y} \in \mathcal{L}$, $\alpha_W(\mathbf{y}) = \varepsilon_W(\mathbf{y}) \wedge \mathbf{b}$. Indeed, from (1) we see that for any $\mathbf{x}, \mathbf{y} \in \mathcal{L}$, $\delta_W(\mathbf{x}) \leq \mathbf{y} \iff \mathbf{x} \leq \varepsilon_W(\mathbf{y})$. Therefore, since $\delta_W(\mathbf{a}) = \mathbf{a} \leq \mathbf{y}$ we get $\mathbf{a} \leq \varepsilon_W(\mathbf{y})$, which implies $\varepsilon_W(\mathbf{y}) \wedge \mathbf{b} \in \mathcal{L}$; furthermore, $\varepsilon_W(\mathbf{y}) \wedge \mathbf{b} \leq \varepsilon_W(\mathbf{y})$ so $\varepsilon_W(\mathbf{y}) \wedge \mathbf{b} \in E_\mathbf{y}$; finally, as both $\varepsilon_W(\mathbf{y})$ and $\mathbf{b}$ are upper-bounds of $E_\mathbf{y}$, so is $\varepsilon_W(\mathbf{y}) \wedge \mathbf{b}$. Hence, $\varepsilon_W(\mathbf{y}) \wedge \mathbf{b} = \bigvee E_\mathbf{y} = \alpha_W(\mathbf{y})$. By a similar reasoning as in Sect. 2.2, we get the following result.

Proposition 2. *Let $W \in \mathcal{M}_n$ and ε_W be the function defined by (2). Then ε_W is an erosion mapping $\mathcal{L}$ to $\mathcal{L}$ if and only if W is column-0-astic.*

If W is also row-0-astic, then $\varepsilon_W = \alpha_W$ is the adjoint of δ_W, as stated next.

Proposition 3. *Let $W \in \mathcal{M}_n$ and δ_W and ε_W be the functions defined by (1) and (2), respectively. Then $(\varepsilon_W, \delta_W)$ is an adjunction on $\mathcal{L}$ if and only if W is doubly-0-astic. Furthermore, $(\varepsilon_W, \delta_W)$ is an adjunction on $\mathcal{L}$ with δ_W extensive (and ε_W anti-extensive) if and only if W is a CMW matrix.*

Proof. Most of the points have already been addressed above or are straightforward from Proposition 1. To see that δ_W extensive implies $w_{ii} = 0$ for all i, just remark that $w_{ii} < 0$ would imply $\delta_W(\mathbf{e}^{(i)})_i < b = e_i^{(i)}$. □

2.4   Generality of $(\varepsilon_W, \delta_W)$

The dilation δ_W, already introduced in [2,11,15], can be viewed as a generalisation of the non-local and adaptive mathematical morphology [13,14] on signals and images. Each column $W_{:,j}$ of W represents the structuring function corresponding to pixel (or instant) j.

As pointed out in [9,11], the dilations that can be written as matrix-based max-plus products like Eq. (1) are the shift (or *vertical-translation*) invariant ones. However the result stated in [9,11] does not directly apply to our setting where the lattice $\mathcal{L}$ is different from the lattice of scalars which define vertical translation of signal values, usually $\mathbb{R} \cup \{-\infty, +\infty\}$. Still, the same idea holds here with some adaptation, as stated in the next proposition.

Proposition 4. *Let $\delta : \mathcal{L} \to \mathcal{L}$ be a dilation. Then there exists $W \in \mathcal{M}_n$ such that $\delta = \delta_W$ if and only if*

$$\forall \lambda \leq 0, \forall \mathbf{x} \in \mathcal{L}, \quad \delta\big((\lambda + \mathbf{x}) \vee \mathbf{a}\big) = \big(\lambda + \delta(\mathbf{x})\big) \vee \mathbf{a}. \tag{3}$$

In that case, the matrix W whose j-th column is $W_{:,j} = \delta(\mathbf{e}^{(j)}) - \mathbf{b}$ for $1 \leq j \leq n$, is such a representing matrix.

We see that this class of dilations is very broad and covers the most commonly used in morphological image and signal processing: dilations based on structuring functions, possibly non-local, varying in space and non-flat.

Proof (Proposition 4). If $\delta = \delta_W$ for some $W \in \mathcal{M}_n$, then it is straightforward to check that δ verifies Eq. (3).

Conversely, suppose δ verifies Eq. (3). Then we first remark that $\delta(\mathbf{b}) = \mathbf{b}$. Indeed, on the one hand, $\delta(\mathbf{a}) = \mathbf{a}$ as $\mathbf{a} = \bigwedge \mathcal{L}$ and δ is a dilation mapping $\mathcal{L}$ to $\mathcal{L}$. On the other hand, $\delta(\mathbf{a}) = \delta\big((a - b) + \mathbf{b}\big) = (a - b) + \delta(\mathbf{b})$ by Eq. (3). Hence $\mathbf{a} = (a - b) + \delta(\mathbf{b})$ which means that $\delta(\mathbf{b}) = \mathbf{b}$.

As a consequence: for any $i \in [\![1, n]\!]$, there is a $j_i \in [\![1, n]\!]$ such that $\delta(\mathbf{e}^{(j_i)})_i = b$. This is simply because $\mathbf{b} = \bigvee_{1 \leq j \leq n} \mathbf{e}^{(j)}$ so $\mathbf{b} = \delta\big(\bigvee_{1 \leq j \leq n} \mathbf{e}^{(j)}\big) = \bigvee_{1 \leq j \leq n} \delta(\mathbf{e}^{(j)})$, which means that, for any i, $b = \bigvee_{1 \leq j \leq n} \delta(\mathbf{e}^{(j)})_i$ and finally that $\delta(\mathbf{e}^{(j_i)})_i = b$ for some j_i, as the supremum is reached here.

Now, let $\mathbf{x} \in \mathcal{L}$. Then it can be decomposed as $\mathbf{x} = \bigvee_{1 \leq j \leq n} \big[(\lambda_j + \mathbf{e}^{(j)}) \vee \mathbf{a}\big]$ with $\lambda_j = x_j - b \leq 0$. Hence, as δ is a dilation verifying Eq. (3), we get $\delta(\mathbf{x}) = \bigvee_{1 \leq j \leq n} \big[(\lambda_j + \delta(\mathbf{e}^{(j)})) \vee \mathbf{a}\big]$. We now use the result stated just above: for any $i \in [\![1, n]\!]$ there is a $j_i \in [\![1, n]\!]$ such that $\lambda_{j_i} + \delta(\mathbf{e}^{(j_i)})_i = x_{j_i} - b + b = x_{j_i} \geq a$. Therefore, $\bigvee_{1 \leq j \leq n} \big[(\lambda_j + \delta(\mathbf{e}^{(j)})) \vee \mathbf{a}\big] = \bigvee_{1 \leq j \leq n} \lambda_j + \delta(\mathbf{e}^{(j)})$ from which we finally get

$$\delta(\mathbf{x}) = \bigvee_{1 \leq j \leq n} \lambda_j + \delta(\mathbf{e}^{(j)}) = \bigvee_{1 \leq j \leq n} (x_j - b) + \delta(\mathbf{e}^{(j)}) = \bigvee_{1 \leq j \leq n} x_j + [\delta(\mathbf{e}^{(j)}) - \mathbf{b}] \tag{4}$$

which is exactly $W \otimes \mathbf{x}$ for W the matrix with columns $W_{:,j} = \delta(\mathbf{e}^{(j)}) - \mathbf{b}$ for $1 \leq j \leq n$. $\qquad \square$

Note that the dual of Proposition 4 obviously holds: the erosions $\varepsilon : \mathcal{L} \to \mathcal{L}$ which can be written as ε_W for some $W \in \mathcal{M}_n$ are those for which

$$\forall \lambda \geq 0, \forall \mathbf{x} \in \mathcal{L}, \quad \varepsilon\big((\lambda + \mathbf{x}) \wedge \mathbf{b}\big) = \big(\lambda + \varepsilon(\mathbf{x})\big) \wedge \mathbf{b}. \tag{5}$$

To show this it is sufficient to see that ε verifies (5) if and only if the dilation $\delta = \varepsilon(\cdot^c)^c$ verifies (3), and recall that $\delta_W(\cdot^c)^c = \varepsilon_{W^T}(\cdot)$.

2.5   Equivalent Dilations and Erosions

In Proposition 4 we exhibited one possible matrix $W \in \mathcal{M}_n$ that represents a dilation, but this matrix is not unique. In this section we characterise the set of such matrices and show that it is a complete lattice.

Since we are interested in adjunctions $(\varepsilon_W, \delta_W)$, following Proposition 3 we focus on the set of matrices in $\mathcal{M}_n$ that are doubly-0-astic, which we denote by $\mathcal{D}_0(n)$. Let the equivalence relation defined for any two matrices $A, B \in \mathcal{D}_0(n)$ by

$$A \sim B \iff \delta_A = \delta_B \iff \forall \mathbf{x} \in \mathcal{L},\ \delta_A(\mathbf{x}) = \delta_B(\mathbf{x}) \tag{6}$$

and note $\mathcal{C}_W = \{M \in \mathcal{D}_0(n),\ M \sim W\}$ the equivalence class of any $W \in \mathcal{D}_0(n)$. We provide an easy characterisation of $\mathcal{C}_W$ that will show useful in numerical computations of the morphological operators defined earlier. For any $u \in \mathbb{R}_{\max}$ let I_u denote the matrix in $\mathcal{M}_n$ whose coefficients are all equal to u. Then two equivalent matrices are characterised as follows.

Proposition 5. *Let* $M, W \in \mathcal{D}_0(n)$. *Then*

$$M \in \mathcal{C}_W \iff M \vee I_{a-b} = W \vee I_{a-b} \iff \begin{cases} m_{ij} = w_{ij} & \text{if } w_{ij} > a - b \\ m_{ij} \leq a - b & \text{otherwise.} \end{cases} \tag{7}$$

This means that if W has coefficients not larger than $a - b$, these can be set to any value not larger than $a - b$, including $-\infty$, and can therefore be ignored in the computation of $\delta_W(\mathbf{x})$.

Proof (Proposition 5). The second equivalence is just a matter of writing, so we prove the first one. Let us first notice that for any $\mathbf{x} \in \mathcal{L}$, $I_{(a-b)} \otimes \mathbf{x} \leq \mathbf{a}$. Therefore $\forall \mathbf{x} \in \mathcal{L}$, $(W \vee I_{(a-b)}) \otimes \mathbf{x} = (W \otimes \mathbf{x}) \vee (I_{(a-b)} \otimes \mathbf{x}) = W \otimes \mathbf{x}$, since $W \otimes \mathbf{x} \geq \mathbf{a}$, and this holds for M too. Hence, if $M \vee I_{a-b} = W \vee I_{a-b}$, then for any $\mathbf{x} \in \mathcal{L}$, $W \otimes \mathbf{x} = (W \vee I_{a-b}) \otimes \mathbf{x} = (M \vee I_{a-b}) \otimes \mathbf{x} = M \otimes \mathbf{x}$, which means $M \in \mathcal{C}_W$.

Conversely, suppose that $M \sim W$ and that $w_{i_0 j_0} > a - b$ for some $i_0, j_0 \in [\![1, n]\!]$. Let $\mathbf{x} = \mathbf{e}^{(j_0)} \in \mathcal{L}$, i.e. $x_{j_0} = b$ and $x_j = a \ \forall j \neq j_0$. The 0-asticity of W and M implies $(W \otimes \mathbf{x})_{i_0} = b + w_{i_0 j_0}$ and $(M \otimes \mathbf{x})_{i_0} = b + m_{i_0 j_0}$, hence $m_{i_0 j_0} = w_{i_0 j_0}$. We have just shown that $\forall i, j \in [\![1, n]\!]$, $(w_{ij} > a - b \Rightarrow w_{ij} = m_{ij})$ and by symmetry of the equivalence relation $(m_{ij} > a - b \Rightarrow w_{ij} = m_{ij})$, which combined yields $\max(m_{ij},\ a - b) = \max(w_{ij},\ a - b)$. So finally $M \sim W \Rightarrow M \vee I_{a-b} = W \vee I_{a-b}$. $\square$

While it is clear that if $A, B \in \mathcal{C}_W$ then $A \vee B \in \mathcal{C}_W$, the characterisation in Proposition 5 shows that $\mathcal{C}_W$ is also closed under infimum, that is: $A \wedge B \in \mathcal{C}_W$. This has the following straightforward consequence.

Proposition 6. *Let* $W \in \mathcal{D}_0(n)$ *and* $\leq$ *the partial ordering on* $\mathcal{C}_W$ *defined by* $A \leq B \iff A \vee B = B \iff a_{ij} \leq b_{ij} \ \forall i, j \in [\![1, n]\!]$. *Then*

- $(\mathcal{C}_W, \leq)$ *is a complete lattice (with coefficient-wise supremum and infimum);*
- *Its greatest element is* $\overline{W} = W \vee I_{a-b}$;
- *Its smallest element is* $\underline{W}$, *defined by* $\underline{w}_{ij} = \begin{cases} w_{ij} & \text{if } w_{ij} > a - b \\ -\infty & \text{otherwise.} \end{cases}$

2.6   Iterated Operators and Granulometries

In this section, given $W \in \mathcal{D}_0(n)$ and $p \in \mathbb{N}^*$, we focus on the iterated dilations and erosions δ_W^p and ε_W^p, as well as their sup and inf integrations, that we note respectively $D_W^{[p]} \doteq \bigvee_{k=1}^p \delta_W^k$ and $E_W^{[p]} \doteq \bigwedge_{k=1}^p \varepsilon_W^k$. One can easily check that both $(\varepsilon_W^p, \delta_W^p)$ and $(E_W^{[p]}, D_W^{[p]})$ are adjunctions. We note respectively $\gamma_W^{[p]} \doteq \delta_W^p \varepsilon_W^p$ and $G_W^{[p]} \doteq D_W^{[p]} E_W^{[p]}$ their corresponding openings.

Note that if δ_W is extensive, or equivalently if W is a CMW matrix (Proposition 3), then these adjunctions are equal: $(\varepsilon_W^p, \delta_W^p) = (E_W^{[p]}, D_W^{[p]})$. As this is not true in general, both adjunctions are worth studying. In particular, we shall examine whether $(\gamma_W^{[p]})_{p\in\mathbb{N}^*}$ and $(G_W^{[p]})_{p\in\mathbb{N}^*}$ define granulometries, that is to say families of openings that are decreasing with p. The answer is yes and it is a general result that does not depend on the representation of the adjunction.

Proposition 7. *Let (ε, δ) be an adjunction on a complete lattice. For any integer $p \in \mathbb{N}^*$, let us note $\gamma_p = \delta^p \varepsilon^p$ and $G_p = D_p E_p$ the openings associated to the adjunctions $(\varepsilon^p, \delta^p)$ and $\left(E_p = \bigwedge_{1\le k\le p} \varepsilon^k, D_p = \bigvee_{1\le k\le p} \delta^k\right)$, respectively. Then $(\gamma_p)_{p\in\mathbb{N}^*}$ and $(G_p)_{p\in\mathbb{N}^*}$ are granulometries.*

Proof. We first show that the family of openings $(\gamma_p)_{p\ge 1}$ decreases with p, hence a granulometry. This is straightforward by writing $\gamma_{p+1} = \delta^{p+1}\varepsilon^{p+1} = \delta^p \gamma_1 \varepsilon^p \le \delta^p \varepsilon^p = \gamma_p$. Secondly, regarding $(G_p)_{p\ge 1}$, we show $G_{p+1} \le G_p$ by proving that $G_p G_{p+1} = G_{p+1}$. We obtain this by remarking that $D_{p+1} = D_p(id \vee \delta)$, which makes it an invariant of G_p: $G_p D_{p+1} = D_p E_p D_p(id \vee \delta) = D_p(id \vee \delta) = D_{p+1}$. Then we can conclude $G_p G_{p+1} = G_p D_{p+1} E_{p+1} = D_{p+1} E_{p+1} = G_{p+1}$. $\square$

To conclude this section, let us write δ_W^p, ε_W^p, $D_W^{[p]}$ and $E_W^{[p]}$ as dilations and erosions represented by one suitable doubly-0-astic matrix. This will help in their graph interpretation of the next section. The associativity of $\otimes$ yields $\forall \mathbf{x} \in \mathcal{L}$, $\delta_W^p(\mathbf{x}) = W \otimes \ldots \otimes W \otimes \mathbf{x} = W^p \otimes \mathbf{x}$, therefore $\delta_W^p = \delta_{W^p}$. We obtain similarly $\varepsilon_W^p = \varepsilon_{W^p}$. The distributivity of $\otimes$ over $\vee$ yields $D_W^{[p]}(\mathbf{x}) = \bigvee_{k=1}^p \delta_W^k(\mathbf{x}) = \bigvee_{k=1}^p (W^k \otimes \mathbf{x}) = (\bigvee_{k=1}^p W^k) \otimes \mathbf{x}$ therefore $D_W^{[p]} = \delta_{S_p(W)}$, with $S_p(W) \doteq \bigvee_{k=1}^p W^k$. Similarly, $E_W^{[p]} = \varepsilon_{S_p(W)}$. Note that by the same arguments and Proposition 3, we get that $\mathcal{D}_0(n)$ is closed under $\otimes$ and $\vee$.

3   Graph Interpretations

3.1   Weighted Graphs

Let $W \in \mathcal{M}_n$ and $\mathcal{G}(W) = (V, E)$ be a weighted and directed graph containing n vertices whose $n \times n$ adjacency matrix is W, with the convention that $w_{ij} > -\infty$ if and only if $(i, j) \in E$. We now recall that a *path* from vertex i to vertex j in $\mathcal{G}(W)$ is a tuple $\pi = (k_1, \ldots, k_l)$ of vertices such that $k_1 = i$, $k_l = j$, and $(k_p, k_{p+1}) \in E$ for $1 \le p \le l-1$. The *length* of the path, denoted by $\ell(\pi)$, is $l-1$ (the number of its edges). For $p \ge 1$, $\Gamma_{ij}^{(p)}(W)$ denotes the set of paths from i to

j in $\mathcal{G}(W)$ of length p and $\Gamma_{ij}^{(\infty)}(W)$ the set of all paths from i to j. The *weight of a path* $\pi = (k_1, \ldots, k_l)$, denoted by $\omega(\pi)$, is the sum $\omega(\pi) = \sum_{p=1}^{l-1} w_{k_p k_{p+1}}$.

3.2 Iterated Operators

Recall that for $W \in \mathcal{M}_n$ and $p \in \mathbb{N}^*$, W^p is the p-th power of W in the $\otimes$ sense, and $S_p(W)$ is the matrix defined in Sect. 2.6, denoted by S_p here for simplicity. We note respectively $w_{ij}^{(p)}$ and $s_{ij}^{[p]}$ their coefficients. The following result is well known in tropical algebra and graph theory [5,6], and will help interpret the operators defined earlier. It can be proved by induction.

Proposition 8. *Let $W \in \mathcal{M}_n$ and $p \in \mathbb{N}^*$. Then for any $1 \leq i, j \leq n$,*

$$w_{ij}^{(p)} = \max\left\{\omega(\pi), \ \pi \in \Gamma_{ij}^{(p)}(W)\right\} \quad and \quad s_{ij}^{[p]} = \max\left\{\omega(\pi), \ \pi \in \bigcup_{1 \leq k \leq p} \Gamma_{ij}^{(k)}(W)\right\} \tag{8}$$

with the convention $\max(\emptyset) = -\infty$.

The equations in (8) are equivalent to saying that

1. $w_{ij}^{(p)} > -\infty$ (resp. $s_{ij}^{[p]} > -\infty$) if and only if there is at least a path in $\mathcal{G}(W)$ from vertex i to vertex j of length exactly (resp. at most) p;
2. $w_{ij}^{(p)}$ (resp. $s_{ij}^{[p]}$) is the maximal weight over the set of paths from vertex i to vertex j of length exactly (resp. at most) p.

Therefore the graphs $\mathcal{G}(W^p)$ and $\mathcal{G}(S_p)$ have the same set of vertices as the original graph $\mathcal{G}(W)$, but an edge exists between vertices i and j in $\mathcal{G}(W^p)$ (resp. $\mathcal{G}(S_p)$) whenever there is a path of length exactly (resp. at most) p from i to j in $\mathcal{G}(W)$. The weights associated with this new edge are the maximal weights over the corresponding set of paths.

Now if $W \in \mathcal{D}_0(n)$, following Sect. 2.6 we get, for $\mathbf{x} \in \mathcal{L}$ and $i \in [\![1, n]\!]$:

$$\delta_W^p(\mathbf{x})_i = \bigvee_{j \in \mathcal{N}_i^p} \{x_j + w_{ij}^{(p)}\} \ , \quad \varepsilon_W^p(\mathbf{x})_i = \bigwedge_{j \in \check{\mathcal{N}}_i^p} \{x_j - w_{ji}^{(p)}\} \tag{9}$$

and

$$D_W^{[p]}(\mathbf{x})_i = \bigvee_{j \in N_i^p} \{x_j + s_{ij}^{[p]}\} \ , \quad E_W^{[p]}(\mathbf{x})_i = \bigwedge_{j \in \check{N}_i^k} \{x_j - s_{ji}^{[p]}\} \tag{10}$$

where $\mathcal{N}_i^p$ is the set of neighbours of vertex i in $\mathcal{G}(W^p)$ or, equivalently, the set of vertices in $\mathcal{G}(W)$ that can be reached from i through a path of length p; $\check{\mathcal{N}}_i^p = \{j \in \{1, \ldots, n\}, i \in \mathcal{N}_j^p\}$; $N_i^p = \cup_{1 \leq k \leq p} \mathcal{N}_i^k$ and $\check{N}_i^p = \cup_{1 \leq k \leq p} \check{\mathcal{N}}_i^k$. Hence these dilations and erosions are suprema and infima of "penalised" values over extended neighbourhoods induced by the original graph. The penalization is given by the strength of the connection between vertices: the closer the penalising weight to zero, the more the neighbours' value contributes to the result. The fact that we can restrict the supremum and infimum over graph neighbourhoods in (9) and (10) is due to the weight values being $-\infty$ outside these neighbourhoods, hence not contributing to the supremum and infimum.

3.3   Path Interpretation of the Opening $G_W^{[p]}$

The goal of this section is to show that $G_W^{[p]}$ can be interpreted similarly to a path opening [8], in the sense that it preserves bright values that are connected to other bright values forming long enough paths in a graph. We can first remark that for any $\mathbf{x} \in \mathcal{L}$, $i \in [\![1, n]\!]$ and $t \in [a, b]$:

$$G_W^{[p]}(\mathbf{x})_i \geq t \iff \exists j \in N_i^p, \text{ such that } \forall l \in \check{N}_j^p \ x_l \geq t - s_{ij}^{[p]} + s_{lj}^{[p]}, \qquad (11)$$

which is straightforward from the expressions in (10), as $G_W^{[p]} = D_W^{[p]} E_W^{[p]}$. This directly yields

$$G_W^{[p]}(\mathbf{x})_i = \bigvee \left\{ t \in [a, b], \exists j \in N_i^p, \forall l \in \check{N}_j^p, \ x_l \geq t - s_{ij}^{[p]} + s_{lj}^{[p]} \right\}. \qquad (12)$$

In the case of binary weights, i.e. $w_{ij} = 0$ if vertex j is neighbour of i in $\mathcal{G}$ and $w_{ij} = -\infty$ otherwise, which corresponds to a non-weighted graph, then $s_{ij}^{[p]} = s_{lj}^{[p]} = 0$ in (11) and (12). Therefore, if $G_W^{[p]}(\mathbf{x})_i \geq t$, then there is a vertex j which is at most p steps away from i, such that all paths of length at most p and ending in j, including those of length exactly p and passing through i (if they exist), show values larger than t. In the general case, the additional term $-s_{ij}^{[p]} + s_{lj}^{[p]}$ modulates this constraint in function of the strength of the connection of i and the other vertices of $\check{N}_j$, to j.

4   Links to the Max-Plus Spectral Theory

Now we present the consequences and interpretations of some results from the spectral theory in max-plus algebra. We first report definitions from [6] necessary to Theorem 1 (also from [6]). Then we draw the links to our setting and more particularly in the case of a symmetric matrix, corresponding to a non-directed graph. In all this section, $W \in \mathcal{M}_n$.

4.1   General Definitions and Results

Definition 3 (Eigenvector, eigenvalue [6]). *Let $\mathbf{x} \in \mathbb{R}_{\max}^n$ and $\lambda \in \mathbb{R}_{\max}$. Then $\mathbf{x}$ is an* eigenvector *of W with λ as corresponding* eigenvalue *if $W \otimes \mathbf{x} = \lambda \otimes \mathbf{x} = \lambda + \mathbf{x}$. If there exists finite $\mathbf{x}$ and λ solutions to this equation, we say that the eigenproblem is* finitely soluble.

In the graph $\mathcal{G}(W)$, a path $(k_1, \ldots, k_l)$ is called a *circuit* if $k_1 = k_l$. We will note $\mathcal{C}(W)$ the set of all circuits of $\mathcal{G}(W)$. Circuits allow us to distinguish another class of matrices in $\mathcal{M}_n$, called *definite* matrices. They are important to the present framework as they include the doubly-0-astic matrices.

Definition 4 (Definite matrix [6]). *W is said* definite *if $\max_{c \in \mathcal{C}(W)} \omega(c) = 0$. In other words, all the circuits of $\mathcal{G}(W)$ have non positive weights, and at least one circuit c^*, called a* zero-weight circuit, *achieves $\omega(c^*) = 0$.*

To see that if W is row or column-0-astic, then it is definite, it is sufficient to build an increasing path with zero-weight, until one vertex repeats. The path can be initialized with any vertex j_1. Then given the current path $(j_1, \ldots, j_m)$, we extend it by adding a vertex j_{m+1} such that $w_{j_m j_{m+1}} = 0$. This is always possible thanks to the row or column-0-asticity of W. Since there are n distinct vertices in $\mathcal{G}(W)$, an index will repeat after at most n iterations.

Definition 5 (Eigen-node, equivalent eigen-nodes [6]**).** *Let W be a definite matrix. An* eigen-node *is any vertex in $\mathcal{G}(W)$ belonging to a zero-weight circuit. Two eigen-nodes are said* equivalent *if there is a zero-weight circuit passing through both of them.*

In [6], $S_n(W) = \bigvee_{1 \leq k \leq n} W^k$ is denoted by $\Delta(W)$ and called the **metric matrix**. Recall that for $i, j \in [\![1, n]\!]$, $\Delta(W)_{ij}$ is the maximal weight over the set of paths from vertex i to vertex j of length at most n, in $\mathcal{G}(W)$ (Proposition 8). If W is definite, circuits have non-positive weights in $\mathcal{G}(W)$ and therefore any path longer than n can be reduced to a shorter path with non larger weight. Hence, $\Delta(W)_{ij}$ is actually the maximal weight over the set of *all* paths from i to j. This provides an easy characterisation of eigen-nodes for W definite: j is an eigen-node of $\mathcal{G}(W)$ if and only if $\Delta(W)_{jj} = 0$. Furthermore, the j-th column ξ_j of $\Delta(W)$ is a map of the ancestors of j in $\mathcal{G}(W)$. It tells which vertices can reach j and at which cost.

Definition 6 (Fundamental eigenvectors, eigenspace [6]**).** *Let W be a definite matrix. Then a* fundamental eigenvector *of W is any j-th column ξ_j of $\Delta(W)$, where j is an eigen-node. Two fundamental eigenvectors are said* equivalent *if their associated eigen-nodes are equivalent (see Definition 5).*

Let $\mathcal{E} = \{\xi_{i_1}, \xi_{i_2}, \ldots, \xi_{i_k}\}$ be a set of $k \geq 1$ fundamental eigenvectors of W, all pairwise non-equivalent. The set $\mathcal{E}$ is said to be a maximal set of non-equivalent fundamental eigenvectors *if any other fundamental eigenvector of W is equivalent to one of the eigenvectors in $\mathcal{E}$.*

In this case the set $\{\bigvee_{j=1}^{k} x_j + \xi_{i_j}, \mathbf{x} \in \mathbb{R}_{\max}^k\}$ is called the eigenspace *of W and does not depend on $\mathcal{E}$ (see [6], Lemma 24-1).*

Theorem 1 ([6]**).** *Let W be a doubly-0-astic matrix. Then the following statements are valid:*

- *For any fundamental eigenvector ξ_j of W (finite or not), $W \otimes \xi_j = \xi_j$.*
- *The eigenproblem is finitely soluble.*
- *If two fundamental eigenvectors (finite or not) are equivalent, then they are equal.*
- *Any finite eigenvector is associated to the eigenvalue $\lambda = 0$, and lies in the eigenspace of W.*

4.2   Consequences and Interpretations

In general. As said, the results of the previous section apply to our setting since we consider adjunctions represented by doubly-0-astic matrices which are both definite and 0-astic. For $W \in \mathcal{D}_0(n)$, $\Delta(W)$ is also in $\mathcal{D}_0(n)$ and the corresponding opening $\delta_{\Delta(W)}\varepsilon_{\Delta(W)}$ is $G_W^{[n]}$. By definition, $G_W^{[n]}(\mathbf{x})$ projects $\mathbf{x} \in \mathcal{L}$ onto $\delta_{\Delta(W)}(\mathcal{L})$, which is the set $\{\bigvee_{j=1}^n y_j + \xi_j, \mathbf{y} \in \mathcal{L}\}$ of max-plus combinations of columns of $\Delta(W)$. Theorem 1 tells that this decomposition can be split as $G_W^{[n]}(\mathbf{x}) = \mathbf{u} \vee \mathbf{v}$, where $\mathbf{u}$ lies in the eigenspace of W and $\mathbf{v}$ is a max-plus combination of the ξ_j which are not fundamental eigenvectors. This decomposition may be sparser than the original one, as the dimension of the eigenspace of W, i.e. $Card(\mathcal{E})$, can be lower than the number of fundamental eigenvectors.

The case of symmetric $W \in \mathcal{D}_0(n)$. This case corresponds to considering a non-directed graph supporting the signal $\mathbf{x}$. As the adjacency relationship is often based on a symmetrical function on pairs of vertex values, this assumption covers many practical cases (e.g. [2,3]). The main consequence of $W \in \mathcal{D}_0(n)$ symmetric is that *every vertex j is an eigen-node*: for any $j \in [\![1,n]\!]$ there is i such that $w_{ij} = 0 = w_{ji}$ and therefore (j,i,j) is a zero-weight circuit. This entails three other consequences.

First, $\Delta(W)_{jj} = 0$ for every $j \in [\![1,n]\!]$, following the characterisation of eigen-nodes described earlier, which implies that $\delta_{\Delta(W)} = D_W^{[n]}$ is extensive and $\varepsilon_{\Delta(W)} = E_W^{[n]}$ anti-extensive (Proposition 3). Secondly, $W \otimes \xi_j = \xi_j$ for every column ξ_j of $\Delta(W)$, which implies $W^k \otimes \xi_j = \xi_j$ for $1 \leq k \leq n$, hence $\Delta(W) \otimes \xi_j = \xi_j$ and finally $\Delta(W) \otimes \Delta(W) = \Delta(W)$. This means $D_W^{[n]}$ and $E_W^{[n]}$ are idempotent. They are therefore a closing and an opening respectively and $E_W^{[n]} = G_W^{[n]}$, since an adjunction (ε, δ) for which ε is an opening and δ a closing verifies $\varepsilon = \delta\varepsilon$ (and $\delta = \varepsilon\delta$). The third consequence is the following.

Corollary 1. *If $W \in \mathcal{D}_0(n)$ is symmetric, then the set of invariants of $G_W^{[n]}$ is exactly the eigenspace of W.*

When W is symmetric, a maximal set of k non-equivalent fundamental eigenvectors $\{\xi_{i_1}, \xi_{i_2}, \ldots, \xi_{i_k}\}$, $k \leq n$, can be seen as negative distance maps to the k corresponding eigen-nodes $\mathcal{G}(W)$, as they contain the optimal cost (maximal weight) between any vertex and the eigen-nodes[2]. Hence we can picture the aspect of $G_W^{[n]}(\mathbf{x})$, for $\mathbf{x} \in \mathcal{L}$: it is the upper-envelope of the largest vertical translations of these distance maps that are dominated by $\mathbf{x}$. Therefore, adapting $\mathcal{G}(W)$ to $\mathbf{x}$ by well connecting vertices within relevant structures preserves these structures under the filter $G_W^{[n]}$, as shown in [2,3]. In practice, n might be large, such as the number of pixels of an image. Since $(G_W^{[p]})_{1 \leq p \leq n}$ is a granulometry, we know that $G_W^{[n]}$ can be approximated by $G_W^{[p]}$ with increasing p.

---

[2] Note that $\Delta(W)$ is a metric, not exactly between vertices, but between their equivalence classes induced by Definition 5, as all vertices are eigen-nodes when W is symmetric.

5   Conclusion

In this paper we consolidated the basis of the representation of adjunctions by matrices in max-plus algebra. We showed that it is a very flexible framework that generalises many types of morphological adjunctions. In particular, it allows describing precisely the behaviour of iterated operators based on spatially-variant, non-flat structuring functions. This is made possible by their graph interpretation and spectral results in max-plus algebra. Future works shall investigate further the insights that max-plus algebra can bring to mathematical morphology through this framework.

References

1. Akian, M., Bapat, R., Gaubert, S.: Max-plus algebra. In: Handbook of Linear Algebra (Discrete Mathematics and its Applications), vol. 39, pp. 10–14 (2006)
2. Blusseau, S., Velasco-Forero, S., Angulo, J., Bloch, I.: Tropical and morphological operators for signal processing on graphs. In: 25th IEEE International Conference on Image Processing, pp. 1198–1202 (2018)
3. Blusseau, S., Velasco-Forero, S., Angulo, J., Bloch, I.: Adaptive anisotropic morphological filtering based on co-circularity of local orientations. Image Process. Line 12, 111–141 (2022). https://doi.org/10.5201/ipol.2022.397
4. Bouaynaya, N., Charif-Chefchaouni, M., Schonfeld, D.: Theoretical foundations of spatially-variant mathematical morphology part I: binary images. IEEE Trans. Pattern Anal. Mach. Intell. 30(5), 823–836 (2008)
5. Carré, B.A.: An algebra for network routing problems. IMA J. Appl. Math. 7(3), 273–294 (1971)
6. Cuninghame-Green, R.A.: Minimax Algebra, vol. 166. Springer-Verlag, Berlin Heidelberg (1979). https://doi.org/10.1007/978-3-642-48708-8
7. Debayle, J., Pinoli, J.C.: General adaptive neighborhood image processing: Part I: Introduction and theoretical aspects. J. Math. Imaging Vis. 25(2), 245–266 (2006)
8. Heijmans, H., Buckley, M., Talbot, H.: Path openings and closings. J. Math. Imaging Vis. 22(2), 107–119 (2005). https://doi.org/10.1007/s10851-005-4885-3
9. Heijmans, H., Ronse, C.: The algebraic basis of mathematical morphology I. Dilations and erosions. Comput. Vis. Graph. Image Process. 50(3), 245–295 (1990)
10. Lerallut, R., Decencière, E., Meyer, F.: Image filtering using morphological amoebas. Image Vis. Comput. 25(4), 395–404 (2007)
11. Maragos, P.: Chapter two - representations for morphological image operators and analogies with linear operators. Adv. Imag. Electron Phys. 177, 45–187 (2013)
12. Ronse, C.: Why mathematical morphology needs complete lattices. Sig. Process. 21(2), 129–154 (1990)
13. Salembier, P.: Study on nonlocal morphological operators. In: 16th IEEE International Conference on Image Processing, pp. 2269–2272 (2009)
14. Velasco-Forero, S., Angulo, J.: On nonlocal mathematical morphology. In: Hendriks, C.L.L., Borgefors, G., Strand, R. (eds.) ISMM 2013. LNCS, vol. 7883, pp. 219–230. Springer, Heidelberg (2013). https://doi.org/10.1007/978-3-642-38294-9_19

15. Velasco-Forero, S., Angulo, J.: Nonlinear operators on graphs via stacks. In: Nielsen, F., Barbaresco, F. (eds.) GSI 2015. LNCS, vol. 9389, pp. 654–663. Springer, Cham (2015). https://doi.org/10.1007/978-3-319-25040-3_70
16. Verdú-Monedero, R., Angulo, J., Serra, J.: Anisotropic morphological filters with spatially-variant structuring elements based on image-dependent gradient fields. IEEE Trans. Image Process. **20**(1), 200–212 (2011)

Hierarchical and Graph-Based Models, Analysis and Segmentation

A Topological Tree of Shapes

Nicolas Passat[1]([✉])[iD] and Yukiko Kenmochi[2][iD]

[1] Université de Reims Champagne Ardenne, CReSTIC EA 3804, 51100 Reims, France
nicolas.passat@univ-reims.fr
[2] Normandie Univ, UNICAEN, ENSICAEN, CNRS, GREYC, 14050 Caen, France

Abstract. In this article, we enrich the framework of morphological hierarchies with new acyclic graphs and trees. These structures lie at the convergence of hierarchical models and topological descriptors. We define them in the context of digital grey-level imaging. We discuss their links with component-trees, trees of shapes and adjacency trees. This analysis leads to new notions, including a notion of *topological tree of shapes*.

1 Introduction

Many hierarchical, graph-based structures have been defined in the framework of mathematical morphology, especially for designing connected operators [25]. The most popular are trees (i.e. rooted, connected, acyclic graphs). They model finite sets of partitions organized with respect to the refinement order relation. These partitions can be partial. This is the case of the component-tree and its variants [9,24], the level-line tree (a.k.a. tree of shapes) and its variants [3,11]. These partitions can also be total. This is the case of the binary partition tree and its variants [19,23,27] and the hierarchical watershed [13,26]. Other hierarchical structures are directed acyclic graphs (DAGs), e.g. the component-hypertree [15], the component-graph [17], the braid of partitions [8] and the directed component hierarchy [18].

The partitions modeled by these hierarchical structures are composed of connected sets defined with respect to a topology defined on a given space which is generally discrete (e.g. a part of $\mathbb{Z}^n$ [22], a complex on/tesselation of $\mathbb{R}^n$). Hierarchical structures carry intrinsic, topological information. However, these information are often limited and generally not sufficient to perform high-level topological analysis of the modeled images/data. In particular, hierarchical structures are generally less informative than high-level topological invariants/descriptors, e.g. the homology groups/homology persistence [6] or the homotopy type.

In this article, we introduce a new family of hierarchical structures—DAGs and trees, including a new notion of *topological tree of shapes*—dedicated to the modeling of grey-level images. They aim to gather (i) connectedness/intensity information carried by component- (min- and max-) trees [24] and (ii) topological information carried by the adjacency tree, a classical topological invariant [21]. Basically, we will first build a DAG that is composed by the min-tree and max-tree

© Springer Nature Switzerland AG 2022
E. Baudrier et al. (Eds.): DGMM 2022, LNCS 13493, pp. 221–235, 2022.
https://doi.org/10.1007/978-3-031-19897-7_18

of a grey-level image, and we will enrich the nodes of these two trees by the adjacency tree structure at each grey-level, leading to the notion of a graph of valued shapes. Then, we will establish that this graph of valued shape can be simplified in a lossless fashion as a tree structure by discarding some transitive, redundant edges. This will lead to a simpler tree structure called tree of valued shapes. By factorizing some spatially equivalent nodes, we will then define a more compact structure, called the complete tree of shapes. We will establish that this complete tree of shapes can be reduced (in a lossy fashion) in two different ways, leading on the one hand to the usual notion of a tree of shapes and on the other hand to the new notion of a topological tree of shapes. (The chosen terminology of *topological trees of shapes* is justified by the way it is defined; however it will be shown to be different from another homonymous notion previously introduced in the literature.) We will finally evoke the links between these new structures (graph and tree of valued shapes, complete tree of shapes, topological tree of shapes) and usual morphological trees (component-tree, tree of shapes, adjacency tree).

This article is organized as follows. Section 2 provides definitions related to hierarchies and grey-level images. We introduce the notions of graph of valued shapes and tree of valued shapes in Sects. 3 and 4, respectively. Section 5 derives from the tree of valued shapes the two essential notions of this work, namely the complete tree of shapes (that generalizes the tree of shapes) and the topological tree of shapes. In Sect. 6, we discuss on the links that exist between these new notions and well-known morphological hierarchies. We provide concluding remarks in Sect. 7.

2   Basics: Hierarchies and Images

Definition 1 (Hierarchical order). *Let X be a set and $\leq$ be an order on X. We say that $\leq$ is a hierarchical order if $\forall x \in X$ the subset $x^{\uparrow} = \{y \in X \mid x < y\}$ is totally ordered.*

Definition 2 (Hierarchical function). *Let X be a set and $\leq$ a hierarchical order on X. The hierarchical function $\zeta_{\leq} : X \to X$ is defined by $\zeta_{\leq}(x) = \bigwedge_{\leq} x^{\uparrow}$. This function is defined everywhere on X except for the greatest elements of $(X, \leq)$.*

Remark 3. *Let $\lhd$ be the Hasse relation obtained from $\leq$ by reflexive-transitive reduction. $\forall x \in X$ we have*

$$x \lhd \zeta_{\leq}(x) \tag{1}$$

This formula induces an isomorphism between $(X, \lhd)$ and $(X, \zeta_{\leq})$.

Definition 4 (Tree). *Let X be a set and $\leq$ a hierarchical order on X such that $(X, \leq)$ admits a maximum. The Hasse diagram $(X, \lhd)$ is called a tree.*

Let $\mathbb{U}$ be a discrete set endowed with a topological structure which provides the notions of adjacency and connectedness, and where the separation theorem (Jordan-Brouwer) holds.

Remark 5. *In this article we choose $\mathbb{U} = \mathbb{Z}^n$ ($n \geq 2$) and we consider the usual framework of digital topology on binary images [22], with the standard couples of $(2n, 3^n - 1)$ and $(3^n - 1, 2n)$-adjacencies for the foreground and background.*

Let $\mathbb{K}$ be a set of values endowed with a total order $\leqslant_{\mathbb{K}}$. Let $\mathcal{F} : \mathbb{U} \to \mathbb{K}$ be an application. We assume that there exist a finite, nonempty subset $\mathbb{S} \subset \mathbb{U}$ and two values $\perp <_{\mathbb{K}} \top \in \mathbb{K}$ such that for all $\mathbf{x} \in \mathbb{U}$

$$\begin{cases} \mathcal{F}(\mathbf{x}) = \perp & \text{if } \mathbf{x} \notin \mathbb{S} \\ \perp <_{\mathbb{K}} \mathcal{F}(\mathbf{x}) <_{\mathbb{K}} \top & \text{if } \mathbf{x} \in \mathbb{S} \end{cases} \tag{2}$$

We set $\mathbb{V} = \mathcal{F}(\mathbb{S}) \cup \{\perp, \top\}$. It is a finite set that we equip with the total order $\leqslant_{\mathbb{V}}$ induced by $\leqslant_{\mathbb{K}}$.

Remark 6. *The application $\mathcal{F}$ is isomorphic to a grey-level image taking its values in an interval of $\mathbb{Z}$ of size $|\mathbb{V}|$, e.g. $[\![0, |\mathbb{V}| - 1]\!]$. See Fig. 1(a).*

3   Graph of Valued Shapes

3.1   (Valued) Connected Components

We set $\leqslant^{\circ} = \leqslant_{\mathbb{V}}$ and $\leqslant^{\bullet} = \geqslant_{\mathbb{V}}$. Let $v \in \mathbb{V}$. We define the threshold sets of $\mathcal{F}$ at value $v \in \mathbb{V}$ (see Fig. 1(b–i)) as

$$\begin{aligned} \Lambda_v^{\circ}(\mathcal{F}) &= \{\mathbf{x} \in \mathbb{U} \mid v \leqslant^{\circ} \mathcal{F}(\mathbf{x})\} \\ \Lambda_v^{\bullet}(\mathcal{F}) &= \{\mathbf{x} \in \mathbb{U} \mid v <^{\bullet} \mathcal{F}(\mathbf{x})\} \end{aligned} \tag{3}$$

Let $X \subseteq \mathbb{U}$. When X is nonempty, we note $\Pi[X] \subset 2^{\mathbb{U}}$ the partition gathering all the connected components of X. If X is empty, we set $\Pi[X] = \emptyset$.

Let $v \in \mathbb{V}$. We set

$$\begin{aligned} \Xi_v^{\circ} &= \Theta_v^{\circ} \times \{v\} & \Theta_v^{\circ} &= \Pi[\Lambda_v^{\circ}(\mathcal{F})] \\ \Xi_v^{\bullet} &= \Theta_v^{\bullet} \times \{v\} & \text{with} \quad \Theta_v^{\bullet} &= \Pi[\Lambda_v^{\bullet}(\mathcal{F})] \\ \Xi_v &= \Theta_v \times \{v\} & \Theta_v &= \Theta_v^{\circ} \cup \Theta_v^{\bullet} \end{aligned} \tag{4}$$

Remark 7. *We have $\Xi_\top^\circ = \Xi_\bot^\bullet = \emptyset$. For any $v \in \mathbb{V} \setminus \{\bot, \top\}$, we have $\Xi_v^\circ \neq \emptyset$ and $\Xi_v^\bullet \neq \emptyset$. In the sequel, $(\mathbb{U}, \bot)$ and $(\mathbb{U}, \top)$ are considered as a unique element noted ∞. Then, we have $\Xi_\bot^\circ = \{(\mathbb{U}, \bot)\} = \{\infty\} = \{(\mathbb{U}, \top)\} = \Xi_\top^\bullet$ and $\Xi_\bot = \Xi_\top = \{\infty\}$.*

We set

$$
\begin{aligned}
\Xi^\circ &= \bigcup_{v \in \mathbb{V}} \Xi_v^\circ \\
\Xi^\bullet &= \bigcup_{v \in \mathbb{V}} \Xi_v^\bullet \\
\Xi &= \Xi^\circ \cup \Xi^\bullet = \bigcup_{v \in \mathbb{V}} \Xi_v
\end{aligned}
\quad \text{and} \quad
\begin{aligned}
\Theta^\circ &= \bigcup_{v \in \mathbb{V}} \Theta_v^\circ \\
\Theta^\bullet &= \bigcup_{v \in \mathbb{V}} \Theta_v^\bullet \\
\Theta &= \Theta^\circ \cup \Theta^\bullet = \bigcup_{v \in \mathbb{V}} \Theta_v
\end{aligned}
\tag{5}
$$

3.2   Orders on Valued Connected Components

We define the partial orders $\sqsubseteq^\circ$ on Ξ° and $\sqsubseteq^\bullet$ on $\Xi^\bullet$ as

$$
\begin{aligned}
((X, v) \sqsubseteq^\circ (Y, w)) &\Leftrightarrow (X \subseteq Y \wedge w \leqslant^\circ v) \\
((X, v) \sqsubseteq^\bullet (Y, w)) &\Leftrightarrow (X \subseteq Y \wedge w \leqslant^\bullet v)
\end{aligned}
\tag{6}
$$

We define the order $\sqsubseteq^\varphi$ as the union of $\sqsubseteq^\circ$ and $\sqsubseteq^\bullet$, i.e. $P \sqsubseteq^\varphi Q$ iff $P \sqsubseteq^\circ Q$ or $P \sqsubseteq^\bullet Q$.

Remark 8. $\sqsubseteq^\varphi$, $\sqsubseteq^\circ$ *and* $\sqsubseteq^\bullet$ *are hierarchical orders. They admit ∞ as maximum.*

Let $v \in \mathbb{V}$. We define the order $\sqsubseteq^v$ on Ξ_v as

$$
((X, v) \sqsubseteq^v (Y, v)) \Leftrightarrow \tau(X) \subseteq \tau(Y)
\tag{7}
$$

where $\tau : 2^{\mathbb{U}} \to 2^{\mathbb{U}}$ is the hole closing application defined by $\tau(X) = X \cup \bigcup Z$ where $Z \subseteq \Pi[\overline{X}]$ is composed by the *finite* connected components of $\overline{X} = \mathbb{U} \setminus X$.

We define the order $\sqsubseteq^\psi$ on Ξ as $\sqsubseteq^\psi = \bigcup_{v \in \mathbb{V}} \sqsubseteq^v$, i.e. $P \sqsubseteq^\psi Q$ iff $\exists v \in \mathbb{V}$ such that $P \sqsubseteq^v Q$.

Remark 9. $\sqsubseteq^\psi$, *and* $\sqsubseteq^v$ *($v \in \mathbb{V}$) are hierarchical orders. Each ordered set $(\Xi_v, \sqsubseteq^v)$ ($v \in \mathbb{V}$) admits a maximum $(\mathbb{U}_v, v)$ where $\mathbb{U}_v \subseteq \mathbb{U}$ is the unique element of Θ_v which is infinite.*

We note $\lhd^\varphi$ (resp. $\lhd^\circ$, $\lhd^\bullet$) and $\lhd^\psi$ (resp. $\lhd^v$) the Hasse relations associated to $\sqsubseteq^\varphi$ (resp. $\sqsubseteq^\circ$, $\sqsubseteq^\bullet$) and $\sqsubseteq^\psi$ (resp. $\sqsubseteq^v$). The graph $(\Xi, \lhd^\varphi)$ is "similar" to the union of the max- and min-trees, whereas $(\Xi, \lhd^\psi)$ is "similar" to the union of the adjacency trees of each threshold set of $\mathcal{F}$. This will be more formally discussed in Sect. 6.

Remark 10. *Let $P = (X, v), Q = (Y, w) \in \Xi$ such that $P \lhd^\varphi Q$. We have $P, Q \in \Xi^\star$ with $\star =$ either $\circ$ or $\bullet$. In addition, we have $X \subseteq Y$ and $v = \zeta_{\leqslant^\star}(w)$.*

Remark 11. *Let $P = (X, v), Q = (Y, w) \in \Xi$ such that $P \lhd^\psi Q$. We have $(P \in \Xi^\bullet$ and $Q \in \Xi^\circ)$ or $(P \in \Xi^\circ$ and $Q \in \Xi^\bullet)$. In addition, we have $v = w$ and $\tau(X) \in \Pi[\overline{Y}]$.*

We note $\varphi = \zeta_{\sqsubseteq^\varphi}$, $\varphi^\circ = \zeta_{\sqsubseteq^\circ}$, $\varphi^\bullet = \zeta_{\sqsubseteq^\bullet}$, $\psi = \zeta_{\sqsubseteq^\psi}$ and $\psi^v = \zeta_{\sqsubseteq^v}$ $(v \in \mathbb{V})$.

3.3 Definition of the Graph of Valued Shapes

Let $\lhd_\Xi$ be the relation defined as the union of $\lhd^\varphi$ and $\lhd^\psi$.

Definition 12 (Graph of valued shapes). *The graph of valued shapes (or VS-graph, for brief) is the couple $\mathfrak{G}_{VS} = (\Xi, \lhd_\Xi)$.*

Remark 13. *The intersection between $\lhd^\varphi$ and $\lhd^\psi$ is empty. We can then consider $\mathfrak{G}_{VS}$ as $(\Xi, \lhd_\Xi)$ or as $(\Xi, \lhd^\varphi, \lhd^\psi)$ and equivalently as (Ξ, φ, ψ).*

Property 14. $\mathfrak{G}_{VS} = (\Xi, \lhd_\Xi)$ *is a directed acyclic graph.*

We define $\sqsubseteq_\Xi$ as the reflexive-transitive closure of $\lhd_\Xi$.

Remark 15. $(\Xi, \sqsubseteq_\Xi)$ *is an ordered set that admits ∞ as maximum.*

4 Tree of Valued Shapes

4.1 Transitive Reduction of the Graph of Valued Shapes

Let $\blacktriangleleft_\Xi$ be the relation on Ξ defined as the transitive reduction of $\lhd_\Xi$.
    Let $P \in \Xi$. Let us consider the following three equalities

$$\psi(P) = [\varphi \circ \psi \circ \varphi](P) \tag{8}$$
$$\varphi(P) = [\varphi \circ \psi \circ \psi](P) \tag{9}$$
$$\varphi(P) = [\varphi^{|\mathbb{V}|-2} \circ \psi](P) \tag{10}$$

Remark 16. *If P satisfies Eq. (8), then we have $P \lhd_\Xi \psi(P)$ and $P \blacktriangleleft_\Xi \psi(P)$. If P satisfies Eq. (9) or (10), then we have $P \lhd_\Xi \varphi(P)$ and $P \blacktriangleleft_\Xi \varphi(P)$.*

Proposition 17. *Let $P \in \Xi$ be such that $\varphi(P)$ and $\psi(P)$ exist. One of Eqs. (8–10) is satisfied.*

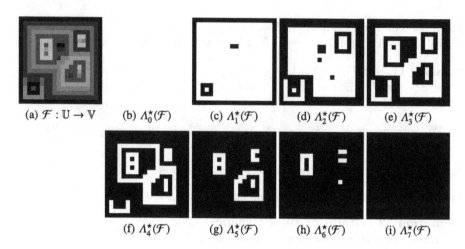

Fig. 1. (a) A grey-level image $\mathcal{F} : \mathbb{U} \to \mathbb{V}$ ($\mathbb{U} = \mathbb{Z}^2$ and $\mathbb{V} = [\![0,7]\!]$). (b–i) The threshold sets $\Lambda_v^\star(\mathcal{F})$ ($\Lambda_v^\circ(\mathcal{F})$ in white; $\Lambda_v^\bullet(\mathcal{F})$ in black), for $v = 0$ (b) to 7 (i).

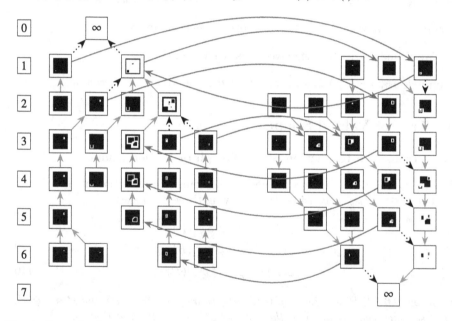

Fig. 2. Tree of valued shapes of the image $\mathcal{F}$ (Fig. 1(a)). The valued connected components are depicted by squares (Ξ° on the left side; $\Xi^\bullet$ on the right side) and are positioned with respect to the threshold value v (see on left), from 0 (top) to 7 (bottom). Red and green arrows correspond to the $\blacktriangleleft_\Xi$ relation. Green and black dotted arrows correspond to the $\lhd^\varphi$ relation. Red arrows are a subset of the $\lhd^\psi$ relation, not fully depicted for the sake of readibility. (Color figure online)

Proof. Let $P = (X, v) \in \Xi$ be such that $\varphi(P)$ and $\psi(P)$ exist. Case 1: $\varphi(P) = (\mathbb{U}, \perp) = \infty$ and $\psi(P) = (\mathbb{U}_v, v)$ (the unique element of $\Pi[\Lambda_v^{\bullet}(\mathcal{F})]$ which is infinite). It is plain that $\varphi^{|\mathbb{V}|-2}((\mathbb{U}_v, v)) = (\mathbb{U}, \top) = \infty = \varphi(P)$, and Eq. (10) then holds. Case 2: $\varphi(P) = (\mathbb{U}, \perp)$ and $\psi(P) \neq (\mathbb{U}_v, v)$. It is plain that $\psi^2(P)$ exists and $\varphi(\psi^2(P)) = (\mathbb{U}, \perp) = \varphi(P)$, thus Eq. (9) holds. Case 3: $\varphi(P) = (\mathbb{U}, \top)$. Since $\psi(P)$ exists (and is finite), it is plain that $\psi^2(P)$ also exists. But then $\varphi(\psi^2(P)) = (\mathbb{U}, \top) = \varphi(P)$, thus Eq. (9) holds. Case 4: $\varphi(P) \neq \infty$. If $\psi(P) = (\mathbb{U}_v, v)$, it is plain that $\psi(\varphi(P)) = (\mathbb{U}_w, w)$ with $(\mathbb{U}_v, v) = \varphi((\mathbb{U}_w, w))$ and Eq. (8) then holds. Let us now suppose that $\psi(P) \neq (\mathbb{U}_v, v)$ and that Eq. (8) does not hold, i.e. $[\varphi \circ \psi \circ \varphi](P) \neq \psi(P)$. Then we have $P \sqsubset^{\psi} \psi^2(P) \sqsubset^{\psi} [\varphi \circ \psi \circ \varphi](P)$. Now, let us suppose that $\varphi(\psi^2(P)) \neq \varphi(P)$. Then we have $\varphi(P) \sqsubset^{\psi} \varphi(\psi^2(P))$ and it comes $\varphi(P) \sqsubset^{\psi} \psi(\varphi(P)) \sqsubset^{\psi} \varphi(\psi^2(P))$ in contradiction with the Jordan theorem. Thus Eq. (9) holds. ∎

If follows from Proposition 17 that for any $P \in \Xi$ such that both $\psi(P)$ and $\varphi(P)$ exist we have $P \blacktriangleleft_{\Xi} \psi(P)$ or $P \blacktriangleleft_{\Xi} \varphi(P)$. Since $(\Xi, \sqsubseteq_{\Xi})$ admits a maximum (namely ∞), for each $P \in \Xi$, $P \neq \infty$, we have either $P \blacktriangleleft_{\Xi} \psi(P)$ or $P \blacktriangleleft_{\Xi} \varphi(P)$. The following property derives from these facts.

Property 18. *Let $P \in \Xi$.*

- *If $\varphi(P)$ is defined and $\psi(P)$ is not, then $P \blacktriangleleft_{\Xi} \varphi(P)$.*
- *If $\psi(P)$ is defined and $\varphi(P)$ is not, then $P \blacktriangleleft_{\Xi} \psi(P)$.*
- *If $\varphi(P)$ and $\psi(P)$ are defined, then either $P \blacktriangleleft_{\Xi} \varphi(P)$ or $P \blacktriangleleft_{\Xi} \psi(P)$.*

Remark 19. *The transitive reduction from $\mathfrak{G}_{\mathrm{VS}} = (\Xi, \lhd_{\Xi})$ to $(\Xi, \blacktriangleleft_{\Xi})$ is a lossless compression. The graph $\mathfrak{G}_{\mathrm{VS}} = (\Xi, \lhd_{\Xi})$ can be reconstructed from $(\Xi, \blacktriangleleft_{\Xi})$.*

4.2 Definition of the Tree of Valued Shapes

Property 20. *$(\Xi, \blacktriangleleft_{\Xi})$ is a tree. Equivalently, $\sqsubseteq_{\Xi}$ is a hierarchical order on Ξ.*

Definition 21 (Tree of valued shapes). *The tree of valued shapes (or VS-tree, for brief) is the couple $\mathfrak{T}_{\mathrm{VS}} = (\Xi, \blacktriangleleft_{\Xi})$. See Fig. 2.*

5 Complete Tree of Shapes and Topological Tree of Shapes

5.1 Spatial Compression: From the Tree of Valued Shapes to the Complete Tree of Shapes

Let $\pi_{\Theta} : \Xi \to \Theta$ be the function defined by $\pi_{\Theta}((X, v)) = X$. Let $\sim_{\Theta}$ be the equivalence relation on Ξ defined by

$$P \sim_{\Theta} Q \Leftrightarrow \pi_{\Theta}(P) = \pi_{\Theta}(Q) \tag{11}$$

Property 22. *The function $\tilde{\pi}_{\Theta} : \Xi/\sim_{\Theta} \to \Theta$ defined by $\tilde{\pi}_{\Theta}([P]_{\sim_{\Theta}}) = \pi_{\Theta}(P)$ is a bijection.*

Remark 23. *Based on the above property, we identify $\Xi/\sim_\Theta$ and Θ. More precisely, for any $P = (X, v) \in \Xi$, we identify $[P]_{\sim_\Theta}$ and X. In particular, we have $[\infty]_{\sim_\Theta} = \{\infty\} \in \Xi/\sim_\Theta$ and it is identified to $\mathbb{U} \in \Theta$.*

Property 24. *Let $K \in \Xi/\sim_\Theta$. Let $\sqsubseteq_K$ be the order induced by $\sqsubseteq_\Xi$ on K. Then $(K, \sqsubseteq_K)$ is a totally ordered set.*

For any $K \in \Xi/\sim_\Theta$, we note $\langle K \rangle_\Theta = \bigwedge_{\sqsubseteq_\Xi} K$ and $\langle\!\langle K \rangle\!\rangle_\Theta = \bigvee_{\sqsubseteq_\Xi} K$. We note $\rho_\Xi = \zeta_{\sqsubseteq_\Xi}$.

Remark 25. *From Property 24, it follows that $\forall p \in [\![1, |K| - 1]\!]$ we have $\rho_\Xi^p(\langle K \rangle_\Theta) \in K$. In particular, we have $\rho_\Xi^{|K|-1}(\langle K \rangle_\Theta) = \langle\!\langle K \rangle\!\rangle_\Theta$ whereas $\rho_\Xi^{|K|}(\langle K \rangle_\Theta) \notin K$.*

Proposition 26. *Let $K \in \Xi/\sim_\Theta$, $K \neq \{\infty\}$. Let $P = \rho_\Xi^{|K|}(\langle K \rangle_\Theta)$. We have $P = \langle [P]_{\sim_\Theta} \rangle_\Theta$.*

Proof. Let $P = \rho_\Xi^{|K|}(\langle K \rangle_\Theta) = \rho_\Xi(\langle\!\langle K \rangle\!\rangle_\Theta)$, with $\langle\!\langle K \rangle\!\rangle_\Theta = (X, v)$ and $P = (Y, w)$. In particular, we have $\langle\!\langle K \rangle\!\rangle_\Theta \blacktriangleleft_\Xi P$ and thus $\langle\!\langle K \rangle\!\rangle_\Theta \lhd_\Xi P$. Let $Q = (Y, u) = \langle [P]_{\sim_\Theta} \rangle_\Theta$. Case 1: $\langle\!\langle K \rangle\!\rangle_\Theta \lhd^\varphi P$. This implies $X \subseteq Y$ and $w \leqslant^\star v$ (with $\star = $ either $\circ$ or $\bullet$). As $P \notin K$, we have $X \subset Y$, and it follows that $w <^\star v$. We have $w \leqslant^\star u$. If $u \neq w$, then we have $w <^\star u$ and it follows that $\langle\!\langle K \rangle\!\rangle_\Theta \sqsubset^\varphi Q \sqsubset^\varphi P$, and thus $\langle\!\langle K \rangle\!\rangle_\Theta \ntriangleleft^\varphi P$: a contradiction. Then, we have $u = w$, and it follows that $P = \langle [P]_{\sim_\Theta} \rangle_\Theta$. Case 2: $\langle\!\langle K \rangle\!\rangle_\Theta \lhd^\psi P$. We have $X \in$ either Θ_v° or $\Theta_v^\bullet$ (for instance, Θ_v°; the same reasoning holds with $\Theta_v^\bullet$) and $\varphi(\langle\!\langle K \rangle\!\rangle_\Theta)$ exists. Let $R = (Z, t) = \varphi(\langle\!\langle K \rangle\!\rangle_\Theta)$. We have $X \subseteq Z$, and since $R \notin K$, it comes $X \subset Z$. Let us suppose that $P \neq Q$. Then, we have $S = (Y, t) \in [P]_{\sim_\Theta}$. It follows that $\tau(X) = \tau(Z)$. Consequently, we have $R \lhd^\psi S$. But then, we obtain $\langle\!\langle K \rangle\!\rangle_\Theta \lhd^\psi R \lhd^\psi S \lhd^\varphi P$, in contradiction with $\langle\!\langle K \rangle\!\rangle_\Theta \blacktriangleleft_\Xi P$. Then, we have $P = \langle [P]_{\sim_\Theta} \rangle_\Theta$. ∎

For any $K \in \Xi/\sim_\Theta$, we consider $\langle K \rangle_\Theta \in \Xi$ as canonical element, and we identify $\langle K \rangle_\Theta = (X, v)$ with $X \in \Theta$.

Let $\sqsubseteq_\Theta$ be the order on $\{\langle K \rangle_\Theta \mid K \in \Xi/\sim_\Theta\} \subseteq \Xi$—and equivalently on Θ—induced by $\sqsubseteq_\Xi$.

We note $\kappa_\Theta = \tilde{\pi}_\Theta^{-1}$. Let $\rho_\Theta : \Theta \to \Theta$ be the function defined by

$$\rho_\Theta(X) = \pi_\Theta(\rho_\Xi^{|K|}(\langle K \rangle)) \tag{12}$$

with $K = \kappa_\Theta(X)$.

Remark 27. *From the above property, we have $\rho_\Theta = \zeta_{\sqsubseteq_\Theta}$. We note $\blacktriangleleft_\Theta$ the relation on Θ associated to ρ_Θ, namely the Hasse relation of $\sqsubseteq_\Theta$.*

Definition 28 (Complete tree of shapes). *The complete tree of shapes (or CS-tree, for brief) is the couple $\mathfrak{T}_{\mathrm{CS}} = (\Theta, \blacktriangleleft_\Theta)$. See Fig. 3 (left).*

The following proposition directly derives from the above results.

Proposition 29. *The equivalence relation $\sim_\Theta$ induces a decreasing homeomorphism from $\mathfrak{T}_{VS} = (\Xi, \blacktriangleleft_\Xi)$ to $\mathfrak{T}_{CS} = (\Theta, \blacktriangleleft_\Theta)$.*

Remark 30. *The homeomorphism from $\mathfrak{T}_{VS} = (\Xi, \blacktriangleleft_\Xi)$ to $\mathfrak{T}_{CS} = (\Theta, \blacktriangleleft_\Theta)$ is a lossless compression. The tree $\mathfrak{T}_{VS} = (\Xi, \blacktriangleleft_\Xi)$ can be fully reconstructed from $\mathfrak{T}_{CS} = (\Theta, \blacktriangleleft_\Theta)$.*

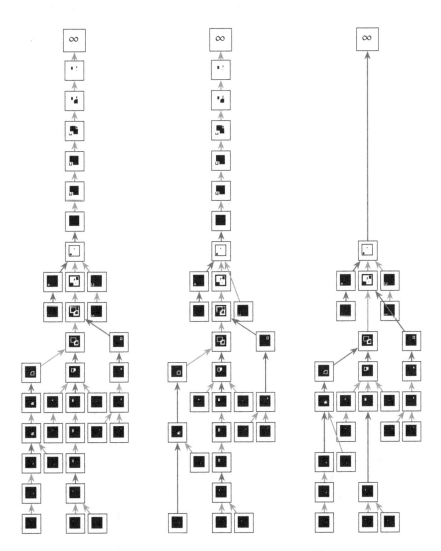

Fig. 3. From left to right: (1) the complete tree of shapes; (2) the topological tree of shapes; (3) the tree of shapes of the image $\mathcal{F}$ of Fig. 1. The complete tree of shapes (1) derives from the reduction of the tree of valued shapes of Fig. 2. Green arrows are originated from the $\lhd^\varphi$ relation. Red arrows are originated from the $\lhd^\psi$ relation. (Color figure online)

5.2 Topological Compression: From the Tree of Valued Shapes to the Topological Tree of Shapes

Let $X, Y \in \mathbb{U}$, with $Y \subset X$. We aim to characterize the preservation of topological properties by a decreasing transformation from X to Y. A frequent strategy is to consider the notion of homotopic transformation. In particular, if there exists a (decreasing) homotopic transformation from X to Y, then X and Y have the same homotopy type. However, this is hardly tractable in 3D [10] and in higher dimensions. Then we consider a weaker topological invariant induced by the notion of strongly deletable set [20]. More precisely, if $X \setminus Y$ is strongly deletable, then the inclusion relation induces a bijection between the (foreground and background) connected components of X and those of Y.

Remark 31. *If* $\mathbb{U} = \mathbb{Z}^2$, *the notion of strongly deletable set is equivalent to the notion of simple set [14]. This implies that if* $X \setminus Y$ *is strongly deletable, then* X *and* Y *have the same homotopy type and* Y *is obtained from* X *by a decreasing homotopic transformation defined as the iterative removal of a sequence of simple points [4].*

Let $P, Q \in \Xi$ be such that $\rho_\Xi(P) = Q$. If $\rho_\Xi^{-1}(\{Q\}) = \{P\}$ and $\pi_\Theta(Q) \setminus \pi_\Theta(P)$ is a strongly deletable set, then we note $Q \searrow P$.

Remark 32. *If* $Q \searrow P$, *then we have* $\rho_\Xi(P) = \varphi(P)$.

Proposition 33. *Let* $P \in \Xi$. *Let* $A = \varphi^{-1}(\{\varphi(P)\}) \cup \psi^{-1}(\{\varphi(P)\})$. *Let* $B = \{\varphi(P)\} \cup \psi^{-1}(\{P\})$. *We have* $\varphi(P) \searrow P$ *iff the restriction* $\varphi_{|A} : A \to B$ *is a bijection.*

Proof. Let $X = \pi_\Theta(P)$. By definition, X is connected, i.e. $\Pi[X] = \{X\}$. The set $\Pi[\overline{X}]$ is composed by one infinite set $X_0 = \mathbb{U} \setminus \tau(X)$ and $k \geq 0$ sets X_i $(1 \leq i \leq k)$ such that $\{X_i\}_{i=1}^k = \tau(\pi_\Theta(\psi^{-1}(\{P\})))$. Let $Y = \pi_\Theta(\varphi(P))$. By definition, Y is connected, i.e. $\Pi[Y] = \{Y\}$. The set $\Pi[\overline{Y}]$ is composed by one infinite set $Y_0 = \mathbb{U} \setminus \tau(Y)$ and $l \geq 0$ sets Y_j $(1 \leq j \leq l)$ such that $\{Y_j\}_{j=1}^l = \tau(\pi_\Theta(\psi^{-1}(\{\varphi(P)\})))$. Let $D = Y \setminus X$. Let us suppose that $\varphi(P) \searrow P$. Then, we have $\varphi^{-1}(\{\varphi(P)\}) = \{P\}$, i.e. φ is bijective between $\varphi^{-1}(\{\varphi(P)\})$ and $\{P\}$. Since D is deletable we have $k = l$ and (up to reindexing), for any $i \in [\![0, k]\!]$, $Y_i \subseteq X_i$. For each $i \in [\![0, k]\!]$, there exist $\widehat{P}_i = (\widehat{X}_i, v) \in \psi^{-1}(\{P\})$ such that $X_i = \tau(\widehat{X}_i)$ and $\widehat{Q}_i = (\widehat{Y}_i, w) \in \psi^{-1}(\{\varphi(\{P\})\})$ such that $Y_i = \tau(\widehat{Y}_i)$. We have $Y_i \subseteq X_i$ and then $\tau(\widehat{Y}_i) \subseteq \tau(\widehat{X}_i)$. We set $D_i = D \cap \widehat{X}_i$. We have $\tau(\widehat{X}_i \setminus D_i) = \tau(\widehat{X}_i) \setminus D_i = \tau(\widehat{Y}_i)$. It follows that $\widehat{Y}_i \subseteq \widehat{X}_i$, and φ is then bijective between $\psi^{-1}(\{\varphi(P)\})$ and $\psi^{-1}(\{P\})$. Let us suppose that φ is bijective between $\varphi^{-1}(\{\varphi(P)\})$ and $\{\varphi(P)\}$. Then both $P = \varphi(P) \setminus D$ and $\varphi(P)$ are connected and $P \subset \varphi(P)$. Let us suppose that φ is bijective between $\psi^{-1}(\{\varphi(P)\})$ and $\psi^{-1}(\{P\})$. The function $\tau \circ \pi_\Theta$ is a bijection between $\psi^{-1}(\{P\})$ (resp. $\psi^{-1}(\{\varphi(P)\})$) and $\{X_i\}_{i=1}^k$ (resp. $\{Y_j\}_{j=1}^l$). It follows that $\varphi(P) \searrow P$. ∎

Let $\sim_H$ be the equivalence relation on Ξ defined as the reflexive-transitive-symmetric closure of $\searrow$.

Remark 34. *We have* $[\infty]_{\sim_H} = \{\infty\} \in \Xi/\sim_H$.

Property 35. *Let* $K \in \Xi/\sim_H$. *Let* $\sqsubseteq_K$ *be the order induced by* $\sqsubseteq_\Xi$ *on* K. $(K, \sqsubseteq_K)$ *is a totally ordered set.*

For any $K \in \Xi/\sim_H$, we note $\langle K \rangle_H = \bigwedge_{\sqsubseteq_\Xi} K$ and $\langle\!\langle K \rangle\!\rangle_H = \bigvee_{\sqsubseteq_\Xi} K$.

Remark 36. *From these results, it follows that* $\forall p \in [\![1, |K| - 1]\!]$ *we have* $\rho_\Xi^p(\langle K \rangle_H) \in K$. *In particular, we have* $\rho_\Xi^{|K|-1}(\langle K \rangle_H) = \langle\!\langle K \rangle\!\rangle_H$ *whereas* $\rho_\Xi^{|K|}(\langle K \rangle_H) \notin K$.

Property 37. *Let* $K \in \Xi/\sim_H$, $K \neq \{\infty\}$. *Let* $P = \rho_\Xi^{|K|}(\langle K \rangle_H)$. *We have* $P = \langle [P]_{\sim_H} \rangle_H$.

For any $K \in \Xi/\sim_H$, we consider $\langle K \rangle_H \in \Xi$ as canonical element, and we identify $\langle K \rangle_H = (X, v)$ with $X \in \Theta$. We set $H = \Xi/\sim_H$.

Let $\sqsubseteq_H$ be the order on $\{\langle K \rangle_H \mid K \in H\} \subseteq \Xi$—and equivalently on H— induced by $\sqsubseteq_\Xi$.

Let $\rho_H : H \to H$ be the function defined by

$$\rho_H(K) = [\rho_\Xi^{|K|}(\langle K \rangle)]_{\sim_H} \tag{13}$$

We define the relation $\blacktriangleleft_H$ on H, induced by the relation $\blacktriangleleft_\Xi$ on Ξ by $K \blacktriangleleft_H \rho_H(K)$.

Definition 38 (Topological tree of shapes). *The topological tree of shapes (or TS-tree, for brief) is the couple* $\mathfrak{T}_{\mathrm{TS}} = (H, \blacktriangleleft_H)$. *See Fig. 3 (centre).*

The following proposition directly derives from the two above properties.

Proposition 39. *The equivalence relation* $\sim_H$ *induces a decreasing homeomorphism from* $\mathfrak{T}_{\mathrm{VS}} = (\Xi, \blacktriangleleft_\Xi)$ *to* $\mathfrak{T}_{\mathrm{TS}} = (H, \blacktriangleleft_H)$.

Remark 40. *The homeomorphism from* $\mathfrak{T}_{\mathrm{VS}} = (\Xi, \blacktriangleleft_\Xi)$ *to* $\mathfrak{T}_{\mathrm{TS}} = (H, \blacktriangleleft_H)$ *is a topologically lossless but a geometrically lossy compression. The structure of the tree* $\mathfrak{T}_{\mathrm{VS}} = (\Xi, \blacktriangleleft_\Xi)$ *can be fully reconstructed from* $\mathfrak{T}_{\mathrm{TS}} = (\Theta, \blacktriangleleft_\Theta)$, *but not the shapes of its nodes.*

6   Links with Other Trees

The graph of valued shapes $\mathfrak{G}_{\mathrm{VS}}$ presents a DAG structure, similarly to other morphological hierarchies, e.g. the component-graph [17], the directed component hierarchy [18] or the braid of partitions [8]. $\mathfrak{G}_{\mathrm{VS}}$ is also organized via two kinds of relations, similarly to the component-hypertree [15] and the directed component hierarchy [18] (where the initial order can be split into two distinct orders).

But, contrary to these morphological hierarchies, $\mathfrak{G}_{\mathrm{VS}}$ can be modeled as a tree structure, namely $\mathfrak{T}_{\mathrm{VS}}$. This should open the way to efficient construction strategies, compared, e.g. to the component-hypertree [12], the component-graph [16] or the braid of partitions [29], whose construction remains complex and/or costly.

Beyond these considerations, the graph of valued shapes and the induced trees (trees of valued shapes, complete tree of shapes) also allow to unify various morphological trees.

With the notations introduced in Sect. 3, the max-tree (resp. min-tree) [24] of $\mathcal{F}$ is defined as $\mathfrak{T}_{\mathrm{max}} = (\Theta_{\mathrm{max}}, \blacktriangleleft_{\mathrm{max}})$ (resp. $\mathfrak{T}_{\mathrm{min}} = (\Theta_{\mathrm{min}}, \blacktriangleleft_{\mathrm{min}})$) with $\Theta_{\mathrm{max}} = \Theta^{\circ}$ (resp. $\Theta_{\mathrm{min}} = \Theta^{\bullet}$) and $\blacktriangleleft_{\mathrm{max}}$ (resp. $\blacktriangleleft_{\mathrm{min}}$) is the Hasse relation induced by the restriction of $\subseteq$ on Θ_{max} (resp. Θ_{min}).

Proposition 41. *There is a decreasing homeomorphism from the subgraph* $(\Xi^{\circ}, \lhd^{\circ})$ *(resp. $(\Xi^{\bullet}, \lhd^{\bullet})$) of $\mathfrak{G}_{\mathrm{VS}}$ to the max-tree $\mathfrak{T}_{\mathrm{max}} = (\Theta_{\mathrm{max}}, \blacktriangleleft_{\mathrm{max}})$ (resp. the min-tree $\mathfrak{T}_{\mathrm{min}} = (\Theta_{\mathrm{min}}, \blacktriangleleft_{\mathrm{min}})$).*

Proof. The proof is similar to that of Proposition 29 (Properties 22, 24, 26) by considering Ξ° and Θ° (resp. $\Xi^{\bullet}$ and $\Theta^{\bullet}$) instead of Ξ and Θ.   ∎

The adjacency tree [21] of a binary set $X \subset \mathbb{U}$ is defined as $\mathfrak{T}_{\mathrm{adj}} = (\Theta_{\mathrm{adj}}(X), \blacktriangleleft_{\mathrm{adj}})$ where $\Theta_{\mathrm{adj}}(X) = \Pi[X] \cup \Pi[\overline{X}]$ and $\blacktriangleleft_{\mathrm{adj}}$ is the Hasse relation induced by the "surrounding" order relation on $\Theta_{\mathrm{adj}}(X)$.

Proposition 42. *Let $v \in \mathbb{V}$. The subgraph $(\Theta_v, \lhd^v)$ of $\mathfrak{G}_{\mathrm{VS}}$ is isomorphic to the adjacency-tree $\mathfrak{T}_{\mathrm{adj}} = (\Theta_{\mathrm{adj}}(\Lambda_v^{\circ}(\mathcal{F})), \blacktriangleleft_{\mathrm{adj}})$.*

Proof. This proposition directly derives from the equivalence of the definitions.   ∎

The tree of shapes [11] of $\mathcal{F}$ is defined as $\mathfrak{T}_{\mathrm{shape}} = (\Theta_{\mathrm{shape}}, \blacktriangleleft_{\mathrm{shape}})$ where $\Theta_{\mathrm{shape}} = \tau(\Theta)$ and $\blacktriangleleft_{\mathrm{shape}}$ is the Hasse relation induced by $\subseteq$ on Θ_{shape}. See Fig. 3 (right).

Proposition 43. *There is a decreasing homeomorphism from the tree $(\Theta, \blacktriangleleft_{\Theta})$ to the tree of shapes $\mathfrak{T}_{\mathrm{shape}} = (\Theta_{\mathrm{shape}}, \blacktriangleleft_{\mathrm{shape}})$.*

Proof. The proof is similar to that of Proposition 29 (Properties 22, 24 and Proposition 26) by considering Θ instead of Ξ and the equivalence relation on Θ defined by $X \sim_S Y \Leftrightarrow \tau(X) = \tau(Y)$.   ∎

Remark 44. *In [28], the notion of a topological monotonic tree was introduced, where "monotonic tree" has the same meaning as "tree of shapes". However, the topological monotonic tree of [28] is indeed different from our topological tree of shapes. Unformally, the difference between both structures lies in the fact that our topological tree of shapes relies on a topological equivalence relation between the nodes of the complete tree of shapes, whereas the topological monotonic as defined in [28] relies on a similar equivalence relation between the external border* *of the nodes. From [28], there is a decreasing homeomorphism from the tree of shapes to the topological monotonic tree. It can be proved that there is also a decreasing homeomorphism from our topological tree of shapes to the topological monotonic tree. These links are not formalized here by lack of room; they will be more deeply investigated in our further works.*

The following result derives from the above propositions and properties.

Property 45. *We have*

$$|H| \leq |\Theta| \leq |\Xi| \leq (|\mathbb{S}| + 1).|\mathbb{V}| \tag{14}$$

$$|\Theta_{\text{shape}}| \leq |\Theta| = |\Theta_{\text{max}}| + |\Theta_{\text{min}}| - 1 \leq 2.|\mathbb{S}| \tag{15}$$

7   Concluding Remarks and Perspectives

This article gathers some preliminary results about the notions of graph/tree of valued shapes and complete/topological tree of shapes. These notions shed a new light on well known morphological hierarchies, namely the component-tree and the tree of shapes. In particular they allow to unify and extend these notions and to link them to topological invariants related to the adjacency tree, the deletable sets and—under favourable hypotheses—the homotopy type. We believe that these structures constitute a promising subject of research in the framework of morphological hierarchies. At this stage, our introductive study focused only on the structural side of these notions.

Our perspective works will also consider the algorithmic aspects, in particular the way to build these structures efficiently. Due to their strong links with the component-tree, it is possible to propose first, naive strategies to build the graph and then the tree of valued shapes from the min- and max-trees [2], and then to derive the complete and topological trees of shapes. It is also possible to start from the construction of the tree of shapes [7] to derive the same structures. However, such approaches, although tractable, are not optimal, and seeking dedicated construction algorithms makes sense.

We initially designed the graph of valued shapes by "mixing" the min-/max-trees and adjacency trees with a precise idea in mind. Our purpose was to develop conceptual tools that would allow one to carry out the topological analysis of objects in non-binary paradigms (e.g. for grey-level images or fuzzy modeling). In this regard, our next step will be to investigate the links that exist between these new structures and frameworks developed in topological analysis, especially with respect to grey-scale topology [5] and to homology persistence and Morse theory [1].

More generally, we also believe that these new structures could be useful for developing approaches dedicated e.g. to homotopic morphology, topological compression or topological comparison.

Acknowledgements. This work was supported by the *Centre National de la Recherche Scientifique* (IEA Project DiTopAM, INS2I Acc-CH-EC TopIA) and by the French *Agence Nationale de la Recherche* (Grants ANR-18-CE23-0025, ANR-20-CE23-0022).

References

1. Boutry, N., Géraud, T., Najman, L.: An equivalence relation between morphological dynamics and persistent homology in n-D. In: Lindblad, J., Malmberg, F., Sladoje, N. (eds.) DGMM 2021. LNCS, vol. 12708, pp. 525–537. Springer, Cham (2021). https://doi.org/10.1007/978-3-030-76657-3_38
2. Carlinet, E., Géraud, T.: A comparative review of component tree computation algorithms. IEEE T Image Proc. **23**, 3885–3895 (2014)
3. Carlinet, E., Géraud, T.: MToS: a tree of shapes for multivariate images. IEEE T Image Proc. **24**, 5330–5342 (2015)
4. Couprie, M., Bertrand, G.: New characterizations of simple points in 2D, 3D, and 4D discrete spaces. IEEE T Pattern Anal. **31**, 637–648 (2009)
5. Couprie, M., Nivando Bezerra, F., Bertrand, G.: Topological operators for grayscale image processing. J. Electron. Imaging **10**, 1003–1015 (2001)
6. Edelsbrunner, H., Letscher, D., Zomorodian, A.: Topological persistence and simplification. Discrete Comput. Geom. **28**, 511–533 (2002)
7. Géraud, T., Carlinet, E., Crozet, S., Najman, L.: A quasi-linear algorithm to compute the tree of shapes of nD images. In: Hendriks, C.L.L., Borgefors, G., Strand, R. (eds.) ISMM 2013. LNCS, vol. 7883, pp. 98–110. Springer, Heidelberg (2013). https://doi.org/10.1007/978-3-642-38294-9_9
8. Kiran, B.R., Serra, J.: Braids of partitions. In: Benediktsson, J.A., Chanussot, J., Najman, L., Talbot, H. (eds.) ISMM 2015. LNCS, vol. 9082, pp. 217–228. Springer, Cham (2015). https://doi.org/10.1007/978-3-319-18720-4_19
9. Kurtz, C., Naegel, B., Passat, N.: Connected filtering based on multivalued component-trees. IEEE T Image Proc. **23**, 5152–5164 (2014)
10. Malgouyres, R., Francés, A.R.: Determining whether a simplicial 3-complex collapses to a 1-complex is NP-complete. In: Coeurjolly, D., Sivignon, I., Tougne, L., Dupont, F. (eds.) DGCI 2008. LNCS, vol. 4992, pp. 177–188. Springer, Heidelberg (2008). https://doi.org/10.1007/978-3-540-79126-3_17
11. Monasse, P., Guichard, F.: Fast computation of a contrast-invariant image representation. IEEE T Image Proc. **9**, 860–872 (2000)
12. Morimitsu, A., Passat, N., Luz Alves, W.A., Hashimoto, R.F.: Efficient component-hypertree construction based on hierarchy of partitions. Pattern Recogn. Lett. **135**, 30–37 (2020)
13. Najman, L., Schmitt, M.: Geodesic saliency of watershed contours and hierarchical segmentation. IEEE T Pattern Anal. **18**, 1163–1173 (1996)
14. Passat, N., Mazo, L.: An introduction to simple sets. Pattern Recogn. Lett. **30**, 1366–1377 (2009)
15. Passat, N., Naegel, B.: Component-hypertrees for image segmentation. In: Soille, P., Pesaresi, M., Ouzounis, G.K. (eds.) ISMM 2011. LNCS, vol. 6671, pp. 284–295. Springer, Heidelberg (2011). https://doi.org/10.1007/978-3-642-21569-8_25
16. Passat, N., Naegel, B., Kurtz, C.: Component-graph construction. J. Math. Imaging Vis. **61**, 798–823 (2019). https://doi.org/10.1007/s10851-019-00872-5
17. Passat, N., Naegel, N.: Component-trees and multivalued images: structural properties. J. Math. Imaging Vis. **49**, 37–50 (2014). https://doi.org/10.1007/s10851-013-0438-3
18. Perret, B., Cousty, J., Tankyevych, O., Talbot, H., Passat, N.: Directed connected operators: asymmetric hierarchies for image filtering and segmentation. IEEE T Pattern Anal. **37**, 1162–1176 (2015)

19. Randrianasoa, J.F., Kurtz, C., Desjardin, E., Passat, N.: Binary partition tree construction from multiple features for image segmentation. Pattern Recogn. **84**, 237–250 (2018)
20. Ronse, C.: A topological characterization of thinning. Theor. Comput. Sci. **43**, 31–41 (1986)
21. Rosenfeld, A.: Adjacency in digital pictures. Inf. Control **26**, 24–33 (1974)
22. Rosenfeld, A.: Digital topology. Am. Math. Mon. **86**, 621–630 (1979)
23. Salembier, P., Garrido, L.: Binary partition tree as an efficient representation for image processing, segmentation, and information retrieval. IEEE T Image Proc. **9**, 561–576 (2000)
24. Salembier, P., Oliveras, A., Garrido, L.: Anti-extensive connected operators for image and sequence processing. IEEE T Image Proc. **7**, 555–570 (1998)
25. Salembier, P., Serra, J.: Flat zones filtering, connected operators, and filters by reconstruction. IEEE T Image Proc. **4**(8), 1153–1160 (1995)
26. Santana Maia, D., Cousty, J., Najman, L., Perret, B.: Characterization of graph-based hierarchical watersheds: theory and algorithms. J. Math. Imaging Vis. **62**, 627–658 (2020). https://doi.org/10.1007/s10851-019-00936-6
27. Soille, P.: Constrained connectivity for hierarchical image decomposition and simplification. IEEE T Pattern Anal. **30**, 1132–1145 (2008)
28. Song, Y., Zhang, A.: Monotonic tree. In: Braquelaire, A., Lachaud, J.-O., Vialard, A. (eds.) DGCI 2002. LNCS, vol. 2301, pp. 114–123. Springer, Heidelberg (2002). https://doi.org/10.1007/3-540-45986-3_10
29. Tochon, G., Dalla Mura, M., Veganzones, M.A., Géraud, T., Chanussot, J.: Braids of partitions for the hierarchical representation and segmentation of multimodal images. Pattern Recogn. **95**, 162–172 (2019)

Component-Tree Simplification Through Fast Alpha Cuts

Michael H. F. Wilkinson$^{(\boxtimes)}$

Bernoulli Institute for Mathematics, Computer Science, and Artifical Intelligence,
University of Groningen, P.O. Box 407, 9700 AK Groningen, The Netherlands
m.h.f.wilkinson@rug.nl

Abstract. Tree-based hierarchical image representations are commonly
used in connected morphological image filtering, segmentation and multi-
scale analysis. In the case of component trees, filtering is generally based
on thresholding single attributes computed for all the nodes in the tree.
Alternatively, so-called shapings are used, which rely on building a component tree of a component tree to filter the image. Neither method is
practical when using vector attributes. In this case, more complicated
machine learning methods are required, including clustering methods. In
this paper I present a simple, fast hierarchical clustering algorithm based
on cuts of α-trees to simplify and filter component trees.

Keywords: Connected filters · Component trees · α-trees ·
Clustering · Algorithms

1 Introduction

Connected filters [1–3] have found many uses in image processing and analysis,
and many different types of filters and multi-scale tools have been developed
since the introduction of the first connected filters in the form of openings by
reconstruction [4], and area openings [5,6]. Many methods are built using hierarchical image representations in the form of tree structures [7–10], for a recent
review see [11]. In the grey-scale case, much work has been done on attribute
filters [12], in particular using tree structures variously known as component
trees [9], min-trees and max-trees [7]. They have also found use for multi-scale
analysis, e.g. through pattern spectra [13] or morphological profiles [14].

This paper will focus on component trees, which are trees containing the
connected components of threshold sets of a grey-scale image. Each node represents a single connected component, and usually contains some scalar attribute
value like area or elongation to characterise the component. Filtering is done
by applying some threshold to the attributes, and removing nodes that have
attribute values lower than the threshold.

Selecting the "right" threshold for attribute filtering is not an easy task.
Apart from simple trial and error, only a few papers address this issue systematically. Jones [9] notes that threshold on attributes could be chosen automatically

© Springer Nature Switzerland AG 2022
E. Baudrier et al. (Eds.): DGMM 2022, LNCS 13493, pp. 236–247, 2022.
https://doi.org/10.1007/978-3-031-19897-7_19

by traversing the tree from leaf to root, and choosing thresholds at points where the attribute value changes abruptly. Thresholds could be chosen per branch, or globally in the tree. In [15], various automatic grey-scale thresholding methods are studied, but these do not always take the topology of the tree into account.

In the case of increasing attributes like area, filtering essentially boils down to pruning the tree such that the remaining leaves have an area larger than some threshold. For non-increasing attributes, such as elongation or perimeter, several other filtering strategies have been proposed [7,13]. Xu et al. [16] introduced the idea of tackling non-increasing scalar attributes in max-trees by computing a max-tree of a max-tree. Because a max-tree of an image with scalar attributes on the nodes is just a node-weighted graph, it is fairly trivial to compute a secondary max-tree of this primary max-tree. The idea is to filter this secondary max-tree, reconstituting the primary max-tree based on the filtering results, and then generating a filtered image from the filtered primary max-tree. The resulting filters were dubbed "shapings". Though interesting new results were obtained, it remains hard to envisage the precise effect of filtering a max-tree based on a secondary max-tree in this way.

None of the above approaches are suitable for so-called vector-attribute filtering [17], in which each connected component has a feature vector instead of a single scalar value. In this paper we will extend the idea of Xu et al. [16] to the vector-attribute case. Rather than building a max-tree of a max-tree, which requires a total order on the attributes, I construct an α-tree [10] of the max-tree. For α-trees, which derive from partition hierarchies described in [18], no total order is needed, which is why they are suitable for graphs with vectorial weights. The aim is to create a hierarchy of simplifications of the input component tree, each of which contains only the nodes at which large transitions in attribute vectors occur. These ideally contain the most essential information in the tree.

In the rest of the paper, I will focus on max-trees, although the method described will work on other tree structures as well. I will first discuss attribute filters and max-trees, and hierarchical clustering using α-trees, and the principles behind the method. I will then present a fast algorithm to generate simplified max-trees at any level of the α-trees, by cutting at a particular dissimilarity threshold α, and where necessary correct the attribute values in the case of attributes that depend on grey-scale content. An algorithm for pattern spectra based on α-cuts is also presented, along with a discussion of the computational complexity. A simple experiment showing the effect of applying α-cuts to pattern spectra is presented, followed by a discussion future and plans for future work are given in the final section.

2   Attribute Filters and Component Trees

Breen and Jones [12] introduced attribute filters, which are attribute thinnings in the non-increasing, anti-extensive case we will focus on here. In the binary case they remove connected foreground components that do not meet some non-increasing criterion T.

Definition 1. *The binary attribute thinning Φ^T of set X with criterion T is given by*

$$\Phi^T(X) = \{x \in X \mid T(\Gamma_x(X))\} \tag{1}$$

where Φ_T is the trivial thinning with non-increasing criterion T, and Γ_x is the connectivity opening at point x. The latter returns the connected component to which x belongs if $x \in X$ and $\emptyset$ otherwise. As can be seen, only those points x that are members of a connected component that meets criterion T are retained.

Vector-attribute filtering [17] was introduced as an extension to attribute filtering. Rather than computing a single attribute, a vector of attributes is computed for each node. As it would be impractical to set thresholds for each vector, Urbach et al. [17] proposed using thresholds to distances to some collection of prototypes to detect objects in images. Later Naegel et al. [19] used Mahalanobis distances to a set of prototypes for segmentation of dermatological images. Formally, vector-attribute thinnings can be defined as follow:

Definition 2. *The vector-attribute thinning $\Phi^{\vec{\tau}}_{\vec{r},\epsilon}$ of X with respect to a reference vector $\vec{r}$ and using vector-attribute $\vec{\tau}$ and scalar value ϵ is given by*

$$\Phi^{\vec{\tau}}_{\vec{r},\epsilon}(X) = \{x \in X \mid T^{\vec{\tau}}_{\vec{r},\epsilon}(\Gamma_x(X))\}. \tag{2}$$

The criterion $T^{\vec{\tau}}_{\vec{r},\epsilon}$ is defined as

$$T^{\vec{\tau}}_{\vec{r},\epsilon}(C) = \rho(\vec{\tau}(C), \vec{r}) > \epsilon. \tag{3}$$

with ρ some metric or dissimilarity function.

The above definitions can be generalised to grey scale by the usual threshold superposition method [21], and implemented using max-trees in the antiextensive case [7]. Max-tree nodes represent the connected foreground components of threshold sets at all threshold levels in the image. The connected components of the threshold levels are referred to as *peak components*. A simple example is shown in Fig. 1. Each node may be assigned one or more attributes,

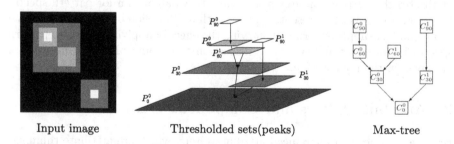

Input image   Thresholded sets(peaks)   Max-tree

Fig. 1. A simple grey-scale image, the foreground components of each threshold set, also know as *peak components*, and the resulting component tree, which is referred to as a *max-tree* in this case. Figure from [20].

and filtering the image applies the attribute criterion to all of the nodes, and reconstituting the image based on which nodes are preserved and which removed. In the case of an increasing criterion T, i.e. if $C \subseteq D$ then $T(C) \leq T(D)$, filtering corresponds to pruning the tree. In the non-increasing case, more complex strategies are used [7,13,16].

3   Hierarchical Clustering of Max-Trees

Previously, the use of cluster analysis on max-tree nodes has been explored in [22]. In that case, nodes were clustered purely based on the attribute vectors, completely disregarding the tree structure. Here I propose to use hierarchical clustering of nodes taking the tree structure into account explicitly, through the use of α-trees [10] of max-trees.

Ouzounis and Soille [10] introduced term α-tree as a way of representing hierarchies of α-connected components (α-CCs) of images, suitable for vector images such as colour and hyperspectral images. The α-CCs of an edge-weighted graph are connected subgraphs of maximal extent such that there exists a path within the α-CCs between each pair elements, such that the edge weights in the path are all smaller than or equal to some threshold α [18]. An α-tree can be created on any graph with vector weights on the vertices by assigning weights to the edges between any two vertices, based on some dissimilarity measure. This allows α-trees to be built on any image, whereas max-trees are restricted to those cases where a total order can be imposed upon the pixel values. It has been shown that α-trees are equivalent to min-trees of an edge-weighted graph [23], and the computational complexity of building one is therefore equivalent to that of building a max-tree.

If we have a max-tree of an image, with vector attributes on the nodes, we cannot readily compute a max-tree of this max-tree to compute a shaping [16]. However, we can obviously compute an α-tree, using any of the existing algorithms. This does not make use of the fact that the max-tree is a tree, not a general graph, which means that there is just a single shortest path connecting any two nodes, and any longer path (taking detours to the root) must traverse the edges in the shortest path. This in turn means that the dissimilarity δ of the edge linking a node to its parent forms a boundary between two α-CCs for any $\alpha < \delta$. In the following I will discuss the special case of α-trees, and in particular α-cuts of max-trees.

Let us assign a weight δ on the edge between current node and its parent. I will refer to these edges and weights as parent edges and parent weights respectively. The weights can be computed using some dissimilarity measure $\rho : \mathbb{R}^n \times \mathbb{R}^n \to \mathbb{R}$, assuming n-dimensional attribute vectors on each of the nodes.

Definition 3. *An α-connected component of any tree with weights δ on the edges is a subtree of maximal extent containing no edges with $\delta > \alpha$.*

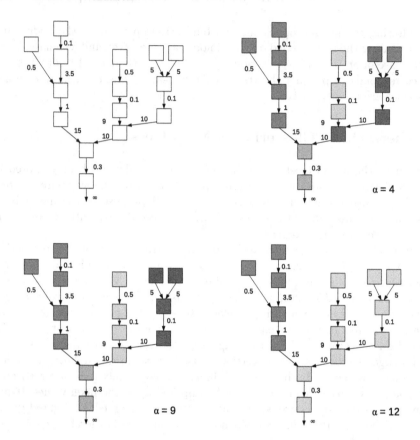

Fig. 2. Hierarchical clustering of a max-tree (top left) using increasing α-cuts. At α = 15 the tree would become a single connected component

It can readily be seen that the root element of any α-connected component must have a parent weight δ > α on its parent-edge. Furthermore, these root nodes uniquely identify the α-connected components of the tree. Indeed, to determine which α-connected component any node with parent weight δ ≤ α belongs to we simply need to find its α-parent.

Definition 4. *The α-parent of a node with parent weight δ ≤ α is the nearest ancestor with parent weight δ > α.*

Thus, the nodes with parent weights δ > α can be seen as the canonical elements of the nodes at level α in the α-tree of the max-tree. An example of a series of α-cuts of a simple max-tree is shown in Fig. 2. The assignment of α-parents and reduction to a simplified max-tree is shown in Fig. 3.

4   Algorithm

We assume that we have a max-tree built using any one of many available algorithms [24]. Without loss of generality, we can assume the nodes are stored in an array `node`. For the moment we assume each node has a vector of attributes `attr` which do not depend on grey level, like area or various moment invariants used previously [13,25]. Each node also contains an index `par` to its parent, and a field `delta` containing the dissimilarity measure between its attribute vector and that of its parent in the max-tree. Finally, we add a field `alpha_par` to each node, which is initialized to be equal to that of the `par` field. To simplify the algorithm, the root node is initialized to have a maximal value of its `delta` field and both its `par` and `alpha_par` point to root node itself.

The algorithm to compute an α-cut of a max-tree now boils down to the following steps:

1. Create an index array `Index` of max-tree nodes, sorted in increasing order of their `delta` field.
2. For all nodes in the max-tree set the `alpha_par` field to its α-parent.

The first step ensures we can easily select all the roots of the α-CCs for a given value of α, simply by using e.g. binary search in the index array to find the first node in which `node[index[i]].delta` $> \alpha$.

The latter step ensures the α-parents of the tree are properly set. This can be achieved in linear time by calling function `find_alpha_par` shown in Algorithm 1 for each node of the tree. Each call to this function follows the root path until the α-parent has been found, and sets all the `alpha_par` field along the root path. This means that subsequent calls that explore the same root path will essentially yield a shortcut to the correct α-parent. Note that before the initial construction of the α-cut, each `alpha_par` was set to the value of the `par` field, which is the correct value for $\alpha = 0$. Therefore, the algorithm follows the usual root paths in the max-tree initially, and should process at most all the nodes in the max-tree once. Once a particular α-cut has been computed, and we wish to compute a new cut with $\alpha' > \alpha$, we need only call `find_alpha_par` for the nodes with `node[index[i]].delta` $> \alpha'$, and these calls would only traverse those nodes with `node[index[i]].delta` $> \alpha$.

The result of calling the above algorithm for $\alpha = 4$ on the max-tree from Fig. 2 is shown in Fig. 3. It also shows how limiting the tree to only the nodes with `node[index[i]].delta` $> \alpha$ yields a simplified version of the max-tree.

4.1   Attributes of α-CCs of Max-Trees

Until now, we have only considered the structure of the simplified trees, but not the attributes. If we consider all "flat" attributes, i.e. those that only depend of the shape of the peak component, but not the grey-levels within it, nothing needs to be done, as each node contains all the information pertaining to that shape. In the case of non-flat attributes, i.e. those that depend on the grey-level

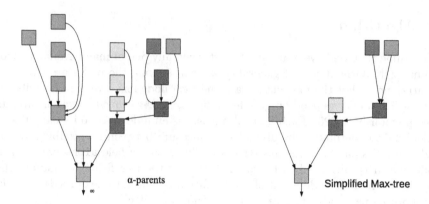

Fig. 3. The result of the find_alpha_par function for $\alpha = 4$: (left) alpha_par pointers of max-tree, (right) Simplified max-tree.

content of the node, not just its shape, some extra post-processing needs to be done, if we want to represent the properties of each partition of the max-tree from the computed attributes. Here I will restrict myself to the attributes based on the first and second moments of the grey level distribution, although the method can be extended to higher orders, and in principle also to certain other attributes.

We assume that each node stores the sum of grey levels and sum of squared grey levels of all the pixels within the peak component in fields SumGrey and SumGreySquare respectively. A field Gval stores the grey value of each node.

The process traverses the tree in increasing grey level order. Whenever it finds a node such that node[i].delta $> \alpha$ it calls function correct_alpha_par on that node. This looks up the α-parent, and subtracts the sum of (squared) grey levels of the current node from the sum of (squared) grey levels of the α-parent.

Algorithm 1. The find_alpha_par function. Note that MTnode is a max-tree node struct type, and *node is the array representing the max-tree.

```
index find_alpha_par(float alpha, MTnode *node, index current){
  index alpha_par = node[current].alpha_par;

  if (node[alpha_par].delta <= alpha)
    node[current].alpha_par = find_alpha_par( alpha, node, alpha_par );

  return node[current].alpha_par;
}
```

It then adds the product of the area of the current node and the (squared) grey level of the α-parent to the sum of the (squared) grey levels of the α-parent. This effectively clips off the contribution of the current node to these sums in its α-parent at the level of the latter's grey level. The algorithm is shown in Algorithm 2.

Algorithm 2. The `correct_alpha_par` function

Precondition: `node[current].delta> alpha`

```
void correct_alpha_par ( MTnode *node, index current ){
  index alpha_par = node[current].alpha_par;

  node[alpha_par].SumGrey =
    node[alpha_par].SumGrey  - node[current].SumGrey
    + node[current].area*node[alpha_par].Gval;
  node[alpha_par].SumGreySquare =
    node[alpha_par].SumGreySquare - node[current].SumGreySquare
    + node[current].area * node[alpha_par].Gval * node[alpha_par].Gval;

}
```

After application of this algorithm, attributes like power [26] and volume (or flux) [27] can be computed in the usual way from these corrected sums. Once the attributes have been computed, we can in principle filter the original max-tree based on only the simplified α-cut version of the tree, by only applying a criterion T to the nodes with $\delta > \alpha$, and for all other nodes copy the decision made for their α-parent. Likewise, granulometries based on α-cuts might also reveal more structure, as single objects are not smeared out over a range of attributes.

4.2 Pattern Spectra

Computation of pattern spectra using alpha-cuts of max-trees can be done with a very minor adaptation of the original code from [13], as shown in Algorithm 3. After α-parents have been assigned, and the array containing the pattern spectrum has been set to zero, we compute each node's contribution to the total sum of grey levels in the image, add it to the appropriate bin of the pattern spectrum. The only difference with the original algorithm is the if-statement in Algorithm 3. If $\delta > \alpha$ the node is also a node in the simplified tree, and we add the flux to its bin in the spectrum in the usual way. If not, we add it to its α-parent.

Algorithm 3. The `alpha_cut_pattern_spectrum` function

Precondition: `find_alpha_par` has been applied to all nodes in the max-tree array `node`. `BinFunc` computes the bin in the spectrum to which a node should be assigned.

```
void alpha_cut_pattern_spectrum ( MTnode *node,
                                  greyval *Spectrum
                                  float   alpha){

  Set all elements of Spectrum to zero

  for all node[i] except root {
     par = node[i].parent;
     flux = (node[i].Gval - node[par].Gval) * node[i].area;

     if (node[i].delta <= alpha)
        bin = BinFunc(node, node[i].alpha_par );
     else
        bin = BinFunc(node, i);

     Spectrum[bin] = Spectrum[bin] + flux;
  }
}
```

4.3   Computational Complexity

Assuming the computation of the dissimilarities δ between node and parent is independent of the number of node N, the initial step of computing the dissimilarities is $O(N)$, as each node need only inspect its own parent, and compute a single value. Sorting the edges by δ is simply $O(N \log N)$. Note, however that this sorting is not necessary in all cases. It is useful if we want to choose α as a percentile of the distribution of δ values. The `find_alpha_par` function of Algorithm 1 is essentially the same as the restitution stage of regular max-tree filtering, which is also linear in N [7]. Indeed, the entire process (without sorting) is essentially the same as that of the entire filtering phase of a max-tree, which in practise is between 1 and 5% of the total compute time. The building phase is the costly phase. By contrast, if we explicitly built an α-tree of the max-tree this is equivalent to building a min-tree of N items [23] which is evidently more costly, both computationally and in terms of memory use.

The complexity of `correct_alpha_par` in Algorithm 2 is independent of N, so applying it in grey-level order to the entire tree is $O(N \log N)$ if the nodes are not sorted in grey level order by the max-tree building algorithm, and $O(N)$ if they are. This too is quite similar to the compute load of filtering an existing max-tree. Likewise, the complexity of Algorithm 3 is $O(N)$, assuming `BinFunc` is independent of N (which is usually the case).

5   Some Initial Results

Algorithm 1 and 3 were implemented in C, within the code base for 2D pattern spectra of [13], and applied to a 621 × 501 image of a diatom from the ADIAC data set [28]. The δ value was computed as

$$\delta(C) = \frac{\text{area}(\text{parent}(C)) - \text{area}(C)}{\text{area}(\text{parent}(C))} \tag{4}$$

Timings revealed small differences in timings between computation of pattern spectra with and without α-cuts, using a desktop PC with an Intel® Core™ i7-6700 CPU at 3.40 GHz. Running 100 iterations of the code with and without α-cuts resulted in a difference of 58 ms on average (or 0.58 ms for a single iteration), out of a total compute time of 929 ms. Thus, roughly 6% of compute time is spent on the computation of δ values, α-parents, and the modifications needed for the α-cut pattern spectra. There seems to be a slight decreasing trend as a function of α, with 64 ms required at $\alpha = 0$, and 52 ms at $\alpha = 0.8$. The average compute time of 0.58 ms is much smaller than building an α-tree of the same image (not its max-tree), which took around 110 ms. Given that the image has 311,121 pixels, vs the max-tree having 70,296 nodes, we can roughly estimate the required time for computing an α-tree of the max-tree as around 24 ms, or some 40× slower than computing a single α-cut.

Figure 4 shows the resulting spectra for $\alpha = 0$, 0.2, and 0.4. These show a pattern reminiscent of a skewed butterfly, in particular at $\alpha = 0$. The left-hand "wing" mainly represents the structures within the diatom, whereas the right-hand side, with larger areas, mainly shows the structure in the background. By increasing α, the flux in these background structures almost all become focused in the top right corner of the spectrum. Changes on the left are subtler, suggesting the detail in the diatom cell is preserved. Much more extensive tests are needed to draw any further conclusions.

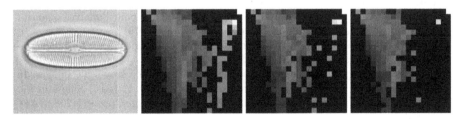

Fig. 4. Pattern spectra of a diatom image for α-cuts at 0, 0.2, and 0.4

6   Conclusions

I have presented a simple algorithm to compute α-cuts of component trees, which are horizontal cuts through α-trees of component trees. These provide an easily tunable simplification of the component trees. Apart from selecting a value of α

and searching for the appropriate location in the index array, we could simply just use the top 10% of the nodes in term of their δ value. A simple slider in an interactive tool would readily allow finding the appropriate level of simplification. Tree simplification could also simplify the analysis and visualization of these trees. Besides, it is hoped that this will allow better selection of meaningful nodes and their attribute vectors for training of machine learning methods. This in turn could lead to better integration of machine learning methods with morphological connected filtering.

Obviously, α-trees and level-line trees can be simplified in the same way, as the presented algorithm for α-cuts carries over without modification to these tree structure, although some modifications would be needed for the grey-scale attribute correction. In principle binary partition trees could be processed this way, if you allow for the fact that the resulting simplified tree might no longer be binary. It should even be possible to extend the technique to the distributed component graph used for distributed computing of attribute filters [20,29], as each local modified component tree contains all the data necessary to compute any filtering or analysis step.

In the near future we will apply this method to detection of important structures in CT, MRI and PET scans, and to detection and analysis of astronomical objects. Given the speed of the simplification method, we aim at building interactive tools for adaptation of the α values to the task at hand.

References

1. Salembier, P., Wilkinson, M.H.F.: Connected operators: a review of region-based morphological image processing techniques. IEEE Signal Process. Mag. **26**(6), 136–157 (2009)
2. Salembier, P., Serra, J.: Flat zones filtering, connected operators, and filters by reconstruction. IEEE Trans. Image Proc. **4**, 1153–1160 (1995)
3. Heijmans, H.J.A.M.: Connected morphological operators for binary images. Comput. Vis. Image Underst. **73**, 99–120 (1999)
4. Klein, J.C.: Conception et réalisation d'une unité logique pour l'analyse quantitative d'images. Ph.D. thesis, Nancy University, France (1976)
5. Cheng, F., Venetsanopoulos, A.N.: An adaptive morphological filter for image processing. IEEE Trans. Image Proc. **1**, 533–539 (1992)
6. Vincent, L.: Morphological area openings and closings for grey-scale images. In: O, Y.L., Toet, A., Foster, D., Heijmans, H.J.A.M., Meer, P. (eds.) Shape in Picture: Mathematical Description of Shape in Grey-level Images, pp. 197–208. NATO (1993)
7. Salembier, P., Oliveras, A., Garrido, L.: Anti-extensive connected operators for image and sequence processing. IEEE Trans. Image Proc. **7**, 555–570 (1998)
8. Salembier, P., Garrido, L.: Binary partition tree as an efficient representation for image processing, segmentation and information retrieval. IEEE Trans. Image Proc. **9**(4), 561–576 (2000)
9. Jones, R.: Connected filtering and segmentation using component trees. Comput. Vis. Image Underst. **75**, 215–228 (1999)
10. Ouzounis, G.K., Soille, P.: The Alpha-Tree Algorithm. Publications Office of the European Union (2012)

11. Bosilj, P., Kijak, E., Lefèvre, S.: Partition and inclusion hierarchies of images: a comprehensive survey. J. Imaging **4**(2), 33 (2018)
12. Breen, E.J., Jones, R.: Attribute openings, thinnings and granulometries. Comput. Vis. Image Underst. **64**(3), 377–389 (1996)
13. Urbach, E.R., Roerdink, J.B.T.M., Wilkinson, M.H.F.: Connected shape-size pattern spectra for rotation and scale-invariant classification of gray-scale images. IEEE Trans. Pattern Anal. Mach. Intell. **29**, 272–285 (2007)
14. Wilkinson, M.H.F., Pesaresi, M., Ouzounis, G.K.: An efficient parallel algorithm for multi-scale analysis of connected components in gigapixel images. ISPRS Int. J. Geo-Inf. **5**(3), 22 (2016)
15. Kiwanuka, F.N., Wilkinson, M.H.F.: Automatic attribute threshold selection for blood vessel enhancement. In: Proceedings of the 20th International Conference Pattern Recognition, pp. 314–2317 (2010)
16. Xu, Y., Géraud, T., Najman, L.: Connected filtering on tree-based shape-spaces. IEEE Trans. Pattern Anal. Mach. Intell. **38**(6), 1126–1140 (2016)
17. Urbach, E.R., Boersma, N.J., Wilkinson, M.H.F.: Vector-attribute filters. In: Mathematical Morphology: 40 Years on, Proceedings International Symposium Mathematical Morphology (ISMM), Paris, 18–20 April 2005, pp. 95–104 (2005)
18. Soille, P.: Constrained connectivity and connected filters. IEEE Trans. Pattern Anal. Mach. Intell. **30**(7), 1132–1145 (2008)
19. Naegel, B., Passat, N., Boch, N., Kocher, M.: Segmentation using vector-attribute filters: methodology and application to dermatological imaging. In: Proceedings of the International Symposium on Mathematical Morphology (ISMM), pp. 239–250 (2007)
20. Gazagnes, S., Wilkinson, M.H.F.: Distributed component forests in 2-D: hierarchical image representations suitable for tera-scale images. Int. J. Pattern Recogn. Artif. Intell. **33**(11), 1940012 (2019)
21. Maragos, P., Ziff, R.D.: Threshold decomposition in morphological image analysis. IEEE Trans. Pattern Anal. Mach. Intell. **12**(5), 498–504 (1990)
22. Kiwanuka, F.N., Wilkinson, M.H.F.: Cluster based vector attribute filtering. In: Benediktsson, J.A., Chanussot, J., Najman, L., Talbot, H. (eds.) ISMM 2015. LNCS, vol. 9082, pp. 277–288. Springer, Cham (2015). https://doi.org/10.1007/978-3-319-18720-4_24
23. Soille, P., Najman, L.: On morphological hierarchical representations for image processing and spatial data clustering. In: Köthe, U., Montanvert, A., Soille, P. (eds.) WADGMM 2010. LNCS, vol. 7346, pp. 43–67. Springer, Heidelberg (2012). https://doi.org/10.1007/978-3-642-32313-3_4
24. Carlinet, E., Géraud, T.: A comparative review of component tree computation algorithms. IEEE Trans. Image Proc. **23**(9), 3885–3895 (2014)
25. Westenberg, M.A., Roerdink, J.B.T.M., Wilkinson, M.H.F.: Volumetric attribute filtering and interactive visualization using the max-tree representation. IEEE Trans. Image Proc. **16**, 2943–2952 (2007)
26. Young, N., Evans, A.N.: Psychovisually tuned attribute operators for pre-processing digital video. IEE Proc. Vis. Image Signal Process. **150**(5), 277–286 (2003)
27. Tushabe, F.B.: Extending attribute filters to color processing and multi-media applications. Ph.D. thesis, University of Groningen (2010)
28. du Buf, H., et al.: Diatom identification: a double challenge called ADIAC. In: Proceedings ICIAP, Venice, pp. 734–739 (1999)
29. Gazagnes, S., Wilkinson, M.H.F.: Distributed connected component filtering and analysis in 2D and 3D tera-scale data sets. IEEE Trans. Image Proc. **30**, 3664–3675 (2021)

Component Tree Loss Function: Definition and Optimization

Benjamin Perret[(✉)][iD] and Jean Cousty[iD]

LIGM, Univ Gustave Eiffel, CNRS, ESIEE Paris, 77454 Marne-la-Vallée, France
{benjamin.perret,jean.cousty}@esiee.fr

Abstract. In this article, we propose a method for designing loss functions based on component trees that can be optimized by gradient descent algorithms and are therefore usable in conjunction with recent machine learning approaches such as neural networks. The nodes of this tree are the connected components of the upper level sets of an image and the leaves represent the regional maxima (or regional minima if the dual tree is considered) of the image, *i.e.*, connected sets of bright pixels surrounded by darker pixels. The proposed loss function is thus defined at the level of connected components rather than at the level of individual pixels, which allows for the optimization of higher semantic level quantities such as topological features. We show how the altitudes associated with the nodes of such hierarchical image representations can be differentiated with respect to the values of the image pixels. This property is used to design a generic loss function that can select or discard image maxima based on various attributes, such as extinction values based on the contrast or the size of the maxima. The possibilities of the proposed method are demonstrated on simulated and real image filtering.

Keywords: Max-tree · Connected filters · Topological loss · Continuous optimization · Mathematical morphology

1 Introduction

Component-trees are hierarchical image representations that are used to perform connected image analysis and filtering [13,19]. In such methods, an image is seen as the collection of the connected components of its level sets, thus offering a representation based on elements of higher semantic level, connected components instead of pixels, to design new image analysis methods. These approaches provide efficient solutions to many image analysis problems such as feature detection [8,23], segmentation [7,13,18,19,22], or object detection [11,20].

However, those methods, based on topological decompositions, do not play well with recent machine learning approaches such as neural networks as their combinatorial nature is, at first sight, not well suited to optimization strategies based on gradient descent. In this context, some authors have recently proposed topological loss functions [4,5,9,12] that enable to enforce topological constraints

ⓒ Springer Nature Switzerland AG 2022
E. Baudrier et al. (Eds.): DGMM 2022, LNCS 13493, pp. 248–260, 2022.
https://doi.org/10.1007/978-3-031-19897-7_20

in continuous optimization frameworks using notions coming from the persistent homology theory. It has also been shown that hierarchies of segmentations can also be used in such context with the introduction of an ultrametric layer [3].

In this article, we propose a novel approach to use component trees, and more specifically max-trees, within continuous optimization methods. This approach is based on the observation that, in such trees, the altitude of a node (the level of the level-set where it first appears) is directly linked to the value of some pixels of the image. Hence, we study how we can back-propagate any slight modification of the altitude of a node of the tree to a slight modification of the initial image. We then design a component tree loss function that enforces the presence of a prescribed number of maxima in the image based on maxima measures. We study how extinction values [21], maxima measures notably used in mathematical morphology to define hierarchical watersheds [6,17], can be used to modify the behavior of the proposed loss function. Finally, the method has been implemented in Pytorch with the hierarchical graph processing library Higra [16] and we provide preliminary results demonstrating the use of the proposed approach on simulated and real images.

This article is organized as follows. The definition of max-trees is recalled in Sect. 2. Then, Sect. 3 presents how max-trees can be used in gradient descent algorithms and formalizes the optimization problem which we address. In Sect. 4, we define a component tree loss function used for maxima selection in the max-trees and we introduce different maxima measures. The experiments are presented in Sect. 5. Finally, Sect. 6 concludes the work and gives some perspectives.

2   Max-Trees

In this section, we recall the definition of max-trees [13,19] which is based on the decomposition of every upper thresholds of an image into connected components.

In the following, the image domain is represented by a finite nonempty set $V = \{v_i\}_{i \in [\![1,n]\!]}$ of cardinality n. The elements of V are called *pixels*. Given any vector $\mathbf{v}$ of $\mathbb{R}^m$ with $m \in \mathbb{N}^+$, the i-th component of $\mathbf{v}$ is denoted $\mathbf{v}_i$. An *image* is represented by a vector $\mathbf{f} \in \mathbb{R}^n$ and, for any $i \in [\![1,n]\!]$, f_i is called *the value of the pixel v_i*. Note that any image can be represented as a vector by choosing an arbitrary ordering of the pixels (*e.g.*, a raster scan for 2d images) and that this choice does not change the results of the proposed method. In order to simplify notations, when we have a vector $\mathbf{f} \in \mathbb{R}^n$ and an element x of a family $\{x_i\}_{i \in [\![1,n]\!]}$ indexed from 1 to n, there exists a single integer $k \in [\![1,n]\!]$ such that $x = x_k$ and we will write $\mathbf{f}_x$ instead of $\mathbf{f}_k$.

Let X be a subset of V, the set of connected components of X is denoted by $\mathcal{CC}(X)$ where connected components may be defined by any appropriate mean: *e.g.*, by path connectivity in a graph. In this article, all the examples involving 2d images are based on a classical 8-adjacency relation on a regular square grid of pixels. Let $\mathbf{f} \in \mathbb{R}^n$ be an image, the set of connected components of $\mathbf{f}$, denoted by $\mathcal{CC}(\mathbf{f})$, is defined by $\mathcal{CC}(\mathbf{f}) = \bigcup_{\lambda \in \mathbb{R}} \{\mathcal{CC}([\mathbf{f}]_\lambda)\}$ where, for any $\lambda \in \mathbb{R}$, $[\mathbf{f}]_\lambda$ is *the upper level set of* $\mathbf{f}$ *of level* λ: $[\mathbf{f}]_\lambda = \{v_i \in V \mid \mathbf{f}_i \geq \lambda\}$. Note that the set $\mathcal{CC}(\mathbf{f})$ is finite and can thus be indexed by integers: $\mathcal{CC}(\mathbf{f}) = \{C_i\}_{i \in [\![1,m]\!]}$, where m is the number of connected components of $\mathbf{f}$. Let C_i in $\mathcal{CC}(\mathbf{f})$, the *altitude of* C_i is defined as the largest level λ in $\mathbb{R}$ such that C_i is a connected component of the upper level set of $\mathbf{f}$ at level λ: *i.e.*, $\max\{\lambda \in \mathbb{R} \mid C_i \in \mathcal{CC}([\mathbf{f}]^\lambda)\}$.

Let $\mathbf{f} \in \mathbb{R}^n$ be an image. The *max-tree* $\mathrm{MT}(\mathbf{f})$ *of* $\mathbf{f}$ is the pair $(\{C_i\}_{i \in [\![1,m]\!]}, \mathbf{a})$ where $\{C_i\}$ is the set of connected components of $\mathbf{f}$ and where $\mathbf{a}$ is a vector of $\mathbb{R}^m$ such that $\mathbf{a}_i$ is equal to the altitude of C_i. The first element of the pair, denoted by $\mathrm{MT}_1(\mathbf{f})$, is called the *hierarchy of* $\mathrm{MT}(\mathbf{f})$. The second element of the pair, denoted by $\mathrm{MT}_2(\mathbf{f})$, is called the *altitude vector of* $\mathrm{MT}(\mathbf{f})$. An example is given in Fig. 1. An element of the hierarchy $\mathcal{H}$ of $\mathrm{MT}(\mathbf{f})$ is called *a node of* $\mathcal{H}$. The node V includes every node of $\mathcal{H}$ and is called the *root*. Let C_i and C_j be two distinct nodes of $\mathcal{H}$. We say that C_i is an ancestor of C_j if C_j is included in C_i. Furthermore, if C_i is an ancestor of C_j, we say that C_i *is a parent of* C_j and that C_j *is a child of* C_i if any ancestor $C_k \neq C_i$ of C_j is also an ancestor of C_i. Any non-root node C_i of $\mathcal{H}$ has a unique parent which is denoted by $\mathrm{par}(C_i)$. The set of children of a node C_i of $\mathcal{H}$ is denoted by $\mathrm{Ch}(C_i)$. A node C_i of $\mathcal{H}$ is called a *leaf* if it has no child. There is a bijection between the leaf nodes of the hierarchy of $\mathrm{MT}(\mathbf{f})$ and the (regional) maxima of $\mathbf{f}$. For any node C_i of $\mathcal{H}$, a pixel v in C_i that is not contained in any child of C_i is called *a proper pixel of* C_i. Any element v of V is a proper pixel of a unique node C_i denoted $\mathrm{par}(v)$.

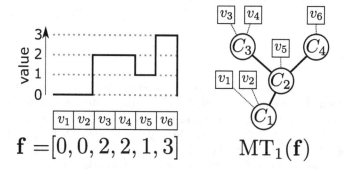

$$\mathbf{f} = [0, 0, 2, 2, 1, 3] \qquad \mathrm{MT}_1(\mathbf{f})$$

Fig. 1. Max-tree example. The left figure shows a 1d image $\mathbf{f} \in \mathbb{R}^6$ defined on the domain $v_1, \ldots, v_6$. Each of the four level sets at levels 0, 1, 2, and 3, has a single connected component $C_1, \ldots, C_4$. Those components are the nodes of hierarchy $MT_1(\mathbf{f})$ (circles) shown on the right image. The plain lines represent the parent relations between nodes. The proper elements of each node are depicted by squares and the dashed lines show the parent relation between those proper elements and their respective nodes, for example we have $\mathrm{par}(v_4) = C_3$. The altitude vector $\mathbf{a} = \mathrm{MT}_2(\mathbf{f})$ of the max-tree of $\mathbf{f}$ is equal to $[0, 1, 2, 3]$, meaning for example that the altitude of the node C_2 is equal to 1. The two maxima of $\mathbf{f}$ correspond to the leaf nodes C_3 and C_4 of the hierarchy $\mathrm{MT}_1(\mathbf{f})$.

3   Optimization with Differentiable Max-Trees

In this section, we first study how the altitude vector of a max-tree can be sub-differentiated with respect to the pixel values, then we state the general formulation of the optimization problem which is addressed.

Differentiable Max-Trees. Trees, as combinatorial structures, are generally not suited to gradient-based optimization. However, in max-trees, the altitude of a component is mapped to the value of some pixels of the base image: its proper pixels. Then, intuitively, a small modification of the values of those proper pixels will not change the hierarchy associated with the max-tree of the image and will produce the exact same modification of the altitude of the corresponding node of the hierarchy.

Property 1. *Let* $\mathbf{f} \in \mathbb{R}^n$ *be an image. Let* $\epsilon \in \mathbb{R}^n$ *such that* $\mathrm{MT}_1(\mathbf{f}) = \mathrm{MT}_1(\mathbf{f} + \epsilon)$. *Then, for any node* C_i *of* $\mathrm{MT}_1(\mathbf{f})$, *the altitude of* C_i *in* $\mathrm{MT}(\mathbf{f} + \epsilon)$ *is equal to* $\mathbf{a}_i + \epsilon_j$ *where* $\mathbf{a}_i$ *is the altitude of* C_i *and where* v_j *is any proper pixel of* C_i.

This property indicates that the Jacobian of the function MT_2 can be written as the matrix composed of the indicator column vectors giving the index of the node associated with any pixel of V by the parent mapping (its proper elements):

$$\frac{\partial \mathrm{MT}_2(\mathbf{f})}{\partial \mathbf{f}} = \left[\mathbb{1}_{\mathrm{par}(v_1)}, \ldots, \mathbb{1}_{\mathrm{par}(v_n)} \right], \tag{1}$$

where $\mathbb{1}_{C_k}$ is the column vector of $\mathbb{R}^m$ equals to 1 in position k, and 0 elsewhere. In a back-propagation algorithm, this means that if we have an error measure e and we have already computed $\frac{\partial e}{\partial \mathbf{a}}$, *i.e.*, how the altitude vector $\mathbf{a} = \mathrm{MT}_2(\mathbf{f})$ of the max-tree of $\mathbf{f}$ should be modified in order to minimize e, we can then back-propagate through MT with the chain rule $\frac{\partial e}{\partial \mathbf{f}} = \frac{\partial \mathbf{a}}{\partial \mathbf{f}} \frac{\partial e}{\partial \mathbf{a}}$ leading to the simple formula $\left(\frac{\partial e}{\partial \mathbf{f}} \right)_i = \left(\frac{\partial e}{\partial \mathbf{a}} \right)_{\mathrm{par}(i)}$ telling how $\mathbf{f}$ should be modified to minimize e.

For example, the transpose of the Jacobian of the altitude vector $\mathrm{MT}_2(\mathbf{f})$ of the max-tree shown in Fig. 1 is equal to

$$\begin{array}{c} \\ \mathbf{a}_1 \\ \mathbf{a}_2 \\ \mathbf{a}_3 \\ \mathbf{a}_4 \end{array} \begin{array}{cccccc} \mathbf{f}_1 & \mathbf{f}_2 & \mathbf{f}_3 & \mathbf{f}_4 & \mathbf{f}_5 & \mathbf{f}_6 \\ \left(\begin{array}{cccccc} 1 & 1 & 0 & 0 & 0 & 0 \\ 0 & 0 & 0 & 0 & 1 & 0 \\ 0 & 0 & 1 & 1 & 0 & 0 \\ 0 & 0 & 0 & 0 & 0 & 1 \end{array} \right). \end{array}$$

This matrix indicates, how the image $\mathbf{f}$ should be modified in order to reflect a modification of the altitude vector $\mathbf{a}$ of the nodes of the max-tree of $\mathbf{f}$. For example, in order to increase the altitude $\mathbf{a}_3$ of the component C_3 by a small value ϵ, one must increase the value of $\mathbf{f}_3$ and $\mathbf{f}_4$ by this same value ϵ.

Optimization Problem. We now state a general formulation of the optimization problem that we want to solve. Let $\mathbf{y} \in \mathbb{R}^n$ be an image representing an observation. We are interested in solving the following optimization problem:

$$\underset{\mathbf{f} \in \mathbb{R}^n}{\mathrm{minimize}} \; J(\mathbf{f}; \mathbf{y}), \quad \text{s.t. } 0 \le \mathbf{f} \le 1, \tag{2}$$

where J is a differentiable cost function involving the altitude vector $MT_2(\mathbf{f})$. As this altitude vector $MT_2(\mathbf{f})$ is differentiable with respect to the image $\mathbf{f}$, a local optimum of the above problem can be found by gradient descent algorithm. The constraint to keep the image between the values 0 and 1 will prevent it to shrink (resp. grow) towards $-\infty$ (resp. ∞) when we try to remove (resp. increase) some image features. A similar effect could be achieved by imposing margins on how much a feature can shrink or grow.

4   Maxima Loss

In the following, we study how to define a component tree loss imposing a topological criterion, by prescribing how many maxima should be present in the result. The proposed approach relies on two features characterizing the maxima:

- *a measure of saliency*: increasing this measure should reinforce the maximum and decreasing it should make it disappear; and
- *a measure of (relative) importance*: which provides a ranking of the maximum to identify those that should be reinforced and those which should disappear.

We first introduce a loss function to *select* a given number of maxima and to *discard* the others according to these two measures. Then, we present several measures to assess the importance and the saliency of maxima.

4.1   Ranked Selection Loss

Assume that the hierarchy $\mathcal{H}$ of $MT(\mathbf{f})$ contains k maxima $\{M_i\}_{i=[\![1,k]\!]}$ (its leave nodes). Let $\ell \in \mathbb{N}^+$ be a target number of maxima. Let $\mathbf{sm} \in \mathbb{R}^{+k}$ and $\mathbf{im} \in \mathbb{R}^{+k}$ represent respectively a saliency and an importance measure on the maxima $\{M_i\}$. Let $p \in \mathbb{R}^{+*}$ and $q \in \mathbb{R}^{+*}$ be 2 strictly positive numbers controlling the growth/shrink pressure applied on the maxima. The ranked selection function will seek to maximize the saliency of the ℓ maxima with the largest importance values and decrease the saliency of the others:

$$J_{p,q}(\mathbf{sm}, \mathbf{im}; \ell) = -\sum_{i=1}^{i\leq\ell} \mathbf{sm}_{\mathbf{r}_i}^p + \sum_{i=\ell+1}^{i\leq k} \mathbf{sm}_{\mathbf{r}_i}^q \text{ with } \mathbf{r} = \mathrm{argsort}(\mathbf{im}), \qquad (3)$$

where argsort is the function that associates any vector $\mathbf{v}$ of $\mathbb{R}^k$ with a permutation vector $\mathbf{r} \in [\![1, n]\!]^k$ sorting the elements of $\mathbf{v}$ in decreasing order, *i.e.*, such that for any i, j in $[\![1, n]\!]$, we have $i < j \Rightarrow v_{\mathbf{r}_i} \geq v_{\mathbf{r}_j}$.

4.2   Maxima Measures

We now define maxima measures that will be used as saliency and/or importance measures in the previous loss function. Recall that the leaves of the max-tree of an image $\mathbf{f}$ corresponds to the maxima of this image and assume that the

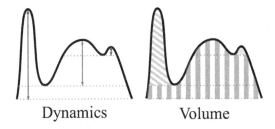

Dynamics   Volume

Fig. 2. Illustration of the dynamics and the volume associated with the maxima of a 1d function. The dynamics of a maximum is equal to the difference of altitude between the top of the maximum and the closest level that contains another maximum of greater altitude. Similarly, the volume of the maximum is equal to the surface between the top of the maximum and the closest level that contains another maximum of greater volume. With the dynamics, the most important maximum is the sharp peak on the left while, with the volume, the most important maximum is the large mount in the middle.

hierarchy $\mathcal{H}$ of $\mathrm{MT}(\mathbf{f})$ contains k maxima $\{M_i\}_{i=[\![1,k]\!]}$; a measure on $\{M_i\}$ is then a positive vector of $\mathbb{R}^k$.

Maxima Altitude: A simple way to measure the importance and the saliency of a maximum is to look at its altitude, *i.e.*, to the altitude $\mathbf{a}_M$ of the maximum M of $\mathcal{H}$. The altitude of the maxima of $\mathbf{f}$ is denoted $\mathbf{alt}(\mathbf{f})$.

Extinction Values: Extinction values are classical maxima measures known for their robustness [21]. Given a family of image filters $\{\sigma_k\}_k$ whose activity increases with k (for any $k_1 \leq k_2$, we have $\sigma_{k_1} \geq \sigma_{k_2}$). The extinction value of a maximum M of $\mathbf{f}$ is equal to the smallest k such that M is not contained in any maximum of $\sigma_k(\mathbf{f})$. Typical examples of extinction values are the *dynamics* and the *volume*. The dynamics and volume associated with the maxima of a function are illustrated in Fig. 2. In the following, we will show how extinction values can be defined and computed based on the max-tree representation.

Any extinction value relies on an increasing attribute measuring the importance of regional maxima. Such attribute will be represented by a vector $\mathbf{v} \in \mathbb{R}^m$ associating a value to each node of the max-tree such that for any two nodes C_i and C_j, $C_i \subseteq C_j \Rightarrow \mathbf{v}_i \leq \mathbf{v}_j$. The idea to compute the extinction value of a maximum for the attribute $\mathbf{v}$ is then to find, for any maximum M, the closest ancestor of M that contains another maximum whose attribute value is greater than the one of M: this node is called the saddle node associated with M for $\mathbf{v}$.

Formally, let N be a node of $\mathcal{H}$ and let A be an ancestor of N. There exists a single branch rooted in A that contains the node N, the child of A in this branch is denoted by $\mathrm{ch}(A)_{\to N}$; in other words $\mathrm{ch}(A)_{\to N}$ is the only child C of A that contains N. The saddle node associated with the node N for the attribute $\mathbf{v}$, denoted by $\mathrm{saddle}_{\mathbf{v}}(N)$, is the closest ancestor A of N such that there exists a child C of A with $\mathbf{v}_{\mathrm{ch}(A)_{\to N}} < \mathbf{v}_C$. If no such ancestor exists, then the saddle node of N for $\mathbf{v}$ is defined as the empty-set $\emptyset$. The base node associated with the node N for the attribute $\mathbf{v}$, denoted by $\mathrm{base}_\mathbf{v}(N)$, is then equal to $\mathrm{ch}(\mathrm{saddle}_\mathbf{v}(N))_{\to N}$ if $\mathrm{saddle}_\mathbf{v}(N) \neq \emptyset$ and the root of the max-tree otherwise.

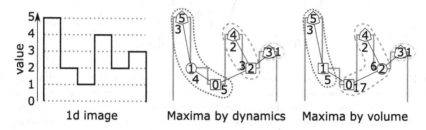

Fig. 3. Regional maximum associated with each maximum for the dynamics and the volume. We consider a 1d image on the left with 3 maxima. The second (resp. third) figure shows how the 1d image is decomposed into 3 regional maxima according to the dynamics (resp. the extinction by volume). In those two images, we see the max-tree of the 1d image where the nodes are depicted by circles, squares and hexagons. The label of each node corresponds to its index, and the blue value beside it corresponds to its attribute value: its depth for the dynamics and its volume for the extinction by volume. The hexagons are the maxima, *i.e.*, the leaves of the tree. Each maximum is associated with a branch of the tree, circled in red (dotted line), green (dashed line) and orange (dot-dash line) whose base node is depicted by a square. For example, for the dynamics (resp. the extinction by volume), the maximum of index 5 extends to its base node 0 (reps. 1); its dynamics (resp. extinction by volume) is thus equal to 5, *i.e.*, the attribute value of this base node. (Color figure online)

Thus, each regional maximum extends from its top node M to its base node $\mathrm{base}_\mathbf{v}(M)$, and the saddle node $\mathrm{saddle}_\mathbf{v}(M)$ is the first ancestor of M that belongs to another maximum according to the attribute $\mathbf{v}$. The extinction value of the maximum M for the attribute $\mathbf{v}$ is then defined as the attribute value of its base node $\mathbf{v}_{\mathrm{base}_\mathbf{v}(M)}$, see Fig. 3. Note that the definition of the extinction value for the attribute $\mathbf{v}$ is just a value selection process in a vector, as in a max-pooling layer, and it can thus be used in the definition of a loss function suitable for gradient descent optimization.

In the following, we consider two different attributes: the depth, whose associated extinction value is usually called the dynamics, and the volume. Both attributes will be defined as a function of the max-tree altitudes, so that any error on the extinction value can be translated as an error on those altitudes.

Let N be a node of $\mathcal{H}$, the depth of N, denoted by $\mathrm{depth}(N)$, is defined by:

$$\mathrm{depth}(N) = \max\{\mathbf{a}_C, C \in \mathcal{H} \mid C \subset N\} - \mathbf{a}_{\mathrm{par}(N)}. \tag{4}$$

The depth of N is thus equal to the difference between the largest altitude in the subtree rooted in N and the altitude of the parent of N. The extinction values of the maxima of $\mathcal{H}$ for the attribute depth is called the dynamics and will be denoted by $\mathbf{dyn}$, see Fig. 3.

Let N be a node of $\mathcal{H}$, the volume of N, denoted by vol(N), is defined recursively by:

$$\text{vol}(N) = |N| \cdot \left(\mathbf{a}_N - \mathbf{a}_{\text{par}(N)}\right) + \sum_{C \in \text{Ch}(N)} \text{vol}(C), \tag{5}$$

where $|N|$ denotes the cardinal of N, *i.e.*, the number of pixels in the node N. The volume of N is thus equal to the volume of the cylinder defined by the node N and its parent, plus the volume of its children. The extinction values of the maxima of $\mathcal{H}$ for the attribute vol is called the extinction value by volume and will be denoted by **vol**, see Fig. 3.

Effect of Modifying the Saliency: The ranked selection loss (3) will try to increase the saliency measure of the selected maxima and decrease the one of the others. In order to better understand how this will affect the result, we propose to study how a single maximum is modified when we try to increase/decrease its saliency according to one of the proposed saliency measures.

In the case of the saliency based on the maxima altitudes, this effect is simple, as increasing (resp. decreasing) the altitude of a maximum simply means increasing (resp. decreasing) the altitude of the leaf node that corresponds to this maximum in the max-tree.

With the dynamics, the saliency of a maximum is determined by the altitude of the leaf node that corresponds to this maximum in the max-tree and by the altitude of the saddle node associated with this leaf for the depth attribute. In this case, increasing (resp. decreasing) the dynamics of a maximum means increasing (resp. decreasing) the altitude of its leaf node and decreasing (resp. increasing) the altitude of the saddle node (see Fig. 4). Note that the altitudes of all the nodes between the leaf node and the saddle node are not modified.

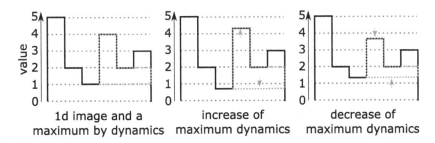

1d image and a
maximum by dynamics

increase of
maximum dynamics

decrease of
maximum dynamics

Fig. 4. Effect of increasing or decreasing the saliency measure of a maximum based on the dynamics **dyn**. The first figure shows a 1d image and a regional maximum for the dynamics (see Fig. 2). The second (resp third) figure shows the effect of increasing (resp decreasing) the measure for this maximum.

Finally, with the extinction value by volume, the saliency of a maximum is determined by the altitudes of all the nodes in the branch going from the leaf node that corresponds to this maximum in the max-tree to the saddle node associated with this leaf for the volume attribute. In this case, increasing (resp. decreasing) the extinction value by volume of a maximum means increasing (resp. decreasing) the altitudes of all the nodes in the branch going from the leaf node to the base node and decreasing (resp. increasing) the altitude of the saddle node (see Fig. 5).

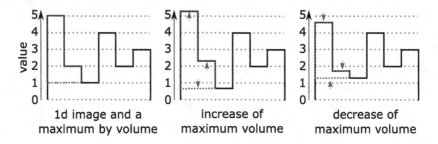

Fig. 5. Effect of increasing or decreasing the saliency measure of a maximum based on the extinction value by volume **vol**. The first figure shows a 1d image and a regional maximum for the volume extinction value (see Fig. 2). The second (resp third) figure shows the effect of increasing (resp decreasing) the measure for this maximum.

Note that in both cases, modifying the saliency of a single maximum usually preserves the ordering of the nodes in the tree; an inversion between the top node and the base node may still happen when a maximum collapses. However, when the saliency of several maxima is modified at the same time, it becomes more probable that the ordering of the nodes in the tree changes, leading to more complex topological modifications in the image domain.

5   Experiments

We demonstrate the behavior of the proposed method and the various maxima measures on a simulated image, and we show how it can be combined with classical loss functions to process real images. The method is implemented using the library Higra [16] for hierarchical graph analysis in combination with the continuous optimization framework Pytorch [15]. In all the experiments, an Adam optimizer [14] is used and the input image $\mathbf{y}$ is used as the initial solution. A Jupyter notebook containing the presented experiments is available online[1].

Simulated Image Filtering: The effect of the optimization of the component tree loss $J_{1,2}(\mathbf{sm}, \mathbf{im}; 2)$ with the proposed importance and saliency maxima measures is demonstrated on a simulated image in Fig. 6. The test image contains four maxima with different altitudes, contrast, and volumes. We can see

---

[1] https://www.esiee.fr/~perretb/notebooks/Component_Tree_Loss.zip.

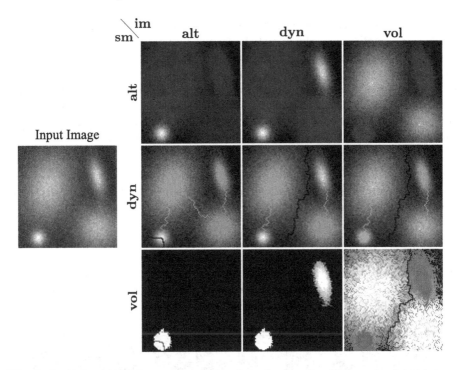

Fig. 6. Optimization of the component tree loss $J_{1,2}(\mathbf{sm}, \mathbf{im}; 2)$ on a simulated image with the objective of selecting 2 maxima for various combinations of saliency measures (**sm**) and importance measures (**im**).

that the altitude measure is not robust to noise and fails to select perceptually significant maxima: both are located in the bottom left blob. On the other hand, the two measures based on extinction values, the dynamics and the volume, both manage to select significant maxima: with the dynamics, the two brightest maxima are selected (bottom left and the top right blobs) while with the volume, the two largest maxima are selected (top left and bottom right blobs). Regarding saliency measures, we can see that the optimization of the saliency based on maxima altitudes leads to increasing the altitudes of the top node of the selected maxima and to raising discarded maxima. The optimization of the dynamics saliency measure is more complex, as increasing/decreasing the dynamics of the maxima involves increasing/decreasing the altitude of its top node and decreasing/increasing the altitudes of its saddle node: this leads to the creation of "bridges" between some maxima. Finally, the optimization of the volume saliency has a more global effect on the maxima as, contrarily to the dynamics, its definition involves the altitudes of all the nodes between the base and the top nodes of a maximum.

In order to work with the saliency measure based on volume, we have observed that raising the power q to 2 inside the loss function can help to erase the unwanted maxima whose contribution to the loss value and hence to the gradient tends to be smaller than the one of the largest maxima.

Note that optimizing J with the dynamics used both as the saliency and the importance measure of maxima is similar [2] to optimizing the barcode length of the connected components used in the "topological loss" based on persistent homology [4, 5, 9, 12]. However, as our approach does not require computing the full persistence diagram associated with the image at each iteration of the optimization algorithm, we observe that it is faster than a classical implementation of the topological loss[2].

Real Image Filtering: Finally, in Fig. 7, we show how the proposed loss function can be combined with classical loss functions used in image analysis: here we optimize the term

$$||\mathbf{f} - \mathbf{y}||_2^2 + \lambda_1 J_{1,1}(\mathbf{dyn}(\mathbf{f}), \mathbf{dyn}(\mathbf{f}), 1) + \lambda_2||\nabla \mathbf{f}||_2^2, \qquad (6)$$

which combines our loss based on the max-tree to enforce the presence of a single maximum with a L2 data attachment term and a total variation regularization term. We can see that we are able to successfully reconnect the different branches of the neurite.

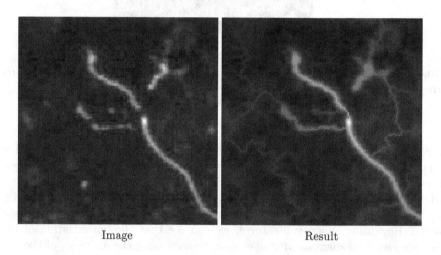

Image   Result

Fig. 7. Reconnection of a neurite image with a combination of the proposed loss J_r to enforce a single maximum, a L2 data attachment term and a TV2 regularization term.

6   Conclusion

We have proposed a continuous optimization framework based on the hierarchical image representation called the max-tree. We showed how it can be used to

---
[2] https://github.com/bruel-gabrielsson/TopologyLayer.

design a component tree loss, *i.e.*, a regularization term, enabling to select or discard maxima in an image based on various measures. This approach can be generalized immediately to other hierarchical representations, such as the min-tree or the tree-of-shapes [1,10]. In future works, we plan to explore more general component tree loss functions based on such hierarchical representations and their use in supervised learning methods involving deep networks.

Acknowledgements. This work was supported by the French ANR grant ANR-20-CE23-0019.

References

1. Ballester, C., Caselles, V., Monasse, P.: The tree of shapes of an image. In: ESAIM: Control, Optimisation and Calculus of Variations, vol. 9, pp. 1–18 (2003)
2. Boutry, N., Géraud, T., Najman, L.: An equivalence relation between morphological dynamics and persistent homology in 1D. In: Burgeth, B., Kleefeld, A., Naegel, B., Passat, N., Perret, B. (eds.) ISMM 2019. LNCS, vol. 11564, pp. 57–68. Springer, Cham (2019). https://doi.org/10.1007/978-3-030-20867-7_5
3. Chierchia, G., Perret, B.: Ultrametric fitting by gradient descent. In: NeurIPS, pp. 3181–3192 (2019)
4. Clough, J., Byrne, N., Oksuz, I., Zimmer, V., Schnabel, J., King, A.: A topological loss function for deep-learning based image segmentation using persistent homology. IEEE TPAMI (2020)
5. Clough, J.R., Oksuz, I., Byrne, N., Schnabel, J.A., King, A.P.: Explicit topological priors for deep-learning based image segmentation using persistent homology. In: Chung, A.C.S., Gee, J.C., Yushkevich, P.A., Bao, S. (eds.) IPMI 2019. LNCS, vol. 11492, pp. 16–28. Springer, Cham (2019). https://doi.org/10.1007/978-3-030-20351-1_2
6. Cousty, J., Najman, L.: Incremental algorithm for hierarchical minimum spanning forests and saliency of watershed cuts. In: Soille, P., Pesaresi, M., Ouzounis, G.K. (eds.) ISMM 2011. LNCS, vol. 6671, pp. 272–283. Springer, Heidelberg (2011). https://doi.org/10.1007/978-3-642-21569-8_24
7. Dalla Mura, M., Benediktsson, J., Waske, B., Bruzzone, L.: Morphological attribute profiles for the analysis of very high resolution images. IEEE TGRS **48**(10), 3747–3762 (2010)
8. Donoser, M., Bischof, H.: Efficient maximally stable extremal region (MSER) tracking. In: IEEE CVPR, vol. 1, pp. 553–560 (2006)
9. Gabrielsson, R., Nelson, B., Dwaraknath, A., Skraba, P.: A topology layer for machine learning. In: AISTATS, pp. 1553–1563. PMLR (2020)
10. Géraud, T., Carlinet, E., Crozet, S., Najman, L.: A quasi-linear algorithm to compute the tree of shapes of nD images. In: Hendriks, C.L.L., Borgefors, G., Strand, R. (eds.) ISMM 2013. LNCS, vol. 7883, pp. 98–110. Springer, Heidelberg (2013). https://doi.org/10.1007/978-3-642-38294-9_9
11. Girshick, R., Donahue, J., Darrell, T., Malik, J.: Rich feature hierarchies for accurate object detection and semantic segmentation. In: IEEE CVPR, pp. 580–587 (2014)
12. Hu, X., Li, F., Samaras, D., Chen, C.: Topology-preserving deep image segmentation. In: NeurIPS, pp. 5657–5668 (2019)

13. Jones, R.: Connected filtering and segmentation using component trees. CVIU **75**(3), 215–228 (1999)
14. Kingma, D., Ba, J.: Adam: a method for stochastic optimization. In: Bengio, Y., LeCun, Y. (eds.) ICLR (2015)
15. Paszke, A., et al.: Pytorch: an imperative style, high-performance deep learning library. In: NeurIPS, pp. 8026–8037 (2019)
16. Perret, B., Chierchia, G., Cousty, J., Guimarães, S., Kenmochi, Y., Najman, L.: Higra: hierarchical graph analysis. SoftwareX **10**, 100335 (2019)
17. Perret, B., Cousty, J., Guimaraes, S., Maia, D.: Evaluation of hierarchical watersheds. IEEE TIP **27**(4), 1676–1688 (2017)
18. Robic, J., Perret, B., Nkengne, A., Couprie, M., Talbot, H.: Self-dual pattern spectra for characterising the dermal-epidermal junction in 3D reflectance confocal microscopy imaging. In: Burgeth, B., Kleefeld, A., Naegel, B., Passat, N., Perret, B. (eds.) ISMM 2019. LNCS, vol. 11564, pp. 508–519. Springer, Cham (2019). https://doi.org/10.1007/978-3-030-20867-7_39
19. Salembier, P., Oliveras, A., Garrido, L.: Anti-extensive connected operators for image and sequence processing. IEEE TIP **7**(4), 555–570 (1998)
20. Teeninga, P., Moschini, U., Trager, S., Wilkinson, M.: Statistical attribute filtering to detect faint extended astronomical sources. Math. Morphol.-Theory Appl. **1** (2016)
21. Vachier, C., Meyer, F.: Extinction value: a new measurement of persistence. In: IEEE Workshop on Nonlinear Signal and Image Processing, vol. 1, pp. 254–257 (1995)
22. Xu, Y., Carlinet, E., Géraud, T., Najman, L.: Hierarchical segmentation using tree-based shape spaces. IEEE TPAMI **39**(3), 457–469 (2016)
23. Xu, Y., Monasse, P., Géraud, T., Najman, L.: Tree-based Morse regions: a topological approach to local feature detection. IEEE TIP **23**(12), 5612–5625 (2014)

Fast and Effective Superpixel Segmentation Using Accurate Saliency Estimation

Felipe Belém[1,3]($\boxtimes$) , Isabela Borlido[2] , Leonardo João[1] ,
Benjamin Perret[3] , Jean Cousty[3] , Silvio J. F. Guimarães[2] ,
and Alexandre Falcão[1]

[1] University of Campinas, Campinas, Brazil
{felipe.belem,leonardo.joao,afalcao}@ic.unicamp.br
[2] Pontifical Catholic University of Minas Gerais, Belo Horizonte, Brazil
isabela_borlido@hotmail.com, sjamil@pucminas.br
[3] LIGM, Univ. Gustave Eiffel, CNRS, F-77454 Marne-la-Vallée, France
{felipe.belem,benjamin.perret,jean.cousty}@esiee.fr

Abstract. *Superpixels through Iterative CLEarcutting* (SICLE) is a
recently proposed framework for superpixel segmentation. SICLE consists of three steps: (i) seed oversampling; (ii) superpixel generation; and
(iii) seed removal; such that, after step (i), steps (ii) and (iii) are repeated
until a desired number of superpixels is obtained. Such pipeline showed
effective and efficient multiscale superpixel segmentation. Furthermore,
if an object is desired, it is possible to improve delineation by providing its probable location, often called saliency. While classical methods estimate object saliency by contrast-based criteria, recent ones use
deep-learning strategies for accurate estimation. SICLE shows robustness
for low-quality saliency estimations, but it struggles to effectively take
advantage of the high-quality ones. In this work, we propose a generalization of its path-cost function and seed removal criterion (steps (ii) and
(iii), respectively), adapting SICLE to a given saliency map. By choice of
a binary parameter, SICLE can take advantage of low- and high-quality
saliency maps for better segmentation. Results show that, by exploiting
the accurate information of the saliency map, our improved SICLE version surpasses state-of-the-art methods in traditional delineation metrics
while requiring only two iterations for segmentation, being significantly
faster than its predecessor and SLIC.

Keywords: Superpixel segmentation · Object saliency map · Image
foresting transform

The authors thank the Conselho Nacional de Desenvolvimento Científico e Tecnológico
– CNPq – (Universal 407242/2021-0, PQ 303808/2018-7, 310075/2019-0), the Fundação
de Amparo a Pesquisa do Estado de Minas Gerais – FAPEMIG – (PPM-00006-18), the
Fundação de Amparo a Pesquisa do Estado de São Paulo – FAPESP – (2014/12236-1)
and the Coordenação de Aperfeiçoamento de Pessoal de Nível Superior – CAPES –
Finance code 001 (COFECUB 88887.191730/2018-00) for the financial support.

É. Baudrier et al. (Eds.): DGMM 2022, LNCS 13493, pp. 261–273, 2022.
https://doi.org/10.1007/978-3-031-19897-7_21

1 Introduction

Superpixels represent homogeneous regions that contain a perceptual meaning and provide more information than pixels. Although some authors raise compactness and regularity as indicators of high-quality superpixel segmentation, boundary adherence, efficiency, and controllable quantity of superpixels, are indispensable for any method [23,25]. From that, several applications such as object tracking [10], semantic segmentation [32], and image classification [21] exploit their properties.

Classical approaches differ in their strategy for superpixel generation and generally consider only color and spatial position to measure superpixel similarity, without any prior information. We may cite *Simple Linear Iterative Clustering* (SLIC) [1] as an example of a clustering-based method given its adapted K-means strategy in a 5-dimensional feature space. Conversely, graph-based approaches, such as *Entropy Rate Superpixels* (ERS) [16], have higher boundary adherence but they are often slow. Also, *Superpixel Hierarchy* (SH) [26] and *Waterpixels* [17] are effective hierarchical graph-based examples whose drawback is error propagation to coarser scales. Finally, *Dynamic and Iterative Spanning Forest* (DISF) [6] is a path-based method that applies oversampling and iteratively generates superpixels on refined seed sets.

Using local information without any prior or high-level knowledge may be insufficient to obtain a good delineation in images with complex characteristics, such as textured or noisy images [27]. More recent approaches circumvent this drawback with Deep Learning architectures [14,28] or by including high-level information, such as texture [30] and gradient mask [27]. However, they present moderate delineation, and their constraints are the regular grid shape on standard convolution operations for deep-learning-based methods, high computational time, and the lack of superpixel groundtruth [28].

Although using saliency in segmentation is not a novel strategy [12], it has not been thoroughly exploited for generating superpixels until recently. In [31], the authors proposed a SLIC-based algorithm that uses a saliency map based on the Fourier Transform for generating more superpixels in textured regions. The *Object-based DISF* (ODISF) [3] method is another example that extends DISF for incorporating object saliency maps. However, since the map's influence is not controllable in both, higher-quality saliency may not promote higher-quality delineation. Conversely, *Object-based ISF* (OISF) [4,5] overcomes this issue by allowing user control over the saliency influence during delineation, but it is slow and highly sensitive to incorrect estimations.

A recent proposal named *Superpixels through Iterative CLEarcutting* (SICLE) [8] generalizes ODISF for allowing user control over the number of iterations for segmentation, being more efficient than SLIC at generating superpixels in different experiments. The SICLE pipeline is composed of three steps: (i) seed oversampling; (ii) superpixel generation by the *Image Foresting Transform* (IFT) [11]; and (iii) object-based seed removal. After step (i), steps (ii) and (iii) are performed until obtaining a desired number of superpixels. By using object information only in the last step, SICLE delineation performance is robust to

Original   SICLE [8]

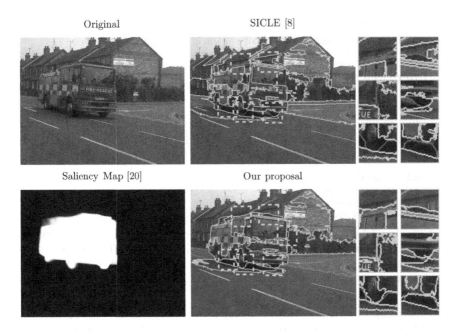

Saliency Map [20]   Our proposal

Fig. 1. Comparison between SICLE and our improved version for accurate maps considering 25 superpixels. Red lines indicate object boundaries, whereas cyan ones, superpixel borders. Yellow rectangles indicate delineation errors that our approach overcame. (Color figure online)

incorrect estimations, contrasting with OISF. However, similarly to ODISF, it cannot improve its delineation performance for higher-quality saliency maps.

One may argue that an object-based method should exploit the prior object location information with respect to its quality. For high-quality information, the approximation to the object boundaries can assist its delineation. Conversely, although low-quality information poorly estimates the object boundaries, it presents valuable information on its location. Therefore, in this work, we propose a generalization of SICLE's path-cost function and object-based seed removal criteria (*i.e.*, steps (ii) and (iii), respectively) for exploiting low- and high-quality saliency maps to improve segmentation results by choice of a binary parameter. Experimental results show that our proposal, named $SICLE^\alpha$, is robust to low-quality maps and improves delineation in the case of high-quality ones, as exemplified in Fig. 1. Moreover, by exploiting the accurate estimation of the object boundaries, $SICLE^\alpha$ achieves higher precision in step (iii), thus requiring only two iterations for effective object delineation. Given both, our method surpasses state-of-the-art methods in terms of efficiency and effectiveness, considering classical evaluation metrics.

This paper is organized as follows. The related definitions are presented in Sect. 2 and our proposal is described in Sect. 3. Section 4 presents the experiments, with an ablation study, and qualitative and quantitative evaluation. Finally, the conclusion and future work are presented in Sect. 5.

2   Theoretical Background

In this section, we present the theoretical background for our work. In Sect. 2.1, we discuss basic concepts in image and graphs and, subsequently, in Sect. 2.2, we present the *Image Foresting Framework* (IFT) [11].

2.1   Image and Graphs

An *image* I is a pair $\langle \mathcal{P}, \mathbf{F} \rangle$ in which $\mathbf{F}(p) \in \mathbb{R}^m$ maps the *features* of every *picture element* (*i.e.*, pixel) $p \in \mathcal{P} \subset \mathbb{Z}^2$, given $m \in \mathbb{N}^*$. I is either *colored* or *grayscale* whenever $m > 1$ or $m = 1$, respectively. An *object saliency map* $O = \langle \mathcal{P}, \mathbf{O} \rangle$ is an instance of the latter in which $\mathbf{O}(p) \in [0, 1]$ maps p to its probability of belonging to an object of interest (*i.e.*, saliency). Finally, for a set of pixels $X \subseteq \mathcal{P}$, we may compute its *mean feature* $\overline{\mathbf{F}}(X) = \sum_{x \in X} \mathbf{F}(x)/|X|$ and *mean saliency* $\overline{\mathbf{O}}(X) = \sum_{x \in X} \mathbf{O}(x)/|X|$.

From I, we may build a *directed graph* (*i.e.*, digraph) $G = \langle \mathcal{V}, \mathcal{A} \rangle$ so that $\mathcal{V} \subseteq \mathcal{P}$ contains its *vertices* and $\mathcal{A} \subset \mathcal{V}^2$ its *arcs*. The existence of an arc $\langle x, y \rangle \in \mathcal{A}$ indicates that x is *adjacent* to y. Often, $\mathcal{A}$ is defined by the 8-adjacents of every pixel $x \in \mathcal{P}$, such that $\mathcal{A} = \{\langle x, y \rangle : \|x - y\|_2 \leq \sqrt{2}\}$. In this work, $\mathcal{A}$ holds no self-loops nor parallel edges (*i.e.*, G is a *simple graph*).

A *(directed) path* $\rho = \langle v_1, \ldots, v_k \rangle$ is a sequence of $k \in \mathbb{N}^*$ distinct adjacent vertices (*i.e.*, $\langle v_i, v_i + 1 \rangle \in \mathcal{A}$ for $i < k$). If $k = 1$, ρ is said to be *trivial*, and *non-trivial* otherwise. In ρ, we term v_i as the *predecessor* of v_{i+1} and the *successor* v_{i-1} given $1 < i < k$. Moreover, we may exhibit the *root* v_1 and the *terminus* v_k of ρ either by $\rho_{v_1 \leadsto v_k}$ or by ρ_{v_k} whenever v_1 is irrelevant for the context. For instance, $\rho_y = \rho_x \odot \langle x, y \rangle$ denotes the path ρ_y resultant from concatenating ρ_x with $\langle x, y \rangle$.

2.2   Image Foresting Transform

The *Image Foresting Transform* (IFT) [11] is a framework whose effectiveness in object delineation has been reported in several works [3,6,9,13]. When a set of representative vertices (*i.e.*, seeds) $\mathcal{S} \subset \mathcal{V}$ is provided, the algorithm builds trees with optimum path-cost from their seed $s \in \mathcal{S}$ to any $p \in \mathcal{V} \setminus \mathcal{S}$ through path concatenation.

We can estimate the cost of an arc $\langle x, y \rangle \in \mathcal{A}$ by an *arc-cost function* $\mathbf{w}_*(x, y) \in \mathbb{R}$ and, likewise, the cost of any path ρ_x can be computed by an *path-cost function* $\mathbf{f}_*(\rho_x) \in \mathbb{R}_+$. As an example, the $\mathbf{f}_{max}$ (Eq. 1) is commonly used in IFT-based methods due to its effectiveness [3,6] in delineating objects:

$$\mathbf{f}_{max}(\langle x \rangle) = \begin{cases} 0, & \text{if } x \in \mathcal{S} \\ +\infty, & \text{otherwise} \end{cases}$$

$$\mathbf{f}_{max}(\rho_x \odot \langle x, y \rangle) = \max \{\mathbf{f}_{max}(\rho_x), \mathbf{w}_*(x, y)\}$$

$$(1)$$

If $\mathbf{f}_*(\rho_x) \leq \mathbf{f}_*(\tau_x)$, considering $\tau_x \in \mathscr{P}$ to be any other path reaching x within the set $\mathscr{P}$ of all possible paths in G, then ρ_x is *optimum*.

The IFT minimizes a *cost map* $\mathbf{C}(x) = \min_{\rho_x \in \mathscr{P}} \{\mathbf{f}_*(\rho_x)\}$ by assigning an optimum path ρ_x from a seed to $x \in \mathcal{V} \setminus \mathcal{S}$. Simply put, the method builds trees whose paths to non-seed vertices are more closely connected to its seed than to any other using a generalization of the Dijkstra's shortest-path algorithm. Even if $\mathbf{f}_*$ is not smooth, it still exhibits properties suitable for segmentation [18]. While minimizes $\mathbf{C}$, the algorithm builds an acyclic map $\mathbf{P}$ (*i.e.*, *predecessor map*) which assigns x either to its predecessor defined in ρ_x or to a distinctive marker $\blacktriangle \notin \mathcal{V}$, whenever x is the root of ρ_x and, thus, of $\mathbf{P}$. As one may see, it is possible to map x to its root $\mathbf{R}(x) \in \mathcal{S}$ recursively through $\mathbf{P}$. Furthermore, by assuming $s = \mathbf{R}(x)$, we can map every vertex to its *optimum-path tree* $\mathbf{T}(s) \subset \mathcal{V}$ by $\mathbf{T}(s) = \{t : \mathbf{R}(t) = s\}$. In this work, every superpixel is an optimum-path tree rooted in a seed.

3   Methodology

In this section, we review *Superpixels through Iterative CLEarcutting* (SICLE) [8] alongside our proposed evolutions to better take advantage of high quality saliency estimations. Briefly, each section refers to a specific SICLE step, given that our contributions reside on the last ones: (i) seed oversampling; (ii) superpixel generation; and (iii) object-based seed removal (*i.e.*, Sects. 3.1, 3.2 and 3.3, respectively).

3.1   Seed Oversampling

Being a seed-based method, the first SICLE step consists in selecting a set $\mathcal{S} \subset \mathcal{V}$ of N_0 initial seeds for generating N_f superpixels, given $N_0, N_f \in \mathbb{N}^*$. However, differently from most approaches, it *oversamples* (*i.e.*, $N_0 \gg N_f$) for improving the probability of selecting the seeds that promote accurate object delineation (*i.e.*, *relevant*). In this strategy, the aim is to remove the irrelevant ones until reaching N_f seeds in the final iteration (see Sect. 3.3). In this work, we argue no need for seed selection improvement given the reported loss of efficiency when object saliency maps are considered [7] and the seed relevance redundancy premise [8]. Consequently, the central strategy for delineation improvement relies on maintaining the relevant seeds throughout the iterations.

3.2   Superpixel Generation

For generating superpixels, SICLE uses the seed-restricted IFT version. Furthermore, although several path-cost and arc-cost functions have been proposed [6,9,24], it opts for the $\mathbf{f}_{max}$ and a root-based arc-cost estimation $\mathbf{w}_{root}(x,y) = \|\mathbf{F}(\mathbf{R}(x)) - \mathbf{F}(y)\|_2$ due to its reported effectiveness in superpixel delineation [8].

As one may note, such arc-cost function does not consider any object information. The authors in [8] justify such option in SICLE mainly on the existence of incorrect estimations in the map, deteriorating the object delineation. And although it resulted in a more robust performance for any map, it also led to the inability to improve its delineation for state-of-the-art estimators. First, let $\alpha \in \{0, 1\}$ be a user-defined "trustiness" switch of the saliency map's object boundary approximation. When $\alpha = 1$, the user judges that the map's borders are sufficiently accurate for assisting in delineation, due to its closeness to the object's boundaries. Otherwise, it may set $\alpha = 0$ for avoiding incorrect estimations within the map, preventing segmentation degradation, while still exploiting the object location information for improving the SICLE performance. For both cases, it is expected that the object is known beforehand and it was properly located by the saliency estimator. Then, to achieve each property when desired, we propose a generalization $\mathbf{w}_*^{\alpha}(x, y) = (\mathbf{w}_*(x, y))^{1+\alpha \cdot \|\mathbf{O}(\mathbf{R}(x)) - \mathbf{O}(y)\|_1}$. Note that, aside from not requiring optimization, $\mathbf{w}_*^{\alpha} = \mathbf{w}_*$ when $\alpha = 0$ since it discards the influence of the saliency difference. Finally, in contrast to the arc-cost function proposed in [4], the magnitude of the saliency influence in $\mathbf{w}_*^{\alpha}$ is significantly smaller, leading to a lighter impact by eventual incorrect estimations.

3.3   Seed Removal

In SICLE, N_f superpixels are obtained after successively removing $N_0 - N_f$ seeds from $\mathcal{S}$, requiring at most $\Omega \in \mathbb{N}^* > 1$ iterations. At each iteration $i \in \mathbb{N} < \Omega$, $\mathbf{M}(i) = \max\{(N_0)^{1-\omega \cdot i}, N_f\}$, given $\omega = 1/(\Omega - 1)$, most irrelevant seeds are removed, while the remaining ones are perpetuated for testing their relevance in the next iteration $i + 1$.

For each seed $s \in \mathcal{S}$, its relevance $\mathbf{V}_*(s) \in \mathbb{R}_+$ is estimated based on the characteristics of its superpixel $\mathbf{T}(s)$ resultant from the last IFT execution. As an example, one may opt for a size- and contrast-based criterion $\mathbf{V}_{sc}(s)$ for accurate selection of relevant seeds irrespective of whether a map is provided [8]. First, we define the color gradient between two superpixels rooted in $s, t \in \mathcal{S}$ by $\mathbf{G}_{\mathbf{F}}(s, t) = \left\|\overline{\mathbf{F}}(\mathbf{T}(s)) - \overline{\mathbf{F}}(\mathbf{T}(t))\right\|_2$. Moreover, it is possible to define the *adjacents* of s by $\mathbf{A}(s) = \{t : \exists \langle x, y \rangle \in \mathcal{A}\}$ considering $t \in \mathcal{S}$, $x \in \mathbf{T}(s)$, $y \in \mathbf{T}(t)$ and $s \neq t$. Finally, from both, we can compute the relevance of s by $\mathbf{V}_{sc}(s) = \frac{|\mathbf{T}(s)|}{|\mathcal{V}|} \cdot \min_{\forall t \in \mathbf{A}(s)} \{\mathbf{G}_{\mathbf{F}}(s, t)\}$ using the aforementioned criterion.

However, when a saliency map is given, the seed relevance is linked to its object proximity: the farther the superpixel, the more irrelevant it is, even though it is considered relevant by its non-object-based criterion [8]. Thus, in SICLE, every criterion is subjected to an object-based weighting factor. Similarly to $\mathbf{G}_{\mathbf{F}}(s, t)$, we define $\mathbf{G}_{\mathbf{O}}(s, t) = \left\|\overline{\mathbf{O}}(\mathbf{T}(s)) - \overline{\mathbf{O}}(\mathbf{T}(t))\right\|_2$ as the saliency gradient between the superpixels of s and t. Then, the object-based relevance of s is measured by $\mathbf{V}_{obj}(s) = \mathbf{V}_*(s) \cdot \max\{\overline{\mathbf{O}}(\mathbf{T}(s)), \max_{\forall t \in \mathbf{A}(s)} \{\mathbf{G}_{\mathbf{O}}(s, t)\}\}$.

One can see that such function favors seeds near the object border depicted on the map, promoting competition in crucial regions for delineation. Moreover, by favoring those within it, SICLE populates the regions incorrectly estimated as object parts, diminishing the influence of such error through competition. However, when a high-quality map is provided, not only are its borders more accurate but promoting competition in internal object borders minimally impacts its exterior boundaries. In such case, one may favor the seeds nearby the object rather than those within it. Thus, and similarly to $\mathbf{w}_*^\alpha$, we generalize SICLE's object-based seed relevance function for assessing such properties whenever one of them is requested: $\mathbf{V}_{obj}^\alpha = \mathbf{V}_*(s) \cdot \max\{(1-\alpha) \cdot \overline{\mathbf{O}}(\mathbf{T}(s)), \max_{\forall\, t \in \mathbf{A}(s)} \{\mathbf{G}_\mathbf{O}(s,t)\}\}$. Similarly to $\mathbf{w}_{obj}^\alpha$, when $\alpha = 0$, the tree's saliency is also considered as a relevant feature, resulting in $\mathbf{V}_{obj}^\alpha = \mathbf{V}_{obj}$. Otherwise, only the trees near the saliency borders are favored in the computation.

4   Experimental Results

In this section, we present the experimental framework for analyzing and evaluating the proposed method. We first describe the experimental setup in Sect. 4.1 and subsequently perform an ablation study in Sect. 4.2. Lastly, we present a quantitative and qualitative analysis in Sect. 4.3.

4.1   Experimental Setup

We selected three datasets for assessing the performance of all methods. Given that the most used segmentation evaluation dataset [2] is contour-driven, it is not applicable when a single object is desired. Conversely, our selection tries to assess different delineation difficulties for distinct objects, while offering a broad perspective on the methods' performance in their primary goal: generating superpixels. For handling different objects, we opt for the popular *Extended Complex Saliency Scene Dataset* (ECSSD) [22], which contains 1000 natural images with complex objects and backgrounds. On the other hand, the thin object legs in *Insects* [18] (130 images) offers a proper delineation challenge. Similarly, the *Liver* [24] dataset contains 40 CT slices of the human liver whose smooth boundaries are difficult to detect. We selected, by random, 30% and 70% of each dataset for optimization and testing, respectively. Finally, we considered the U^2-Net [20], fine-tuned with its default parameters, for generating the object saliency maps.

As baselines, we chose the following state-of-the-art methods based on their speed and accuracy: (i) SLIC [1][1]; (ii) SH [26][2]; (iii) ERS [16][3]; (iv) OISF [5][4]. By selecting such baselines, we assess the major properties for superpixel segmentation: speed and object delineation. Thus, although deep-learning-based methods with promising results have been proposed, more research is required for surpassing the performance of classical algorithms [8,15,27,29,33]. As initial

---

[1] https://www.epfl.ch/labs/ivrl/research/slic-superpixels/.

[2] https://github.com/semiquark1/boruvka-superpixel.

[3] https://github.com/mingyuliutw/EntropyRateSuperpixel.

[4] https://github.com/LIDS-UNICAMP/OISF.

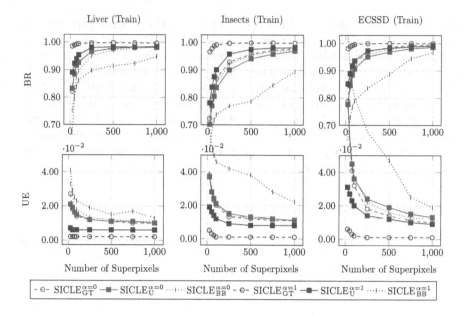

Fig. 2. Impacts of object saliency map quality on SICLE$^\alpha$.

setting for SICLE$^\alpha$, we used the default recommendation [8]:(i) random over-sampling with $N_0 = 3000$; (ii) $\mathbf{f}_{max}$; (iii) $\mathbf{V}_* = \mathbf{V}_{sc}$; and (iv) $\Omega = 5$. Our code is publicly available online[5]. For measuring their performances, we used the *Boundary Recall* (BR) [23] and the *Under-segmentation Error* (UE) [19] due to their expressiveness [23]. While the former measures the ratio between object boundaries and superpixel borders (*i.e.*, higher is better), the latter estimates errors from superpixel "leakings" (*i.e.*, lower is better).

4.2 Ablation Study

We first analyzed the impacts of the saliency map quality on SICLE$^\alpha$, as shown in Fig. 2. In this experiment, we considered a representative of a poor, a state-of-the-art, and an ideal estimator: (i) object's minimum bounding box (BB); (ii) U^2-Net (U); and (iii) ground-truth (GT); respectively. We highlight that the GT is only considered in this experiment. By setting $\alpha = 1$, our improved SICLE improves its segmentation proportionally to the saliency map quality (*i.e.*, the better the map, the better the delineation). Note that SICLE$^\alpha$ performance deteriorates when changing from GT to U maps, indicating that although highly accurate, the latter is not ideal. Still, by improving the saliency incorporation, our approach significantly improves when $\alpha = 1$, especially considering UE. Finally, by simply setting $\alpha = 0$ when a poor quality map is provided, SICLE$^\alpha$ becomes robust against saliency errors.

---

[5] https://github.com/LIDS-UNICAMP/SICLE.

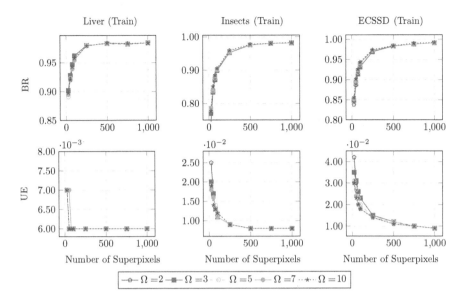

Fig. 3. Impacts of the maximum number of iterations on SICLE$^\alpha$ considering $\alpha = 1$ and the U^2-Net estimator.

Our second experiment analyses if the proposed method assists in reducing the number of iterations for segmentation. From the curves in Fig. 3, we see that SICLE$^\alpha$ manages to achieve its top performance requiring only two iterations and increasing Ω does not lead to improvements when $\alpha = 1$. We argue that our improved seed relevance criteria accurately select N_f relevant seeds in only one iteration, requiring one more for promoting effective object delineation. For that, we set $\Omega = 2$ whenever $\alpha = 1$.

4.3 Quantitative and Qualitative Analysis

Our last experiment (Fig. 4) compared our improved SICLE$^\alpha$ against the baselines. In terms of BR, SICLE$^\alpha$ managed to surpass all methods significantly, especially for $N_f = 200$ superpixels. Given its discrepant performance compared to OISF, we can argue that our approach best exploits the saliency information for segmentation. Moreover, SICLE$^\alpha$ presented better delineations in Insects than ERS, the best method known in such dataset. Regarding UE, our improvements reduced the superpixel leaking significantly, leading to on par results with OISF, which often presents the lowest values in several works [3–5].

Table 1 shows the average speed performance of all methods in the ECSSD dataset on a 64-bit Intel(R) Core(TM) i7-4790S PC with CPU speed of 3.20 GHz. As one can see, even though SICLE$^\alpha$ is O $(|\mathcal{V}| \log |\mathcal{V}|)$, it is the fastest method amongst all. For instance, SLIC and SH are O $(|\mathcal{V}|)$, but they perform burdensome operations for obtaining a single segmentation. While the former executes a strict number of iterations (*e.g.*, 10), the latter computes the whole hierarchy.

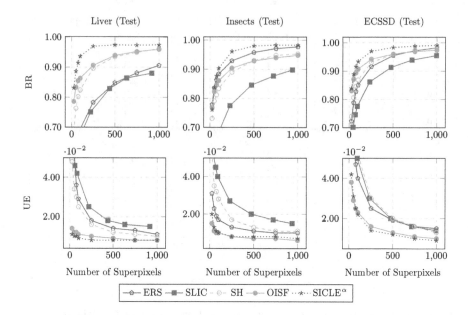

Fig. 4. Performance comparison between our approach, considering $\alpha = 1$ against state-of-the-art methods.

Table 1. Average speed performance (in seconds) on the ECSSD dataset considering $\alpha = 1$ for SICLE$^\alpha$. The best value for each N_f is depicted in bold.

N_0	SLIC	ERS	SH	OISF	SICLE$^\alpha$
25	0.537 ± 0.028	0.913 ± 0.085	0.758 ± 0.028	1.987 ± 0.367	$\mathbf{0.279 \pm 0.037}$
100	0.540 ± 0.029	0.952 ± 0.093	0.756 ± 0.026	1.252 ± 0.236	$\mathbf{0.279 \pm 0.030}$
750	0.541 ± 0.029	1.027 ± 0.109	0.756 ± 0.027	0.849 ± 0.163	$\mathbf{0.296 \pm 0.037}$

In contrast, SICLE$^\alpha$ surpasses both speed and delineation by benefitting from our improvements, leading to only two iterations. Lastly, it is straightforward to obtain an object-based multiscale segmentation on the fly from SICLE$^\alpha$ [8].

Finally, the superior performance of SICLE$^\alpha$ can be exemplified by Fig. 5. Note that, by setting $\alpha = 1$, our improved method can correct the errors for $\alpha = 0$, achieving top object delineation and surpassing all baselines. As indicated by the yellow rectangles, SICLE$^\alpha$ best exploits the saliency information and best approximates the object borders, especially when compared to other object-based methods like OISF.

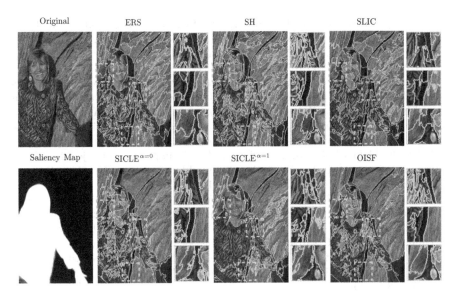

Fig. 5. Comparison between SICLE$^\alpha$ and state-of-the-art methods considering 100 superpixels. Red lines indicate object boundaries, whereas cyan ones, superpixel borders. Yellow rectangles indicate delineation errors that our approach overcame. (Color figure online)

5   Conclusion and Future Work

This work proposes SICLE$^\alpha$, an improved version of the state-of-the-art object-based method *Superpixels through Iterative CLEarcutting* (SICLE) by generalizing its path-cost function and seed removal criterion. Our proposal may promote robustness for low-quality saliency maps or may improve its effectiveness and efficiency in delineation for high-quality ones through a single and intuitive parameter. Results show that SICLE$^\alpha$ surpasses state-of-the-art methods regarding popular metrics while being the fastest one in all datasets considered. For future work, we intend to extend SICLE$^\alpha$ for video supervoxel segmentation and study its performance for interactive image segmentation.

References

1. Achanta, R., Shaji, A., Smith, K., Lucchi, A., Fua, P., Süsstrunk, S.: SLIC superpixels compared to state-of-the-art superpixel methods. Trans. Pattern Anal. Mach. Intell. **34**(11), 2274–2282 (2012)
2. Arbelaez, P., Maire, M., Fowlkes, C., Malik, J.: Contour detection and hierarchical image segmentation. Trans. Pattern Anal. Mach. Intell. **33**(5), 898–916 (2011)
3. Belém, F., Cousty, J., Perret, B., Guimarães, S., Falcão, A.: Towards a simple and efficient object-based superpixel delineation framework. In: 34th Conference on Graphics, Patterns and Images (SIBGRAPI), pp. 346–353 (2021)

4. Belém, F., Guimarães, S.J.F., Falcão, A.X.: Superpixel segmentation by object-based iterative spanning forest. In: Vera-Rodriguez, R., Fierrez, J., Morales, A. (eds.) CIARP 2018. LNCS, vol. 11401, pp. 334–341. Springer, Cham (2019). https://doi.org/10.1007/978-3-030-13469-3_39
5. Belém, F., Guimarães, S., Falcão, A.: Superpixel generation by the iterative spanning forest using object information. In: 33rd Conference on Graphics, Patterns and Images (SIBGRAPI), pp. 22–28 (2020). workshop of Thesis and Dissertations
6. Belém, F., Guimarães, S., Falcão, A.: Superpixel segmentation using dynamic and iterative spanning forest. Signal Process. Lett. **27**, 1440–1444 (2020)
7. Belém, F., Melo, L., Guimarães, S., Falcão, A.: The importance of object-based seed sampling for superpixel segmentation. In: 32nd Conference on Graphics, Patterns and Images (SIBGRAPI), pp. 108–115 (2019)
8. Belém, F., Perret, B., Cousty, J., Guimarães, S., Falcão, A.: Efficient multiscale object based superpixel framework. arXiv preprint, pp. 1 10 (2022)
9. Borlido Barcelos, I., Belém, F., Miranda, P., Falcão, A.X., do Patrocínio, Z.K.G., Guimarães, S.J.F.: Towards interactive image segmentation by dynamic and iterative spanning forest. In: Lindblad, J., Malmberg, F., Sladoje, N. (eds.) DGMM 2021. LNCS, vol. 12708, pp. 351–364. Springer, Cham (2021). https://doi.org/10.1007/978-3-030-76657-3_25
10. Conze, P.H., Tilquin, F., Lamard, M., Heitz, F., Quellec, G.: Unsupervised learning-based long-term superpixel tracking. Image Vis. Comput. **89**, 289–301 (2019)
11. Falcão, A., Stolfi, J., Lotufo, R.: The image foresting transform: theory, algorithms, and applications. Trans. Pattern Anal. Mach. Intell. **26**(1), 19–29 (2004)
12. Fehri, A., Velasco-Forero, S., Meyer, F.: Prior-based hierarchical segmentation highlighting structures of interest. Math. Morphol.-Theory Appl. **3**(1), 29–44 (2019)
13. Galvão, F., Falcão, A., Chowdhury, A.: RISF: recursive iterative spanning forest for superpixel segmentation. In: 31st Conference on Graphics, Patterns and Images (SIBGRAPI), pp. 408–415 (2018)
14. Jampani, V., Sun, D., Liu, M., Yang, M., Kautz, J.: Superpixel sampling networks. In: 18th European Conference on Computer Vision (ECCV), pp. 352–368 (2018)
15. Kang, X., Zhu, L., Ming, A.: Dynamic random walk for superpixel segmentation. IEEE Trans. Image Process. **29**, 3871–3884 (2020)
16. Liu, M., Tuzel, O., Ramalingam, S., Chellappa, R.: Entropy rate superpixel segmentation. In: 24th Conference on Computer Vision and Pattern Recognition (CVPR), pp. 2097–2104 (2011)
17. Machairas, V., Faessel, M., Cárdenas-Peña, D., Chabardes, T., Walter, T., Decenciere, E.: Waterpixels. IEEE Trans. Image Process. **24**(11), 3707–3716 (2015)
18. Mansilla, L., Miranda, P.: Oriented image foresting transform segmentation: connectivity constraints with adjustable width. In: 29th Conference on Graphics, Patterns and Images (SIBGRAPI), pp. 289–296 (2016)
19. Neubert, P., Protzel, P.: Superpixel benchmark and comparison. In: Forum Bildverarbeitung, vol. 6, pp. 1–12 (2012)
20. Qin, X., Zhang, Z., Huang, C., Dehghan, M., Zaiane, O., Jagersand, M.: U2-net: going deeper with nested u-structure for salient object detection. Pattern Recogn. **106**, 107404 (2020)
21. Sellars, P., Aviles-Rivero, A.I., Schönlieb, C.B.: Superpixel contracted graph-based learning for hyperspectral image classification. IEEE Trans. Geosci. Remote Sens. **58**(6), 4180–4193 (2020)

22. Shi, J., Yan, Q., Xu, L., Jia, J.: Hierarchical image saliency detection on extended CSSD. Trans. Pattern Anal. Mach. Intell. **38**(4), 717–729 (2015)
23. Stutz, D., Hermans, A., Leibe, B.: Superpixels: an evaluation of the state-of-the-art. Comput. Vis. Image Underst. **166**, 1–27 (2018)
24. Vargas-Muñoz, J., Chowdhury, A., Alexandre, E., Galvão, F., Miranda, P., Falcão, A.: An iterative spanning forest framework for superpixel segmentation. Trans. Image Process. **28**(7), 3477–3489 (2019)
25. Wang, M., Liu, X., Gao, Y., Ma, X., Soomro, N.Q.: Superpixel segmentation: a benchmark. Signal Process. Image Commun. **56**, 28–39 (2017)
26. Wei, X., Yang, Q., Gong, Y., Ahuja, N., Yang, M.: Superpixel hierarchy. Trans. Image Process. **27**(10), 4838–4849 (2018)
27. Wu, J., Liu, C., Li, B.: Texture-aware and structure-preserving superpixel segmentation. Comput. Graph. **94**, 152–163 (2021)
28. Yang, F., Sun, Q., Jin, H., Zhou, Z.: Superpixel segmentation with fully convolutional networks. In: 33rd Conference on Computer Vision and Pattern Recognition (CVPR) (2020)
29. Yu, Y., Yang, Y., Liu, K.: Edge-aware superpixel segmentation with unsupervised convolutional neural networks. In: 28th International Conference on Image Processing (ICIP), pp. 1504–1508 (2021)
30. Yuan, Y., Zhang, W., Yu, H., Zhu, Z.: Superpixels with content-adaptive criteria. IEEE Trans. Image Process. **30**, 7702–7716 (2021)
31. Zhang, J., Aviles-Rivero, A.I., Heydecker, D., Zhuang, X., Chan, R., Schönlieb, C.B.: Dynamic spectral residual superpixels. Pattern Recogn. **112**, 107705 (2021)
32. Zhao, W., Fu, Y., Wei, X., Wang, H.: An improved image semantic segmentation method based on superpixels and conditional random fields. Appl. Sci. **8**(5), 837 (2018)
33. Zhu, L., et al.: Learning the superpixel in a non-iterative and lifelong manner. In: 34th Conference on Computer Vision and Pattern Recognition (CVPR), pp. 1225–1234 (2021)

Join, Select, and Insert: Efficient Out-of-core Algorithms for Hierarchical Segmentation Trees

Josselin Lefèvre[1,2(✉)], Jean Cousty[1], Benjamin Perret[1], and Harold Phelippeau[2]

[1] LIGM, Univ Gustave Eiffel, CNRS, ESIEE Paris, F-77454 Marne-la-Vallée, France
{josselin.lefevre,jean.cousty,benjamin.perret}@esiee.fr
[2] Thermo Fisher Scientific, Bordeaux, France
harold.phelippeau@thermofisher.com

Abstract. Binary Partition Hierarchies (BPH) and minimum spanning trees are fundamental data structures involved in hierarchical analysis such as quasi-flat zones or watershed. However, classical BPH construction algorithms require to have the whole data in memory, which prevent the processing of large images that cannot fit entirely in the main memory of the computer. To cope with this problem, an algebraic framework leading to a high level calculus was introduced allowing an out-of-core computation of BPHs. This calculus relies on three operations: *select*, *join*, and *insert*. In this article, we introduce three efficient algorithms to perform these operations providing pseudo-code and complexity analysis.

1 Introduction

Hierarchies of partitions are versatile representations that have proven useful in many image analysis and processing problems. In this context, binary partition hierarchies [14] (BPH) built from altitude ordering and associated minimum spanning trees are key structures for several (hierarchical) segmentation methods: in particular it has been shown [2,11] that such hierarchies can be used to efficiently compute quasi-flat zone (also referred as α-trees) hierarchies [2,10] and watershed hierarchies [2,9]. Efficient algorithms for building BPHs on standard size images are well established, but, with the constant improvement of acquisition systems comes a dramatic increase in image resolutions, which can reach several terabytes in size. In such case, it becomes impossible to put a single image in the main memory of a standard workstation and classical algorithms for BPHs stop working. This creates the need for scalable algorithms to construct BPHs in an out-of-core manner to handle images that cannot fit in memory.

In [4,6,8], the authors investigate distributed memory algorithms to compute min and max trees for terabytes images. In [5], computation of minimum spanning trees of streaming images is considered. A parallel algorithm for the computation of quasi-flat zones hierarchies has been proposed in [7]. Finally, the authors of [1] recently proposed massively parallel algorithms for the computation of max-trees on GPUs. All these work rely on a common idea which is to

© Springer Nature Switzerland AG 2022
E. Baudrier et al. (Eds.): DGMM 2022, LNCS 13493, pp. 274–286, 2022.
https://doi.org/10.1007/978-3-031-19897-7_22

work independently on small pieces of the space, "join" the information found on adjacent pieces, and "insert" this joint information into other pieces.

In a previous work [3], the authors specifically addressed the problem of computing a BPH under the out-of-core constraint, *i.e.*, when the objective is to minimize the amount of memory required by the algorithms. To do so, they introduced an algebraic framework formalizing the distribution of a hierarchy over a partition of the space together with three algebraic operations acting on BPHs: *select, join*, and *insert*. They showed that, when a causal partition of the space is considered, it is possible to compute the distribution of a BPH using these three operations by browsing the different regions of the partition only twice (once in a forward pass and once in a backward pass) and by requiring to have only the information about two adjacent regions in the main memory at any step of the algorithm. However, no efficient algorithm has been proposed for the three operations *select, join*, and *insert*.

In this work, we propose efficient implementations for these operations. The proposed algorithms rely on a particular data structure to represent *local* hierarchies which is designed to efficiently search and browse the nodes of the hierarchy and to store only the necessary and sufficient information required locally to compute the distribution of the BPH. We give algorithms, with their pseudo-code, for the three operations whose time complexity is either linear or linearithmic. In order to ease the presentation, we consider the particular case of 2d images, modelled as 4-adjacency graphs, but the method can be easily extended to any regular graph. The implementation of the method in C++ and Python based on the hierarchical graph processing library Higra [12] is available online https://github.com/PerretB/Higra-distributed.

This article is organized as follows. Section 2 gives the definition of BPH. Section 3 recalls the notion of the distribution of a hierarchy and the calculus method that can be used to compute such distribution over a causal partition of the space. Section 4 explains the proposed data structures. Section 5 presents the algorithms for the three operations *select, join*, and *insert*. Finally, Sect. 6 concludes the work and gives some perspectives.

2   Binary Partition Hierarchy by Altitude Ordering

In this section, we first remind the definitions of hierarchy of partitions. Then we define the binary partition hierarchy by altitude ordering using the edge-addition operator [3] and we recall the bijection existing between the regions of this hierarchy and the edges of a minimum spanning tree of the graph.

Let V be a set. A *partition of* V is a set of pairwise disjoint subsets of V. Any element of a partition is called a *region* of this partition. The *ground* of a partition $\mathbf{P}$, denoted by $gr(\mathbf{P})$, is the union of the regions of $\mathbf{P}$. A partition whose ground is V is called a *complete partition of* V. Let $\mathbf{P}$ and $\mathbf{Q}$ be two partitions of V. We say that $\mathbf{Q}$ is a *refinement of* $\mathbf{P}$ if any region of $\mathbf{Q}$ is included in a region of $\mathbf{P}$. A *hierarchy on* V is a sequence $(\mathbf{P}_0, \ldots, \mathbf{P}_\ell)$ of partitions of V such that, for any λ in $\{0, \ldots, \ell - 1\}$, the partition $\mathbf{P}_\lambda$ is a refinement of $\mathbf{P}_{\lambda+1}$.

Let $\mathcal{H} = (\mathbf{P}_0, \ldots, \mathbf{P}_\ell)$ be a hierarchy. The integer ℓ is called the *depth of* $\mathcal{H}$ and, for any λ in $\{0, \ldots, \ell\}$, the partition $\mathbf{P}_\lambda$ is called the λ-*scale of* $\mathcal{H}$. In the following, if λ is an integer in $\{0, \ldots, \ell\}$, we denote by $\mathcal{H}[\lambda]$ the λ-scale of $\mathcal{H}$. For any λ in $\{0, \ldots, \ell\}$, any region of the λ-scale of $\mathcal{H}$ is also called a *region of* $\mathcal{H}$. The hierarchy $\mathcal{H}$ is *complete* if $\mathcal{H}[0] = \{\{x\} \mid x \in V\}$ and if $\mathcal{H}[\ell] = \{V\}$. We denote by $\mathcal{H}_\ell(V)$ the set of all hierarchies on V of depth ℓ, by $\mathcal{P}(V)$ the set of all partitions on V, and by $2^{|V|}$ the set of all subsets of V.

In the following, the symbol ℓ stands for any strictly positive integer.

We define a graph as a pair $G = (V, E)$ where V is a finite set and E is composed of unordered pairs of distinct elements in V. Each element of V is called a vertex of G, and each element of E is called an edge of G. The Binary Partition Hierarchy (BPH) by altitude ordering relies on a total order on E, denoted by $\prec$. Let k in $\{1, \ldots, \ell\}$, we denote by $u_k^\prec$ the k-th element of E for the order $\prec$. Let u be an edge in E, the *rank of* u *for* $\prec$, denoted by $r^\prec(u)$, is the unique integer k such that $u = u_k^\prec$. We then define the *update of a hierarchy* $\mathcal{H}$ *with respect to an edge* $\{x, y\}$, denoted by $\mathcal{H} \oplus \{x, y\}$: with k the rank of $\{x, y\}$, $\mathcal{H} \oplus \{x, y\}[\lambda]$ remains unchanged for any λ in $\{0, k-1\}$ while, for any λ in $\{k, \ldots, \ell\}$, we have $(\mathcal{H} \oplus \{x, y\})[\lambda] = \mathcal{H}[\lambda] \setminus \{R_x, R_y\} \cup \{R_x \cup R_y\}$ where R_x (resp. R_y) denotes the region of $\mathcal{H}[\lambda]$ containing x (resp. y). Let $E' \subseteq E$ and let $\mathcal{H}$ be a hierarchy. We set $\mathcal{H} \boxplus E' = \mathcal{H} \oplus u_1 \oplus \ldots \oplus u_{|E'|}$ where $E' = \{u_1, \ldots, u_{|E'|}\}$. The binary operation $\boxplus$ is called the *edge-addition*. Thanks to this operation, we can define formally the BPH for $\prec$. Let X be a set, we denote by $\perp_X$ the hierarchy defined by $\perp_X [\lambda] = \{\{x\} \mid x \in X\}$, for any λ in $\{0, \ldots \ell\}$. The BPH for $\prec$, denoted by $\mathcal{B}^\prec$ is the hierarchy $\perp_X \boxplus E$.

Let $\mathcal{B}^\prec$ be a binary partition by altitude ordering, $\mathcal{R}$ be a region of $\mathcal{B}^\prec$ and $\mathcal{R}^\star$ be the set of non-leaf regions of $\mathcal{B}^\prec$. The *rank* of $\mathcal{R}$, denoted by $r(\mathcal{R})$, is the lowest integer λ such that $\mathcal{R}$ is a region of $\mathcal{B}^\prec[\lambda]$. We consider the map μ from $\mathcal{R}^\star$ in E such that, for any non-leaf region $\mathcal{R}$ of $\mathcal{B}^\prec$, we have $\mu^\prec(\mathcal{R}) = u_{r(\mathcal{R})}^\prec$. We say that $\mu^\prec(\mathcal{R})$ is the *building edge* of $\mathcal{R}$. Building edges of the binary partitions hierarchy defines a minimum spanning tree of an edge-weighted graph. In Fig. 1, $\mathcal{Y}$ is the BPH built on the 4-adjacency graph B. Non-leaf nodes of $\mathcal{Y}$ correspond to the edges of the minimum spanning tree of B (dashed edges).

3   Distributed Hierarchies of Partitions on Causal Partition

In this section, we recall the definition of the distribution of a BPH on a sliced graph and the principle of its calculus in an out-of-core manner. Intuitively, distributing a hierarchy consists in splitting it into a set of smaller trees such that: 1) each smaller tree corresponds to a *selection* of a sub part of whole tree that intersects a slice of the graph and 2) the initial hierarchy can be reconstructed by "gluing" those smaller trees.

Let V be a set. The operation *sel* is the map from $2^{|V|} \times \mathcal{P}(V)$ to $\mathcal{P}(V)$ which associates to any subset X of V and to any partition $\mathbf{P}$ of V the subset sel$(X, \mathbf{P})$ of $\mathbf{P}$ which contains every region of $\mathbf{P}$ that contains an element of X. The

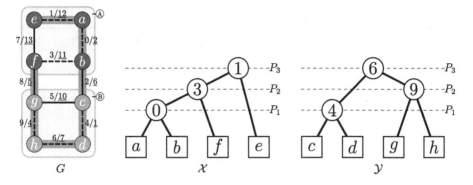

Fig. 1. G a 4-adjacency graph divided into two slices A and B (respectively blue and green). Each edge of G is associated with a pair (*index*, *weight*). The two slices are separated by their common neighborhood *i.e.* edges 8 and 2. Hierarchy $\mathcal{X}$ (respectively $\mathcal{Y}$) is the BPH built on A (respectively B). Indices associated with non-leaf nodes of the BPHs correspond to the indices of their corresponding building edges represented by dashed edges in G. We can note that MSTs built on slices (dashed edges) are not sub-trees of the complete MST (shown as shadow). In consequence, edge 3 is part of the hierarchy $\mathcal{X}$ when it should not be.

operation *select* is the map from $2^{|V|} \times \mathcal{H}_\ell(V)$ in $\mathcal{H}_\ell(V)$ which associates to any subset X of V and to any hierarchy $\mathcal{H}$ on V the hierarchy select $(X, \mathcal{H}) = (\text{sel}(X, \mathcal{H}[0]), \dots, \text{sel}(X, \mathcal{H}[\ell]))$.

We are then able to define the distribution of a hierarchy thanks to *select*. Let V a set, let $\mathbf{P}$ be *a complete partition on V* and let $\mathcal{H}$ be a hierarchy on V. The distribution of $\mathcal{H}$ over $\mathbf{P}$ is the set $\{\text{select}(R, \mathcal{H}) \mid R \in \mathbf{P}\}$ and for any region R of $\mathbf{P}$, select $(R, \mathcal{H})$ is called a *local hierarchy (of $\mathcal{H}$ on R)*.

The calculus introduced in [3] aims to compute the distribution of a BPH over a partition of the space. In this article, we consider the special case of a 4 adjacency graph representing a 2d image that can be divided into slices, and we are interested in computing a distribution of the BPH over those slices. It should be noted that this is not a limiting factor, and the method can easily be adapted to any regular grid graph.

Let h and w be two integers representing the height and the width of an image. In the following, the set V is the Cartesian product of $\{0, \dots, h-1\} \times \{0, \dots, w-1\}$. Thus, any element x of V is a pair $x = (x_i, x_j)$ such that x_i and x_j are the coordinates of x. In the 4-adjacency grid, the set of all edges E is equal to $\{\{x, y\} \in V \mid |x_i - y_i| + |x_j - y_j| \leq 1\}$. Let k be a positive integer, the causal partition of V is the sequence $(S_0, \dots, S_k)$ such that for any t in $\{0, \dots, k\}$, $S_t = \{(i, j) \in V \mid t \times \frac{w}{k} \leq i < (t+1) \times \frac{w}{k}\}$. Each element of this partition is called a *slice*. The set of vertices at the interface between two neighbor slices A and B and belonging to A is noted $\gamma_B^\bullet(A)$. The major advantage of considering this partition over a regular graph is that each subset of V or E can be computed on the fly efficiently from a computational and memory point of view.

Algorithm 1: Out-of-core binary partition hierarchy [3].

Data: A graph (V, E), a total order $\prec$ on E, and a causal partition $(S_0, \ldots, S_k)$ of V

Result: $\{\mathcal{B}_0^\downarrow, \ldots, \mathcal{B}_k^\downarrow\}$: the distribution of the BPH $\mathcal{B}_V^\prec$ over $\{S_0, \ldots, S_k\}$.

1 $\mathcal{B}_0^\uparrow := \mathcal{B}_{S_0}^\prec$ // call PlayingWithKruskal algorithm

2 **foreach** i **from** 1 **to** k **do**   // Causal traversal of the slices

3   Call PlayingWithKruskal algorithm to compute $\mathcal{B}_{S_i}^\prec$

4      $\mathcal{M}_i^\uparrow := \text{join}\left(\text{select}\left(\gamma_{S_i}^\bullet(S_{i-1}), \mathcal{B}_{i-1}^\uparrow\right), \text{select}\left(\gamma_{S_{i-1}}^\bullet(S_i), \mathcal{B}_{S_i}^\prec\right)\right)$

5      $\mathcal{B}_i^\uparrow := \text{insert}(\text{select}\left(\gamma_{S_{i-1}}^\bullet(S_i), \mathcal{M}_i^\uparrow\right), \mathcal{B}_{S_i}^\prec)$

6 $\mathcal{B}_k^\downarrow := \mathcal{B}_k^\uparrow;\ \mathcal{M}_k^\downarrow := \mathcal{M}_k^\uparrow$

7 **foreach** i **from** $k-1$ **to** 0 **do**   // Anticausal traversal of the slices

8      $\mathcal{B}_i^\downarrow := \text{insert}(\text{select}\left(\gamma_{S_{i+1}}^\bullet(S_i), \mathcal{M}_{i+1}^\downarrow\right), \mathcal{B}_i^\downarrow)$

9   **if** $i > 0$ **then** $\mathcal{M}_i^\downarrow := \text{insert}(\text{select}\left(\gamma_{S_{i-1}}^\bullet(S_i), \mathcal{B}_i^\downarrow\right), \mathcal{M}_i^\uparrow)$

Given this causal partition, Algorithm 1 allows computing the local hierarchies of the BPH of the complete graph on each slice. This algorithm can be divided in two parts: causal and anti-causal traversal of the slices. Each of these parts relies on the same idea. First, start with the causal traversal. Given a causal partition of V into $k+1$ slices, for any i in $\{1, \cdots, k\}$ compute the BPH on S_i with a call to the algorithm presented in [11] hereafter called PlayingWithKruskal (line 3). Then, select the part of this hierarchy containing the vertices adjacent to the previous slice and join it with the part of the hierarchy associated to the previous slice containing the vertices adjacent to the current slice, leading to the "merged" hierarchy denoted by $\mathcal{M}_i^\uparrow$ (line 4). The merged hierarchy is then inserted in the BPH which gives $\mathcal{B}_i^\uparrow$ (line 5). The hierarchies $\mathcal{B}_i^\uparrow$ associated to slice i misses the information located in slices of higher indices, and consequently only the last local hierarchy $\mathcal{B}_k^\uparrow$ is correct *i.e.* $\mathcal{B}_k^\uparrow = \text{select}(S_k, \mathcal{H})$. In order to compute the valid distribution, and after having spread information in the causal direction, information must be back propagated in the reverse anti-causal direction so that each local hierarchy is enriched with the global context (lines 7 to 9).

4   Data Structures

In this section, we present the data structures used in the algorithms defined in the following sections. These data structures are designed to contain only the necessary and sufficient information so that we never need to have all the data in the main memory at once. The data structure representing a local hierarchy assumes that the nodes of the hierarchy are indexed in a particular order and relies on three "attributes": 1) a mapping of the indices from the local context (a given slice) to the global one (the whole graph) noted $\mathcal{H}.\texttt{map}$, 2) the parent array denoted by $\mathcal{H}$.par encoding the parent relation between the tree nodes, and 3) an array $\mathcal{H}$.weights giving, for each non-leaf-node of the tree, the weight of its corresponding building edge.

More precisely, given a binary partition hierarchy $\mathcal{H}$ with n regions, every integer between 0 and $n-1$ is associated to a unique region of $\mathcal{H}$. Moreover, this indexing of the regions of $\mathcal{H}$ follows a topological order such that: 1) any leaf region is indexed before any non-leaf region; 2) two leaf regions $\{x\}$ and $\{y\}$ are sorted with respect to an arbitrary order on the element V, called the *raster scan order* of V. Thus $\{x\}$ has an index lower than $\{y\}$ if x is before y with respect to the raster scan order; and 3) two non-leaf regions are sorted according to their rank, *i.e.*, the order of their building edges for $\prec$. This order can be seen as an extension of the order $\prec$ on E to the set $V \cup E$ that enables 1) to efficiently browse the nodes of a hierarchy according to their scale of appearance in the hierarchy and 2) to efficiently match regions of V with the leaves of the hierarchy. By abuse of notation, this extended order is also denoted by $\prec$ in the following.

To keep track of the global context, a link between the indices in the local tree and the global indices in the whole graph is stored in the form of an array map which associates: 1) to the index i of any leaf region R, the vertex x of the graph G such that $R = \{x\}$, *i.e.* map[i]=x; and 2) to the index i of any non-leaf region R, its building edge, *i.e.* map[i]=$\mu^{\prec}(R)$.

The parent relation of the hierarchy is stored thanks to an array par such that par[i]=j if the region of index j is the parent of the region of index i.

The binary partition hierarchy is built for a particular ordering $\prec$ of the edges of G. In practice, this ordering is induced by weights computed over the edges of G. To this end, we store an array weights of $|\mathcal{R}^\star(\mathcal{H})|$, *i.e.* the number of non-leaf regions, elements such that, for every region R in $\mathcal{R}^\star(\mathcal{H})$ of index i, weights[i] is the weight of the building edge $\mu^{\prec}(R)$ of region R. The edges can then be compared according to the following total order induced by the weights: we set $u \prec v$ if the weight of u is less than the one of v or if u and v have equal weights but u comes before v with respect to the raster scan order.

5   Algorithms

Select. In this part, we give an algorithm to compute the result of the *select* operation. This operation consists in "selecting" the part of a given hierarchy intersecting a subset of the space. In Algorithm 1, *select* takes as input a set of vertices located at the "border" of a slice and a hierarchy in order to obtain a smaller "border hierarchy".

Select algorithm proceeds in 3 steps:

1. *Lines 3–7.* Mark any leaf-node of $\mathcal{H}$ that corresponds to an element of X, *i.e.* any leaf-region $\{x\}$ with $x \in X$;
2. *Lines 8–9.* Traverse the hierarchy from leaves to root and mark any node that is a parent of a marked node;
3. *Lines 11–17.* Build the hierarchy $\mathcal{S}$ whose nodes are only marked nodes of $\mathcal{H}$.

Algorithm 2: SELECT

Data: $\mathcal{H}$: a hierarchy, X: set of selected nodes st. $X \subset gr(\mathcal{H})$
Result: $\mathcal{S}$: the hierarchy select $(X, \mathcal{H})$

1 Initialize an array mark to false for every region of $\mathcal{H}$
2 $i := 0; j := 0$ // i iterates over X and j over the leaves of $\mathcal{H}$
3 **while** $i < |X|$ **and** $j < |\mathcal{H}.leaves|$ **do**
4 | **if** $X[i] = \mathcal{H}.map[j]$ **then**
5 | | $mark[j] := true$
6 | | $i := i + 1$
7 | $j := j + 1$

8 **foreach** n **from** 0 **to** $|\mathcal{H}| - 1$ **do**
9 | **if** $mark[n] = true$ **then** $mark[\mathcal{H}.\text{par}[n]] := true$

10 $n_{\mathcal{S}} := 0$
11 **foreach** n **from** 0 **to** $|\mathcal{H}| - 1$ **do**
12 | **if** $mark[n] = true$ **then**
13 | | $\mathcal{S}.\text{par}[n_{\mathcal{S}}] := \mathcal{H}.\text{par}[n]$
14 | | $\mathcal{S}.\text{map}[n_{\mathcal{S}}] := \mathcal{H}.\text{map}[n]$
15 | | **if** $n \in \mathcal{R}^*(\mathcal{H})$ **then**
16 | | | $\mathcal{S}.\text{weight}[n_{\mathcal{S}} - |X|] := \mathcal{H}.\text{weight}[n_{\mathcal{S}} - |\mathcal{H}.leaves|]$

17 | $n_{\mathcal{S}} := n_{\mathcal{S}} + 1$

18 **return** $\mathcal{S}$

In Algorithm 2 we assume that X is sorted and that $X \subset gr(\mathcal{H})$, which is always the case in Algorithm 1. For each element $X[i]$ of X, we search for the index j of a leaf of $\mathcal{H}$ mapped to $X[i]$, *i.e.* such that $\mathcal{H}.\text{map}[j] = X[i]$. To this end, it is necessary to make a traversal of the leaves of $\mathcal{H}$. As mentioned before, the leaves correspond to the first indices by construction. The first step can then be performed in linear time with respect to the number of leaf-regions of $\mathcal{H}$. The second step consist in traversing the whole hierarchy from leaves to root in order to mark every region of $\mathcal{H}$ which belongs to select $(X, \mathcal{H})$ *i.e.* regions parent of a marked one. The complexity of this step is therefore linear with respect to the number of regions of $\mathcal{H}$. Finally, the last step boils down to extracting the hierarchy select $(X, \mathcal{H})$ from the marked nodes. For this a new hierarchy is created by traversing $\mathcal{H}$ again. As the traversal is done by increasing order of index, the properties relating to the weights of the building edges and order of appearance of regions are preserved. The complexity of this last step is linear with respect to the number of regions of $\mathcal{H}$. Thus, Algorithm 2 has a linear $O(n)$ complexity, where n is the number of regions of $\mathcal{H}$.

Join. Formally the *join of* $\mathcal{X}$ *and* $\mathcal{Y}$, denoted by $join(\mathcal{X}, \mathcal{Y})$, is the hierarchy defined by $join(\mathcal{X}, \mathcal{Y}) = (\mathcal{X} \sqcup \mathcal{Y}) \boxplus F$, where F is the common neighborhood of the grounds of $\mathcal{X}$ and of $\mathcal{Y}$, and $\sqcup$ denotes the supremum on hierarchies (see [13]). Intuitively, this operation merges two hierarchies according to their common neighborhood, that is the set of edges linking their grounds. In [7], the authors proposed an algorithm that can be used to successively add edges of the common neighborhood. Intuitively, to add an edge, the hierarchy is updated while climbing the branches associated with the edge extremities. The worst-case complexity is then linear with respect to the size of the hierarchy for adding a single edge. Thus the overall complexity of such join procedure would be $O(k \times n)$ where n is the size of the hierarchies and k is the number of edges in the common neighborhood. In this section, we drop the multiplicative dependency in the size of the neighborhood at the cost of introducing a sorting of F and we present an algorithm whose complexity is quasi-linear with respect to the size n of the hierarchies and linearithmic with respect to the number k of edges in F.

Algorithm 3: JOIN

Data: $\mathcal{X}$ and $\mathcal{Y}$: two hierarchies, F common neighborhood of $gr(\mathcal{X})$ and $gr(\mathcal{Y})$.
Result: A collection $Q_D = \text{join}(\mathcal{X}, \mathcal{Y})$

1 **foreach** *node n_i of* $\mathcal{X}$ **do** $Q_D.\text{MakeSet}(i)$
2 **foreach** *node n_i of* $\mathcal{Y}$ **do** $Q_D.\text{MakeSet}(i + |\mathcal{X}.leaves|)$
3 aDescendent$(\mathcal{X}, 0)$
4 aDescendent$(\mathcal{Y}, |\mathcal{X}.leaves|)$
5 $F := \text{sort}(F)$
6 $i_1 := |\mathcal{X}.leaves|$; $i_2 := |\mathcal{Y}.leaves|$; $i_3 := 0$
7 **while** $i_1 < |\mathcal{X}|$ *or* $i_2 < |\mathcal{Y}|$ *or* $i_3 < |F|$ **do**
8   **if** $F[i_3] \prec \mathcal{X}.map[i_1]$ *and* $F[i_3] \prec \mathcal{Y}.map[i_2]$ **then**
9          $(x, y) := F[i_3]$; $m := F[i_3]$; $w := weight(F[i_3])$; $i_3 \mathrel{+}= 1$
10   **else if** $\mathcal{X}.map[i_1] \prec \mathcal{Y}.map[i_2]$ **then**
11         $(x, y) := \mathcal{X}.\text{desc}[i_1]$; $m := \mathcal{X}.map[i_1]$; $w := \mathcal{X}.weight[i_1]$; $i_1 \mathrel{+}= 1$
12   **else**
13         $(x, y) := \mathcal{Y}.\text{desc}[i_2]$; $m := \mathcal{Y}.map[i_2]$; $w := \mathcal{Y}.weight[i_2]$; $i_2 \mathrel{+}= 1$
14     $c_x := Q_D.\text{FindCanonical}(x)$; $c_y := Q_D.\text{FindCanonical}(y)$
15   **if** $c_x ! = c_y$ **then**
16         $n := Q_D.\text{Union}(c_x, c_y)$; $Q_D.map[n] := m$
17         $Q_D.\text{weight}[n - (|\mathcal{X}.leaves| + |\mathcal{Y}.leaves|)] := w$

A detailed presentation of the proposed algorithm is given in Algorithm 3 which calls auxiliary functions presented in Algorithm 4. Intuitively, in order to compute the join of two hierarchies $\mathcal{X}$ and $\mathcal{Y}$, Algorithm 3 consists in "emulating" PlayingWithKruskal algorithm on the graph obtained from (i) the edges associated to the non-leaf nodes of $\mathcal{X}$ and of $\mathcal{Y}$ and (ii) the edges in the common neighborhood F of $\mathcal{X}$ and $\mathcal{Y}$. Therefore, all these edges are considered in increasing order with respect to $\prec$ and, for each edge, it is decided if this edge must be considered or not in the creation process of the join hierarchy. The decision is made based on the potential creation of a cycle if this edge were added during the minimum spanning tree creation process. We can thus see on the Fig. 2 that the node 3 has been added to $\mathcal{X}$ by construction but that it is then discarded during the construction of the joined hierarchy. Potential-cycles creation is efficiently checked with Tarjan Union-Find data structures as in Kruskal's algorithm. A main observation can be made to highlight the difference between the situation encountered in the contexts of join algorithm and PlayingWithKruskal algorithms: in the context of join, some edges, which are associated to the nodes of the hierarchies $\mathcal{X}$ and $\mathcal{Y}$, are made of vertices that do not belong to the underlying space (*i.e.*, the common neighborhood of the slices supporting the grounds of $\mathcal{X}$ and $\mathcal{Y}$). When such edge is found, the standard algorithm can be shortcut leading to a modified version of the PlayingWithKruskal auxiliary functions presented in Algorithm 4. Compared to original functions, the only change is the insertion of the if test at line 9. This test detects the edges for which a shortcut must occur based on an attribute called desc. This attribute is pre-computed for every node of $\mathcal{X}$ and of $\mathcal{Y}$ by the auxiliary function aDescendent. Overall, the following steps are performed in Algorithm 3:

- *Lines* 1–2. Initialize the Union-find data structures;
- *Lines* 3–4. Compute the attribute desc for both $\mathcal{X}$ and $\mathcal{Y}$;
- *Lines* 5 *to* 13. Browse the edges in increasing order. Observe that it implies sorting the edges in the common neighborhood F of $\mathcal{X}$ and $\mathcal{Y}$ in increasing order for $\prec$ (non-leaf-nodes of $\mathcal{X}$ and $\mathcal{Y}$ are already sorted by construction);
- *Lines* 15–17 Apply PlayingWithKruskal steps, calling the modified version of the auxiliary functions.

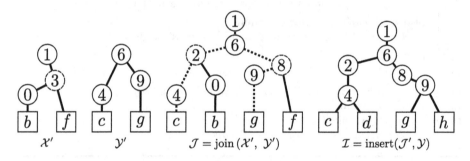

Fig. 2. The hierarchy $\mathcal{J}$ is build by computing the join over $\mathcal{X}' = \text{select}(\{c, h\}, \mathcal{X})$ and $\mathcal{Y}' = \text{select}(\{d, i\}, \mathcal{Y})$ (border trees computed from BPHs of Fig. 1). We can see that the node 3 of $\mathcal{X}'$ not longer appear in the joint tree in the favor of nodes corresponding to the common neighborhood of the grounds of $\mathcal{X}$ and $\mathcal{Y}$ *i.e.* nodes 2 and 8. That is to say, that by taking into account the topological order on the edges of the MSTs associated with the border trees and the common neighborhood, 3 does not belongs to the BPH. $\mathcal{I}$ is then build by inserting the hierarchy $\mathcal{J}' = \text{select}(\{c, g\}, \mathcal{J})$ into the hierarchy $\mathcal{Y}$. Given the Definition 16 of 1, $\mathcal{I}$ is a "correct" local hierarchy for the tile B.

The first step complexity is linear with respect to the number of elements of $gr(\mathcal{X}) \cup gr(\mathcal{Y})$. The second step uses the auxiliary function aDescendent to compute attributes desc for both $\mathcal{X}$ and $\mathcal{Y}$. It should be noted that this

---

Algorithm 4: AUXILIARY FUNCTIONS FOR *join* ALGORITHM

```
// The functions called hereafter on Q_T and Q_BT are those described in [11]
1  Procedure Q_D.MakeSet(q)
2  |  Q_D.Root[q] := q; Q_BT.MakeSet(q); Q_T.MakeSet(q)
3
4  Function Q_D.FindCanonical(q)
5  |  return Q_T.FindCanonical(q)
6
7  Function Q_D.Union(c_x, c_y)
8  |  t_u := Q_D.Root[c_x]; Q_BT.par[t_u] := Q_BT.size
9  |  if c_y = -1 then  Q_D.Root[c_x] := Q_BT.size
10 |  else
11 |  |  t_v := Q_D.Root[c_y]; Q_BT.par[t_v] := Q_D.size
12 |  |  c := Q_T.Union(c_x, c_y); Q_D.Root[c] := Q_BT.size
13 |  end
14 |  Q_BT.MakeSet(Q_BT.size)
15 |  return Q_D.size - 1
16
17 Function aDescendent(H: a hierarchy, s: shift)
      // set the attribute H.desc: an array that maps to each node n of H two leaves
         that are descendants of the two children of n.
18 |  foreach node n of H do  H.desc[n]:= (-1, -1)
19 |  foreach leaf node n of H do  H.desc[n].first:= n + s
20 |  foreach non-root non-leaf node n of H in increasing order for ≺ do
21 |  |  p := H.par[n]
22 |  |  if H.desc[p].first= -1 then H.desc[p].first:= H.desc[n].first
23 |  |  else H.desc[p].second:= H.desc[n].first
24 |  end
```

---

last function takes a parameter *shift* which allows to index the leaves of the second hierarchy after those of the first. For each node of the two hierarchies, an attribute is computed during a leaves to root traversal which gives a linear complexity with respect to the number of regions of each hierarchy. Third step requires to sort the edges of F with respect to $\prec$ before browsing the edges which implies a complexity of $O(k \times log(k) + |\mathcal{X}| + |\mathcal{Y}|)$ with k the number of edges in F. The fourth step is equivalent to PlayingWithKruskal algorithm in terms of complexity. Then, its complexity is $O(m \times \alpha(n))$ where m is sum of the number of edges in F and the number of non-leaf nodes of $\mathcal{X}$ and $\mathcal{Y}$, where n is the number of leaf nodes in $\mathcal{X}$ and $\mathcal{Y}$ and where $\alpha()$ is the inverse Ackermann function which grows sub-logarithmically.

Insert. In this part, we present an algorithm to compute the hierarchy $\mathcal{Z} = insert(\mathcal{X}, \mathcal{Y})$. We assume that $\mathcal{X}$ is *insertable* in $\mathcal{Y}$ i.e. for any λ in $\{0, \ldots, \ell\}$, for any region Y of $\mathcal{Y}[\lambda]$, Y is either included in a region of $\mathcal{X}[\lambda]$ or is included in $V \setminus gr(\mathcal{X}[\lambda])$. This assumption holds true at each call to *insert* in Algorithm 1. The *insertion of* $\mathcal{X}$ *into* $\mathcal{Y}$ is the hierarchy $\mathcal{Z}$, such that, for any λ in $\{0, \ldots, \ell\}$, $\mathcal{Z}[\lambda] = \mathcal{X}[\lambda] \cup \{R \in \mathcal{Y}[\lambda] \mid R \cap gr(\mathcal{X}[\lambda]) = \emptyset\}$. Algorithm 5, presented hereafter, computes the insertion $\mathcal{Z}$ of $\mathcal{X}$ into $\mathcal{Y}$. From a high level point of view, it proceeds in two main steps:

Algorithm 5: INSERT

Data: $\mathcal{X}$ and $\mathcal{Y}$: two hierarchies such that $\mathcal{X}$ insertable in $\mathcal{Y}$.
Result: $\mathcal{Z}$: the hierarchy insert$(\mathcal{X}, \mathcal{Y})$

1 $x := 0$; $y := 0$; $z := 0$; // indices for the nodes/regions of $\mathcal{X}$, $\mathcal{Y}$, and $\mathcal{Z}$
2 Initialize an array InZ of $|\mathcal{Y}|$ Booleans to true (resp. to false) for every leaf (resp. non-leaf) region of $\mathcal{Y}$
3 **while** $x < |\mathcal{X}|$ *or* $y < |\mathcal{Y}|$ **do**
4   **if** $x < |\mathcal{X}|$ *and* $y < |\mathcal{Y}|$ *and* $\mathcal{X}.map[x] = \mathcal{Y}.map[y]$ **then**
      // Duplicate region (x,y) found in $\mathcal{X}$ and $\mathcal{Y}$, keep (and renumber) it in $\mathcal{Z}$
5        $C_{\mathcal{X}\to\mathcal{Z}}[x] := z$; $C_{\mathcal{Y}\to\mathcal{Z}}[y] := z$; $C_{\mathcal{Z}\to\mathcal{X},\mathcal{Y}}[z] := (x,y)$
6        $x += 1$; $y += 1$; $z += 1$
7   **else if** $\mathcal{Y}.map[y] \prec \mathcal{X}.map[x]$ **then**
8   **if** $InZ[y] = true$ **then** // Keep (en renumber) region y in $\mathcal{Z}$
9          $InZ[\mathcal{Y}.\mathrm{par}[y]] := true$
10         $C_{\mathcal{Y}\to\mathcal{Z}}[y] := z$; $C_{\mathcal{Z}\to\mathcal{X},\mathcal{Y}}[z] := (-1, y)$
11         $y += 1$; $z += 1$
12   **else** $y += 1$ // Discard region y from $\mathcal{Z}$
13   **else** // Keep (and renumber) region x in $\mathcal{Z}$
14       $C_{\mathcal{X}\to\mathcal{Z}}[x] := z$; $C_{\mathcal{Z}\to\mathcal{X},\mathcal{Y}}[z] := (x, -1)$
15       $x += 1$; $z += 1$

16 $\mathcal{Z} :=$ initialize a tree structure with $n_{\mathcal{Z}} = z$ nodes
17 **foreach** z *from 0 to* $n_{\mathcal{Z}}$ **do**
18    $(x, y) := C_{\mathcal{Z}\to\mathcal{X},\mathcal{Y}}[r]$
19   **if** $x \neq -1$ **then**
20       $\mathcal{Z}.map[z] := \mathcal{X}.map[x]$
21   **if** $[x] = \mathcal{X}.root$ **then** $\mathcal{Z}.\mathrm{par}[z] := x$
22   **else** $\mathcal{Z}.\mathrm{par}[z] := C_{\mathcal{X}\to\mathcal{Z}}[\mathcal{X}.\mathrm{par}[x]]$
23   **if** $z \geq |\mathcal{X}.leaves|$ **then**
        $\mathcal{Z}.weight[z - |\mathcal{X}.leaves|] := \mathcal{X}.weight[x - |\mathcal{X}.leaves|]$
24   **else**
25       $\mathcal{Z}.map[z] := \mathcal{Y}.map[y]$
26   **if** $y = \mathcal{Y}.root$ **then** $\mathcal{Z}.\mathrm{par}[z] := y$
27   **else** $\mathcal{Z}.\mathrm{par}[z] := C_{\mathcal{Y}\to\mathcal{Z}}[\mathcal{Y}.\mathrm{par}[y]]$
28   **if** $z \geq |\mathcal{Y}.leaves|$ **then**
        $\mathcal{Z}.weight[z - |\mathcal{Y}.leaves|] := \mathcal{Y}.weight[x - |\mathcal{Y}.leaves|]$

---

1. *Lines* 3–15. Identify and renumber the regions of $\mathcal{X}$ and $\mathcal{Y}$ that belong to $\mathcal{Z}$ and store the correspondences between the new number of the regions in $\mathcal{Z}$ and the indices of the initial regions in $\mathcal{X}$ and $\mathcal{Y}$. It can be observed that this step is necessary since a region of $\mathcal{Z}$ can be duplicated in both $\mathcal{X}$ and $\mathcal{Y}$ and that some regions of $\mathcal{Y}$ are discarded from $\mathcal{Z}$. In order to perform this

step, the regions of $\mathcal{X}$ and $\mathcal{Y}$ are simultaneously browsed in increasing order for $\prec$. The correspondences between the regions of the hierarchies are stored in three arrays: $C_{\mathcal{X} \to \mathcal{Z}}$, $C_{\mathcal{Y} \to \mathcal{Z}}$, and $C_{\mathcal{Z} \to \mathcal{X}, \mathcal{Y}}$;

2. *Lines* 17–28. Build the parenthood relation (`par`) of the hierarchy $\mathcal{Z}$ using the parenthood relation of the hierarchies $\mathcal{X}$ and $\mathcal{Y}$ and the correspondences between the regions of the hierarchies. At the same time, we also build the attributes `map` and `weight` associated to $\mathcal{Z}$.

During the first step, each region of the two hierarchies $\mathcal{X}$ and $\mathcal{Y}$ is considered once and processed with a limited number of constant-time instructions. Thus, the overall time complexity of Lines 3–15 is linear with respect to the number of nodes of $\mathcal{X}$ and $\mathcal{Y}$. The worst-case complexity of the second step is also linear with respect to the number of nodes of $\mathcal{X}$ and $\mathcal{Y}$ since $\mathcal{Z}$ contains at most all regions of each of hierarchy. Thus, the overall complexity of Algorithm 5 is $O(|\mathcal{X}| + |\mathcal{Y}|)$.

6   Conclusion

In this article, we proposed efficient and easily implementable algorithms for the three algebraic operations on hierarchies *select*, *join*, and *insert*. These algorithms rely on a particular data structure to represent local hierarchies in order to achieve linear or linearithmic time complexity while limiting the amount of information required in the main memory. Thanks to these contributions it is now possible to efficiently implement the calculus scheme proposed in [3] for the out-of-core computation of BPHs. In future works, we plan to study the time and memory consumption of the proposed algorithms in practice and to develop efficient algorithms to process the distribution of a BPH in order to obtain a completely out-of-core pipeline for seeded watershed segmentation.

Acknowledgements. This work was supported by the French ANR grant ANR-20-CE23-0019.

References

1. Carlinet, E., Blin, N., Lemaitre, F., Lacassagne, L., Geraud, T.: Max-tree computation on GPUs. IEEE TPDS (2022)
2. Cousty, J., Najman, L., Perret, B.: Constructive links between some morphological hierarchies on edge-weighted graphs. In: Hendriks, C.L.L., Borgefors, G., Strand, R. (eds.) ISMM 2013. LNCS, vol. 7883, pp. 86–97. Springer, Heidelberg (2013). https://doi.org/10.1007/978-3-642-38294-9_8
3. Cousty, J., Perret, B., Phelippeau, H., Carneiro, S., Kamlay, P., Buzer, L.: An algebraic framework for out-of-core hierarchical segmentation algorithms. In: Lindblad, J., Malmberg, F., Sladoje, N. (eds.) DGMM 2021. LNCS, vol. 12708, pp. 378–390. Springer, Cham (2021). https://doi.org/10.1007/978-3-030-76657-3_27
4. Gazagnes, S., Wilkinson, M.H.F.: Distributed connected component filtering and analysis in 2d and 3d tera-scale data sets. IEEE Trans. Image Process. **30**, 3664–3675 (2021)

5. Gigli, L., Velasco-Forero, S., Marcotegui, B.: On minimum spanning tree streaming for hierarchical segmentation. PRL **138**, 155–162 (2020)
6. Götz, M., Cavallaro, G., Geraud, T., Book, M., Riedel, M.: Parallel computation of component trees on distributed memory machines. TPDS **29**(11), 2582–2598 (2018)
7. Havel, J., Merciol, F., Lefèvre, S.: Efficient tree construction for multiscale image representation and processing. JRTIP **16**(4), 1129–1146 (2019). https://doi.org/10.1007/s11554-016-0604-0
8. Kazemier, J.J., Ouzounis, G.K., Wilkinson, M.H.F.: Connected morphological attribute filters on distributed memory parallel machines. In: Angulo, J., Velasco-Forero, S., Meyer, F. (eds.) ISMM 2017. LNCS, vol. 10225, pp. 357–368. Springer, Cham (2017). https://doi.org/10.1007/978-3-319-57240-6_29
9. Meyer, F.: The dynamics of minima and contours. In: Maragos, P., Schafer, R.W., Butt, M.A. (eds.) ISMM, vol. 5, pp. 320 336. Springer, Boston (1996). https://doi.org/10.1007/978-1-4613-0469-2_38
10. Meyer, F., Maragos, P.: Morphological scale-space representation with levelings. In: Nielsen, M., Johansen, P., Olsen, O.F., Weickert, J. (eds.) Scale-Space 1999. LNCS, vol. 1682, pp. 187–198. Springer, Heidelberg (1999). https://doi.org/10.1007/3-540-48236-9_17
11. Najman, L., Cousty, J., Perret, B.: Playing with Kruskal: algorithms for morphological trees in edge-weighted graphs. In: Hendriks, C.L.L., Borgefors, G., Strand, R. (eds.) ISMM 2013. LNCS, vol. 7883, pp. 135–146. Springer, Heidelberg (2013). https://doi.org/10.1007/978-3-642-38294-9_12
12. Perret, B., Chierchia, G., Cousty, J., Guimarães, S.J.F., Kenmochi, Y., Najman, L.: Higra: hierarchical graph analysis. SoftwareX **10**, 100335 (2019)
13. Ronse, C.: Partial partitions, partial connections and connective segmentation. JMIV **32**(2), 97–125 (2008). https://doi.org/10.1007/s10851-008-0090-5
14. Salembier, P., Garrido, L.: Binary partition tree as an efficient representation for image processing, segmentation, and information retrieval. TIP **9**(4), 561–576 (2000)

Graph-Based Image Segmentation with Shape Priors and Band Constraints

Caio de Moraes Braz, Luiz Felipe D. Santos, and Paulo A.V. Miranda$^{(\boxtimes)}$ (iD)

Institute of Mathematics and Statistics, University of São Paulo,
05508–090 São Paulo, SP, Brazil
{caiobraz,lfdolabela,pmiranda}@ime.usp.br

Abstract. In this work, we describe an efficient algorithm, with proof of correctness, for finding an optimal binary segmentation of an image such that the indicated object satisfies a novel high-level prior, called the Band constraint (B), which is the extension of a recent shape prior, called Local Band constraint (LB), to its limiting case with radius tending to infinity. Unlike the LB constraint, the new algorithm can be applied directly to the original image graph saving memory. In our theoretical investigations, we discuss the theoretical relationship of the new B constraint with the Boundary Band (BB) constraint, formerly known as Geodesic Band constraint. Finally, we experimentally conduct a template rotation invariance study of the B constraint within the Oriented Image Foresting Transform framework in region adjacency graphs, when applied to natural images with templates by Gielis geometric equation.

Keywords: Band constraint · Shape constraints · Oriented Image Foresting Transform

1 Introduction

Image segmentation can be interpreted as a graph partition problem subject to hard constraints, such as seed pixels selected in the image for object recognition, by modelling neighborhood relations of picture elements. In this work, we are interested in fast seed-based methods in graphs to efficiently deal with large amounts of data, but which must also be versatile enough to support the inclusion of high-level constraints from prior object knowledge. We intend to develop new methods, where shape constraints are used as a priori knowledge to circumvent problems due to weak edges and the presence of multiple objects with similar color/intensity profiles. The proposed methods, developed in the strong formalism of graphs, could also be used as an additional layer in a segmentation pipeline, guaranteeing the theoretical establishment of the formal properties of the generated objects, increasing the robustness of the obtained results.

Thanks to Conselho Nacional de Desenvolvimento Científico e Tecnológico – CNPq – (Grant 407242/2021-0, 313087/2021-0, 465446/2014-0), CAPES (88887.136422/2017-00) and FAPESP (2014/12236-1, 2014/50937-1).

É. Baudrier et al. (Eds.): DGMM 2022, LNCS 13493, pp. 287–299, 2022.
https://doi.org/10.1007/978-3-031-19897-7_23

By restricting our scope of shape constraints to optimization methods on graphs aiming at globally optimal results, we have the following early works. In [4, 12], for each execution of the min-cut/max-flow algorithm, the shape prior is considered as a soft constraint, only penalizing segmentations that differ from a given binary template of the expected object. Soft constraints are also considered in [8], but the template is estimated as a fuzzy map from a training set of aligned binary shapes using kernel PCA. In [4], the shape prior is embedded in the edge weights of neighboring pixels in the graph (pairwise term), being the penalty value proportional to a distance function from the boundary of the template, while in [8, 12] the shape prior is embedded in the weights of arcs interconnecting pixels with terminal nodes (unary term). Increasing the penalty value in order to obtain a hard shape constraint in these methods from [4, 12] is not useful, as it would lead to a segmentation practically identical to the given binary template, disregarding all image information. Hedgehog Shape Prior [6, 7] is closer to the objectives of our work, since it corresponds to the definition of a genuine hard shape constraint. Compact Shape Prior [3] also defines a class of shapes by a hard constraint, but it is very restrictive and does not generalize to highly variable shapes, having a very limited scope of applications. The star convexity [5] is another example, which can be interpreted as a visibility constraint. That is, an object is star convex in relation to a center point c, if for every point p in the object, p is visible to c via the line segment connecting them, which must also be part of the object (the background is an obstruction to the "light" emitted by p to the observer c). Some shape constraints demand more sophisticated algorithms or more sophisticated graph constructions, such as the Boundary Band (BB) constraint [1] and Local Band (LB) constraint [10], respectively. The solution subject to BB constraint from [1] allows the segmentation to follow a pre-established template of shapes, with variances within a range of permitted deformations around an arbitrary scale. On the other hand, LB is less sensitive to the seed/template positioning as it is more flexible to adapt locally to image characteristics.

In this work, we present an efficient algorithm to compute an optimal segmentation, with proof of correctness, which can be applied directly to the original image graph, to handle the limiting case of LB with infinite radius, denoted as the Band (B) constraint. The solution from [10] via the usage of an expanded graph would not be feasible in this case, due to the excessive number of arcs. We also discuss how to create shape templates by Gielis equation, which can simulate many natural shapes, such as, diatoms, eggs, cross sections of plants, snowflakes and starfish [11] and present experiments to handle rotation variations.

2   Background

An image can be interpreted as a directed graph $G = \langle \mathcal{N}, \mathcal{A} \rangle$ whose nodes in $\mathcal{N}$ are the image pixels in its image domain $\mathcal{I} \subset Z^n$, and whose arcs, elements of $\mathcal{A}$, are the ordered pixel pairs $\langle s, t \rangle$ of vertices that are adjacent, that is, spatially close (e.g., 4-neighborhood, or 8-neighborhood, in case of 2D images).

We use $t \in \mathcal{A}(s)$ and $\langle s, t \rangle \in \mathcal{A}$ to indicate that t is adjacent to s. We will usually assume also that our image graph G is arc-weighted, that is, that each arc $\langle s, t \rangle \in \mathcal{A}$ has a fixed weight $\omega(s, t) \in [-\infty, \infty]$ (e.g., $\omega(s, t) = |I(t) - I(s)|$ for a single channel image with values given by $I(t)$). An arc-weighted digraph will be denoted as $G = \langle \mathcal{N}, \mathcal{A}, \omega \rangle$. A digraph G is symmetric if, for all $\langle s, t \rangle \in \mathcal{A}$, the pair $\langle t, s \rangle$ is also an arc of G. Note that in the symmetric graphs we can still have $\omega(s, t) \neq \omega(t, s)$. In this work, all considered graphs are symmetric and connected.

2.1   Oriented Image Foresting Transform (OIFT)

In the case of binary segmentation (object/background), we consider two non-empty disjoint seed sets, $\mathcal{S}_1$ and $\mathcal{S}_0$ ($\mathcal{S}_1 \cap \mathcal{S}_0 = \emptyset$), containing pixels selected inside the object $\mathcal{O}$ and in its exterior, respectively. A label, $L(t) = 1$ for all $t \in \mathcal{S}_1$ and $L(t) = 0$ for all $t \in \mathcal{S}_0$, is propagated to all unlabeled pixels during the OIFT algorithm [9]. For a label map $L \colon \mathcal{N} \to \{0, 1\}$ the object $\mathcal{O}$ identified with it is defined as the set $L^{-1}(1)$, where $L^{-1}(i) := \{t \in \mathcal{N} \colon L(t) = i\}$.

There are two important classes of energy formulations within the Generalized Graph Cut framework, the Max-Min[1] and Min-Sum optimizers [2]. OIFT and ORFC algorithms are Max-Min optimizers while the min cut/max-flow algorithm is a Min-Sum optimizer. The segmentation by OIFT gives a global optimum solution by maximizing the graph-cut measure $\varepsilon_{\min}$ (Eq. 1) subject to the seed constraints [9], among all segmentations satisfying these constraints.

$$\varepsilon_{\min}(L) = \min_{\langle s,t \rangle \in \mathcal{A} \mid L(s) > L(t)} \omega(s, t) \tag{1}$$

2.2   Closely Related Shape Constraints

Let $C \colon \mathcal{N} \to [0, \infty)$ be a fixed vertex cost function associated with an image digraph $G = \langle \mathcal{N}, \mathcal{A} \rangle$. The values $C(t)$ can be based on templates of shapes as discussed in Sect. 4.1, which will also be considered for evaluation in Sect. 5. The *boundary* of object $\mathcal{O}$ is defined as $\mathrm{bd}(\mathcal{O}) = \{t \in \mathcal{O} \colon \exists s \in \mathcal{A}(t) \text{such that} s \notin \mathcal{O}\}$.

Definition 1 (Boundary Band constraint (BB)). For $\Delta > 0$, an object $\mathcal{O}$ is BB_Δ (satisfies Boundary Band constraint with band size Δ) provided $C(t) < C(s) + \Delta$ for all $t \in \mathcal{O}$ and $s \in \mathrm{bd}(\mathcal{O})$.

As a consequence of Definition 1, we have that $\mathrm{bd}(\mathcal{O})$ is contained in the band $\{s \in \mathcal{N} \colon C(s) \in (m - \Delta, m]\}$, where $m = \max\{C(t) \colon t \in \mathcal{O}\}$. In particular, we have $|C(s) - C(t)| < \Delta$ for all $s, t \in \mathrm{bd}(\mathcal{O})$. Consequently, this regularizes the shape of $\mathrm{bd}(\mathcal{O})$, see [1]. Therefore, the idea of BB is to establish a maximum possible variation of the cost C between the boundary points $\mathrm{bd}(\mathcal{O})$ of the object $\mathcal{O}$ to be segmented. This is expected to prevent the generated segmentation to be irregular in relation to the C-level sets [1].

---

[1] Min-Max optimizer is a dual equivalent problem.

Definition 2 (Local Band constraint (LB)). For $\Delta, R > 0$ and a cost map $C \colon \mathcal{N} \to [0, \infty)$, a pixel $t \in \mathcal{O}$ is LB^R_Δ (satisfies Local Band constraint with band size Δ and parameter R) provided $C(t) < C(s) + \Delta$ for all $s \in \mathcal{N} \setminus \mathcal{O}$ such that $\|s - t\| \le R$. An object $\mathcal{O}$ is LB^R_Δ provided every $t \in \mathcal{O}$ is LB^R_Δ.

In this definition, the symbol $\|\cdot\|$ denotes the standard Euclidean L_2 norm on $\mathcal{N} \subset \mathbb{Z}^2$. In other words, if $\mathcal{O}$ is LB^R_Δ, then for any pair of pixels s and t such that $\|s - t\| \le R$ and $C(t) - C(s) \ge \Delta$, we have that $t \in \mathcal{O}$ implies $s \in \mathcal{O}$. The set of arcs $\{\langle p, q \rangle \in \mathcal{N} \times \mathcal{N} \colon \|p - q\| \le R \ \& \ C(p) \ge C(q) + \Delta\}$ representing the LB constraint forms a *Directed Acyclic Graph* (DAG). Several of these arcs that represent the LB constraint may actually be redundant and we can apply a *transitive reduction* to eliminate them from this DAG (Fig. 1).

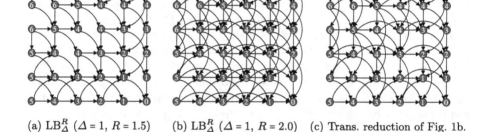

(a) LB^R_Δ ($\Delta = 1$, $R = 1.5$)   (b) LB^R_Δ ($\Delta = 1$, $R = 2.0$)   (c) Trans. reduction of Fig. 1b.

Fig. 1. (a-b) The LB arcs for LB^R_Δ (with $\Delta = 1$) for increasing radius values. The values $C(t)$ are indicated inside the nodes. (c) The transitive reduction of (b).

3   Theoretical Relationships Between Shape Constraints

To relate Local Band constraint to Boundary Band constraint from [1], the following notion of *Local Boundary Band constraint (LBB)* was introduced in [10]:

Definition 3 (Local Boundary Band (LBB)). For $\Delta, R > 0$ and a cost map $C \colon \mathcal{N} \to [0, \infty)$, a pixel $t \in \mathcal{O}$ is LBB^R_Δ (satisfies Local Boundary Band Constraint with band size Δ and parameter R) provided $C(t) < C(s) + \Delta$ for all $s \in \mathrm{bd}(\mathcal{O})$ such that $\|s - t\| \le R$. An object $\mathcal{O}$ is LBB^R_Δ provided every $t \in \mathcal{O}$ is LBB^R_Δ.

The following properties can be established for these shape constraints.

Properties. For $\Delta, \delta, R, r > 0$, we have:

(P1) If $\delta \le \Delta$, then $\mathcal{O}$ is $\mathrm{LBB}^R_\delta \implies \mathcal{O}$ is LBB^R_Δ,

(P2) If $r \le R$, then $\mathcal{O}$ is $\mathrm{LBB}^R_\Delta \implies \mathcal{O}$ is LBB^r_Δ,

(P3) If $\delta \leq \Delta$, then $\mathcal{O}$ is $\mathrm{LB}_\delta^R \implies \mathcal{O}$ is LB_Δ^R,

(P4) If $r \leq R$, then $\mathcal{O}$ is $\mathrm{LB}_\Delta^R \implies \mathcal{O}$ is LB_Δ^r.

Proof. These properties follow immediately from Definitions 2 and 3.

Note that neither of the statements "$\mathcal{O}$ is LB_Δ^R" and "$\mathcal{O}$ is LBB_Δ^R" implies the other. Nevertheless, they are closely related, for small/appropriate values of r and δ according to the following proposition from [10].

Proposition 1. *Let* $r = \max_{\langle s,t \rangle \in \mathcal{A}} \|s - t\|$ *and* $\delta = \max_{\langle s,t \rangle \in \mathcal{A}} |C(t) - C(s)|$. *If* $\Delta, R > 0$ *and* $\mathcal{O}$ *is* LB_Δ^{R+r}, *then* $\mathcal{O}$ *is* $\mathrm{LBB}_{\Delta+\delta}^R$.

Note however that the converse of this statement may not be true. The relationship in the opposite direction normally requires extra assumptions on map C and $\mathcal{O}$, as exemplified in the proposed proposition below.

Proposition 2. *Let* $\mathcal{O}$ *be a star-convex object with respect to center* c *and* $C(t) = \|t - c\|$ *be the Euclidean distance from* c. *If* $\Delta, R > 0$ *and* $\mathcal{O}$ *is* LBB_Δ^R, *then* $\mathcal{O}$ *is* LB_Δ^R.

Proof. Let $B(t) = \{s \in \mathcal{N} : \|t - s\| \leq R\}$ be the closed ball centered at t with radius R. Consider a boundary pixel $b \in \mathrm{bd}(\mathcal{O})$ inside $B(t)$, such that $C(b) = \|b - c\| = d_{min}$, where $d_{min} = \min_{x \in \mathrm{bd}(\mathcal{O}) \cap B(t)} C(x)$, as indicated in Fig. 2a. Since $\mathcal{O}$ is LBB_Δ^R, we know that $C(t) < C(b) + \Delta = d_{min} + \Delta$. To prove that $\mathcal{O}$ is LB_Δ^R, we must show that the following condition holds:

(∗) $C(u) \geq d_{min}$ for all $u \in (\mathcal{N} \setminus \mathcal{O}) \cap B(t)$.

Note that this condition given by (∗) implies that $C(t) < d_{min} + \Delta \leq C(u) + \Delta$.
For each $u \in (\mathcal{N} \setminus \mathcal{O}) \cap B(t)$, we have two cases to consider:

1. The line segment $\overline{uc}$ with endpoints u and c intercepts a pixel $p \in \mathrm{bd}(\mathcal{O}) \cap B(t)$ (Fig. 2a). In this case, we have $C(p) < C(u)$. By the definition of d_{min}, we know that $d_{min} \leq C(p)$. Hence we have that $d_{min} \leq C(p) < C(u)$.
2. The line segment $\overline{uc}$ with endpoints u and c intercepts a pixel $p \in \mathrm{bd}(B(t))$ and $p \notin \mathcal{O}$ (Fig. 2b). In this case, moving from p through the circular arc $\overset{\frown}{pr}$ towards the point r, which is the ball's pixel closest to c, there will always be a point $q \in \mathrm{bd}(\mathcal{O})$ and $C(q) \leq C(p) \leq C(u)$. Note that this is true since $\mathcal{O}$ is assumed to be a star-convex object with respect to center c, so that the segment $\overline{tc}$ is always contained within the object and $r \in \mathcal{O}$. By the definition of d_{min}, we know that $d_{min} \leq C(q)$. Hence we have that $d_{min} \leq C(q) \leq C(u)$.

Although Proposition 2 requires a star-convex object with respect to center c and $C(t) = \|t - c\|$, in practice, we have close results by similar arguments when we consider the Geodesic Star Convexity [5] and the cost map C is the geodesic length (i.e., $\psi_{sum}(\langle v_0, \ldots, v_\ell \rangle) := \sum_{1 \leq j \leq \ell} \|v_{j-1} - v_j\|$), from a compact and connected set of seeds $\mathcal{S}_1$ in $G = \langle \mathcal{N}, \mathcal{A} \rangle$.

Similar to establishing BB_Δ as the limiting case of LBB_Δ^R, as $R \to \infty$, we can also take the limit of LB_Δ^R as $R \to \infty$, which results in the following definition.

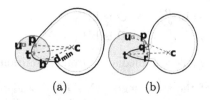

(a)   (b)

Fig. 2. Schematic illustration of Proposition 2. (a) The line segment $\overline{uc}$ with endpoints u and c intercepts a pixel $p \in \text{bd}(\mathcal{O}) \cap B(t)$. (b) The line segment $\overline{uc}$ with endpoints u and c intercepts a pixel $p \in \text{bd}(B(t))$ and $p \notin \mathcal{O}$.

Definition 4 (Band constraint (B)). For $\Delta > 0$, an object $\mathcal{O}$ is B_Δ (satisfies Band constraint with band size Δ) provided it is LB_Δ^∞, that is, when $C(t) < C(s) + \Delta$ for all $t \in \mathcal{O}$ and $s \in \mathcal{N} \setminus \mathcal{O}$.

Now let's consider the limiting cases of Propositions 1 and 2, as $R \to \infty$. For Proposition 1, we have that if $\Delta > 0$ and $\mathcal{O}$ is B_Δ, then $\mathcal{O}$ is $\text{BB}_{\Delta+\delta}$. Under the assumptions of Proposition 2, we have that $\mathcal{O}$ is $\text{BB}_\Delta \implies \mathcal{O}$ is B_Δ. Note also that the limiting case of Property (P1), as $R \to \infty$, implies that $\mathcal{O}$ is $\text{BB}_\Delta \implies \mathcal{O}$ is $\text{BB}_{\Delta+\delta}$. Therefore the Boundary Band constraint (BB) and the Band constraint (B) are strongly correlated.

4   The Band Constraint Algorithm

Now let's consider the limit case of the segmentation by OIFT subject to the LB constraint, as $R \to \infty$. In this case, the algorithm from [10] would become unfeasible due to the excessive amount of arcs created for the expanded graph, even considering the transitive reduction. Here we present an alternative and efficient algorithm (Algorithm 1) for solving this particular case. Algorithm 1 corresponds to a modified OIFT algorithm, to ensure the B constraint throughout its iterations, being more elegant and simpler than the solution of BB [1], as it uses a single auxiliary procedure.

Algorithm 1 – SEGMENTATION BY OIFT SUBJECT TO THE B CONSTRAINT

INPUT: Symmetric edge-weighted image digraph $G = \langle \mathcal{N}, \mathcal{A}, \omega \rangle$ and non-empty disjoint seed sets S_0 and S_1, cost map $C \colon \mathcal{N} \to [0,\infty)$, and $\Delta > 0$.

OUTPUT: The label map $L \colon \mathcal{N} \to \{0,1\}$.

AUXILIARY: Initially empty sets Q_0, Q_1, Q_x and Q, and an array of status $S \colon \mathcal{N} \to \{0,1\}$, where $S(t) = 1$ for processed nodes and $S(t) = 0$ for unprocessed nodes. The value $V(t)$ represents a potential penalty that a change of $L(t)$ would contribute to $\varepsilon_{\min}(L)$.

1. **For each $t \in \mathcal{N}$, do**
2. | *Set $S(t) \leftarrow 0$ and $V(t) \leftarrow \infty$;*
3. | *Insert t in Q_x;*
4. | **If $t \in S_0$, then**

5. $\llcorner$ $V(t) \leftarrow -\infty$, $L(t) \leftarrow 0$, *and insert t in Q;*
6.   **If** $t \in \mathcal{S}_1$, **then**
7. $\llcorner$ $V(t) \leftarrow -\infty$, $L(t) \leftarrow 1$, *and insert t in Q.*
8.   **While** $Q \neq \emptyset$ **do**
9.   *Remove s from Q such that V(s) is minimum;*
10.   *Set* $S(s) \leftarrow 1$;
11.   *Propagate(s,G,Q,V,L,S) by Algorithm 2;*
12.   *Remove s from* Q_x;
13.   **If** $L(s) = 0$, **then**
14.   *Insert s in* Q_0;
15.   **While** $Q_x \neq \emptyset \wedge (\max_{a \in Q_x} C(a) - \min_{b \in Q_0} C(b) \geq \Delta)$ **do**
16.   *Remove t from* Q_x *such that C(t) is maximum;*
17.   *Insert t in* Q_0 *and set* $L(t) \leftarrow 0$;
18.   *Set* $S(t) \leftarrow 1$;
19.   **If** $t \in Q$, **then** *remove t from Q.*
20. $\llcorner$ *Propagate(t,G,Q,V,L,S) by Algorithm 2.*
21.   **Else If** $L(s) = 1$, **then**
22.   *Insert s in* Q_1;
23.   **While** $Q_x \neq \emptyset \wedge (\max_{a \in Q_1} C(a) - \min_{b \in Q_x} C(b) \geq \Delta)$ **do**
24.   *Remove t from* Q_x *such that C(t) is minimum;*
25.   *Insert t in* Q_1 *and set* $L(t) \leftarrow 1$;
26.   *Set* $S(t) \leftarrow 1$;
27.   **If** $t \in Q$, **then** *remove t from Q.*
28. $\llcorner$ *Propagate(t,G,Q,V,L,S) by Algorithm 2.*
29. *Return L.*

Algorithm 2 – PROPAGATE

INPUT: Pixel $s \in \mathcal{N}$, graph $G = \langle \mathcal{N}, \mathcal{A}, \omega \rangle$, set Q, cost map V, label map
             $L: \mathcal{N} \to \{0,1\}$, and the array of status $S: \mathcal{N} \to \{0,1\}$, where $S(t) = 1$
             for processed nodes and $S(t) = 0$ for unprocessed nodes.
OUTPUT: The updated priority queue Q and the updated maps V and L.
AUXILIARY: Variable *tmp*.

1. **For each** $\langle s,t \rangle \in \mathcal{A}$ such that $S(t) = 0$ **do**
2.   **If** $L(s) = 1$, **then** $tmp \leftarrow \omega(s,t)$.
3.   **Else** $tmp \leftarrow \omega(t,s)$;
4.   **If** $tmp < V(t)$, **then**
5.   *Set* $V(t) \leftarrow tmp$ *and* $L(t) \leftarrow L(s)$.
6. $\llcorner$ **If** $t \notin Q$, **then** *insert t in Q.*

Theorem 1. *Let* $G = \langle \mathcal{N}, \mathcal{A}, \omega \rangle$ *be a symmetric edge-weighted image digraph with* $\omega: \mathcal{A} \to \mathbb{R}$. *Let L be a segmentation returned by Algorithm 1 applied to G, non-empty disjoint seed sets* $\mathcal{S}_1$ *and* $\mathcal{S}_0$, *cost map* $C: \mathcal{N} \to [0, \infty)$, *and parameter* $\Delta > 0$. *Assume that* $\mathcal{S}_1$ *and* $\mathcal{S}_0$ *are* B_Δ-*consistent, that is, that*

$(\star)$ *there exists a labeling satisfying seeds and* B_Δ *constraints.*

Then L satisfies seeds and B_Δ *constraints and maximizes the energy* $\varepsilon_{\min}$, *given by (1) w.r.t. G, among all segmentations satisfying these constraints.*

Proof. During the computation of Algorithm 1, three sets of pixels are indirectly defined, a growing object $\mathcal{O}' = \{t \in \mathcal{N}: S(t) = 1 \wedge L(t) = 1\}$, a growing background $\mathcal{B}' = \{t \in \mathcal{N}: S(t) = 1 \wedge L(t) = 0\}$ and an unprocessed zone $\mathcal{U}' = \{t \in \mathcal{N}: S(t) = 0\}$, initially given by $\mathcal{N}$, which is constantly reducing in size. From these sets, two candidate objects $\mathcal{O}_1 = \mathcal{O}'$ and $\mathcal{O}_2 = \mathcal{N} \setminus \mathcal{B}'$ are defined.

At each iteration of the main loop at Lines 9–10, Algorithm 1 always modifies the candidate object ($\mathcal{O}_1$ or $\mathcal{O}_2$) that has the lowest energy (given by $V(s)$ at Line 9), through the arc with the worst cost at its cut boundary. This is an event that should not be undone, since it is the only change of a pixel that can improve the candidate object with the worst energy. Changing other pixels will not improve it. However, this change may lead to a violation of the Band constraint for the corresponding candidate object ($\mathcal{O}_1$ or $\mathcal{O}_2$).

(V1) Violations for $\mathcal{O}_1$: The candidate object $\mathcal{O}_1$ is modified by the acquisition of s at Lines 9–10. Violations can only occur when $C(s) \geq C(t) + \Delta$ for some $t \in \mathcal{N} \setminus \mathcal{O}_1$. Since s cannot be modified, the only way to correct this violation is by adding t to $\mathcal{O}_1$ (Lines 25–26). This new insertion of t cannot in turn generate other unforeseen violations in a chain of events since $C(s) > C(t)$. However, there may be multiple pixels in a condition similar to that of node t with respect to s. They are handled in a similar way through the loop of Lines 23–28.

(V2) Violations for $\mathcal{O}_2$: The candidate object $\mathcal{O}_2$ is modified by the removal of s at Lines 9–10. Violations can only occur when $C(t) \geq C(s) + \Delta$ for some $t \in \mathcal{O}_2$. Since s cannot be modified, the only way to correct this violation is by removing t from $\mathcal{O}_2$ (Lines 17–18). This new removal of t cannot in turn generate other unforeseen violations in a chain of events since $C(t) > C(s)$. However, there may be multiple pixels in a condition similar to that of node t with respect to s. They are handled in a similar way through the loop of Lines 15–20.

Algorithm 1 has the following invariant:

Proposition 3. *At the end of each iteration of the main loop (Lines 8–28), we have that $\mathcal{O}_1$ and $\mathcal{O}_2$ satisfy the Band constraint (i.e., $\mathcal{O}_1$ is B_Δ and $\mathcal{O}_2$ is B_Δ) and $\mathcal{O}_1 \subseteq \mathcal{O}_2$.*

Condition $\mathcal{O}_1 \subseteq \mathcal{O}_2$ is guaranteed by construction of Algorithm 1[2].

$\mathcal{O}_1$ and $\mathcal{O}_2$ satisfying the Band constraint can be proved by induction. The base case is assured by the fact that $\mathcal{S}_1$ and $\mathcal{S}_0$ are assumed to be B_Δ-consistent. Therefore, in the first iterations of the main loop, the seeds will be processed

---

[2] Note that, in Lines 9–10, since s was removed from set Q on Line 9, we can conclude that it previously had $S(s) = 0$. Therefore, prior to the execution of Lines 9–10, we had $s \notin \mathcal{O}_1$ and $s \in \mathcal{O}_2$. During the execution of Lines 9–10, if $L(s) = 1$ then s is inserted into $\mathcal{O}_1$, but it was already in $\mathcal{O}_2$. On the other hand, if $L(s) = 0$ then s is removed from $\mathcal{O}_2$, but $s \notin \mathcal{O}_1$. A pixel t can also be removed from $\mathcal{O}_2$ on Lines 17–18, but according to Line 16, this pixel t was previously in the set Q_x, which stores unprocessed pixels, so t was not in $\mathcal{O}_1$. A pixel t can also be inserted into $\mathcal{O}_1$ on Lines 25–26, but according to Line 24, this pixel t was previously in the set Q_x, which stores unprocessed pixels, so $t \in \mathcal{O}_2$ since $t \notin \mathcal{B}'$.

and the handling of violations V1 and V2, as explained previously, will establish the base case, being the processing order of seeds irrelevant. For the induction step, we assume that $\mathcal{O}_1$ is B_Δ and $\mathcal{O}_2$ is B_Δ before executing the body of the main loop and we prove that they still satisfy the Band constraint at the end of the loop. The induction step is guaranteed by the handling of violations V1 and V2, as explained previously, in Lines 23–28 and Lines 15–20 of Algorithm 1 respectively. However, there are two important issues that need to be discussed.

(i) Lines 23–28 only consider pixels t from the set $Q_x = \mathcal{U}'$, whereas from the discussion of V1 we should actually consider $t \in \mathcal{N} \setminus \mathcal{O}_1 = \mathcal{U}' \cup \mathcal{B}'$.
(ii) Lines 15–20 only consider pixels t from the set $Q_x = \mathcal{U}'$, whereas from the discussion of V2 we should actually consider $t \in \mathcal{O}_2 = \mathcal{U}' \cup \mathcal{O}_1$.

For case (i), we must prove that there are no violations of the B constraint of $\mathcal{O}_1$ for $t \in \mathcal{B}'$ during Lines 23–28, and therefore, there is no need to test pixels from this set $\mathcal{B}'$. We can carry out a proof by contradiction, assume that there is $t \in \mathcal{B}'$ such that $C(s) \geq C(t) + \Delta$. Note that $t \in \mathcal{B}'$ implies that $t \notin \mathcal{O}_2$. Since s corresponds to a pixel removed from Q in Line 9, before computing Line 10, it previously had $S(s) = 0$ and therefore we have $s \in \mathcal{O}_2$. However, $t \notin \mathcal{O}_2$, $s \in \mathcal{O}_2$ and $C(s) \geq C(t) + \Delta$ indicate a violation of the B constraint for $\mathcal{O}_2$, leading to a contradiction, since $\mathcal{O}_2$ is B_Δ by the induction hypothesis.

For case (ii), we must prove that there are no violations of the B constraint of $\mathcal{O}_2$ for $t \in \mathcal{O}_1$ during the computation of Lines 15–20, and therefore, there is no need to test pixels from this set $\mathcal{O}_1$. We can carry out a proof by contradiction, assume that there is $t \in \mathcal{O}_1$ such that $C(t) \geq C(s) + \Delta$. Since s corresponds to a pixel removed from Q in Line 9, before computing Line 10, it previously had $S(s) = 0$ and therefore we have $s \notin \mathcal{O}_1$. However, $t \in \mathcal{O}_1$, $s \notin \mathcal{O}_1$ and $C(t) \geq C(s) + \Delta$ indicate a violation of the B constraint for $\mathcal{O}_1$, leading to a contradiction, since $\mathcal{O}_1$ is B_Δ by the induction hypothesis.

Following the invariant of Proposition 3 and the fact that the set $Q_x = \mathcal{U}'$ is always decreasing in size, in the end, all pixels will have been removed from set Q and objects $\mathcal{O}_1$ and $\mathcal{O}_2$ will have converged to the same set of pixels that satisfy the B constraint. A global maximum of the energy function $\varepsilon_{\min}$ (Eq. 1) subject to the B constraint is guaranteed by the energy competition between the two candidate objects $\mathcal{O}_1$ and $\mathcal{O}_2$ throughout the process, always carrying out operations strictly necessary to improve the object with lower energy and conserving the one with higher energy.

For implementation purposes, the sets Q_0, Q_1, Q_x and Q must be implemented using appropriate priority queue data structures in order to support the efficient removal of their extreme value elements (maximum or minimum). The queues Q_0, Q_1 and Q_x represent the sets $\mathcal{B}'$, $\mathcal{O}'$ and $\mathcal{U}'$, respectively. Regarding the computational complexity, if a binary heap is used for the queues, then Algorithm 1 can be implemented in $O((m + n) \log n)$, where $n = |\mathcal{N}|$ and $m = |\mathcal{A}|$.

4.1   Shape Templates

In [4,13], the shape prior is constructed from a fixed instance of the target object taken as a reference, by calculating the *Euclidean Distance Transform* (EDT) in relation to its contour. In order to adapt this scheme for use as the cost map C, we must consider a signed distance function (assuming negative values for distances within the object as in [13]) and then subtract its most negative value from it to shift the brightness range to consider only positive values. However, level curves farther from the initial contour suffer deformations, with their salient parts becoming rounded on larger scales. Therefore, this is only viable when we have a good a priori idea of an appropriate scale for the object of interest.

To create a cost template invariant to scale changes of a given predefined shape, it is necessary that its scale variations relative to a fixed reference point keep the smaller scale shapes entirely contained within the upscaled shapes. Star-convex shapes meet this constraint with respect to their star centers. Many natural shapes can be described by the Gielis equation and its generalizations, including shapes of diatoms, eggs, cross sections of plants, snowflakes and starfish [11]. Gielis equation in polar coordinates can be written as a function $r: [0, 2\pi) \rightarrow \mathbb{R}$, given by Eq. 2, where $r(\varphi)$ and φ are the polar radius and the angle between the straight line where the polar radius lies and the x-axis, respectively.

$$r(\varphi) = \left(\left| \frac{1}{A} \cos \left(\frac{m}{4} \varphi \right) \right|^{n_2} + \left| \frac{1}{B} \sin \left(\frac{m}{4} \varphi \right) \right|^{n_3} \right)^{-1/n_1} \tag{2}$$

where n_1, n_2 and n_3 are real constants, and m is a positive integer. Given that $r(\varphi) > 0$ and we have a single point on the curve for each angle φ in the interval $[0, 2\pi)$, we can conclude that all shapes described by Eq. 2 are star convex that can be used immediately to create scale invariant templates for the B constraint. Algorithm 3 computes a template C of shapes given by Gielis equation that is appropriate for the B constraint.

Algorithm 3 – TEMPLATE BY GIELIS EQUATION

INPUT: The parameters of the Gielis equation (A, B, n_1, n_2, n_3, and m), the coordinates (c_x, c_y) of the model center, and dimensions T_x and T_y.

OUTPUT: The template (cost map) $C: \mathcal{N} \rightarrow [0, \infty)$, where $\mathcal{N} \subset \mathbb{Z}^2$ and for $(x, y) \in \mathcal{N}$ we have $0 \leq x < T_x$ and $0 \leq y < T_y$.

1.   Set $s_{min} \leftarrow \infty$ and $d_{min} \leftarrow \infty$;
2.   **For each** $(x, y) \in \mathcal{N}$ such that $x = 0 \vee x = T_x - 1 \vee y = 0 \vee y = T_y - 1$, **do**
3.   Set $d_x \leftarrow x - c_x$ and $d_y \leftarrow y - c_y$;
4.   Compute $\varphi \leftarrow \arctan2 (d_y, d_x)$; **If** $\varphi < 0$, **then** $\varphi \leftarrow \varphi + 2\pi$.
5.   Compute $r \leftarrow$ GielisEquation $(\varphi, A, B, n_1, n_2, n_3, m)$;
6.   Set $d \leftarrow \sqrt{d_x^2 + d_y^2}$ and $s \leftarrow d/r$;
7.   **If** $s < s_{min} \vee (s = s_{min} \wedge d < d_{min})$ **then**
8.   Set $s_{min} \leftarrow s$ and $d_{min} \leftarrow d$.
9.   **For each** $p = (x, y) \in \mathcal{N}$, **do**
10.   Set $d_x \leftarrow x - c_x$ and $d_y \leftarrow y - c_y$;
11.   Compute $\varphi \leftarrow \arctan2 (d_y, d_x)$; **If** $\varphi < 0$, **then** $\varphi \leftarrow \varphi + 2\pi$.
12.   Compute $r \leftarrow$ GielisEquation $(\varphi, A, B, n_1, n_2, n_3, m)$;
13.   Set $d \leftarrow \sqrt{d_x^2 + d_y^2}$ and $s \leftarrow d/r$;
14.   Set $C(p) \leftarrow$ Round $((s/s_{min}) \cdot d_{min})$.
15. Return C.

In Lines 4 and 11 of Algorithm 3, $\text{arctan2}(d_y, d_x)$ is used to compute φ by the two-argument arctangent with a range of $(-\pi, \pi]$. Gielis function $r(\varphi)$ by Eq. 2 is then computed on Lines 5 and 12. The d/r ratios in s from Lines 6 and 13 are used to indicate different scale values for the template, which must be normalized and quantized into an appropriate range as done on Line 14 for the digital cost map C. To this end, the first loop is used to find an appropriate normalization scale value s_{min} by inspecting the directions along the image border.

We still need to find the proper translation and rotation of the template in relation to the desired object in the image. In [4], they are estimated based on the given object markers. However, accurately estimating rotation based on object markers requires a reasonable amount of markers preferably distributed close to the object boundary. In [12], a sequence of intermediate segmentations is produced, starting with a segmentation free of shape constraints and the penalties for shape constraints are gradually increased, with translation and rotation being estimated between iterations based on the previous segmentation of the sequence. This strategy, however, will only work in easier cases, when a reasonable approximation of the object can be obtained for the first segmentations of the sequence. In this work, we focus on the more formal aspects of shape constraints, avoiding the usage of such heuristic procedures as much as possible.

5   Experimental Results

We tested the method for a fixed translation of the template, by aligning the zero-valued pixel of C with the center of mass of the object seeds, and rotations were handled by exhaustive search with angle increments of $5°$. In order for this search to become viable, we used a graph of superpixels with an average area of 20 pixels for each superpixel and $\omega(s, t) = \|I(t) - I(s)\|$, where $I(t)$ is the mean CIELAB color of superpixel t. We used a template C by Algorithm 3 of a leaf modeled via Eq. 2 to segment plant leaves (Fig. 3a). To study the proposed treatment of rotations, we repeated the experiments for dozens of images (with 640×480 pixels) in different orientations achieving similar results (see the accuracy by the Dice coefficient in Figs. 3b-d). The execution time was 487.31 ms per image on an Intel Core i5-10210U CPU 1.60 GHz $\times$ 8 laptop.

(a)   (b) Dice: 98.92%   (c) Dice: 98.48%   (d) Dice: 98.31%

Fig. 3. (a) Shape template by Gielis equation with $A = 10^5$, $B = 10^4$, $n_1 = 2$, $n_2 = 1$, $n_3 = 1$, and $m = 2$. (b-d) Results by Algorithm 1 with $\Delta = 100$ and rotation handling.

6 Conclusion

As pointed out in [10], in applications where you want to obtain a good segmentation result even in the presence of uncertainties in the template positioning, LB constraint is the recommended method. However, our experimental results indicate that if the objective is to execute the method several times varying the position of the template, as we did for the rotation, in order to identify its best positioning, B (Algorithm 1) becomes a more interesting option, as its sensitivity can help us to isolate the best position, without the need to recreate the graph for each new position of the template as required by LB. As future work we intend to test the developed shape constraints in the segmentation of three-dimensional magnetic resonance imaging and computed tomography.

References

1. de Moraes Braz, C., Miranda, P.A.: Image segmentation by image foresting transform with geodesic band constraints. In: 2014 IEEE International Conference on Image Processing, pp. 4333–4337. IEEE (2014). https://doi.org/10.1109/ICIP.2014.7025880

2. Ciesielski, K., Udupa, J., Falcão, A., Miranda, P.: A unifying graph-cut image segmentation framework: algorithms it encompasses and equivalences among them. In: Proceedings of SPIE on Medical Imaging: Image Processing, vol. 8314 (2012)

3. Das, P., Veksler, O.: Semiautomatic segmentation with compact shapre prior. In: The 3rd Canadian Conference on Computer and Robot Vision, pp. 28–28 (2006)

4. Freedman, D., Zhang, T.: Interactive graph cut based segmentation with shape priors. In: 2005 IEEE Computer Society Conference on Computer Vision and Pattern Recognition. CVPR'05, vol. 1, pp. 755–762. IEEE (2005)

5. Gulshan, V., Rother, C., Criminisi, A., Blake, A., Zisserman, A.: Geodesic star convexity for interactive image segmentation. In: Proceedings of Computer Vision and Pattern Recognition, pp. 3129–3136 (2010)

6. Isack, H., Veksler, O., Sonka, M., Boykov, Y.: Hedgehog shape priors for multi-object segmentation. In: 2016 IEEE Conference on Computer Vision and Pattern Recognition (CVPR), pp. 2434–2442 (2016)

7. Isack, H.N., Boykov, Y., Veksler, O.: A-expansion for multiple "hedgehog" shapes. CoRR abs/1602.01006 (2016), https://arxiv.org/abs/1602.01006

8. Malcolm, J., Rathi, Y., Tannenbaum, A.: Graph cut segmentation with nonlinear shape priors. In: 2007 IEEE International Conference on Image Processing, vol. 4, pp. IV-365–IV-368 (2007). https://doi.org/10.1109/ICIP.2007.4380030

9. Miranda, P., Mansilla, L.: Oriented image foresting transform segmentation by seed competition. IEEE Trans. Image Process. 23(1), 389–398 (2014)

10. de Moraes Braz, C., Miranda, P.A.V., Ciesielski, K.C., Cappabianco, F.A.M.: Optimum cuts in graphs by general fuzzy connectedness with local band constraints. J. Math. Imaging Vis. 62(5), 659–672 (2020). https://doi.org/10.1007/s10851-020-00953-w

11. Shi, P., Ratkowsky, D.A., Gielis, J.: The generalized Gielis geometric equation and its application. Symmetry 12(4), 645 (2020). https://doi.org/10.3390/sym12040645

12. Vu, N., Manjunath, B.S.: Shape prior segmentation of multiple objects with graph cuts. In: IEEE Conference on Computer Vision and Pattern Recognition, pp. 1–8 (2008)
13. Zhang, T., Freedman: tracking objects using density matching and shape priors. In: Proceedings Ninth IEEE International Conference on Computer Vision, vol. 2, pp. 1056–1062 (2003)

Differential Oriented Image Foresting Transform Segmentation by Seed Competition

Marcos A. T. Condori⬤ and Paulo A. V. Miranda$^{(\boxtimes)}$⬤

Institute of Mathematics and Statistics, University of São Paulo,
São Paulo, SP 05508-090, Brazil
{mtejadac,pmiranda}@ime.usp.br

Abstract. The Image Foresting Transform (IFT) is a graph-based framework to develop image operators based on optimum connectivity between a root set and the remaining nodes, according to a given path-cost function. Oriented Image Foresting Transform (OIFT) was proposed as an extension of some IFT-based segmentation methods to directed graphs, enabling them to support the processing of global object properties, such as connectedness, shape constraints, boundary polarity, and hierarchical constraints, allowing their customization to a given target object. OIFT lies in the intersection of the Generalized Graph Cut and the General Fuzzy Connectedness frameworks, inheriting their properties. Its returned segmentation is optimal, with respect to an appropriate graph cut measure, among all segmentations satisfying the given constraints. In this work, we propose the Differential Oriented Image Foresting Transform (DOIFT), which allows multiple OIFT executions for different root sets, making the processing time proportional to the number of modified nodes. Experimental results show considerable efficiency gains over the sequential flow of OIFTs in image segmentation, while maintaining a good treatment of tie zones. We also demonstrate that the differential flow makes it feasible to incorporate area constraints in OIFT segmentation of multi-dimensional images.

Keywords: Oriented Image Foresting Transform · Image segmentation in directed graphs · Generalized Graph Cut · Differential algorithms

1 Introduction

In graph-based methods, image segmentation can be seen as a graph partition problem between sets of seed pixels. Oriented Image Foresting Transform (OIFT) [14] and Oriented Relative Fuzzy Connectedness (ORFC) [1] are extensions to directed weighted graphs of some methods from the Generalized

Thanks to Conselho Nacional de Desenvolvimento Científico e Tecnológico – CNPq – (Grant 407242/2021-0, 313087/2021-0, 465446/2014-0, 166631/2018-3), CAPES (88887.136422/2017-00) and FAPESP (2014/12236-1, 2014/50937-1).

E. Baudrier et al. (Eds.): DGMM 2022, LNCS 13493, pp. 300–311, 2022.
https://doi.org/10.1007/978-3-031-19897-7_24

Graph Cut (GGC) framework [3], including Fuzzy Connectedness [4] and Watersheds [7]. OIFT generates an optimal cut in the graph according to an appropriate graph cut measure, while having a lower computational complexity compared to the min-cut/max-flow algorithm [2].

OIFT's energy formulation on digraphs makes it a very versatile method, supporting several high-level priors for object segmentation, including global properties such as connectedness [12], shape constraints [15], boundary polarity [13,14], and hierarchical constraints [11], which allow the customization of the segmentation to a given target object.

In interactive region-based segmentation from markers (i.e., set of seeds), the user can add markers to and/or remove markers from previous interactions in order to improve the results. In the context of Image Foresting Transform (IFT) [9], which is based on propagating paths from seeds, instead of starting over the segmentation for each new set of seeds, Differential Image Foresting Transform (DIFT) algorithm [8] can be employed to update the segmentation in a differential manner, by correcting only the wrongly labeled parts of the optimum-path forest in time proportional to the size of the modified regions in the image (i.e., in sublinear time). This greatly increases efficiency, which is crucial to obtain interactive response times in the segmentation of large 3D volumes. However, DIFT [8] requires that the path-cost function be *monotonically incremental* (MI), consequently not supporting the OIFT path-cost functions.

More recently, a novel differential IFT algorithm, named *Generalized DIFT* (GDIFT) [6], has been proposed, which extends the original DIFT algorithm to handle connectivity functions with root-based increases (which can be non-monotonically incremental), avoiding segmentation inconsistencies (e.g., disconnected regions) in applications to superpixel segmentation [10,16]. However, there are still no studies of the differential computation for the case of the OIFT path-cost functions. This work aims to close this gap by testing three alternatives for Differential Oriented Image Foresting Transform (DOIFT). Our experimental results show considerable efficiency gains of the differential flow of DOIFTs over the sequential flow of OIFTs in image segmentation of medical images, while maintaining a good treatment of tie zones for two of the presented solutions. We also demonstrate that the differential flow makes it feasible to incorporate area constraints in OIFT segmentation of multi-dimensional images, which is useful for getting regions of interest in the image with less user interaction.

2   Background

A multi-dimensional and multi-spectral image $\hat{I}$ is a pair $\langle \mathcal{I}, \mathbf{I} \rangle$, where $\mathcal{I} \subset \mathbb{Z}^n$ is the image domain and $\mathbf{I}(t)$ assigns a set of m scalars $I_i(t)$, $i = 1, 2, \ldots, m$, to each pixel $t \in \mathcal{I}$. The subindex i is removed when $m = 1$.

An image can be interpreted as a weighted digraph $G = \langle \mathcal{V}, \mathcal{A}, \omega \rangle$, whose nodes $\mathcal{V}$ are the image pixels in its image domain $\mathcal{I} \subset \mathbb{Z}^n$, and whose arcs are the ordered pixel pairs $\langle s, t \rangle \in \mathcal{A}$. (e.g., 4-neighborhood or 8-neighborhood, in case of 2D images). The digraph G is symmetric if for any of its arcs $\langle s, t \rangle \in \mathcal{A}$, the pair $\langle t, s \rangle$ is also an arc of G. Each arc $\langle s, t \rangle \in \mathcal{A}$ has a weight $\omega(s, t)$, such as a dissimilarity measure between pixels s and t (e.g., $\omega(s, t) = |I(t) - I(s)|$).

For a given image graph $G = \langle \mathcal{V}, \mathcal{A}, \omega \rangle$, a path $\pi = \langle t_1, t_2, \ldots, t_n \rangle$ is a sequence of adjacent pixels (i.e., $\langle t_i, t_{i+1} \rangle \in \mathcal{A}$, $i = 1, 2, \ldots, n - 1$) with no repeated vertices ($t_i \neq t_j$ for $i \neq j$). Other greek letters, such as τ, can also be used to denote different paths. A path $\pi_t = \langle t_1, t_2, \ldots, t_n = t \rangle$ is a path with terminus at a pixel t. When we want to explicitly indicate the origin of the path, the notation $\pi_{s \rightsquigarrow t} = \langle t_1 = s, t_2, \ldots, t_n = t \rangle$ may also be used, where s stands for the origin and t for the destination node. A path is *trivial* when $\pi_t = \langle t \rangle$. A path $\pi_t = \pi_s \cdot \langle s, t \rangle$ indicates the extension of a path π_s by an arc $\langle s, t \rangle$.

A *predecessor map* is a function $P \colon \mathcal{V} \to \mathcal{V} \cup \{nil\}$ that assigns to each pixel t in $\mathcal{V}$ either some other adjacent pixel in $\mathcal{V}$, or a distinctive marker nil not in $\mathcal{V}$, in which case t is said to be a *root* of the map. A *spanning forest* is a predecessor map which contains no cycles, i.e., one which takes every pixel to nil in a finite number of iterations. For any pixel $t \in \mathcal{V}$, a spanning forest P defines a path π_t^P recursively as $\langle t \rangle$ if $P(t) = nil$, and $\pi_s^P \cdot \langle s, t \rangle$ if $P(t) = s \neq nil$.

2.1 Image Foresting Transform (IFT)

The Image Foresting Transform (IFT) algorithm (Algorithm 1) is a generalization of Dijkstra's algorithm for multiple sources (root sets) and more general connectivity functions [5,9]. A *connectivity function* computes a value $f(\pi_t)$ for any path π_t, usually based on arc weights. A path π_t is *optimum* if $f(\pi_t) \leq f(\tau_t)$ for any other path τ_t in G. By taking to each pixel $t \in \mathcal{V}$ one optimum path with terminus at t, we obtain the optimum-path value $V_{opt}^f(t)$, which is uniquely defined by $V_{opt}^f(t) = \min_{\forall \pi_t \text{ in } G} \{f(\pi_t)\}$. The *image foresting transform* (IFT) [9] takes an image graph $G = \langle \mathcal{V}, \mathcal{A}, \omega \rangle$, and a path-cost function f; and assigns one optimum path to every pixel $t \in \mathcal{V}$ such that an *optimum-path forest* P is obtained, i.e., a spanning forest where all paths π_t^P for $t \in \mathcal{V}$ are optimum. However, f must satisfy the conditions indicated in [5], otherwise, the paths π_t^P of the returned spanning forest may not be optimum.

The cost of a trivial path $\pi_t = \langle t \rangle$ is usually given by a handicap value $H(t)$. For example, $H(t) = 0$ for all $t \in \mathcal{S}$ and $H(t) = \infty$ otherwise, where $\mathcal{S}$ is a seed set. The costs for non-trivial paths follow a path-extension rule. For example:

$$f_{\max}(\pi_s \cdot \langle s, t \rangle) = \max\{f_{\max}(\pi_s), \omega(s, t)\} \tag{1}$$

In Algorithm 1, the root map R stores the origin of the paths and the path-cost map V converges to V_{opt}^f, when f satisfies the conditions indicated in [5].

Algorithm 1 – IFT Algorithm

INPUT: Image graph $\langle \mathcal{V}, \mathcal{A}, \omega \rangle$, path-cost function f and an initial labeling function $\lambda \colon \mathcal{V} \to \{0, \ldots, l\}$.

OUTPUT: The label map $L \colon \mathcal{V} \to \{0, \ldots, l\}$, root map $R \colon \mathcal{V} \to \mathcal{V}$, path-cost map $V \colon \mathcal{V} \to \mathbb{R}$ and the spanning forest $P \colon \mathcal{V} \to \mathcal{V} \cup \{nil\}$.

AUXILIARY: Priority queue Q, variable tmp and an array of status $S \colon \mathcal{V} \to \{0, 1\}$, where $S(t) = 1$ for processed nodes and $S(t) = 0$ for unprocessed nodes.

1. **For each** $t \in \mathcal{V}$, **do**
2.   *Set* $S(t) \leftarrow 0$.
3.   *Set* $P(t) \leftarrow nil$, $V(t) \leftarrow f(\langle t \rangle)$, $R(t) \leftarrow t$ *and* $L(t) \leftarrow \lambda(t)$.
4.   **If** $V(t) \neq +\infty$, **then**
5.   *insert* t *in* Q.
6. **While** $Q \neq \varnothing$, **do**
7.   *Remove* s *from* Q *such that* $V(s) = \min_{\forall t \in Q}\{V(t)\}$.
8.   *Set* $S(s) \leftarrow 1$.
9.   **For each** *node* t *such that* $\langle s,t \rangle \in \mathcal{A}$, **do**
10.   **If** $S(t) \neq 1$, **then**
11.   *Compute* $tmp \leftarrow f(\pi_s^P \cdot \langle s,t \rangle)$.
12.   **If** $tmp < V(t)$, **then**
13.   **If** $t \in Q$, **then** *remove* t *from* Q.
14.   *Set* $P(t) \leftarrow s$ *and* $V(t) \leftarrow tmp$.
15.   *Set* $R(t) \leftarrow R(s)$ *and* $L(t) \leftarrow L(s)$.
16.   *Insert* t *in* Q.

2.2 Oriented Image Foresting Transform (OIFT)

In its first version [14], OIFT was built on the IFT framework by considering the following path-cost function in a symmetric digraph with integer weights:

$$f_1^{\vec{\sigma}}(\langle t \rangle) = \begin{cases} -1 & \text{if } t \in \mathcal{S}_1 \cup \mathcal{S}_0 \\ +\infty & \text{otherwise} \end{cases}$$

$$f_1^{\vec{\sigma}}(\pi_{r \rightsquigarrow s} \cdot \langle s,t \rangle) = \begin{cases} \max\{f_1^{\vec{\sigma}}(\pi_{r \rightsquigarrow s}), 2 \times w(s,t) + 1\} & \text{if } r \in \mathcal{S}_1 \\ \max\{f_1^{\vec{\sigma}}(\pi_{r \rightsquigarrow s}), 2 \times w(t,s)\} & \text{if } r \in \mathcal{S}_0 \end{cases} \quad (2)$$

Later, a second version [13] with a better handling of ties was proposed based on the following path-cost function:

$$f_2^{\vec{\sigma}}(\langle t \rangle) = f_1^{\vec{\sigma}}(\langle t \rangle)$$

$$f_2^{\vec{\sigma}}(\pi_{r \rightsquigarrow s} \cdot \langle s,t \rangle) = \begin{cases} w(s,t) & \text{if } r \in \mathcal{S}_1 \\ w(t,s) & \text{otherwise} \end{cases} \quad (3)$$

The segmented object $\mathcal{O}^P$ by OIFT is defined from the forest P computed by Algorithm 1, with $f_2^{\vec{\sigma}}$ (or $f_1^{\vec{\sigma}}$), by taking as object pixels the set of pixels that were conquered by paths rooted in $\mathcal{S}_1$ (i.e., $t \in \mathcal{O}^P$ if and only if $R(t) \in \mathcal{S}_1$).

The functions $f_1^{\vec{\sigma}}$ and $f_2^{\vec{\sigma}}$ are non-monotonically incremental connectivity functions, as described in [13,14]. The optimality of $\mathcal{O}^P$ by OIFT is supported by an energy criterion of cut in graphs involving arcs from object to background pixels $\mathcal{C}(\mathcal{O}^P)$ (outer-cut boundary), according to Theorem 1 from [13,14].

$$\mathcal{C}(\mathcal{O}) = \{\langle s,t \rangle \in \mathcal{A} \mid s \in \mathcal{O} \text{ and } t \notin \mathcal{O}\} \quad (4)$$

$$E(\mathcal{O}) = \min_{\langle s,t \rangle \in \mathcal{C}(\mathcal{O})} w(s,t) \quad (5)$$

Theorem 1 (Outer-cut optimality by OIFT). *For two given sets of seeds* $\mathcal{S}_1$ *and* $\mathcal{S}_0$, *let* $\mathcal{U}(\mathcal{S}_1, \mathcal{S}_0) = \{\mathcal{O} \subseteq \mathcal{V} \mid \mathcal{S}_1 \subseteq \mathcal{O} \subseteq \mathcal{V} \setminus \mathcal{S}_0\}$ *denote the universe of all possible objects satisfying the seed constraints. Any spanning forest* P *computed by Algorithm 1 for function* $f_1^{\vec{\sigma}}$ *(or* $f_2^{\vec{\sigma}}$*) defines a segmented object* $\mathcal{O}^P$ *that maximizes* E *(Eq. 5) among all possible segmentation results in* $\mathcal{U}$. *That is,* $E(\mathcal{O}^P) = \max_{\mathcal{O} \in \mathcal{U}(\mathcal{S}_1, \mathcal{S}_0)} E(\mathcal{O})$.

2.3 Differential Image Foresting Transform (DIFT)

Let a sequence of IFTs be represented as $\langle IFT_{(\mathcal{S}^1)}, IFT_{(\mathcal{S}^2)}, \ldots, IFT_{(\mathcal{S}^n)} \rangle$, where n is the total number of IFT executions on the image. At each execution, the seed set $\mathcal{S}^i$ is modified by adding and/or removing seeds to obtain a new set $\mathcal{S}^{i+1}$. We define a scene $\mathcal{G}^i$ as the set of maps $\mathcal{G}^i = \{P^i, V^i, L^i, R^i\}$, resulting from the ith iteration in a sequence of IFTs.

The DIFT algorithm [6,8] allows to efficiently compute a scene $\mathcal{G}^i$ from the previous scene $\mathcal{G}^{i-1}$, a set $\Delta_{\mathcal{S}^i}^+ = \mathcal{S}^i \setminus \mathcal{S}^{i-1}$ of new seeds for addition, and a set $\Delta_{\mathcal{S}^i}^- = \mathcal{S}^{i-1} \setminus \mathcal{S}^i$ of seeds marked for removal. In the execution flow by DIFT, after the first execution of $IFT_{(\mathcal{S}^1)}$, we have that the scenes $\mathcal{G}^i$ for $i \geq 2$ are calculated based on the scene $\mathcal{G}^{i-1}$, taking advantage of the trees that were computed in the previous iteration, thus reducing the processing time. Hence, we have the following differential flow: $\langle IFT_{(\mathcal{S}^1)}, DIFT_{(\Delta_{\mathcal{S}^2}^+, \Delta_{\mathcal{S}^2}^-, \mathcal{G}^1)}, DIFT_{(\Delta_{\mathcal{S}^3}^+, \Delta_{\mathcal{S}^3}^-, \mathcal{G}^2)}, \ldots, DIFT_{(\Delta_{\mathcal{S}^n}^+, \Delta_{\mathcal{S}^n}^-, \mathcal{G}^{n-1})} \rangle$.

3 Differential OIFT (DOIFT)

Figure 1 shows that the Generalized DIFT (GDIFT) algorithm [6] with $f_2^{\vec{\sigma}}$, to differentially compute the sequence $\langle IFT_{(\mathcal{S}^1)}, IFT_{(\mathcal{S}^2)} \rangle$, where $\mathcal{S}^1 = \mathcal{S}_1^1 \cup \mathcal{S}_0^1 = \{a\} \cup \{i, l\}$ and $\mathcal{S}^2 = \mathcal{S}_1^2 \cup \mathcal{S}_0^2 = \{a\} \cup \{i\}$, may generate a result not predicted by $IFT_{(\mathcal{S}^2)}$ via Algorithm 1. The problem occurs because nodes b and g are initially processed in a given order during the first run of the IFT (Fig. 1b), but later become frontier nodes, i.e., neighboring nodes of removed trees/subtrees (Fig. 1c) that can be reprocessed in a different order than the original (Fig. 1d). Due to the strictly minor inequality of Line 12 of Algorithm 1, in the case of ties in offered costs, we have that the node that first sees its contested neighbor will win the dispute. Therefore, multiple processing orders affect the conquest of neighboring nodes (such as nodes c and f in Fig. 1).

The DIFT algorithms [6,8] do not attempt to address this issue, as they assume that the usage of the "$\leq$" comparison on Line 12 of Algorithm 1 would also be perfectly valid. However, in the case of functions such as $f_2^{\vec{\sigma}}$, in which the cost along the path is not a non-decreasing function, these problems in the processing order of frontier nodes are severely aggravated and can generate solutions that would never be obtained in the sequential flow. To resolve these issues, it would be necessary to explicitly store the processing order of the nodes, to ensure that later, the frontier nodes would be reprocessed in the same previous order. However, in addition to spending more memory, it would be complex to ensure the consistency in maintaining this new map of order over several iterations.

In order to address these issues without compromising the execution time of the algorithms, we chose to develop solutions for the differential OIFT focused only on the issue of generating segmentation labels that are consistent with the sequential flow labeling (consequently ensuring an optimal cut as in Theorem 1), without worrying about minor topology details of the resulting forest, that are irrelevant to the segmentation task.

The first proposed solution is simply to consider the usage of the Generalized DIFT (GDIFT) algorithm from [6] with the f_1^{σ} path-cost function. Note that f_1^{σ} is a function with non-decreasing costs along the path, with cost variations depending only on the root label and the arc weights $\omega(s,t)$ and $\omega(t,s)$, which perfectly fits the conditions required in [6]. Note that problems like the one reported in Fig. 1 do not occur with f_1^{σ}, since there are no cost ties between object and background in this formulation, as they are treated as odd and even numbers, respectively, and the background is always favored.

The second proposed solution is to use Algorithm 2, which considers for each path π_t a lexicographical path-cost function with two components $\langle F_2^{\sigma}(\pi_t), T(\pi_t) \rangle$, where $F_2^{\sigma}(\pi = \langle t_1, \ldots, t_n \rangle) = \max_{i=1,2,\ldots,n} \{ f_2^{\sigma}(\langle t_1, \ldots, t_i \rangle) \}$ and $T(\pi_t)$ is related to the number of maximum valued arcs crossed along the path, aiming at a better handling of tie zones, but we use odd numbers in $T(\pi_t)$ for paths from the background seeds and even numbers for the object, so that there are no ties in the second component between object and background.

Algorithm 2 – ALGORITHM DOIFT

INPUT: Image graph $G = \langle \mathcal{V}, \mathcal{A}, \omega \rangle$, the set Δ_S^+ of seeds for addition, set Δ_S^- of seeds for removal, the maps L, V and P initialized with the result from the previous OIFT/DOIFT execution, and an initial labeling function $\lambda : \Delta_S^+ \to \{0,1\}$ for the new seeds. We consider $V(t) = \langle V_1(t), V_2(t) \rangle$ as we work with lexicographical costs.

OUTPUT: The updated maps L, V and P.

AUXILIARY: Priority queue Q, and variables tmp_1 and tmp_2.

1. *Set $Q \leftarrow \varnothing$.*
2. **If** $\Delta_S^- \neq \varnothing$, **then**
3. $\lfloor$ $(L, V, P, Q) \leftarrow DOIFT\text{-}RemoveSubTrees(G, L, V, P, Q, \Delta_S^-)$
4. **For each** $s \in \Delta_S^+$, **do**
5. $\lfloor$ *Set $L(s) \leftarrow \lambda(s)$, $P(s) \leftarrow nil$, $V(s) \leftarrow \langle -1, L(s)+1 \rangle$*
6. $\lfloor$ **If** $s \notin Q$, **then** *insert s in Q.*
7. **While** $Q \neq \varnothing$, **do**
8.   *Remove s from Q such that $V(s) \overset{lex}{\leq} V(r)$ for all $r \in Q$.*
9.   **For each** *node t such that $\langle s,t \rangle \in \mathcal{A}$*, **do**
10.   *Compute $tmp_1 \leftarrow F_2^{\sigma}(\pi_s^P \cdot \langle s,t \rangle)$.*
11.   **If** $tmp_1 \neq f_2^{\sigma}(\pi_s^P \cdot \langle s,t \rangle)$, **then**
12. $\lfloor$ *Set $tmp_2 \leftarrow V_2(s)$.*

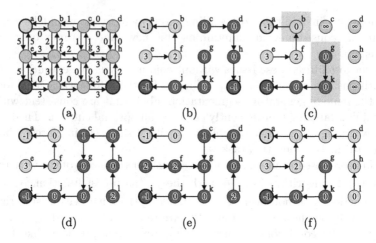

Fig. 1. (a) Input graph with marked seeds $\mathcal{S}_1^1 = \{a\}$ and $\mathcal{S}_0^1 = \{i, l\}$. (b) Initial computed forest by OIFT with $f_2^{\vec{\sigma}}$, assuming node b was processed first than node g. The values within the nodes indicate the costs of the paths and the arrows point to the predecessor of each node. (c) The tree of node l is marked for removal and its nodes are made available for a new dispute between the frontier nodes of neighboring trees (marked with a pink background). (d) A possible result of the differential flow, where the frontier node g was processed first than b, thus gaining c, but leading to a result that cannot be generated by the sequential flow via Algorithm 1. (e–f) The two possible outcomes of sequential flow for $f_2^{\vec{\sigma}}$ with $\mathcal{S}_1^2 = \{a\}$ and $\mathcal{S}_0^2 = \{i\}$.

13.   **Else If** $V_1(s) = tmp_1$, **then**
14.   └ *Set* $tmp_2 \leftarrow V_2(s) + 2$.
15.   **Else , then**
16.   └ *Set* $tmp_2 \leftarrow L(s) + 1$.
17.   **If** $\langle tmp_1, tmp_2 \rangle \overset{lex}{<} V(t)$, **then**
18.   │ *Set* $P(t) \leftarrow s$, $V(t) \leftarrow \langle tmp_1, tmp_2 \rangle$, $L(t) \leftarrow L(s)$.
19.   └ **If** $t \notin Q$, **then** *insert* t *in* Q.
20.   **Else If** $s = P(t)$, **then**
21.   │ **If** $tmp_1 \neq V_1(t)$ *or* $tmp_2 > V_2(t)$ *or* $L(t) \neq L(s)$, **then**
22.   │   │ $(L, V, P, Q) \leftarrow DOIFT\text{-}RemoveSubTrees(G, L, V, P, Q, \{t\})$
23.   └   └ **Break;** #GOTO LINE 8

Procedure DOIFT-RemoveSubTrees in Algorithm 3, releases the entire subtrees, converting its pixels to trivial trees of infinite cost, and transforms all of its neighboring pixels into frontier pixels, inserting them in Q, assuming that the graph is symmetric. It plays the role of both DIFT-RemoveSubTree and DIFT-TreeRemoval from [6], but has been modified to not rely on the use of a root map to save memory.

Algorithm 3 – PROCEDURE DOIFT-REMOVESUBTREES

INPUT: Image graph G, the maps L, V and P, the priority queue Q, and a set $\mathcal{R}$ of roots of the subtrees to be removed.
OUTPUT: The updated maps L, V and P, and the updated priority queue Q.
AUXILIARY: Queue J and a set $\mathcal{F}$.

1. Set $J \leftarrow \varnothing$, $\mathcal{F} \leftarrow \varnothing$.
2. **For each** $t \in \mathcal{R}$, **do**
3.   │   **If** $t \in Q$, **then** *remove t from Q.*
4.   │   *Set $V(t) \leftarrow \langle \infty, \infty \rangle$, $P(t) \leftarrow nil$.*
5.   └   *Insert t in J.*
6. **While** $J \neq \varnothing$, **do**
7.   │   *Remove s from J.*
8.   │   **For each** *node t such that* $\langle s, t \rangle \in \mathcal{A}$, **do**
9.   │   │   **If** $s = P(t)$, **then**
10.   │   │   │   *Insert t in J.*
11.   │   │   │   **If** $t \in Q$, **then** *remove t from Q.*
12.   │   │   └   *Set $V(t) \leftarrow \langle \infty, \infty \rangle$, $P(t) \leftarrow nil$.*
13.   │   │   **Else If** $V(t) \neq \langle \infty, \infty \rangle$ *and* $t \notin Q$, **then**
14.   └   └   └   *Insert t in $\mathcal{F}$.*
15. **While** $\mathcal{F} \neq \varnothing$, **do**
16.   │   *Remove t from $\mathcal{F}$.*
17.   │   **If** $V(t) \neq \langle \infty, \infty \rangle$ *and* $t \notin Q$, **then**
18.   └   └   *Insert t in Q.*

Other differences of Algorithm 2 in relation to GDIFT [6], are the absence of the state map used in [6], which proved to be unnecessary for functions with non-decreasing costs along the paths, as for the lexicographical cost $\langle F_2^{\vec{\sigma}}(\pi_t), T(\pi_t) \rangle$, and modifications to avoid using the root map to save memory. Another difference is the inclusion of Line 23 in Algorithm 2, to immediately break the innermost loop, thus avoiding the repeated processing of part of the neighborhood.

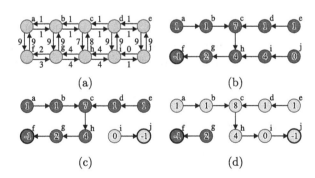

Fig. 2. (a) Input graph. (b) Initial forest by OIFT with $f_2^{\vec{\sigma}}$ for $\mathcal{S}^1 = \{f\}$. (c) The updated result by Algorithm 2, as a new object seed j is inserted, so that $\mathcal{S}^2 = \{f, j\}$. The values within nodes reflect the costs of $f_2^{\vec{\sigma}}$. (d) The correct result by Proposition 1.

The third proposed version of DOIFT is a variant of the second, modified so that disputed nodes with the same cost are given to the first processed neighbor, so as to respect Proposition 1, that will be defined next.

For any function $f(\pi)$, let $F(\pi)$ denote the maximum cost along the path:

$$F(\pi = \langle t_1, \ldots, t_n \rangle) = \max_{i=1,2,\ldots,n} \{f(\langle t_1, \ldots, t_i \rangle)\} \tag{6}$$

Consider the following lemma:

Lemma 1. *Let P be a predecessor map computed by Algorithm 1. For any two paths $\delta_t^P = \langle t_1, t_2, \ldots, t_n = t \rangle$ and $\tau_s^P = \langle s_1, s_2, \ldots, s_m = s \rangle$, defined by P, if $F(\delta_t^P) < F(\tau_s^P)$, then we have that node t was removed before s from Q on Line 7 of Algorithm 1.*

Proof. Let s_k be a node in τ_s^P, such that $f(\langle s_1, \ldots, s_k \rangle) = F(\tau_s^P)$. From Eq. 6, we have that $f(\langle t_1, \ldots, t_i \rangle) \leq F(\delta_t^P)$, $i = 1, 2, \ldots, n$. From the assumptions of Lemma 1, we may conclude that $F(\delta_t^P) < f(\langle s_1, \ldots, s_k \rangle)$. Thus, $f(\langle t_1, \ldots, t_i \rangle) < f(\langle s_1, \ldots, s_k \rangle)$, $i = 1, 2, \ldots, n$.

From the dynamic of execution of Algorithm 1, we know that paths δ_t^P and τ_s^P stored in the map P are gradually computed by the removal from Q of nodes with minimum cost (Line 7). After s_k gets inserted in Q with cost $V(s_k) = f(\langle s_1, \ldots, s_k \rangle)$, it won't be removed from Q before all nodes t_i, $i = 1, 2, \ldots, n$, are consecutively processed in Q, with lower costs $V(t_i) = f(\langle t_1, \ldots, t_i \rangle)$. Therefore, we have that $t = t_n$ is removed prior to s from Q.

From Lemma 1, we can also conclude the following proposition:

Proposition 1. *Let P be a predecessor map computed by Algorithm 1. For any two paths δ_s^P and $\tau_{s'}^P$, $s \neq s'$, defined in P, if $F(\tau_{s'}^P) < F(\delta_s^P)$ and $f(\delta_s^P \cdot \langle s, t \rangle) = f(\tau_{s'}^P \cdot \langle s', t \rangle)$, then we have that $\pi_t^P \neq \delta_s^P \cdot \langle s, t \rangle$.*

Proof. Algorithm 1 will assign t to the first optimum path that reaches it, because of the strict inequality in Line 12. According to Lemma 1, we have that s' leaves Q before s. Consequently, the path $\tau_{s'}^P \cdot \langle s', t \rangle$ is evaluated before $\delta_s^P \cdot \langle s, t \rangle$, offering the same cost (i.e., $f(\delta_s^P \cdot \langle s, t \rangle) = f(\tau_{s'}^P \cdot \langle s', t \rangle)$). Therefore, we have that π_t^P cannot be $\delta_s^P \cdot \langle s, t \rangle$.

Figure 2 discusses the consequences of Proposition 1 in the differential execution of OIFT. Note that Algorithm 2 does not satisfy Proposition 1. To correct this issue, the condition of Line 17 of Algorithm 2 must be changed to a much more complex condition:

$$tmp_1 < V_1(t) \text{ or } (tmp_1 = V_1(t) \text{ and } ((tmp_2 < V_2(t) \text{ and } not H_2) \text{ or } H_1))$$

where X, H_1 and H_2 are boolean variables defined as:

$$X \leftarrow V_1(t) = f_2^{\vec{\sigma}}(\pi_s^P \cdot \langle s, t \rangle) > V_1(s) \text{ and } V_1(t) > V_1(P(t))$$

$$H_1 \leftarrow P(t) \neq nil \text{ and } X \text{ and } V(s) \overset{lex}{<} V(P(t))$$

$$H_2 \leftarrow P(t) \neq nil \text{ and } X \text{ and } V(s) \overset{lex}{>} V(P(t)) \tag{7}$$

With these modifications, we have the third version of the DOIFT algorithm.

4 OIFT with Area/volume Constraint

Let $E_A = \max_{\mathcal{O} \in \mathcal{U}(A, \mathcal{S}_0)} E(\mathcal{O})$ denote the optimum energy value by Eq. 5 of a segmentation by OIFT using set A as internal seeds in Theorem 1. In order to introduce the idea of the incorporation of a size constraint in OIFT, we need first to establish some supporting propositions.

Proposition 2. *The optimum energy $E_{A \cup B}$ among all objects in $\mathcal{U}(A \cup B, \mathcal{S}_0)$, satisfies $E_{A \cup B} = \min\{E_A, E_B\}$.*

Proposition 3. *For a given strongly connected and symmetric digraph G, and sets of seeds $\mathcal{S}_1$ and $\mathcal{S}_0$, such that $\mathcal{S}_1 = \{t\}$ we have that $E_{\{t\}} = V_{opt}^{f_{max}^*}(t)$, where f_{max}^* is the path-cost function from Eq. 1, but being computed in the transpose graph and only from the external seeds in $\mathcal{S}_0$.*

The proofs of Proposition 2 and 3 are given in [12].

Suppose we want to define an optimal object of maximum energy via OIFT but having area/volume below a given threshold. Let's assume that the defined background must be connected to the originally selected background seeds. If the object has an area above the threshold, we can reduce its size by inserting new background seeds in its boundary. In order to apply Propositions 2 and 3, we can temporarily invert the object and background labels, in order to take advantage of the analogous and symmetrical problem. In this complementary problem, the energies of background nodes could be computed by the IFT with f_{max} from the object seeds $\mathcal{S}_1$ in the original graph. In order to get an optimal object, at each iteration we must then select a new background seed at the highest energy node of the object's boundary. We can then repeat this procedure until the area of the resulting object falls below the given threshold. We therefore have a sequence of OIFTs for each new seed inserted that can be calculated faster by DOIFT.

5 Experimental Results

Figure 3 shows the experimental curves for the segmentation of the talus bone using 40 slices from MR images of the foot using a robot user. In the first row, the arc weights were defined as $\omega(s,t) = |I(t) - I(s)|$ and with boundary polarity parameter defined as -50% (see [14]). In the second row, we repeat the experiment but with the arc weights quantized in a smaller range of values, corresponding to a quarter of the original range. $DOIFT_1$ and $OIFT$ (with f_2^{σ} and a heap priority queue) had a performance drop in the second case, due to their worse handling of tie zones. $DOIFT_2$ and $DOIFT_3$ had an accuracy performance consistent with the $OIFT$ with FIFO tie-breaking policy using f_2^{σ}. In case of $OIFT(FIFO)$, we considered a bucket sorting for Q and a binary heap was used for all other cases. Even using a slower queue Q, differential approaches were faster than $OIFT(FIFO)$ with the exception of the first iteration.

We also carried out experiments in a 3D MR image. We consider the accumulated time over the iterations of the automatic seed selection via the area

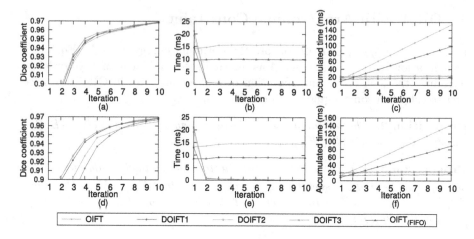

Fig. 3. The mean curves of accuracy, time, and accumulated time.

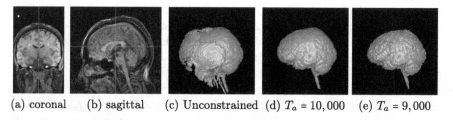

(a) coronal   (b) sagittal   (c) Unconstrained (d) $T_a = 10,000$ (e) $T_a = 9,000$

Fig. 4. Brain segmentation in MR images. Only three markers were selected in the indicated coronal slice.

procedure described in Sect. 4 to segment the brain. We considered a region adjacency graph of supervoxels by [16] with an average of 100 voxels per region. Figure 4 shows the results obtained for different values of the maximum volume threshold T_a, which is expressed in number of supervoxels. Regarding the execution time, for $T_a = 9,000$ we had 1 s for the differential flow by $DOIFT_2$ and 75 s for the sequential flow by OIFT with heap. For $T_a = 10,000$, we had 0.86 s for $DOIFT_2$ and 64 s for OIFT.

6   Conclusion

We have successfully tested different approaches to implement the differential OIFT and its use in implementing an area/volume constraint in OIFT. The use of area constraints can help to improve segmentation considerably, without the need to select multiple markers. As future works we intend to evaluate other applications for DOIFT and to create a hierarchy of OIFT segmentations by varying the area threshold.

References

1. Bejar, H.H., Miranda, P.A.: Oriented relative fuzzy connectedness: theory, algorithms, and its applications in hybrid image segmentation methods. EURASIP J. Image Video Process. **2015**(21) (2015)
2. Boykov, Y., Funka-Lea, G.: Graph cuts and efficient N-D image segmentation. Int. J. Comput. Vision **70**(2), 109–131 (2006)
3. Ciesielski, K., Udupa, J., Falcão, A., Miranda, P.: A unifying graph-cut image segmentation framework: algorithms it encompasses and equivalences among them. In: Proceedings of SPIE on Medical Imaging: Image Processing, vol. 8314 (2012)
4. Ciesielski, K., Udupa, J., Saha, P., Zhuge, Y.: Iterative relative fuzzy connectedness for multiple objects with multiple seeds. Comput. Vision Image Underst. **107**(3), 160–182 (2007)
5. Ciesielski, K.C., Falcão, A.X., Miranda, P.A.V.: Path-value functions for which Dijkstra's algorithm returns optimal mapping. J. Math. Imaging Vision **60**(7), 1025–1036 (2018)
6. Condori, M.A., Cappabianco, F.A., Falcão, A.X., Miranda, P.A.: An extension of the differential image foresting transform and its application to superpixel generation. J. Visual Commun. Image Represent. **71**, 102748 (2020)
7. Cousty, J., Bertrand, G., Najman, L., Couprie, M.: Watershed cuts: minimum spanning forests and the drop of water principle. IEEE Trans. Pattern Anal. Mach. Intell. **31**(8), 1362–1374 (2008)
8. Falcão, A.X., Bergo, F.P.: Interactive volume segmentation with differential image foresting transforms. IEEE Trans. Med. Imaging **23**(9), 1100–1108 (2004)
9. Falcão, A., Stolfi, J., Lotufo, R.: The image foresting transform: theory, algorithms, and applications. IEEE TPAMI **26**(1), 19–29 (2004)
10. Galvão, F.L., Falcão, A.X., Chowdhury, A.S.: RISF: recursive iterative spanning forest for superpixel segmentation. In: 31st SIBGRAPI Conference on Graphics, Patterns and Images (SIBGRAPI), pp. 408–415 (2018)
11. Leon, L.M., Ciesielski, K.C., Miranda, P.A.: Efficient hierarchical multi-object segmentation in layered graphs. Math. Morphol. Theory Appl. **5**(1), 21–42 (2021). https://doi.org/10.1515/mathm-2020-0108
12. Mansilla, L.A.C., Miranda, P.A.V., Cappabianco, F.A.M.: Oriented image foresting transform segmentation with connectivity constraints. In: 2016 IEEE International Conference on Image Processing (ICIP), pp. 2554–2558 (2016)
13. Mansilla, L., Miranda, P.: Image segmentation by oriented image foresting transform: Handling ties and colored images. In: 18th International Conference on Digital Signal Processing, Greece, pp. 1–6 (2013)
14. Miranda, P., Mansilla, L.: Oriented image foresting transform segmentation by seed competition. IEEE Trans. Image Process. **23**(1), 389–398 (2014)
15. de Moraes Braz, C., Miranda, P.A., Ciesielski, K.C., Cappabianco, F.A.: Optimum cuts in graphs by general fuzzy connectedness with local band constraints. J. Math. Imaging Vision **62**, 659–672 (2020)
16. Vargas-Muñoz, J.E., Chowdhury, A.S., Alexandre, E.B., Galvão, F.L., Miranda, P.A.V., Falcão, A.X.: An iterative spanning forest framework for superpixel segmentation. IEEE Trans. Image Process. **28**(7), 3477–3489 (2019)

Discrete Geometry - Models, Transforms, and Visualization

Tangential Cover for 3D Irregular Noisy Digital Curves

Phuc Ngo$^{(\boxtimes)}$ⓘ and Isabelle Debled-Rennesson

Université de Lorraine, CNRS, LORIA, 54000 Nancy, France
{hoai-diem-phuc.ngo,isabelle.debled-rennesson}@loria.fr

Abstract. This paper presents a discrete structure, named *adaptive tangential cover* (ATC), for studying 3D noisy digital curves. The structure relies mainly on the primitive of *blurred segment of width ν* and on the local noise estimator of *meaningful thickness*. More precisely, ATC is composed of maximal blurred segments of different widths deduced from the local noise values estimated at each point of the curve. Two applications of ATC for geometric estimators of 3D noisy digital curves are also presented in the paper. The experimental results demonstrate the efficiency of ATC for analyzing 3D irregular noisy curves.

Keywords: 3D digital curves · Noise estimator · Tangent and curvature estimators

1 Introduction

3D digital curves are often involved in many applications of 3D image processing and computer graphics. For instance, in [12], a system has been designed to model surfaces with collections of 3D curves. In [20], a sketch-based modeling method is proposed to reconstruct 3D curves and to reveal 3D shape information from typical design sketches. In applications of medical image processing [17], 3D digital curves are used for the analysis and can be obtained from a 3D curvilinear skeletonisation [4]. In those applications, the geometric characteristics of the curves play an important role for numerous purposes. In real context, the data present generally noise due to the acquisition process.

In digital geometry, new mathematical definitions of basic geometric objects are introduced to better fit the discrete/digital nature of data to process. In the context of 3D digital curve analysis, the notion of 3D *maximal digital straight segment* [2] has been used to describe the geometric properties of the curves. In particular, the sequence of all maximal segments along a digital curve $\mathcal{C}$, called the *tangential cover*, has been shown to be an efficient tool to study digital curves. Indeed, it is involved in numerous geometric estimators of digital curves: length [2], tangent [18], curvature [3] . . .

However, the tangential covers based on maximal segments are not adapted to noisy or disconnected curves. For this, the notion of *blurred segment of width ν* [5, 16] was proposed to deal with 3D digital curves containing noise or other sources

© Springer Nature Switzerland AG 2022
É. Baudrier et al. (Eds.): DGMM 2022, LNCS 13493, pp. 315–329, 2022.
https://doi.org/10.1007/978-3-031-19897-7_25

of imperfections from real data via the parameter ν. The sequence of maximal blurred segments of width ν along a digital curve is called a ν-*tangential cover*. This structure has been used in different contexts to study and to analyze the geometrical characteristics of noisy curves (*e.g.* [5,15,16]). Nevertheless, in these applications, the width ν needs to be manually adjusted to take into account the noise present on the curves. Furthermore, the structure is not well suited for noise which appears irregularly on the 3D curves.

Based on the works [13,14], the present paper studies a discrete framework for 3D digital curves containing *irregular* noise, and proposes improvements for the 3D approach. More precisely, we present a discrete structure, called *adaptive tangential cover* (ATC), which relies on primitive of blurred segment of width ν. The particularity of ATC is that it contains a sequence of blurred segments with different widths varying in function of noise present along the digital curve. Such adaptive widths are computed thanks to the local noise estimator of *meaningful thickness* [9]. In particular, the ATC can be computed in quasi-linear time and the method is parameter-free. We then show two applications of ATC to geometric estimators: tangent based on λ-MST [18] and curvature based on osculating circle [3] of 3D digital noisy curves. The experimental results show that the proposed method, based on the structure of ATC, is an efficient tool for studying noisy digital curves.

2   Background Notions

In the following, we define a digital curve $\mathcal{C} = (\mathcal{C}_i \in \mathbb{Z}^d)_{i=1..n}$, for $d = \{2,3\}$, as a sequence of discrete points. We then denote $\mathcal{C}_{i,j}$, with $1 \leq i \leq j \leq n$, the set of consecutive points from $\mathcal{C}_i$ to $\mathcal{C}_j$ of $\mathcal{C}$.

Let us first recall some definitions in 2D and then the extensions to 3D since, in this work, the definitions in 3D are computed from the 2D projections.

2.1   2D Blurred Segment of Width ν and Noise Detector

Definition 1 ([19]). *A **2D digital line**, with direction vector $(b,a) \in \mathbb{Z}^2$, $gcd(a,b) = 1$, shift $\mu \in \mathbb{Z}$ and thickness $\omega \in \mathbb{Z}_+$ is defined as the set of points $(x,y) \in \mathbb{Z}^2$ verifying*

$$\mu \leq ax - by < \mu + \omega \tag{1}$$

Such a line is denoted by $\mathcal{D}_2(a, b, \mu, \omega)$.

A *2D digital segment* is a finite and bounded subset of a 2D digital line (see Fig. 1(a)). From the primitive of digital segment, the notion of 2D blurred segment of width ν is proposed and allows for more flexibility in handling noisy data via the width parameter.

Let us consider $\mathcal{S}_2$ a sequence of discrete points of $\mathbb{Z}^2$.

Definition 2 ([6]). *A digital line $\mathcal{D}_2(a, b, \mu, \omega)$ is said to be **bounding** for $\mathcal{S}_2$ if all points of $\mathcal{S}_2$ belong to $\mathcal{D}_2$.*

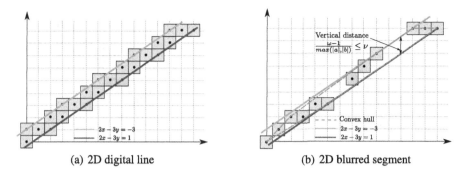

(a) 2D digital line   (b) 2D blurred segment

Fig. 1. Examples of (a) 2D digital line $\mathcal{D}_2(2,3,-3,5)$ and (b) 2D blurred segment of width $\nu = 1.5$ belongs to the optimal digital line $\mathcal{D}_2(2,3,-3,5)$.

Definition 3 ([6]). *A bounding digital line $\mathcal{D}_2(a,b,\mu,\omega)$ of $\mathcal{S}_2$ is said to be* **optimal** *if the value $\frac{\omega-1}{max(|a|,|b|)}$ is minimal, i.e. if its vertical (or horizontal) distance is equal to the vertical (or horizontal) thickness of the convex hull of $\mathcal{S}_2$.*

This definition is illustrated in Fig. 1(b), it leads to the definition of 2D blurred segment.

Definition 4 ([6]). *A sequence $\mathcal{S}_2$ is a* **2D blurred segment of width ν** *if its optimal bounding line has vertical/horizontal distance less than or equal to ν, i.e. $\frac{\omega-1}{max(|a|,|b|)} \leq \nu$.*

From this definition, a linear algorithm of blurred segment recognition is proposed in [16]. The method is based on an incremental growth of convex hull of the points sequentially added to the segment, and the calculation of its vertical and horizontal thickness.

From the primitive of digital straight segment, the noise detector of *meaningful scale* (MS) [7,8] is designed to locally estimate the best scale to analyze a digital curve. This detector is based on the study of the asymptotic properties of the discrete length L of maximal segments. In particular, it has been shown in [11] that the lengths of maximal segments covering a point p located on the boundary of a C^3 real object should be between $\Omega(1/h^{1/3})$ and $O(1/h^{1/2})$ if p is located on a strictly concave or convex part and near $O(1/h)$ elsewhere (where h represents the grid size). This theoretical property, defined on finer and finer grid sizes, was used by taking the opposite approach with the computation of the maximal segment lengths obtained with coarser and coarser grid sizes (from a subsampling process). From the graph of the maximal segment mean lengths $\overline{L}$, obtained at different scales, the noise estimator consists of recognizing the maximal scale for which the lengths follow the previous theoretical behavior.

The method has been extended to the detector of *meaningful thickness* (MT) [9] by using the blurred segment primitive with the scale definition given by the width of the blurred segment. Such a strategy presents the first advantage to be easier to implement without needing to apply different subsamplings and can be

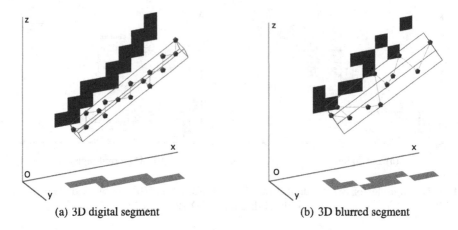

(a) 3D digital segment   (b) 3D blurred segment

Fig. 2. Examples of (a) 3D digital segment of $\mathcal{D}_3(20, -25, 8, -10, -6, 8, 6)$ and (b) 3D blurred segment of width $\nu = 2.5$ belongs to the optimal digital line $\mathcal{D}_3(10, -15, 3, -10, -6, 22, 20)$.

used for non-integer coordinate curves. In this work, we use the MT estimator since it allows to process disconnected curves which is suitable for our framework of studying digital noisy curves.

2.2   3D Blurred Segment of Width ν

Definition 5 ([2]). *A **3D digital line**, with main vector $(a, b, c) \in \mathbb{Z}^3$ such that $a \geq b \geq c > 0$ and $gcd(a, b, c) = 1$, with shifts $\mu, \mu' \in \mathbb{Z}$, and thicknesses $\omega, \omega' \in \mathbb{Z}_+$ is defined as the set of points $(x, y, z) \in \mathbb{Z}^3$ verifying*

$$\begin{cases} \mu \leq cx - az < \mu + \omega \\ \mu' \leq bx - ay < \mu' + \omega' \end{cases} \tag{2}$$

Such a line is denoted by $\mathcal{D}_3(a, b, c, \mu, \mu', \omega, \omega')$.

For the 3D digital lines of coefficients ordered different from $a \geq b \geq c > 0$, they can be obtained by permuting x, y, z as well as their coefficients. From Definition 5, a 3D digital line is bijectively projected into two projection planes as two 2D digital lines. A *3D digital segment* is a finite, bounded and connected subset of a 3D digital line (see Fig. 2(a)).

Definition 6 ([16]). *Let $\mathcal{S}_3$ be a sequence of points of $\mathbb{Z}^3$. $\mathcal{S}_3$ is a **3D blurred segment of width** ν, with main vector $(a, b, c) \in \mathbb{Z}^3$ and $a \geq b \geq c > 0$, if it has an optimal digital line $\mathcal{D}_3(a, b, c, \mu, \mu', \omega, \omega')$ such that*

- *$\mathcal{D}_2(a, b, \mu', \omega')$ is **optimal** for the sequence of projections of points $\mathcal{S}_3$ in the plane Oxy and $\frac{\omega'-1}{max(|a|,|b|)} \leq \nu$;*

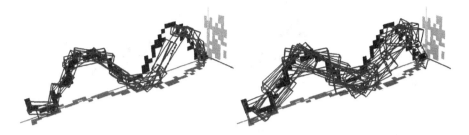

Fig. 3. Example of ν-tangential cover on a noisy digital curve: $\nu = 2$ (left) and $\nu = 4$ (right).

- $\mathcal{D}_2(a, c, \mu, \omega)$ *is **optimal** for the sequence of projections of points* $\mathcal{S}_3$ *in the plane* Oxz *and* $\frac{\omega - 1}{max(|a|, |c|)} \leq \nu$.

An illustration of 3D blurred segment is given in Fig. 2(b).

Let $\mathcal{C} \subset \mathbb{Z}^3$ be a 3D digital curve. Let $BS(i, j, \nu)$ denote the predicate "$\mathcal{C}_{i,j}$ is a blurred segment of width ν".

Definition 7 ([16]). $\mathcal{C}_{i,j} \subset \mathcal{C}$ *is called a **maximal blurred segment of width** ν, noted* $MBS(i, j, \nu)$, *iff* $BS(i, j, \nu)$, $\neg BS(i, j + 1, \nu)$ *and* $\neg BS(i - 1, j, \nu)$.

In [16], an incremental and linear algorithm is proposed to recognize a 3D maximal blurred segment (MBS) of width ν for a sequence of points. The main idea of the algorithm is to add simultaneously the 2D points in the projection planes as far as two of them are valid *i.e.*, we can add more point to the 2D MBS. Then, the 3D MBS is computed from the two corresponding 2D MBS of width ν projected onto the two basic planes. An algorithm for decomposing a digital curve $\mathcal{C}$ into 3D MBS of width ν is also presented in [16] with a complexity of $\mathcal{O}(n \log^2 n)$ for n the number of points in $\mathcal{C}$. The obtained structure is called (3D) ν-*tangential cover*. Some examples are given in Fig. 3.

Let $MBS_\nu(\mathcal{C}) = \{MBS_i(B_i, E_i, \nu)_{i=1..m}\}$ denote the ν-tangential cover of $\mathcal{C}$. By construction, we obtain the following property.

Property 1 ([16]). *Let* $MBS_\nu(\mathcal{C})$ *be the* ν-*tangential cover of* $\mathcal{C}$. *Then, we have* $MBS_\nu(\mathcal{C}) = \{MBS(B_1, E_1, \nu), MBS(B_2, E_2, \nu), \ldots, MBS(B_m, E_m, \nu)\}$ *such that* $B_1 < B_2 < \ldots < B_m$ *and* $E_1 < E_2 < \ldots < E_m$.

3   Adaptive Tangential Cover

The notion of ν-tangential cover has been proved to be an efficient tool to study digital noisy curves [5,15,16]. However, the parameter ν needs to be manually adjusted by the user. In this section, we present a 3D discrete structure, named *adaptive tangential cover* (ATC). It is composed of 3D MBS of different widths according to the amount of noise present in the curve. Before detailing the 3D ATC and its construction, let us first describe the blurred segment recognition algorithm used for building the 3D ATC.

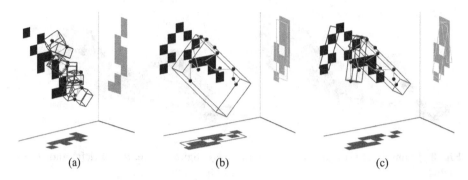

Fig. 4. Recognition of 3D blurred segment of width $\nu = 2$. (a) Result obtained when considering the bijectivity of the functional planes. (b) Result obtained when considering the two of the three projections are 2D blurred segments of width $\nu = 2$. (c) Result obtained when considering the three projections are 2D blurred segments of width $\nu = 2$ and relaxing the bijective condition.

3.1 Blurred Segment Recognition

It must be recalled that, by definition, a 3D digital segment is bijectively projected into two projection planes as two 2D digital segments. From this property, the algorithm for 3D straight segment recognition [2] is performed by considering the projections in the basic planes, namely Oxy, Oxz and Oyz. In particular, a 3D digital segment is said *valid* if **at least two** of its projections are 2D digital segments and the points bijectively projected on the two planes, these planes are called *functional planes*. An algorithm for decomposing a digital curve into 3D straight segments is as well proposed. We refer the readers to [2,5] for more details of the algorithm.

The recognition algorithm has been extended in [16] for the 3D blurred segment primitive by using the 2D blurred segment recognition in the projection planes. Similarly, a 3D blurred segment must be *valid* for at least two of its projections in the three basic planes together with the condition of two functional planes.

In this work, we are interested in 3D noisy digital curves, and in particular, the notion of 3D blurred segments of width ν for studying such curves. Although the 3D blurred segment recognition algorithm proposed in [16] works well for 3D noisy curves and its decomposition, we remarked various degeneracies that require us to modify the recognition algorithm. Firstly for the two functional planes, in case the amount of noise is important, this condition often fails, the algorithm therefore provides the short segments as illustrated in Fig. 4(a). This can bias the results of the geometric estimators on the curves. Secondly, a 3D blurred segment must have at least two of its projections being verified as 2D blurred segments. An example is given in Fig. 4(b) for 3D blurred segment of width $\nu = 2$. One can observe that the 3D segment has its two projections on the planes Oxy and Oxz being two valid 2D blurred segments. However, the obtained 3D blurred segment is far to what we would like to have as a segment.

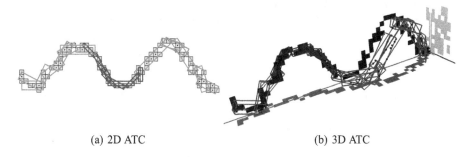

(a) 2D ATC   (b) 3D ATC

Fig. 5. Examples of ATC on 2D et 3D noisy digital curves.

More regular segments as in Fig. 4(c) would be more preferable and relevant for analyzing this sequence of points. It must be mentioned that these issues are generally due to the nature of noisy data which makes discrete points irregularly distributed along the curve.

To overcome these limitations of the method, we propose some modifications in the recognition algorithm of 3D blurred segment. In particular, we **relax the constraints of valid and functional planes**. More precisely, for a given digital curve $\mathcal{C} \subset \mathbb{Z}^3$, a sequence $\mathcal{C}_{i,j}$ of $\mathcal{C}$ is a 3D blurred segment of width ν iff the three projections of $\mathcal{C}_{i,j}$ on the basic planes are all 2D blurred segments of width ν. The characteristics of the optimal 3D digital line $\mathcal{D}_3(a, b, c, \mu, \mu', \omega, \omega')$ of $\mathcal{C}_{i,j}$ is then computed from the two projections of 2D blurred segments having the longest lengths (in term of Euclidean distance). The rest of the algorithm stays the same as in [16].

3.2 Adaptive Tangential Cover Construction

As previously mentioned, in the ν-tangential cover, the width parameter ν needs to be manually adjusted. Furthermore, the width ν is globally set for all MBS composing the ν-tangential cover. This approach works generally well when noise is uniformly distributed on the curve, but it is less relevant to local and irregular noisy curves.

In 2D, to overcome this issue of appropriate width ν, the structure of *adaptive tangential cover* (ATC) is introduced in [13,14]. Such a structure is designed to capture the local noise on a digital curve by adjusting the width of 2D MBS in accordance with the amount of irregular noise present along the curve. In other words, the ATC is composed of MBS with different widths varying in function of detected noise present in the curve. In particular, the MBS has bigger widths at noisy zones, and smaller widths in zones with less or no noise (see Fig. 5(a)).

Definition 8 ([13]). *Let $\mathcal{C} = (\mathcal{C}_i)_{1 \leq i \leq n}$ be a digital curve. Let $\eta = (\eta_i)_{1 \leq i \leq n}$ be the vector of noise level associated to each $\mathcal{C}_i$ of $\mathcal{C}$. Let $MBS(\mathcal{C}) = \{MBS_{\nu_k}(\mathcal{C})\}$*

be the set of ν_k-tangential covers for $\nu_k \in \eta$. An **adaptive tangential cover** of $\mathcal{C}$, noted $ATC(\mathcal{C})$, is the sequence of maximal blurred segments: $ATC(\mathcal{C}) = \{MBS_j = MBS(B_j, E_j, \nu) \in MBS(\mathcal{C}) \mid \nu = \max\{\eta_t \mid t \in [\![B_j, E_j]\!]\}\}$ such that $MBS_j \nsubseteq MBS_i$ for $i \neq j$, where $[\![a, b]\!]$ is the integer interval between a and b, including both.

Still in [13], an algorithm is proposed to build the ATC of a 2D digital curve $\mathcal{C}$. The algorithm is composed of two steps: (1) labeling the points with the noise level vector η and (2) building the $ATC(\mathcal{C})$ with MBS of width from the obtained labels.

It should be mentioned that the definition of ATC is given in 2D, but still valid for 3D. A primary extension in 3D of ATC is also presented in [13]. The method is however restricted to noisy curves $\mathcal{C} \subset \mathbb{Z}^3$ having **two functional planes**.

Regarding the noise vector η for each point of $\mathcal{C}$, the method in [13] performs separately the noise detection – with MT estimator – on the projected 2D curves in the two functional planes, and then chooses the maximum noise level to assign to each corresponding 3D point. Although this strategy works well for 3D noisy curves containing few noise (see Fig. 5(b)), it becomes less performance for important noisy cases. One common problem is that the high noise level is quickly propagated along the constructed ATC. As a consequence, it creates the segments of big widths encompassing the significant details of the input curve (see Fig. 6(a)). Furthermore, in general, the 3D curves do not always have only two functional planes. We refer the reader to [13] for more details. In this work, we keep the idea of the ATC algorithm in [13]. We will, however, make some changes in the method for a more efficient construction of 3D ATC, and allow to handle general curves – without the constraint of two functional planes. More precisely, the noise estimator is performed on the three projected curves in the basic planes. For a better local fitting of the input curve $\mathcal{C}$, **the median value is used, instead of the maximum** as in [13], for the noise associated to each 3D point of $\mathcal{C}$. Indeed, it allows to prevent the strong increment of the width of blurred segments when the noise becomes important. A comparison of the two strategies on a noisy curve is given in Fig. 6. We can observe that the 3D ATC with maximal values contains the segments of big widths where the noise is important. It causes the lost of the curve characteristic at certain places and may bias the geometric estimators of the curve. On the other hand, the 3D ATC with median allows a closer approximation of the curve.

In the following, we use the blurred segment recognition described in Sect. 3.1. The 3D ATC construction is given in Algorithm 1 with modifications explained previously. Examples of 3D ATC with the method are given in Fig. 7. Note that, by definition (Definition 8) and construction, the 3D ATC satisfies also Property 1.

In [16], it is shown that the ν-TC of a 3D digital curve $\mathcal{C}$ can be computed in $O(n \log^2 n)$ for n the number of points in $\mathcal{C}$. The construction of ATC is based on the ν-TC for different widths ν obtained for the MT estimator. The number ν-TC to be computed is equal to the size of ϑ – the number of noise levels present

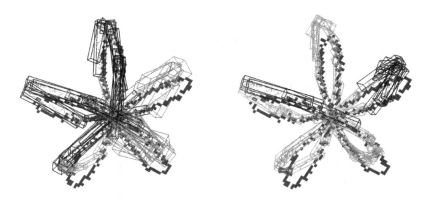

Fig. 6. Illustration of 3D ATC with the max (left) and median (right) value for combing noise.

Algorithm 1: Calculation of 3D adaptive tangential cover.

Input   : $\mathcal{C} = (\mathcal{C}_i)_{i=1\ldots n}$ input 3D digital curve,
                $\eta = (\eta_i)_{i=1\ldots n}$ noise level of each $\mathcal{C}_i$ with median MT of the
    projected planes
                $\vartheta = \{\nu_k \mid \nu_k \in \eta\}$ ordered set of η, and
                $MBS(\mathcal{C}) = \{MBS_{\nu_k}(\mathcal{C})\}_{k=1\ldots m}$ sets of MBS of $\mathcal{C}$ for each width
    value $\nu_k \in \vartheta$
Output   : $ATC(\mathcal{C})$ the 3D adaptive tangential cover of $\mathcal{C}$
Variable : $\gamma = (\gamma)_{i=1\ldots n}$ vector of labels to each $\mathcal{C}_i$

1 **begin**
2    $ATC(\mathcal{C}) = \emptyset$; $\gamma_i = \eta_i$ for $i \in [\![0, n-1]\!]$
   // Step 1: Label each point $\mathcal{C}_i$ of $\mathcal{C}$
3   **foreach** $\nu_k \in \vartheta$ **do**
4   **foreach** $MBS(B_i, E_i, \nu_k) \in MBS_{\nu_k}(\mathcal{C})$ **do**
5          $\alpha = \max\{\eta_i \mid i \in [\![B_i, E_i]\!]\}$
6   **if** $\alpha = \nu_k$ **then** $\gamma_i = \nu_k$ for $i \in [\![B_i, E_i]\!]$

   // Step 2: Calculate the ATC of $\mathcal{C}$ with MBS of width from γ
7   **foreach** $\nu_k \in \vartheta$ **do**
8   **foreach** $MBS(B_i, E_i, \nu_k) \in MBS_{\nu_k}(\mathcal{C})$ **do**
9   **if** $\exists \gamma_i$, *for* $i \in [\![B_i, E_i]\!]$, *such that* $\gamma_i = \nu_k$ **then**
10          $ATC(\mathcal{C}) = ATC(\mathcal{C}) \cup \{MBS(B_i, E_i, \nu_k)\}$

---

in $\mathcal{C}$ – and $|\vartheta| \ll n$. In other words, the complexity of Algorithm 1 for computing a 3D ATC is also $O(n \log^2 n)$.

4   Applications: Tangent and Curvature Estimators

Geometric properties of curves are important characteristics to be exploited in geometry processing. In particular, tangent and curvature are among the important properties to describe a curve. In this section, we present two applications of ATC for geometric estimators at each point of 3D noisy digital curves. The ATC can be used to improve the precision of existing tangent and curvature estimators for irregular noisy curves.

To evaluate the proposed estimators, we consider two following curves: Flower and Astroid. The first curve is a *smooth* curve with nearly the same mean-error along the curve, while the second is a *sharp 4-pointed* curve. These two curves are defined, respectively, by the following parametric equations:

$$\begin{cases} x = 2\cos(t)\cos(5t) \\ y = 2\sin(t)\cos(5t) \quad (3) \\ z = \cos^2(5t) \end{cases} \quad \text{and} \quad \begin{cases} x = \cos^3(t) \\ y = \sin^3(t) \quad (4) \\ z = t \end{cases}$$

These curves are discretized using the class `NaiveParametricCurve` `Digitizer3D` of DGtal [1] to obtain 3D connected curve $\mathcal{C}$. To generate noise for $\mathcal{C}$, a random process with a uniform distribution is considered to change one coordinate of points of $\mathcal{C}$ one unit, ± 1. Thirty curves with random noise are generated for the error measurements between the theoretical tangent/curvature and the estimated ones on these curves.

4.1 Discrete Tangent

In [18], a discrete tangent estimator, called λ-*maximal segment tangent* (λ-MST), has been proposed for *regular* digital curves. It is based on the tangential cover with maximal straight segments. The method is an extension of the algorithm presented in [11] for 2D curves. In particular, the estimator is a simple parameter-free method and has multi-grid convergent properties [10,11].

The λ-MST can be easily applied to ATC with blurred segment primitive, and thus allows to handle noise. More precisely, the tangent at $\mathcal{C}_k \in \mathcal{C}$ is obtained via the definition of a *pencil of maximal blurred segments* $\mathcal{P}(\mathcal{C}_k) = \{MBS_i = MBS(B_i, E_i, .), \text{with } B_i \leq k \leq E_i\}$. From this, a notion of eccentricity was introduced in order to distribute weights on all the segments covering $\mathcal{C}_k$. More formally, the eccentricity is defined as

$$e_i(\mathcal{C}_k) = \begin{cases} ||\mathcal{C}_k - \mathcal{C}_{E_i}||_2^2/L_i & \text{if } MBS_i \in \mathcal{P}(\mathcal{C}_k) \\ 0 & \text{otherwise} \end{cases} \text{, with } L_i = ||\mathcal{C}_{E_i} - \mathcal{C}_{B_i}||_2^2 \quad (5)$$

Then, the tangent direction $\mathbf{t}(\mathcal{C}_k)$ at $\mathcal{C}_k$ is computed as a weighted combination of the direction vectors $\mathbf{t}_i$ of the maximal blurred segments $MBS_i \in \mathcal{P}(\mathcal{C}_k)$ covering $\mathcal{C}_k$ by

$$\mathbf{t}(\mathcal{C}_k) = \frac{\sum_{MBS_i \in \mathcal{P}(\mathcal{C}_k)} \lambda(e_i(\mathcal{C}_k)) \frac{\mathbf{t}_i}{||\mathbf{t}_i||_2}}{\sum_{MBS_i \in \mathcal{P}(\mathcal{C}_k)} \lambda(e_i(\mathcal{C}_k))} \quad (6)$$

where λ is a mapping function defined from $[0, 1]$ to $\mathbb{R}^+$ such that $\lambda(0) = \lambda(1) = 0$ and $\lambda > 0$ elsewhere and λ needs to satisfy convexity/concavity property [11]. In this paper, the C^2 function $\lambda(x) = 64(-x^6 + 3x^5 - 3x^4 + x^3)$ is used. Furthemore, thanks to Property 1, the pencil $\mathcal{P}(\mathcal{C}_k)$, for $\mathcal{C}_k \in \mathcal{C}$, can be easily computed from $ATC(\mathcal{C})$.

Figure 7 shows a visual result of the tangent estimator λ-MST with ATC on two noisy curves. Further results are shown in Table 1 in which we compare the λ-MST estimator using ATC and ν-tangential covers with $\nu = 1...5$ (as there are 5 noise levels detected in the thirty input curves). Different error measures are considered: mean and maximal error and standard deviation. The result shows that the combination of λ-MST with ATC improves globally tangent estimation on digital noisy curves; it has the best mean error measures, and the other measures are every closed to the best ones as well.

Table 1. Error measures of tangent estimator λ-MST on the digitized curves of Flower (Eq. 3) and Astroid (Eq. 4) with random noise added.

Curves		Error of estimated tangents					
		ATC_{MT}	1-TC	2-TC	3-TC	4-TC	5-TC
Flower	Mean error	**0.220012**	0.369261	0.222374	0.283572	0.320565	0.343145
	Std. dev.	0.214158	0.250986	**0.193687**	0.211761	0.225935	0.23631
	Max error	1.26362	1.26657	**1.25397**	1.29662	1.29617	1.30408
Astroid	Mean error	**0.293318**	0.379116	0.295132	0.302137	0.309408	0.344594
	Std. dev.	**0.216065**	0.285028	0.234184	0.222423	0.231388	0.24023
	Max error	1.06834	1.22489	1.13035	1.06979	**1.04051**	1.05925

4.2 Discrete Curvature

In the study of geometric characteristics of 3D digital curves, Coeurjolly and Svensson have proposed in [3] the calculation of discrete curvature based on the osculating circle with maximal straight segments. Inspired by this idea, Nguyen and Debled-Rennesson have presented in [15] the discrete curvature estimator using 3D maximal blurred segments. The proposed method can be naturally applied to the structure of 3D ATC. More precisely, from the 3D ATC of an input curve $\mathcal{C} = (\mathcal{C}_i)_{i=1...n}$, we compute the pencil $\mathcal{P}(\mathcal{C}_k)$ for each point $\mathcal{C}_k \in \mathcal{C}$. Let $\mathcal{C}_l$ (resp. $\mathcal{C}_r$) be the left (resp. right) end point of $\mathcal{P}(\mathcal{C}_k)$. The discrete curvature at $\mathcal{C}_k$ is estimated using the radius $R_\nu(\mathcal{C}_k)$ of the osculating circle of the triangle formed by three points $\mathcal{C}_l$, $\mathcal{C}_k$ and $\mathcal{C}_r$

$$R_\nu(\mathcal{C}_k) = \frac{s_1 s_2 s_3}{\sqrt{(s_1 + s_2 + s_3)(s_1 + s_2 - s_3)(s_1 - s_2 + s_3)(s_2 + s_3 - s_1)}} \qquad (7)$$

with $s_1 = \|\overrightarrow{\mathcal{C}_k \mathcal{C}_r}\|$, $s_2 = \|\overrightarrow{\mathcal{C}_k \mathcal{C}_l}\|$ et $s_3 = \|\overrightarrow{\mathcal{C}_l \mathcal{C}_r}\|$. The curvature of $\mathcal{C}_k$ is then calculated by

$$c_\nu(\mathcal{C}_k) = \frac{s}{R_\nu(\mathcal{C}_k)} \qquad (8)$$

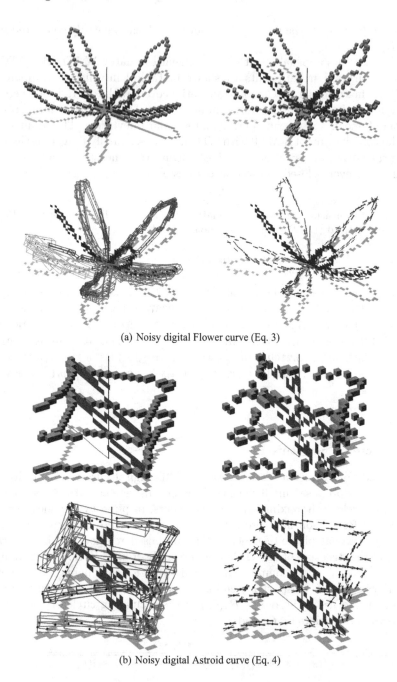

(a) Noisy digital Flower curve (Eq. 3)

(b) Noisy digital Astroid curve (Eq. 4)

Fig. 7. Tangent estimator λ-MST with ATC on noisy digital curves. In each figure (a) and (b), the ATC of the input curve is on the left, the result of estimated tangents is on the right: red segments are expected theoretical tangents and blue segments are estimated tangents. (Color figure online)

where $s = signe(det(\overrightarrow{\mathcal{C}_k\mathcal{C}_r}, \overrightarrow{\mathcal{C}_k\mathcal{C}_l}))$ indicating the concavity or convexity of $\mathcal{C}$ at $\mathcal{C}_k$.

Table 2 presents the results of the proposed curvature estimator using 3D ATC and ν-tangential covers with $\nu = 1...5$ on the error measures of mean and maximal error and standard deviation. The estimator with ATC does not always give the best result but we obtain fairly accurate curvature estimation comparing to the ν-tangential covers.

Table 2. Error measures of curvature estimator on the digitized curves of Flower (Eq. 3) and Astroid (Eq 4) with random noise added.

Curves		Error of estimated curvatures					
		ATC_{MT}	1-TC	2-TC	3-TC	4-TC	5-TC
Flower	Mean error	0.042239	0.192657	0.043795	**0.041935**	0.045646	0.048014
	Std. dev.	0.118326	0.173881	**0.108671**	0.125055	0.130469	0.131886
	Max error	1.05404	1.906708	**0.977172**	1.08214	1.11167	1.11805
Astroid	Mean error	0.026844	0.202735	0.043616	**0.024496**	0.024548	0.028414
	Std. dev.	**0.015149**	0.176366	0.058435	0.016437	0.016609	0.017491
	Max error	0.439742	0.828724	0.349773	0.08767	**0.078456**	0.078618

5 Conclusion

From the studies in [13,14], this paper presents different improvements of the structure of *adaptive tangential cover* (ATC) for 3D noisy digital curves. ATC is a discrete structure composed of a sequence of maximal blurred segment of width ν adjusted according to the amount of noise present on the curve. Due to the nature of noisy data, 3D blurred segment recognition is also modified by considering all three projection planes. From this, a parameter-free and quasi-linear algorithm is proposed to compute the structure of ATC.

Two applications of ATC are also presented for tangent and curvature estimators of noisy curves. The proposed structure opens numerous perspectives for studying and analyzing 3D noisy curves such as 3D curve segmentation or approximation, convex and concave detection, dominant point detection in 3D ... The source code of the proposed method for constructing 3D ATC, based on DGtal library [1], is available at the *GitHub* repository: https://github.com/ngophuc/ATC_3D.

In further works, we also would like to study the aspect of multi-grid convergent estimators of 3D noisy digital curves using the proposed structure of ATC. Furthermore, we would like to elaborate a more efficient computation of ATC without calculation of all ν-tangential covers. Other perspective consists of studying a 3D noise estimator for ATC instead of using the combination of MT estimators on the three projection planes.

Acknowledgment. The authors would like to thank Hugo Ambrozik for his work during a master internship at LORIA which motivated the writing of this article.

References

1. DGtal: Digital Geometry tools and algorithms library. http://libdgtal.org
2. Coeurjolly, D., Debled-Rennesson, I., Teytaud, O.: Segmentation and length estimation of 3D discrete curves. In: Bertrand, G., Imiya, A., Klette, R. (eds.) Digital and Image Geometry. LNCS, vol. 2243, pp. 299–317. Springer, Heidelberg (2001). https://doi.org/10.1007/3-540-45576-0_18
3. Coeurjolly, D., Svensson, S.: Estimation of curvature along curves with application to fibres in 3D images of paper. In: Bigun, J., Gustavsson, T. (eds.) SCIA 2003. LNCS, vol. 2749, pp. 247–254. Springer, Heidelberg (2003). https://doi.org/10.1007/3-540-45103-X_34
4. Couprie, M., Bertrand, G.: New characterizations of simple points in 2D, 3D and 4D discrete spaces. IEEE PAMI **31**(4), 637–648 (2009)
5. Debled-Rennesson, I.: Eléments de géométrie discrète vers une etude des structures discrètes bruitées,: habilitation à Diriger des Recherches. Université Henri Poincaré - Nancy I, France (2007)
6. Debled-Rennesson, I., Feschet, F., Rouyer-Degli, J.: Optimal blurred segments decomposition of noisy shapes in linear time. Comput. Graph. **30**(1), 30–36 (2006)
7. Kerautret, B., Lachaud, J.O.: Meaningful scales detection along digital contours for unsupervised local noise estimation. IEEE PAMI **34**(12), 2379–2392 (2012)
8. Kerautret, B., Lachaud, J.O.: Meaningful scales detection: an unsupervised noise detection algorithm for digital contours. Image Process. Line **4**, 98–115 (2014)
9. Kerautret, B., Lachaud, J.O., Said, M.: Meaningful thickness detection on polygonal curve. In: Pattern Recognition Applications and Methods, pp. 372–379 (2012)
10. Lachaud, J.-O., Vialard, A., de Vieilleville, F.: Analysis and comparative evaluation of discrete tangent estimators. In: Andres, E., Damiand, G., Lienhardt, P. (eds.) DGCI 2005. LNCS, vol. 3429, pp. 240–251. Springer, Heidelberg (2005). https://doi.org/10.1007/978-3-540-31965-8_23
11. Lachaud, J.O., Vialard, A., de Vieilleville, F.: Fast, accurate and convergent tangent estimation on digital contours. Image Vision Comput. **25**(10), 1572–1587 (2007)
12. Nealen, A., Igarashi, T., Sorkine, O., Alexa, M.: Fibermesh: designing freeform surfaces with 3D curves. ACM Trans. Graph. **26**(3), 41 (2007)
13. Ngo, P., Debled-Rennesson, I., Kerautret, B., Nasser, H.: Analysis of noisy digital contours with adaptive tangential cover. J. Math. Imaging Vis. **59**(1), 123–135 (2017)
14. Ngo, P., Nasser, H., Debled-Rennesson, I., Kerautret, B.: Adaptive tangential cover for noisy digital contours. In: Normand, N., Guédon, J., Autrusseau, F. (eds.) DGCI 2016. LNCS, vol. 9647, pp. 439–451. Springer, Cham (2016). https://doi.org/10.1007/978-3-319-32360-2_34
15. Nguyen, T.P., Debled-Rennesson, I.: Curvature and torsion estimators for 3D curves. In: Bebis, G., et al. (eds.) ISVC 2008. LNCS, vol. 5358, pp. 688–699. Springer, Heidelberg (2008). https://doi.org/10.1007/978-3-540-89639-5_66
16. Nguyen, T.P., Debled-Rennesson, I.: On the local properties of digital curves. Int. J. Shape Model. **14**(2), 105–125 (2008)
17. Postolski, M., et al.: Reliable airway tree segmentation based on hole closing in bronchial walls, vol. 57, pp. 389–396 (2009)

18. Postolski, M., Janaszewski, M., Kenmochi, Y., Lachaud, J.O.: Tangent estimation along 3D digital curves. In: International Conference on Pattern Recognition, pp. 2079–2082 (2012)

19. Reveillès, J.P.: Géométrie discrète, calculs en nombre entiersgorithmique, et al.: thèse d'état. Université Louis Pasteur, Strasbourg (1991)

20. Xu, B., Chang, W., Sheffer, A., Bousseau, A., McCrae, J., Singh, K.: True2form: 3d curve networks from 2D sketches via selective regularization. ACM Trans. Graph. **33**(4) (2014)

A Curious Invariance Property of Certain Perfect Legendre Arrays: Stirring Without Mixing

Timothy Petersen[1]([✉]), Matthew Ceko[2], David Paganin[2], and Imants Svalbe[2]

[1] Monash Centre for Electron Microscopy, Monash University, Melbourne, VIC 3800, Australia
timothy.petersen@monash.edu.au
[2] School of Physics and Astronomy, Monash University, Melbourne, VIC 3800, Australia

Abstract. The persistent interest in applications for Legendre arrays is motivated by their delta-like periodic autocorrelation property and their therefore perfectly flat-magnitude Fourier spectrum. Applying discrete affine transforms scrambles such perfect arrays by a random shuffling of their element locations, whilst preserving their original delta-like autocorrelation. We show here that certain slightly different, but new Legendre arrays, under quite general affine rotations, always map exactly back onto themselves, with a sign change of ± 1. These arrays, here called L_p, have 4-fold symmetry and are invariant to discrete rotations. We prove this invariance property is general for L_p arrays of prime size $p \times p$, where $p = 4N - 1$ for integer N. However, rotational invariance in L_p where $p = 4N + 1$ occurs for a restricted range of discrete angles. These arrays are also equal to their own discrete Fourier transform, for a standard Fourier normalisation. Each invariant rotation 'randomly' stirs the locations of all the array elements, whilst never mixing the positions of those elements of L_p with opposite sign.

Keywords: Rotational invariance · Prime sized perfect autocorrelation arrays · Discrete Radon projection

1 Constructing Rotation Invariant Legendre Arrays

In cryptographic or communication applications [1–4], the use of affine transforms to 'randomise' a seed array may be used to improve its security to avoid unwanted intrusion, whilst preserving the favourable array correlation metrics [5]. Otherwise, affine mappings can be used to extend a family of arrays where each array has equally strong autocorrelation and where all members have equally low cross-correlation [4].

A classic Legendre sequence [6], with elements L(i), of prime length p, assigns the value 0 to its first element, at i = 0. Then, for $1 \leq i \leq (p - 1)$, L(i) assigns $+1$ to the $(p - 1)/2$ element locations i that *are* quadratic residues modulo p, and -1 to the $(p - 1)/2$ element locations i that are *not* quadratic residues modulo p.

The new $p \times p$ Legendre arrays L_p, for prime p, presented here are (up to a change of sign), invariant to any affine rotation r : s, for any $p = 4N - 1$, where r,s and N are positive integers. For $p = 4N + 1$, the invariance applies only for rotations r : s

© Springer Nature Switzerland AG 2022
É. Baudrier et al. (Eds.): DGMM 2022, LNCS 13493, pp. 330–340, 2022.
https://doi.org/10.1007/978-3-031-19897-7_26

where $(r^2 + s^2)$ modp is not zero. Methods are given here to construct these transpose-symmetric and perfect $p \times p$ Legendre arrays in real space, via their Fourier phase in the frequency domain, or directly through the periodic discrete Finite Radon Transform (FRT) projections of L_p. The invariance of the L_p arrays here under discrete Fourier transformation (FFT) mirrors that of the constituent Legendre sequences, as has been observed in prior 2D perfect array constructions [7]. An example for $p = 11$ is shown in Fig. 1.

$$
\begin{pmatrix}
0 & 1 & 1 & 1 & 1 & 1 & 1 & 1 & 1 & 1 & 1 \\
1 & -1 & 1 & -1 & -1 & 1 & 1 & -1 & -1 & 1 & -1 \\
1 & 1 & -1 & -1 & 1 & -1 & -1 & 1 & -1 & -1 & 1 \\
1 & -1 & -1 & -1 & 1 & 1 & 1 & 1 & -1 & -1 & -1 \\
1 & -1 & 1 & 1 & -1 & -1 & -1 & -1 & 1 & 1 & -1 \\
1 & 1 & -1 & 1 & -1 & -1 & -1 & -1 & 1 & -1 & 1 \\
1 & 1 & -1 & 1 & -1 & -1 & -1 & -1 & 1 & -1 & 1 \\
1 & -1 & 1 & 1 & -1 & -1 & -1 & -1 & 1 & 1 & -1 \\
1 & -1 & -1 & -1 & 1 & 1 & 1 & 1 & -1 & -1 & -1 \\
1 & 1 & -1 & -1 & 1 & -1 & -1 & 1 & -1 & -1 & 1 \\
1 & -1 & 1 & -1 & -1 & 1 & 1 & -1 & -1 & 1 & -1
\end{pmatrix}
\begin{pmatrix}
-1 & -1 & -1 & -1 & -1 & -1 & -1 & -1 & -1 & -1 & -1 \\
-1 & -1 & -1 & -1 & -1 & -1 & -1 & -1 & -1 & -1 & -1 \\
-1 & -1 & -1 & -1 & -1 & -1 & -1 & -1 & -1 & -1 & -1 \\
-1 & -1 & -1 & -1 & -1 & -1 & -1 & -1 & -1 & -1 & -1 \\
-1 & -1 & -1 & -1 & -1 & -1 & -1 & -1 & -1 & -1 & -1 \\
-1 & -1 & -1 & -1 & -1 & 120 & -1 & -1 & -1 & -1 & -1 \\
-1 & -1 & -1 & -1 & -1 & -1 & -1 & -1 & -1 & -1 & -1 \\
-1 & -1 & -1 & -1 & -1 & -1 & -1 & -1 & -1 & -1 & -1 \\
-1 & -1 & -1 & -1 & -1 & -1 & -1 & -1 & -1 & -1 & -1 \\
-1 & -1 & -1 & -1 & -1 & -1 & -1 & -1 & -1 & -1 & -1 \\
-1 & -1 & -1 & -1 & -1 & -1 & -1 & -1 & -1 & -1 & -1
\end{pmatrix}
$$

Fig. 1. L_{11} displayed as a matrix (left). The 2D FFT of L_{11} is exactly the same as the array L_{11}. The delta-like periodic autocorrelation is shown on the right.

This paper is laid out as follows: Sect. 1.1 describes Legendre sequences in terms of quadratic residues, as necessary background for the construction of 2D Legendre arrays in Sect. 1.2, including their affine rotations and projections. Section 1.3 defines discrete rotations through an affine transform, with the rotation invariance detailed in Sect. 1.4. Alternative FRT and Fourier constructions of 2D Legendre arrays are described in Sect. 1.5. Section 2.1 presents examples of the new arrays constructed here and Sect. 2.2 shows how the invariant affine rotations shuffle signed elements without mixing them. The Appendix shows that the FRT construction method can be used to make pairs or sets of arrays that have close-to-perfect autocorrelation but have exactly zero cross-correlation.

1.1 Legendre Sequences and Quadratic Residues

Quadratic residues provide a simple way to build Legendre sequences of length p. Sets of quadratic residues, Q_p, are defined as

$$
Q_p = \left\{ q_p(i) \right\}, \text{ where } q_p(i) = i^2 \bmod p, \{0 \le i \le p-1\}. \tag{1}
$$

Clearly $i = 0$ is a special case. By definition, $q_p(p - i) = q_p(i)$. For odd primes p, there are $(p + 1)/2$ unique elements in Q_p, including 0, with symmetric values for i and $(p - i)$. This symmetry ensures the remaining $(p - 1)/2$ integers $\underline{q}_p(i)$ are $\underline{not}$ members of the set Q_p, but instead form the complementary set $\underline{Q}_p$ of non-quadratic residue integers over the interval 1 to $(p - 1)$; then $Q_p + \underline{Q}_p = \{0 : p - 1\}$. The sum over all p Legendre sequence element values is then 0. The Legendre symbol, $(i \backslash p) \in \{0, 1, -1\}$ is often used as shorthand for $q_p(i)$ and $\underline{q}_p(i)$.

The periodic autocorrelation of L, written here as L⊗L, is delta-like, with peak value $(p - 1)$, all $(p - 1)$ off-peak values being constant, at -1, and zero mean. One measure to quantify the delta-correlation of a sequence is by the 'merit factor' [5], defined as the square of the correlation peak over the sum of all squared off-peak correlation elements (infinite for perfect arrays but finite for L). The symmetry of the element assignments in Q_p, and hence L, mirrors the periodic structure of the Fourier frequencies, f_k. The sequence L is Nyquist bandwidth limited over $f_k = 0$ to $(p - 1)/2$, with complex conjugate symmetry, where $F(k) = F(p - k)^*$ for real valued sequences. The (normalised) Fourier transform magnitudes, $|F_k|$, of any prime length Legendre sequence are then $F_0 = 0$, and $|F_k| = \sqrt{p}$ for $1 \le k \le (p - 1)/2$. Their flat Fourier spectrum, $|F_k|$, being constant for $k > 0$, is commensurate with the delta-like autocorrelation, and endows L with the epithet as being one member of a diverse class known as 'perfect' periodic sequences [8] (with zero or constant off-peak periodic correlation elements).

It is peripheral, but noteworthy, that the *aperiodic* autocorrelation of L is not 'perfect', having non-uniform off-peak values that vary in size for sequences that are cyclic shifts of the original sequence. The closest-to-perfect aperiodic autocorrelation metrics for (zero-padded) Legendre sequences occur when the original sequence of length p is cyclically shifted by round $(p/4)$ elements [6]. In contrast, Huffman sequences [9] have the optimal delta-like autocorrelation metrics under aperiodic conditions, and are as flat spectrally as is possible for any finite length sequences [10].

1.2   2D Legendre Arrays

A 2D Legendre array of prime size $p \times p$ can be constructed by forming the outer product (Kronecker tensor product) of a 1D Legendre sequence L with the transpose of the same sequence. The leading zero of Legendre sequences results in the first row and column of the tensor product consisting only of zeroes. The array value at location $(0, 0)$ is left as zero. To maintain the zero sum of L, each of the resulting zero-valued elements from $(1 : p - 1)$ along the first row need to be set to $+1$ and the same elements in the first column need to be set to -1 (or vice-versa). Given the symmetry of the quadratic residues between points (i, j) and $(p - i, p - j)$, all 2D Legendre arrays have $180°$ rotational symmetry.

An alternative, slightly different, but new Legendre array, termed L_p in this paper, of size $p \times p$, can be constructed by assigning $+1$ to element locations (i, j) where $(i^2 + j^2)$ mod p is a quadratic residue ($q_p(i) \in Q_p$), and assigned value -1 if not ($q_p(i) \in Q_p$). The location $(0, 0)$ retains the value 0. The elements $1 : (p - 1)$ along the first row and first column are now all assigned the value $+1$ (as $i^2 + 0^2$ is always $q_p(i)$). This slightly different array L_p is transpose symmetric. Arrays made this way using $p = 4N - 1$ (for any positive integer N) have zero sum and now possess 4-fold rotational symmetry. The exact 4-fold symmetry of L_p becomes geometrically evident when the array is circularly shifted to locate the sole 0-valued entry at the centre of the $p \times p$ array.

For primes $p = 4N+1$, to construct zero-sum 'perfect' arrays using the latter method, all of the elements from $1 : (p - 1)$ on either the first column or the first row must be reset to -1 (as done for the outer product method). Those arrays then lose their transpose symmetry. The transpose symmetry can be re-imposed by applying a discrete rotation of 45° (through an affine rotation by 1 : 1, as reviewed in Sect. 1.3). However, those arrays no longer possess 4-fold symmetry. Section 1.5 shows that the 4-fold symmetry of the perfect arrays for $p = 4N + 1$ can be retained by the compromise of relaxing the zero-sum requirement.

1.3 Discrete Affine Rotations

An affine transform, R, applied to image pixels or data $A(x, y)$, reversibly maps the homogeneous coordinates $v = (x, y, 0)$ of A to new locations $v' = (x', y', 0)$ in A' via the matrix product $v' = R[v]$,

$$\begin{pmatrix} x' \\ y' \\ 0 \end{pmatrix} = \begin{pmatrix} a & b & x_0 \\ c & d & y_0 \\ 0 & 0 & 1 \end{pmatrix} \begin{pmatrix} x \\ y \\ 0 \end{pmatrix}. \qquad (2)$$

For a prime-sized 2D array, the six 'free' transform values a, b, c, d and the (cyclic shift) displacements x_0, y_0 that together make up the transform matrix R, contain integers modulo p. Applying the mapping R^{-1}, the inverse of matrix R, (again, with values modulo p), reverses (exactly) the initial transform, so that each location in $A'(x', y')$ maps back to $A(x, y)$ in its original place.

For 'pure' affine rotations by discrete (integer-based) directions r : s, then R has $a = s$, $b = -r$, $c = r$ and $d = s$, with $x_0 = y_0 = 0$. Then the determinant of the matrix R, is $\det(R) = r^2 + s^2$. When $\det(R) = 0$, modulo p, the mapping is null and the transformed array is void.

As p is an odd integer, for any pair of integers r and s such that $r^2 + s^2 = \alpha p$, then p must satisfy $p = 4N + 1$, for positive integers α and N. Importantly, when $p = 4N - 1$, $\det(R)$ modulo p cannot be zero for any affine rotation r : s of a p × p array. The rotation integers r : s need not be relatively prime; $\pm \alpha r : \alpha s$ for integer α, produces the same axis of rotation as $\pm r : s$, with adjacent image locations being separated by vector distance $\alpha^2(r^2 + s^2)$.

In creating an affine-transformed array, $A'(x', y') = R[A(x, y)]$, the modulo p part of the transform matrix operator R (usually) 'randomly' shuffles the element locations along both length p rows and columns of A. The periodic autocorrelations, $A' \otimes A' = A \otimes A$ are identical and remain perfectly delta-like. However, the periodic cross-correlation $A' \otimes A$ will, in general, be very low because of the relative randomisation of array element locations that arises from the affine mapping.

The next section shows that, for the Legendre arrays L_p, the reversible but apparently random shuffling that is produced by any affine rotation $R(r : s)$ never mixes the locations of elements from the set of $+1$ quadratic residue points (Q_p) with the locations of the -1 elements from the set of non-residue points $(\underline{Q}_p)$.

1.4 Affine Rotation of Legendre Arrays L_p

We consider affine rotations, as defined in Eq. 2, applied to a transpose symmetric p × p Legendre array L_p for $p = 4N - 1$. Here all values at the p × p array locations are defined by quadratic residues (the single 0 and the $+1$ points by Q_p, the -1 points by $\underline{Q}_p$). For any general p × p discrete affine rotation by vector r : s, any location (x, y) is mapped to a new, unique location (x′, y′), where

$$x' = xs - yr, \quad y' = xr + ys. \tag{3}$$

The Legendre array element at location $(0, 0)$ is assigned value 0. This point will not be moved by any affine transform R, provided that the cyclic shifts $x_0 = y_0 = 0$. If array location (x, y) is a quadratic residue, we can check if the location of the point (x, y) rotated to point (x′, y′) is also a quadratic residue. It is useful to rephrase Eq. 3 as

$$\left(x'^2 + y'^2\right) = \left(r^2 + s^2\right)\left(x^2 + y^2\right) \tag{4}$$

For any discrete rotation r : s, the new location (x′, y′) will remain as a quadratic residue location iff $\left(r^2 + s^2\right)Q_p$ is $q_p(i)$, an element of Q_p.

Scaling any set Q_p by any of its quadratic residue elements $q_p(i)$, modulo p, produces the same set (by definition). Scaling the set Q_p by a non-quadratic residue, $\underline{q}_p(i)$ produces the complement of the residue set, $\underline{Q}_p$

$$q_p(i)Q_p = Q_p, \quad \underline{q}_p(i)Q_p = \underline{Q}_p. \tag{5}$$

Each and every location (x′, y′) in the four-fold symmetric array L_p is fixed by Q_p or $\underline{Q}_p$. The rotation R(r : s) then maps, via the integer scaling factor $\left(r^2 + s^2\right)$, each array quadratic residue location to another location fixed by Q_p (when $\left(r^2 + s^2\right) \in Q_p$), or else they all map to one of the locations valued -1 (when $\left(r^2 + s^2\right) \in \underline{Q}_p$).

Hence we have now established the curious rotation invariance, R(r : s)[L_p] = $\pm L_p$, as the Legendre array remains unchanged up to a sign. This key result shall be demonstrated through numerical examples in Sect. 2. The periodic cross-correlation of the original array L_p with the affine rotated array $L'_p = R(r : s)[L_p]$, is $L_p' \otimes L_p = \pm L_p \otimes L_p$, which has peak value (p − 1) and all off-peak values −1, or peak −(p − 1) with all off-peak values +1. These affine rotations perform a remarkable, rather curious and decidedly non-random shuffling of the array elements.

1.5 Alternative Methods to Construct 2D Legendre Arrays

Using the Finite Radon Transform. There is a direct 'tomographic' link between affine transforms of an array and sets of periodic discrete projections of the same array, such as obtained through the Finite Radon Transform of a matrix A(x, y) [8, 11, 12]:

$$FRT(A(x, y)) = R(t, m) = \begin{cases} \sum_{y=0}^{p-1} A(\{t + my\} \bmod p, y) \text{ for } 0 \le m \le p - 1 \\ \sum_{y=0}^{p-1} A(x, t) \text{ for } m = p \end{cases} \tag{6}$$

$$FRT^{-1}(R(t, m)) = \left(\sum_{m=0}^{p-1} R(\{x - my\} \bmod p, m) - A_{sum}\right)/p + O(x, y). \tag{7}$$

Within the inverse Finite Radon Transform, FRT^{-1} of Eq. 7, A_{sum} is the sum of all values in the array $A(x, y)$ (which is also the constant sum of each row in the FRT) and the offset $O(x, y)$ has value $R(y, p)/p$ when $x = 0$; $R(0, p)/p$ when $y = 0$; and $R(x, p)/p$ otherwise. An aperiodic discrete projection of any sized array $A(x, y)$, at angle $r : s$, can be obtained by applying the 'cousin' of the periodic FRT, the Mojette transform [13].

The FRT has size $p \times (p + 1)$. Each point $t \times m$ of the FRT is the sum of p values sampled from a $p \times p$ array $A(x,y)$. The mth FRT projection, for $0 \leq m < p$, sums data from p cyclically wrapped points, each shifted from initial point $A(t, 0)$ by m steps in x for each unit increment in y. These points are located along discrete lines $[(km + t) \bmod p, k]$ for $0 \leq k \leq p - 1$. The pth projection is formed from the p sums taken along rows 0 to $(p - 1)$ of $A(x, y)$.

The $p \times p$ arrays L_p, as presented above, can also be constructed using the discrete projections of the FRT. A synthesized set of $(p + 1)$ identical, zero-sum delta-like FRT projections is prepared, each having length p. The first projection element $(t = 0)$ is set to value $(p - 1)$, with all the remaining $(p - 1)$ elements being set to -1. FRT projection m is scaled by $+1$ if $(m^2 - 1)$ is a quadratic residue, modulo p, and by -1 if not. FRT projection m wraps $(m - 1$ times$)$ around the $p \times p$ array for angle $m : 1$.

Each projection at angle $m : 1$ maps to a set of 'nearest neighbour' projected rays that follow a discrete line with an angle $i : j$ that is unique for each m. The $i : j$ angles can be used in place of the geometric ray angle $m : 1$, by scaling the assigned projections by ± 1 where $(i^2 + j^2)$ is or is not a quadratic residue modulo p. The symmetry of both sets of ray angles $m : 1$ and $i : j$ makes the sign for the kth projection the same as for the $(p + 1 - k)^{th}$ projection. The perfect array L_p is recovered by applying the inverse FRT transform to the set of sign-adjusted FRT projections. The (exact) FRT inverse is obtained after normalising the $p \times p$ image obtained from back-projecting the FRT forward-projected sums at the complemented discrete angles, $m' = p - (i^2 + j^2)$.

The array locations that get summed for each discrete FRT projection m at angle $m : 1$ or for the equivalent unique shortest neighbour ray angle $i : j$, are the same set of points, as the spatial array coordinates used to assign $L_p(i, j)$ belong to either Q_p or $\underline{Q}_p$ via $(i^2 + j^2) \bmod p$.

Using the FRT-based method with $p = 4N - 1$ produces the identical transpose and 4-fold symmetric perfect $p \times p$ arrays L_p as presented above. Note that here $p + 1 = 4N - 1 + 1 = 4N$.

For $p = 4N + 1$, the FRT method also produces transpose symmetric and perfect $p \times p$ arrays. However, the resulting $4N + 1$ array L_p now has value 2 at location $(0,0)$ and the array sums to 2p rather than zero. The signs of the $p + 1 = 4N + 2$ projections have 2 more $+1$ entries than -1 entries. The array autocorrelation is then shifted by 4, having peak $p^2 + 3$ with all off-peak values 3. The peak-to-sidelobe ratio for these $4N + 1$ L_p arrays is then changed to $p^2/3 + 1$, with the merit factor $\approx (p^2 + 2)/9$ (both metrics being $p^2 - 1$ for $p = 4N - 1$). The $4N + 1$ FRT-built versions of these perfect arrays do not exhibit invariance to all affine rotations, as observed for the previous methods of construction.

Using the Fast Fourier Transform. The quadratic residue approach to construct L_p also works when applied in the Fourier domain, $F(u, v)$. This is expected, as $A(x, y)$, $FRT[A(x, y)]$ and $F[A(x, y)]$ provide $1 : 1$ mappings of each other, for $p \times p$ data. We set all of the p × p Fourier magnitudes, |F(u, v)| to value p, apart from (0, 0), which is initialised to 0, as for a zero-sum array. We then set the p × p Fourier phases at frequencies (f_u, f_v) to value 0 when $(f_u^2 + f_v^2)$ modp is in Q_p and set the phase to π when in $\underline{Q}_p$. We then apply the inverse Fourier Transform to recover the 2D Legendre array. For p = 4N − 1 we get exactly the same rotationally invariant zero-sum perfect arrays as for the FRT-based method, here $(p^2 + 1)/2$ phases are set to zero and $(p^2 − 1)/2$ phases are set to π.

For the 4N + 1 case, the FFT-built arrays are also transpose and 4-fold symmetric, with sum 2p, as for the FRT method. Of the p^2 Fourier phases, there is an imbalance of 2p − 1 in the number that are set to zero over those set to π, hence the value of the Fourier magnitude at $F = (0, 0)$ gets reset to 2p.

It should be noted that for size p = 4N − 1, L_p is invariant under the discrete Fourier transform such that $F(L_p) = L_p$ (for a certain Fourier normalisation convention). The equality of these 2D arrays may be understood from the choice of Fourier phases (0 and π) that encode signs governed by the same quadratic residue construction described in Sect. 1.2. It is perhaps less obvious that also $F(L_p) = L_p$ (using the same Fourier convention) for the p = 4N + 1 sized L_p arrays built from the FRT.

Invariant rotations for p = 4N + 1 L_p arrays. There are some affine rotations r : s for which the transpose and 4-fold symmetric versions of the 4N + 1 Legendre p × p arrays do map back exactly onto themselves. These correspond to discrete rotation angles r : s where $(r^2 + s^2) \in Q_p$, except for the case when det(R) = $(r^2 + s^2)$ = 0. The non-invariant affine rotations, where $(r^2 + s^2) \in \underline{Q}_p$, result in a close but partial match with the original array, yielding a difference image of mostly zeros, with 2p − 1 values being 2 and one location with value 4.

If the array L_p is shifted (cyclically) by the applied affine rotation matrix R (when x_0 or y_0 are not both zero), the resulting array will match the original after re-aligning the result. This can be done by co-registering the single (0, 0) points of both arrays. If the affine transform R is not a pure rotation, for example if R includes a skew component, then det(R) = ac − bd, the transpose and 4-fold symmetry of the mapping via Q_p is lost and the array is unlikely to be affine invariant.

2 2D Legendre Array Results

2.1 Examples for p = 4N + 1 and p = 4N − 1

We next provide some examples. The array L_{11}, for p = 11 (here p = 4N − 1) is given in Fig. 1, $Q_{11} = \{0, 1, 4, 9, 5, 3\}$. The Fourier phase for F(u, v) of L_{11} is zero everywhere $(f_u^2 + f_v^2)$ mod11 is in Q_{11}, or, equivalently, where L_{11} has the value +1. It has value π where $(f_u^2 + f_v^2)$ mod11 is in $\underline{Q}_{11}$, where L_{11} is −1. The FRT-built array is initially assigned 12 identical projections, [10, −1, −1, −1, −1, −1, −1, −1, −1, −1, −1], one for eachm. We compute a sign for each projection, according to $(m^2 + 1)$ or $(i^2 + j^2) \in Q_{11}$ or $\underline{Q}_{11}$. The number of ± sign changes is equal, as indicated in Fig. 2. L_{11} is invariant to all discrete rotations r : s.

m for p = 11, FRT	0	1	2	3	4	5	6	7	8	9	10	11
angle i of i:j	0	1	2	3	1	-1	1	-1	-3	-2	-1	1
angle j of i:j	1	1	1	1	3	2	2	3	1	1	1	0
residue (i²+j²), 11	1	2	5	10	10	5	5	10	10	5	2	1
residue (m²+1), 11	1	2	5	10	6	4	4	6	10	5	2	1
projection sign	1	-1	1	-1	-1	1	1	-1	-1	1	-1	1

Fig. 2. Angles, residues modulo 11 and projection signs for the 2D Legendre array.

The array L_{11}, for p $=$ 13 (here p $=$ 4N $+$ 1) is shown in Fig. 3, Q_{13} $=$ {0, 1, 4, 9, 3, 12, 10}. The Fourier phase for F(u, v) of L_{13} is zero everywhere $\left(f_u^2 + f_v^2\right)$ is in Q_{13}, where L_{13} has the value $+$ 1. It has value π where $\left(f_u^2 + f_v^2\right)$ is in $\underline{Q}_{13}$, where L_{13} is -1. The FRT-built array is initially assigned 14 identical projections [$\overline{12}, -1, -1,$ $-1, -1, -1, -1, -1, -1, -1, -1, -1, -1, -1$], one for each m. We compute a sign for each projection, according to $(m^2 + 1)$ or $(i^2 + j^2) \in Q_{13}$ or $\underline{Q}_{13}$. There are now an unequal number of ± 1 sign changes, as indicated in Fig. 4; here 8 projection signs are $+1$ but 6 are -1, resulting in the value 2 at (0, 0) in L_{13} after applying the FRT inverse. Affine rotation of L_{13} by 1:3 is invariant as $1^2 + 3^2 = 10$ is in Q_{13}, whilst rotation by 1:2 is not invariant. Affine rotation by 2:3 is not possible here, as R has zero determinant for 13 × 13 arrays.

$$
\begin{pmatrix}
2 & 1 & 1 & 1 & 1 & 1 & 1 & 1 & 1 & 1 & 1 & 1 & 1 \\
1 & -1 & -1 & 1 & 1 & 1 & -1 & -1 & 1 & 1 & 1 & -1 & -1 \\
1 & -1 & -1 & 1 & -1 & 1 & 1 & 1 & 1 & -1 & 1 & -1 & -1 \\
1 & 1 & 1 & -1 & 1 & -1 & -1 & -1 & -1 & 1 & -1 & 1 & 1 \\
1 & 1 & -1 & 1 & -1 & -1 & 1 & 1 & -1 & -1 & 1 & -1 & 1 \\
1 & 1 & 1 & -1 & -1 & -1 & 1 & 1 & -1 & -1 & -1 & 1 & 1 \\
1 & -1 & 1 & -1 & 1 & 1 & -1 & -1 & 1 & 1 & -1 & 1 & -1 \\
1 & -1 & 1 & -1 & 1 & 1 & -1 & -1 & 1 & 1 & -1 & 1 & -1 \\
1 & 1 & 1 & -1 & -1 & -1 & 1 & 1 & -1 & -1 & -1 & 1 & 1 \\
1 & 1 & -1 & 1 & -1 & -1 & 1 & 1 & -1 & -1 & 1 & -1 & 1 \\
1 & 1 & 1 & -1 & 1 & -1 & -1 & -1 & -1 & 1 & -1 & 1 & 1 \\
1 & -1 & -1 & 1 & -1 & 1 & 1 & 1 & 1 & -1 & 1 & -1 & -1 \\
1 & -1 & -1 & 1 & 1 & 1 & -1 & -1 & 1 & 1 & 1 & -1 & -1
\end{pmatrix}
\begin{pmatrix}
3 & 3 & 3 & 3 & 3 & 3 & 3 & 3 & 3 & 3 & 3 & 3 \\
3 & 3 & 3 & 3 & 3 & 3 & 3 & 3 & 3 & 3 & 3 & 3 \\
3 & 3 & 3 & 3 & 3 & 3 & 3 & 3 & 3 & 3 & 3 & 3 \\
3 & 3 & 3 & 3 & 3 & 3 & 3 & 3 & 3 & 3 & 3 & 3 \\
3 & 3 & 3 & 3 & 3 & 3 & 3 & 3 & 3 & 3 & 3 & 3 \\
3 & 3 & 3 & 3 & 3 & 3 & 3 & 3 & 3 & 3 & 3 & 3 \\
3 & 3 & 3 & 3 & 3 & 3 & 172 & 3 & 3 & 3 & 3 & 3 \\
3 & 3 & 3 & 3 & 3 & 3 & 3 & 3 & 3 & 3 & 3 & 3 \\
3 & 3 & 3 & 3 & 3 & 3 & 3 & 3 & 3 & 3 & 3 & 3 \\
3 & 3 & 3 & 3 & 3 & 3 & 3 & 3 & 3 & 3 & 3 & 3 \\
3 & 3 & 3 & 3 & 3 & 3 & 3 & 3 & 3 & 3 & 3 & 3 \\
3 & 3 & 3 & 3 & 3 & 3 & 3 & 3 & 3 & 3 & 3 & 3 \\
3 & 3 & 3 & 3 & 3 & 3 & 3 & 3 & 3 & 3 & 3 & 3
\end{pmatrix}
$$

Fig. 3. L_{13} displayed as a matrix (left). The 2D FFT of L_{13} is exactly the same as the array L_{13}. The delta-like periodic autocorrelation is shown on the right.

m for p = 13, FRT	0	1	2	3	4	5	6	7	8	9	10	11	12	13
angle i of i:j	0	1	2	3	-1	2	-1	1	3	1	-3	-2	-1	1
angle j of i:j	1	1	1	1	3	3	2	2	2	3	1	1	1	0
residue (i²+j²), 13	1	2	5	10	10	0	5	5	0	10	10	5	2	1
residue (m²+1), 13	1	2	5	10	4	0	11	11	0	4	10	5	2	1
projection sign	1	-1	-1	1	1	1	-1	-1	1	1	1	-1	-1	1

Fig. 4. Angles, residues modulo 13 and projection signs for the 2D Legendre array.

The value 2 at location (0, 0) in p $=$ 4N $+$ 1 sized arrays built using the FRT or FFT method can be arbitrarily reset to 0 (or any other value) without changing the rotational invariance properties, however the perfect periodic correlation metrics of the array decline, falling further as the value set at (0, 0) is more distant from $+2$.

2.2 Invariance Demonstrated as Stirring Without Mixing

In Sect. 1.4 it was shown that the 2D Legendre arrays of interest here are invariant to rotation, up to a change of sign. The numerical results of Sect. 2.1 attest to this but it is insightful to visually portray this curious result, as shall be demonstrated in this section.

For affine rotations of arrays L_p, the locations of the class of $+1$ points always remains distinct from those of the class of -1 points. Imagine a stack of three $p \times p$ arrays as coloured (RGB) pixels, perhaps where the $+1$ points of a Legendre array L_p were encoded using different shades of red, and the -1 points were depicted in shades of green. Then applying any affine rotation shuffles the distribution of red and green intensities, but never blends the positions of the red pixels with the green pixels; this is stirring without mixing, as depicted in Fig. 5. Note the internal shuffling of the $+1$ and -1 point locations in L_p for rotation r : s is not the same as for αr : αs, however the ± 1 classes still never mix.

Fig. 5. Three colour-coded 31×31 Legendre arrays, L_p. In the original image on the left, the $+1$ entries of L_p appear shaded in red with an intensity at each pixel that increases from the top centre to the bottom centre. The -1 entries are shown uniformly green. The centre image shows the original L_p array after applying a periodic affine rotation of r : s = 1 : 2 which maps the array back exactly onto itself ($1^2 + 2^2 = 5$, 5 is a quadratic residue $= 6^2$ modulo 31). The image on the right shows the original L_p array after applying a periodic affine rotation of r : s = 2 : 3 ($2^2 + 3^2 = 13$, 13 is not a quadratic residue modulo 31). (Color figure online)

In Fig. 5 the red pixels are shuffled amongst themselves, but never to locations that were and still are occupied by all of the green pixels. Here all of the red pixels have been shuffled to locations that were previously occupied by the green pixels. Both classes of $+1$ pixels and -1 pixels are stirred but never mix.

3   Conclusion

The new $p \times p$ Legendre arrays L_p presented here are (up to a change of sign), invariant to any affine rotation r : s, for any p = 4N − 1. For p = 4N + 1, the invariance applies only for rotations r : s where $(r^2 + s^2)$ modp is not zero but is in Q_p. The effect of more general affine mappings, such as skew operations, remains a topic for further investigation.

The property of invariance to discrete rotations makes these L_p arrays less secure in communication applications. It may also prohibit or restrict the ability to generate families of transpose symmetric arrays based on L_p that retain uniformly-low-cross-correlations (a subject of ongoing work). The current L_p arrays may, however, be beneficial in practical circumstances where this rotation invariance provides greater robustness to orientation-based changes or for symmetrical detector design. For example, Gottesman and Fenimore's modified uniformly redundant arrays [14], based upon quadratic residues, possessed useful two-fold or six-fold symmetry suited for square and hexagonal unimodular arrays, respectively. However, the constraints of periodicity and the accompanying upscale or downscale array size changes that arise from affine rotations (by a factor of $(r^2 + s^2)$ for rotations r : s) would both need to be accommodated.

Appendix

The 1D autocorrelation of each projected view at angle r : s for any 2D array A, mirrors exactly the 1D projected view, at the same angle r : s, of the 2D autocorrelation $A \otimes A$ [13]. The FRT of any p × p array A produces B, an exactly invertible set of $(p + 1)$ discrete projections of length p. The set B of FRT projections of A can be applied in several interesting ways.

When A is a Legendre array, each 1D projection in its FRT set B looks like a delta function (as used here in Sect. 1.5). We can select a partial set B_k of the full FRT set B, that contains k of the (p+1) projections, with otherwise empty data. Applying the inverse FRT to B_k reconstructs an array A_k that has almost-perfect autocorrelation, whilst the projected views of A_k at the 'missing' projection angles are exactly zero. Then sets of partial projections, for example a pair B_1 and B_2, for which $B = B_1 + B_2$, are disjoint. The arrays $A_1 = FRT^{-1}(B_1)$ and $A_2 = FRT^{-1}(B_2)$ will have both $A_1 \otimes A_1$ and $A_1 \otimes A_2$ being delta-like, yet $A_1 \otimes A_2 = 0$, yielding an exactly zero cross-correlation, [15].

The full set of projections in the FRT of a Legendre array can also be cyclically shifted (in either or both angle variable m and projection translate t). Applying the inverse operation to these shifted FRTs makes new, exactly perfect Legendre arrays (at the cost of losing the transpose symmetry of the L_p arrays presented here). Specific patterns of shifts in m can produce large families of distinct Legendre arrays where the periodic cross-correlation between all family members is not zero, but as low as is possible [16]. Changes in the t variable of the FRT produce 'grey' Legendre arrays with a range of integer values beyond -1, 0, and $+1$. The extended alphabet of array values permits the construction of even larger families of arrays with low cross-correlation [4, 17].

References

1. Bomer, L., Antweiler, M.: Perfect three-level and three-phase sequences and arrays. IEEE Trans. Commun. **42**(234), 767–772 (1994)
2. Green, D.H., Smith, M.D., Martzoukos, N.: The linear complexity of polyphase power residue sequences. IEE Proc. Commun. **149**, 195–201 (2002)
3. Green, D.H., Garcia–Perera, L.P.: The linear complexity of related prime sequences. Proc. R. Soc. Lond. Series A Math. Phys. Eng. Sci. **460**(2042), 487–498 (2004)

4. Ceko, M., Svalbe, I., Petersen, T., Tirkel, A.: Large families of "grey" arrays with perfect auto-correlation and optimal cross-correlation. J. Math. Imaging Vis. **61**, 237–248 (2019). https://doi.org/10.1007/s10851-018-0848-3

5. Borwein, P., Choi, K.-K.S., Jedwab, J.: Binary sequences with merit factor greater than 6.34. IEEE Trans. Inf. Theor. **50**(12), 3234–3249 (2004). https://doi.org/10.1109/TIT.2004.838341

6. Golay, M.J.E.: The merit factor of Legendre sequences. IEEE Trans. Inf. Theory **29**(6), 934–936 (1983)

7. Van Schyndel, R.G., Tirkel, A.Z., Svalbe, I.D.: Key independent watermark detection. In: Proceedings IEEE International Conference on Multimedia Computing and Systems, vol. 1, pp. 580–585 (1999). https://doi.org/10.1109/MMCS.1999.779265

8. Ceko, M.: Discrete projection: highly correlated arrays and ghosts for image reconstruction, Thesis, Chapter 3, Sections 3.5.1 to 3.5.3 (2021)

9. Huffman, D.: The generation of impulse-equivalent pulse trains. IRE Trans. Inf. Theory **8**(5), 10–16 (1962)

10. Svalbe, I.D., Paganin, D.M., Petersen, T.C.: Sharp computational images from diffuse beams: factorization of the discrete delta function. IEEE Trans. Comput. Imaging **6**, 1258–1271 (2020)

11. Matúš, F., Flusser, J.: Image representation via a finite Radon transform. IEEE Trans. Pattern Anal. Mach. Intell. **15**(10), 996–1006 (1993)

12. Svalbe, I.: Exact, scaled image rotation using the finite Radon transform. Pattern Recognit. Lett. **32**(9), 1415–1420 (2011)

13. Guédon, J. (ed.).: The Mojette Transform: Theory and Applications. ISTE-Wiley, UK, USA (2009)

14. Gottesman, S.R., Fenimore, E.E.: New family of binary arrays for coded aperture imaging. Appl. Opt. **28**(20), 4344–4352 (1989)

15. Cavy, B., Svalbe, I.: Construction of perfect auto-correlation arrays and zero cross-correlation arrays from discrete projections. In: Barneva, R., Bhattacharya, B., Brimkov, V. (eds.) Combinatorial Image Analysis. IWCIA 2015. LNCS, vol. 9448, pp. 232–243. Springer, Cham (2015). https://doi.org/10.1007/978-3-319-26145-4_17

16. Tirkel, A., Cavy, B., Svalbe, I.: Families of multi-dimensional arrays with optimal correlations between all members. Electron. Lett. **51**(15), 1167–1168 (2015)

17. Svalbe, I.D., Tirkel, A.Z.: Extended families of 2D arrays with near optimal auto and low cross-correlation. EURASIP J. Adv. Sig. Process. **2017**(1), 1–19 (2017). https://doi.org/10.1186/s13634-017-0455-2

A Simple Discrete Calculus for Digital Surfaces

David Coeurjolly[1]([⊠]) [iD] and Jacques-Olivier Lachaud[2] [iD]

[1] Univ Lyon, CNRS, INSA Lyon, UCBL, LIRIS, UMR5205 Villeurbanne, France
david.coeurjolly@liris.cnrs.fr
[2] Université Savoie Mont Blanc, CNRS, LAMA, 73000 Chambéry, France
jacques-olivier.lachaud@univ-smb.fr

Abstract. Computing differential quantities or solving partial derivative equations on discrete surfaces is at the core of many geometry processing and simulation tasks. For digital surfaces in $\mathbb{Z}^3$ (boundary of voxels), several challenges arise when trying to define a discrete calculus framework on such surfaces mimicking the continuous one: the vertex positions and the geometry of faces do not capture well the geometry of the underlying smooth Euclidean object, even when refined asymptotically. Furthermore, the surface may not be a combinatorial 2 manifold even for discretizations of smooth regular shape. In this paper, we adapt a discrete differential calculus defined on polygonal meshes to the specific case of digital surfaces. We show that this discrete differential calculus accurately mimics the continuous calculus operating on the underlying smooth object, through several experiments: convergence of gradient and weak Laplace operators, spectral analysis and geodesic computations, mean curvature approximation and tolerance to non-manifold locii.

Keywords: Discrete calculus · Differential operators · Digital surface

1 Introduction

In many geometry processing and simulation applications, solving variational problems or simulating partial differential equations on the object boundary can be a critical step. Involved shapes are generally represented as discrete meshes. It is thus necessary to have a calculus that operates consistently and accurately on these discrete objects. For embedded graphs or triangular manifold meshes, finite element methods [1,2] or discrete exterior calculus (DEC) [7,8] have played a crucial role in many applied mathematics and geometry processing applications. From a broad perspective, such models provide consistent differential operators to process scalar, vector or tensor functions on meshes or embedded graphs (consistency given by satisfying Stokes' theorem for all discrete elements). These frameworks induce several convergence results for PDE solutions, but with strong hypotheses on the discrete-continuous mapping [10]. On generic non-triangular

---

This work has been partly funded by CoMeDiC ANR-15-CE40-0006 research grant.

É. Baudrier et al. (Eds.): DGMM 2022, LNCS 13493, pp. 341–353, 2022.
https://doi.org/10.1007/978-3-031-19897-7_27

meshes, Virtual Element Methods [20], or ad hoc operators exist [6]. However, these models assume a Euclidean embedding *interpolating* the smooth manifold.

Digital objects and surfaces corresponds to discrete approximation of continuous objects through a discretization process [11], or to partitions in images. In terms of perturbation and stability, digital surfaces, made of isothetic unit squares in $\mathbb{Z}^3$, are very specific: vertices *do not interpolate* the continuous object, and geometric normals poorly approximate the continuous normal bundle. Even worse, the primal quad surface may not be a combinatorial 2-manifold. To design stable geometric estimators, a key ingredient is the digitization grid step h used to represent objects. Hence, a multigrid process, as a function of h, can be designed to relate a digital surface to the underlying smooth surface [14]. Stable geometric estimators with convergence properties can then be obtained (i.e. the estimation converges to the expected one on the smooth manifold as h tends to zero) for various quantities: surface area [14], curvature tensor [12], or even higher order functional such as the Laplace-Beltrami operator [3].

A *combinatorial* DEC can be constructed on digital surfaces, but their specific geometry makes it poorly reflects the continuous calculus. This article presents a *new discrete calculus framework dedicated to digital surfaces*, which relies on two ingredients: (i) a *convergent normal vector field* $\mathbf{u}$ (e.g. the integral invariant normal estimator [12]), which is used to correct the embedding of (ii) a *polygonal differential calculus model* of de Goes *et al.* [6]. Several methods exploits this idea of correcting the embedding with a normal vector field [3,13,14]. Mercat [16] follows this idea with a theory of conformal parametrization and differential operators for digital surfaces restricted to combinatorial 2-manifolds. Our proposal shares some ideas with these works and defines a calculus on generic digital surfaces with a simple per-face construction of the operators.

The paper is organized as follows. We first describe the operator construction from [6] (Sect. 2). Section 3 describes how to correct the geometrical embedding of the surface elements. Finally, we evaluate the performance of the framework on various variational problems (Sect. 4).

2   Polygonal Differential Calculus

We focus here on the formulation proposed by de Goes *et al.* for polygonal surfaces embedded in $\mathbb{R}^3$ [6]. It defines differential operators per face without assumption on the face geometry, which could be non-convex or even non-planar.

Let $\mathcal{M}$ be a mesh with vertices $\mathbf{X}$ and faces $\mathbf{F}$. For a given face f with n_f vertices, we denote by $\mathbf{X}_f$, the vertex positions encoded as a $n_f \times 3$ matrix (the i-th row corresponds to the position of the vertex i of the face f). In this calculus, we consider scalar functions ϕ defined on vertices of $\mathcal{M}$ (see Fig. 1$-(a-b)$). For the face f, we denote by ϕ_f the restriction of ϕ to the face vertices, represented as a vector of size n_f. As all discrete differential estimators will be linear in the vertex positions and the scalar function values defined at vertices, the matrix representation will easily combine operators with matrix products. For instance, the centroid of a face is given by $\mathbf{c}_f := \frac{1}{n_f}\mathbf{X}_f^t \mathbf{1}_f$, with $\mathbf{1}_\mathbf{f} = (1\dots 1)^t$ of size n_f.

First per face local operators are constructed, starting from the gradient. To cope with (non-planar) polygonal faces, the weak form of the gradient is sought (*i.e.* $\int_f \nabla\phi(\mathbf{x})dx$ in the continuous setting) leading to a constant gradient integrated per face. Solving the integral on the discrete structure would require an interpolation scheme on the non-planar face. The key ingredient is to focus on the co-gradient $\nabla\phi^\perp$ that leads to a simpler expression of its weak form using Stoke's theorem:

$$\int_f \nabla\phi^\perp(\mathbf{x})dx = \oint_{\partial f} \phi(\mathbf{x})\mathbf{t}(\mathbf{x})dx,$$

$\mathbf{t}(\mathbf{x})$ being the unit tangent vector at $\mathbf{x} \in \partial f$. When discretizing this form for a polygonal face f, $\mathbf{t}(\mathbf{x})$ corresponds to the (normalized) edge vectors. Now, let us consider a linear function $\phi_f = \mathbf{X}_f\mathbf{s} + \mathbf{1}_f r$ with $\mathbf{s} \in \mathbb{R}^s$ and $r \in \mathbb{R}$, integrating ϕ_f on an edge correspond to averaging the function value at the endpoints. Hence, the discrete operator for the (integrated) co-gradient is the $3 \times n_f$ matrix

$$\mathbf{G}_f^\perp := \mathbf{E}_f^t \mathbf{A}_f,$$

$\mathbf{E}_f$ being the matrix encoding the face edges, and $\mathbf{A}_f$ the operator that averages two consecutive vertex values. To be explicit, $\mathbf{A}_f$ is the $n_f \times n_f$ matrix with $\mathbf{A}^{i,i} = 1/2$, $\mathbf{A}^{i,i+1} = 1/2$ (zero otherwise), and $\mathbf{D}_f$ denotes the $n_f \times n_f$ derivative matrix with $\mathbf{D}^{i,i} = -1$, $\mathbf{D}^{i,i+1} = 1$ (zero otherwise). So $\mathbf{E}_f = \mathbf{D}_f\mathbf{X}_f$.

Given a vector ϕ_f, $\mathbf{E}_f^t\mathbf{A}_f\phi_f$ is a vector of size 3 corresponding to the Euclidean embedding of the integrated co-gradient of ϕ_f on f. For a polygonal face f, the vector area $\mathbf{a}_f$ can be defined as [17]: $\mathbf{a}_f := \frac{1}{2}\sum_{v_i \in f} \mathbf{x}_{v_i} \times \mathbf{x}_{v_{i+1}}$.

The normal vector to f is simply $\mathbf{n}_f := \mathbf{a}_f/a_f$, with $a_f := \|\mathbf{a}_f\|_2$. Now, we can finally get the expression of the gradient as $\nabla\phi^\perp = [\mathbf{n}(\mathbf{x})]\nabla\phi$, where $[\mathbf{n}]$ is the $\pi/2$ rotation matrix such that $[\mathbf{n}]\mathbf{q} = \mathbf{n} \times \mathbf{q}$ for any 3d vector $\mathbf{q}$:

$$\mathbf{G}_f := \frac{1}{a_f}[\mathbf{n}_f]\mathbf{E}_f^t\mathbf{A}_f.$$

The gradient operator is a $3 \times n_f$ matrix outputting a 3d vector in 3d when applied to a scalar vector ϕ_f (see Fig. 1-(c)). Following such per-face construction, de Goes *et al.* define several differential operators listed in Table 1 and represented as matrices acting on scalar functions, vectors, or discrete differential forms (see [7,8] for an introduction). We do not discuss the rationale and the construction of each operator, please refer to [6] for details.

We conclude with the presentation of the Laplace-Beltrami operator Δ, a fundamental tool in many geometry processing applications [15]. To solve global PDE involving this operator (*e.g.* for instance solving a Laplace or Poisson problem $\Delta u = f$), it is interesting to aggregate the per-face operators into a global operator $\mathbf{L}$. We simply sum up the $\mathbf{L}_f$ matrices with a global indexing of the vertices. As every $\mathbf{L}_f$ matrix is negative semi-definite, so is the global Laplace-Beltrami operator. Furthermore, the global operator is very sparse. In opposition with [6], we define it with a negative sign, since usually both the Laplacian and the Laplace-Beltrami are negative operators (with negative eigenvalues).

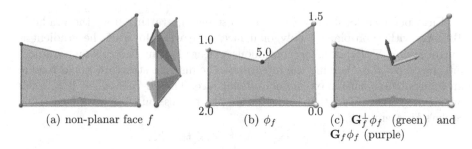

(a) non-planar face f (b) ϕ_f (c) $\mathbf{G}_f^{\perp}\phi_f$ (green) and
$\mathbf{G}_f\phi_f$ (purple)

Fig. 1. Illustration of a non-planar face f (a) equipped with a vertex valued scalar function ϕ_f (b), and the co-gradient and gradient vectors of ϕ_f (c) (Color figure online)

We do not go further into the details of this approach, several results are given in [6] to justify the construction of these operators. For short, they match with classical discrete exterior calculus or finite element ones for triangular meshes [7], and for the polygonal case, they resemble to Virtuel Element Methods operators for 2d structures embedded in $\mathbb{R}^3$ [19]. However, all these relationships between discrete calculus operators and their continuous counterpart on the smooth object only make sense when the discrete structure *interpolates the smooth object*, with a *close normal vector field*. Applying this calculus as is on digital surfaces provide poor results. We need to adapt the surface embedding to correct it.

Table 1. Summary of local per-face operators of polygonal calculus [6].

Operator	Size	Description
$\mathbf{V}_f := \mathbf{E}_f(\mathbf{I}_{3\times3} - \mathbf{n}_f\mathbf{n}_f^t)$	$n_f \times 3$	Flat operator that maps a vector to a 1-form
$\mathbf{U}_f := \frac{1}{a_f}[\mathbf{n}_f](\mathbf{B}_f^t - \mathbf{c}_f\mathbf{1}_f^t)$	$3 \times n_f$	Sharp operator that maps a 1-form to a 3d vector
$\mathbf{P}_f := \mathbf{I}_{n_f\times n_f} - \mathbf{V}_f\mathbf{U}_f$	$n_f \times n_f$	Projection operator acting on a 1-form
$\mathbf{M}_f^0 := \frac{a_f}{n_f}\mathbf{I}_{n_f\times n_f}$	$n_f \times n_f$	Inner product for discrete 0-forms (functions)
$\mathbf{M}_f^1 := a_f\mathbf{U}_f^t\mathbf{U}_f + \lambda\mathbf{P}_f^t\mathbf{P}_f$	$n_f \times n_f$	Inner product for discrete 1-forms (for some $\lambda > 0$)
$\mathrm{Div}_f := -\mathbf{D}_f^t M_f^1$	$n_f \times n_f$	Integrated divergence operator from a 1-form
$\mathbf{L}_f := -\mathbf{D}_f^t M_f^1\mathbf{D}_f$	$n_f \times n_f$	Weak Laplace-Beltrami operator

3   Projected Digital Surface Embedding

As discussed in the introduction, the classical approach is to relate a digital object to its continuous counterpart through the Gauss digitization process [11]. In this setting, many multigrid convergence results have been obtained for various integral and differential quantities such as the length in 2d [4], the surface area in 3d [14], the curvature tensor [12], or the Laplace-Beltrami operator [3]. Among these techniques, the convergent estimation of the normal vector bundle is the cornerstone of many followup results (*e.g.* [3,14]).

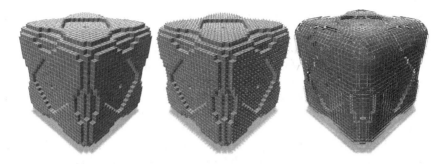

Fig. 2. Input digital surface (left), its estimated normal vectors using [12] (middle), projections of the face vertices onto their respective estimated tangent plane (right).

In the polygonal calculus described above, a key ingredient is that operators are first defined *per-face*, and later summed up globally with the vertex indexing. This suggests that correcting the embedding of the face vertices during the local construction of the operators would lead to globally corrected operators. As a first-order approximation, we propose to define our *projected polygonal calculus* by relying on an external estimation of the face tangent space, and by (implicitly) projecting each face vertex onto this tangent plane when constructing the operators. The idea of incorporating external normal vectors to correct the embedding of a discrete structure follows the idea of Lachaud et al. [13]. Let $\{\mathbf{u}_f\}$ be a normal vector field estimated on the digital surface, for instance using the multigrid convergent approach of [12].

Definition 1 (Embedding operator). *The per-face f projection operator onto the tangent plane orthogonal to $\mathbf{u}_f$ passing through the origin is the 3×3 matrix $\Pi_f := (\mathbf{I}_{3 \times 3} - \mathbf{u}_f \mathbf{u}_f^t)$. Recalling that $\mathbf{X}_f$ is the $n_f \times 3$ matrix of the vertices position of the face f, the new positions are given by $\mathbf{X}_f^* := \mathbf{X}_f \Pi_f$.*

Note that the intercept does not need to be specified as all differential operators are invariant by translation. For illustration purpose, projected faces have been translated to keep their centroid invariant in Fig. 2, but this is meaningless in the calculus. Another observation is that for a given vertex $\mathbf{v}$ in $\mathcal{M}$, its embedding generally differs for its adjacent faces. This follows from the per-face construction of the operators. The global continuity of scalar functions, or their processing (*i.e.* their Laplacian) comes from the global indexing of vertices, and global operators as defined above.

4   Experiments

We demonstrate the interest of the projected calculus model for the processing of scalar functions on digital surfaces. More precisely, we first evaluate the relevance of the new embedding for the gradient computation of a scalar function. Next, we study the Laplace-Beltrami operator and more advanced processing. All operators are implemented in C++ and available in DGtal [18].

Gradient of Scalar Functions. To evaluate the relevance of including external tangent information in the calculus, let us consider the simple case of the gradient evaluation on a quadratic scalar field defined on a tilted digital plane (with normal vector $1/3 \cdot (\sqrt{3}, \sqrt{3}, \sqrt{3})^t$, Fig. 3). In a multigrid setting, we compute the per-face gradient of the scalar function (Euclidean distance in ambient space with respect to the plane center, Fig. 3-(a)), and we compare the expected gradient vector direction (Fig. 3-(b)), to the estimated gradient using the classical polygonal calculus (Fig. 3-(c)), and the projected one (Fig. 3-(d)). Gradient vectors are illustrated from their projection onto the Euclidean plane. The error metric used here is the l_2 norm of the difference between estimated and expected normalized gradient vectors. Figures 3-(e) and (f) detail relative errors. In Fig. 3-(g), we have considered a decreasing grid step h from 1 to 1.5×10^{-2} (3 748 021 faces), demonstrating multigrid convergence of the gradient direction for the projected calculus.

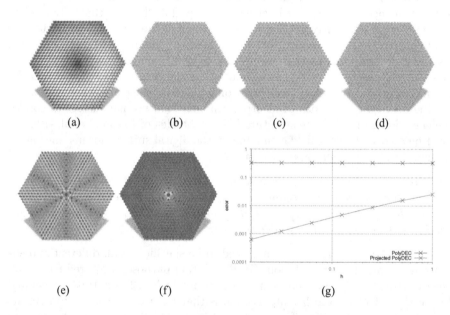

Fig. 3. Convergence of the gradient operator: Input scalar function on a tilted plane (a), expected normalized gradient projected onto the Euclidean plane (b), gradient vectors using the original polygonal calculus with many aliasing artifacts (c), gradient vectors using the projected calculus (d). Figures (e) and (f) details relative errors of gradient estimations (values respectively in $[0, 0.28]$ and $[0, 0.62]$), and (g) illustrates l_2 norm of the error when decreasing the grid step h (from right to left).

Laplacian of a Function. First, let us consider an evaluation of the output of the Laplace-Beltrami function on analytical functions on a domain for which we already know the expected Laplacian of that function. Following the setting in [3], we evaluate $\mathbf{L}$ on a digital sphere with the simple quadratic function

$\phi(\mathbf{x}) := \mathbf{x}_x^2$ on its boundary, and expect $\Delta\phi(\mathbf{x}) = 2 - 6\cos^2(\psi)^2(1 - \cos(\theta)^2)$ in spherical coordinates. Note that Caissard *et al.* [3] achieve strong consistency of the operator (*i.e.* pointwise convergence of the Laplacian values) at the price of a convolution process on the surface elements, leading to a quadratic algorithm for the operator construction, and a non-sparse operator. Clearly it is unlikely that the purely local construction of our projected polygonal calculus provides pointwise convergence. We observe nevertheless that the projection process offers reasonable estimations while being purely local (see Fig. 4).

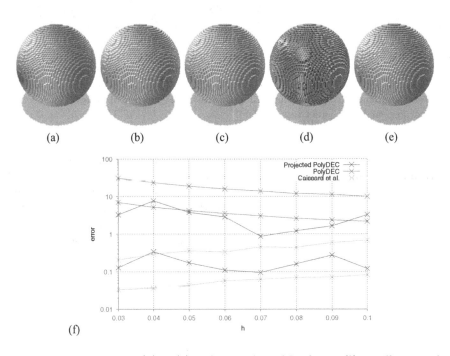

Fig. 4. Given a function $\phi(\mathbf{x})$ in (*a*) and its analytical Laplacian (*b*), we illustrate the result of the pointwise convergence heat kernel based Laplace-Beltrami operator [3] in (*c*). In (*d*) we have the results of the original local polygonal calculus operator, and in (*e*) the result of our projected one. In (*f*) we have performed a multigrid convergence test with decreasing gridstep h. Dashed lines correspond to the maximum absolute error when estimating Δf, the solid ones correspond to mean absolute errors.

Spectral Analysis. We evaluate the stability of the spectrum of the pointwise Laplace-Beltrami operator $(\mathbf{M}^0)^{-1}\mathbf{L}$. Eigenvalues and eigenvectors have a fundamental role in many geometry processing applications [15] as they define a (spectral) basis to represent scalar functions on surfaces. Figure 5 compares the first eigenvectors (corresponding to the largest eigenvalues) of the original polygonal calculus Laplace-Beltrami operator, to the ones we propose. Our embedding operator has a clear positive impact on the smoothness of basis functions.

Fig. 5. First ten eigenvectors of the Laplace-Betrami operator on a digital sphere (and their projection onto the Euclidean sphere): top two rows correspond to the original polygonal calculus, bottom two rows correspond to the projected polygonal calculus we propose (note that due to proximity of eigenvalues and numerical issues, eigenvectors may not appear in the exact same order).

Mean Curvature Estimation. A well-known result related to the Laplace-Beltrami operator is that, if $\mathbf{x}$ is the function associating to any point of $\mathcal{M}$ its coordinates, then $\Delta\mathbf{x} = 2H(\mathbf{x})\mathbf{n}(\mathbf{x})$ with $H(\mathbf{x})$ the mean curvature at $\mathbf{x}$ and $\mathbf{n}(\mathbf{x})$ the normal to $\mathcal{M}$. We check this method for computing the mean curvature with our calculus. We compute the pointwise Laplace-Beltrami of coordinates functions $(X, Y, Z) = \mathbf{x}$ and deduce the mean curvature H by scalar product with the estimated normal $\mathbf{u}$. Otherwise said, the estimated curvature $\hat{H}$ is

$$\hat{H} = \frac{1}{2}(\mathbf{M}^0)^{-1} \left[\mathbf{L}X \ \mathbf{L}Y \ \mathbf{L}Z \right] \mathbf{u}^t,$$

where $\mathbf{u}$ is a $n_f \times 3$ matrix storing per row the corrected normal to the vertex. To account for the fact that the Laplace-Beltrami is only weakly convergent, we can diffuse for a short time the result to simulate a local integration. Figure 6 illustrates this method for approximating the mean curvature, and also confirms that the calculus must be corrected to get meaningful result.

Geodesic Distance Estimation. To evaluate the proposed calculus on a more complex example, let us consider the geodesic distance estimation problem on digital surfaces. We consider the PDE approach of Crane et al. [5] that uses heat diffusion to estimate the geodesic distance function from sources. Sources are defined via a characteristic function u_0 in the domain. The heat method consists in three steps: (i) integrate the heat flow $\dot{u} = \Delta u$ for some fixed time t with initial condition $u(x, 0) = u_0(x)$, (ii) normalize the gradient vector field $d := -\nabla u_t / \|\nabla u_t\|_2$, and (iii) solve the Poisson equation $\Delta\phi = \nabla \cdot d$ ($\nabla \cdot$ denotes the divergence operator on vector fields). The scalar function $\phi - \min_x(\phi)$ is a good approximation of the geodesic distance from the sources to any point [5].

Goursat polynomial surface, digitized at gridstep $h = 0.25$				
Smooth surface	Naive calculus		Projected calculus	
True H	Pointwise $\hat{H}$	Diffused $\hat{H}$	Pointwise $\hat{H}$	Diffused $\hat{H}$
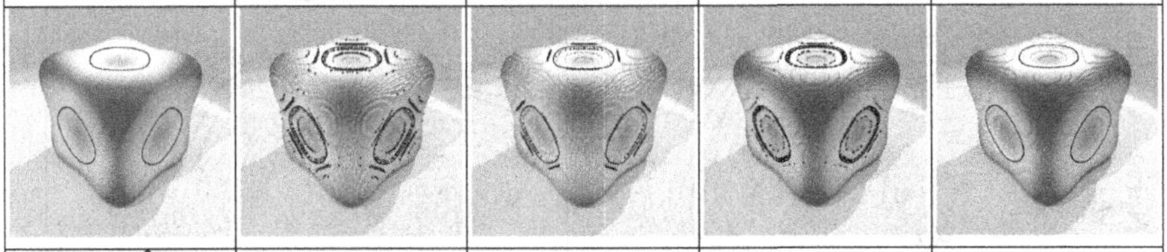				
$\|H - \hat{H}\|_2$	0.059	0.051	**0.025**	**0.003**
$\|H - \hat{H}\|_\infty$	0.197	0.146	**0.155**	**0.022**

Fig. 6. Approximation of mean curvature H with Laplace-Beltrami operator. For the computation of the diffused approximation $\hat{H}$, the diffusion time t is set to $0.04 < \frac{1}{6}h$, so very short. The projected calculus uses integral invariant normal estimator (default parameter $r = 3h^{1/2}$). Projected calculus results are more accurate and stable.

In our discrete calculus framework, the heat diffusion step (i) is obtained with a single backward Euler step for timestep t solving

$$(u_t - u_0)/t = (\mathbf{M}^0)^{-1}\mathbf{L}u_t \qquad \text{or more simply} \qquad (\mathbf{M}^0 - t\mathbf{L})u_t = \mathbf{M}^0 u_0 \,,$$

which involves the global mass matrix $\mathbf{M}^0$ and the weak Laplace-Beltrami $\mathbf{L}$.

Then, denoting w the solution u_t, step (ii) computes per-face the normalized opposite gradient d_f of w_f for face f as $d_f := -\mathbf{G}_f w_f / \|\mathbf{G}_f w_f\|_2$. Bringing back this vector to a one-form with flat operator $\mathbf{V}_f$, the per-face weak divergence is then $\delta_f := \text{Div}_f \mathbf{V}_f d_f$. Vector δ_f is a $n_f \times 1$ vector associated to face f, assigning values to each vertex of f. The global divergence map δ simply consists in summing up per vertex the divergence contribution of each face.

Finally, step (iii) computes the geodesic distance function ϕ by solving the Poisson linear system $\mathbf{L}\phi = \delta$, which must be shifted by $-\min_x(\phi)$ so that the result starts from distance 0.

Figure 7 illustrates geodesic distance computations on a digitized sphere, both with naive calculus and projected calculus (we use integral invariant normal estimations with default parameters). As shown by projecting back distances on the true smooth surface, the projected calculus builds more isotropic isodistance lines while being much more numerically accurate.

Distances on Surfaces with Boundaries. If the digital surface has boundaries, then the heat diffusion as it is written follows Neumann boundary conditions: the Laplace-Beltrami operator $\mathbf{L}$ is built per-face and ignores boundaries. Isodistance lines will tend to be orthogonal to boundaries (see Fig. 8 for an example on a half-sphere). As suggested in [5], we compute a second diffusion in step (i) that assumes Dirichlet zero boundary conditions on the digital surface boundaries. This is done by restricting the linear system $(\mathbf{M}^0 - t\mathbf{L})u_t = \mathbf{M}^0 u_0$ to non-boundary vertices, simply by assuming value 0 for u on boundary vertices.

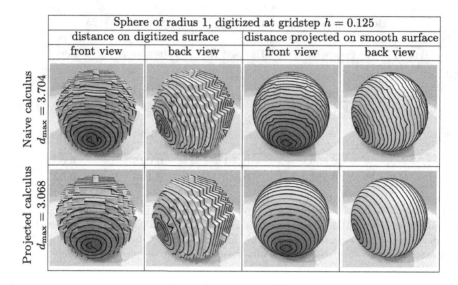

Fig. 7. Distance computation to a source point with geodesic in heat method on a digital sphere. Scale is deep blue (0) to red (max distance) with black isolines every 0.1 and red isolines every 1. The expected max distance is π. Top row displays distances computed with naive polygonal calculus. Bottom row displays distances computed with projected calculus with normals **u** estimated with integral invariant method. All computations made with parameter $\lambda = 0.25$ and heat diffusion time $t = h^2$.

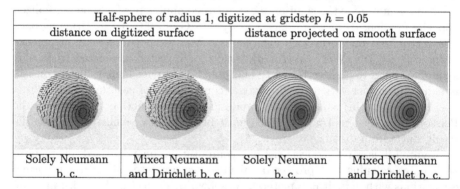

Fig. 8. Distance computation to a source point with geodesic in heat method on a digital surface with boundaries (here a half-sphere). Results are more accurate when mixing two heat diffusion solutions, one with Neumann boundary conditions, one with Dirichlet boundary conditions.

We then average the two solutions to define the vector w that is given to step (ii) of the heat method. Figure 8 confirms the soundness of this approach.

Processing on Non-manifold Surfaces. Since every calculus operator is defined per-face, they are quite oblivious to non-manifold edges and vertices

Fig. 9. Geodesic distance computation on non-manifold surfaces: the projected calculus is quite oblivious to non-manifold parts of the surface, as illustrated on this triple junction surface, digitized coarsely at $h = 1$ (left) and finely at $h = 0.125$ (right).

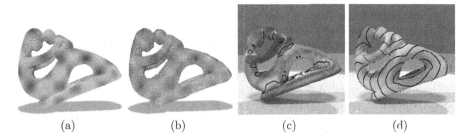

(a)   (b)   (c)   (d)

Fig. 10. Various computations on a more complex voxel object: (from left to right) the gradients of a scalar function, mean curvature and geodesic distances.

of the digital surface. We illustrate this fact on Fig. 9 by computing the geodesic distance to a source point on a digital surface approximating the triple junction of three planes (exact same code as for Fig. 7 and 8).

5   Conclusion

Our contribution is a simple and easily implementable discrete differential calculus for the processing of scalar functions on digital surfaces (see Fig. 10). It relies on changing the natural embedding of the digital surface when constructing per face differential operators, with the use of an external, multigrid convergent, normal vector estimation. Many experiments back up the effectiveness of this new calculus.

Although we demonstrate its interest to solve integrated problems such as Poisson problem (and the geodesic distance estimation is a perfect example of this class), the proposed local construction does not achieve pointwise convergence for second order operators like the Laplace-Beltrami one. Interesting challenges include the design of nonlocal operators, similarly to [3] or [9], or the use of higher order embeddings for the digital surface.

References

1. Arnold, D.N., Falk, R.S., Winther, R.: Differential complexes and stability of finite element methods I. The de Rham complex. In: Arnold, D.N., Bochev, P.B., Lehoucq, R.B., Nicolaides, R.A., Shashkov, M. (eds.) Compatible Spatial Discretizations. The IMA Volumes in Mathematics and its Applications, vol. 142, pp. 23–46. Springer, Heidelberg (2006). https://doi.org/10.1007/0-387-38034-5_2
2. Arnold, D.N., Falk, R.S., Winther, R.: Finite element exterior calculus, homological techniques, and applications. Acta Numer **15**, 1–155 (2006)
3. Caissard, T., Coeurjolly, D., Lachaud, J.O., Roussillon, T.: Laplace-beltrami operator on digital surfaces. J. Math. Imaging Vis. **61**(3), 359–379 (2019). https://doi.org/10.1007/s10851-018-0839-4
4. Coeurjolly, D., Klette, R.: A comparative evaluation of length estimators of digital curves. IEEE Trans. Pattern Anal. Mach. Intell. **26**(2), 252–258 (2004)
5. Crane, K., Weischedel, C., Wardetzky, M.: The heat method for distance computation. Commun. ACM **60**(11), 90–99 (2017). https://doi.org/gcj3hk
6. De Goes, F., Butts, A., Desbrun, M.: Discrete differential operators on polygonal meshes. ACM Trans. Graph. (TOG) **39**(4), 110–1 (2020)
7. Desbrun, M., Hirani, A.N., Leok, M., Marsden, J.E.: Discrete exterior calculus. arXiv preprint math/0508341 (2005)
8. Grady, L.J., Polimeni, J.: Discrete Calculus: Applied Analysis on Graphs for Computational Science. Springer, Heidelberg (2010). https://doi.org/10.1007/978-1-84996-290-2
9. Hildebrandt, K., Polthier, K.: On approximation of the Laplace-Beltrami operator and the Willmore energy of surfaces. In: Computer Graphics Forum, vol. 30, pp. 1513–1520. Wiley Online Library (2011)
10. Hildebrandt, K., Polthier, K., Wardetzky, M.: On the convergence of metric and geometric properties of polyhedral surfaces. Geom. Dedicata. **123**(1), 89–112 (2006). https://doi.org/10.1007/s10711-006-9109-5
11. Klette, R., Rosenfeld, A.: Digital Geometry: Geometric Methods for Digital Picture Analysis. Morgan Kaufmann (2004)
12. Lachaud, J.-O., Coeurjolly, D., Levallois, J.: Robust and convergent curvature and normal estimators with digital integral invariants. In: Najman, L., Romon, P. (eds.) Modern Approaches to Discrete Curvature. LNM, vol. 2184, pp. 293–348. Springer, Cham (2017). https://doi.org/10.1007/978-3-319-58002-9_9
13. Lachaud, J.O., Romon, P., Thibert, B., Coeurjolly, D.: Interpolated corrected curvature measures for polygonal surfaces. In: Computer Graphics Forum (Proceedings of Symposium on Geometry Processing), vol. 39, no. 5 (2020). https://doi.org/gmt2mq
14. Lachaud, J.O., Thibert, B.: Properties of gauss digitized shapes and digital surface integration. J. Math. Imaging Vis. **54**(2), 162–180 (2016)
15. Lévy, B., Zhang, H.: Spectral mesh processing. In: ACM SIGGRAPH 2010 Courses, pp. 1–312 (2010)
16. Mercat, C.: Discrete complex structure on surfel surfaces. In: Coeurjolly, D., Sivignon, I., Tougne, L., Dupont, F. (eds.) DGCI 2008. LNCS, vol. 4992, pp. 153–164. Springer, Heidelberg (2008). https://doi.org/10.1007/978-3-540-79126-3_15
17. Sullivan, J.M.: Curvatures of smooth and discrete surfaces. In: Bobenko, A.I., Sullivan, J.M., Schröder, P., Ziegler, G.M. (eds.) Discrete Differential Geometry. Oberwolfach Seminars, vol. 38, pp. 175–188. Springer, Heidelberg (2008). https://doi.org/10.1007/978-3-7643-8621-4_9

18. The DGtal Project: DGtal (2010). https://dgtal.org
19. Beirão da Veiga, L., Brezzi, F., Cangiani, A., Manzini, G., Marini, L.D., Russo, A.: Basic principles of virtual element methods. Math. Mod. Methods Appl. Sci. **23**(01), 199–214 (2013)
20. Beirão da Veiga, L., Brezzi, F., Marini, L.D., Russo, A.: The hitchhiker's guide to the virtual element method. Math. Mod. Methods Appl. Sci. 24(08), 1541–1573 (2014)

Distance-Driven Curve-Thinning on the Face-Centered Cubic Grid

Gábor Karai[✉][iD]

Department of Image Processing and Computer Graphics, University of Szeged,
Szeged, Hungary
karai@inf.u-szeged.hu

Abstract. Centerline (curve skeleton) is a frequently used skeleton-like shape descriptor for 3D tubular objects. This paper proposes an endpoint-based sequential curve-thinning algorithm (i.e., an iterative object reduction technique to obtain the centerline) for binary objects sampled on the face-centered cubic (FCC) grid. In order to ensure the centeredness of the resulting centerline, the thinning process is driven by distance transform. The introduced method is evaluated on distance maps computed with various distance definitions (i.e., chamfer or Euclidean distance). To the best of our knowledge, the reported algorithm is the very first curve-thinning method for the FCC grid.

Keywords: FCC grid · Distance transform · Skeletonization · Thinning

1 Introduction

Skeletonization provides region-based shape descriptors which represent the general shape of (segmented) digital binary objects [16]. In 3D, there are three types of skeleton-like features: the *medial surface* (*surface skeleton*), the *centerline* (*curve skeleton*), and the *topological kernel*. The medial surface provides an approximation of the continuous 3D skeleton, since it can contain 2D surface patches. The centerline is a line-like 1D representation of objects. In many applications [15], it is a concise representation of tubular 3D objects. The topological kernel is a minimal set of points that is topologically equivalent [11] to the original object.

Distance transform (*DT*) converts a binary picture consisting of feature and nonfeature elements to a picture, where the value of each element gives the distance to the nearest feature element [3]. *Distance-based skeletonization* relies on a distance transform where nonfeature elements form the objects in the input picture to be represented. *Thinning* is another strategy for skeletonization, which is an iterative object reduction in a topology preserving way [12]. Distance-based methods are frequently combined with thinning [1,4,17,24].

Most of the existing 3D skeletonization algorithms act on the cubic grid, in which each point is associated with an element of $\mathbb{Z}^3$. According to our best

© Springer Nature Switzerland AG 2022
É. Baudrier et al. (Eds.): DGMM 2022, LNCS 13493, pp. 354–365, 2022.
https://doi.org/10.1007/978-3-031-19897-7_28

knowledge, no one proposed curve-skeletonization methods on the face-centered cubic (FCC) grid. This non-conventional grid tessellates the 3D Euclidean space into rhombic dodecahedra [9,20]. The importance of the FCC grid shows an upward tendency due to its advantages of geometrical and topological properties [2,5,6,22,23]. In this paper, the very first curve-thinning algorithm acting on the FCC grid is presented.

The rest of this work is organized as follows: Sect. 2 reviews the basic notions of 3D digital topology and distance transform. In Sect. 3, our curve-thinning algorithm is introduced. Some experimental results are shown in Sect. 4, where we compare the algorithm's computation times with different arguments and evaluate the reconstructibility of the original objects. Finally, we round off this work with some concluding remarks in Sect. 5.

2 Basic Notions and Results

The FCC grid is usually denoted by $\mathbb{F}$, whose elements are called *points*. The FCC grid is defined as the following subset of $\mathbb{Z}^3$:

$$\mathbb{F} = \{(x, y, z) \in \mathbb{Z}^3 \mid x + y + z \equiv 0 \ (\text{mod } 2)\}. \tag{1}$$

We make a distinction among the following three types of neighborhood of a point $p = (p_x, p_y, p_z) \in \mathbb{F}$:

$$N_{12}(p) = \{(q_x, q_y, q_z) \in \mathbb{F} \mid (p_x - q_x)^2 + (p_y - q_y)^2 + (p_z - q_z)^2 = 2\}$$
$$N_6(p) = \{(q_x, q_y, q_z) \in \mathbb{F} \mid (p_x - q_x)^2 + (p_y - q_y)^2 + (p_z - q_z)^2 = 4\} \tag{2}$$
$$N_{18}(p) = N_{12}(p) \cup N_6(p)$$

Two points $p, q \in \mathbb{F}$ are i-adjacent if $q \in N_i(p)$ ($i \in \{6, 12, 18\}$), see Fig. 1.

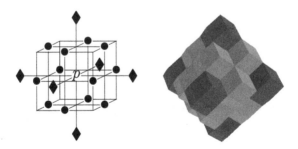

Fig. 1. The 18-neighborhood of point p ($N_{18}(p)$) in $\mathbb{F}$ (left) and its voxel representation (right). $N_{12}(p)$ contains the points marked "•" (lightgray voxels), $N_6(p)$ contains the points marked "♦" (darkgray voxels). Note that unmarked elements of $\mathbb{Z}^3$ are not in $\mathbb{F}$.

Next, we apply the fundamental concepts of digital topology as reviewed by Kong and Rosenfeld [11]. An (m, n) *binary digital picture* on the FCC grid is a quadruple $\mathcal{P} = (\mathbb{F}, m, n, B)$, where each element in $B \subseteq \mathbb{F}$ is called a *black point*, and each point in $\mathbb{F} \setminus B$ is called a *white point*. Furthermore, m and n are the assigned neighborhood relations to all black and white points, respectively. A black point p is called a *border point* if $N_n(p) \setminus B \neq \emptyset$, otherwise p is an *interior point*. Picture $\mathcal{P}$ is *finite* if B contains finitely many black points. In this paper, our attention is focused on $(18, 12)$ pictures.

Since all studied relations are symmetric, their reflexive and transitive closures generate equivalence relations, and their equivalence classes are called *components*. A black component or an *object* is an m-component of B, while a white component is an n-component of $\mathbb{F} \setminus B$. In a finite picture, there is a unique white component called *background*. A finite white component is called a *cavity*.

A *reduction* transforms a binary picture only by changing some black points to white ones (which is referred to as *deletion*). Topology preservation is a major concern in skeletonization. A 3D reduction does *not* preserve topology [10] if

- any object in the input picture is split or is completely deleted,
- any cavity in the input picture is merged with the background or another cavity,
- a cavity is created where there was none in the input picture, or
- a *hole* (that e.g. donuts have) is eliminated, merged with other holes or created.

A *simple point* is a black point whose deletion is a topology preserving reduction [11]. Sequential thinning algorithms traverse the border points of a picture, and focus on the actually visited single point for possible deletion, hence for such algorithms, the deletion of only simple points ensures topology preservation. Theorem 1 states that simpleness is a local property in $\mathbb{F}$ which can be determined by investigating the 18-neighborhood of a point. Figure 2 shows examples for simple and non-simple points.

Theorem 1 [8]. *Let p be a black point in a $(\mathbb{F}, 18, 12, B)$ picture. Then p is a simple point if and only if the following conditions hold:*

1. *Point p is 18-adjacent to exactly one 18-component of $N_{18}(p) \cap B$.*
2. *Point p is 12-adjacent to exactly one 12-component of $N_{12}(p) \setminus B$.*

The most frequently used distance functions for two points $p = (p_x, p_y, p_z)$ and $q = (q_x, q_y, q_z)$ are *neighborhood distances* [13], and the Euclidean distance, $d_e(p, q) = \sqrt{(p_x - q_x)^2 + (p_y - q_y)^2 + (p_z - q_z)^2}$. In $\mathbb{F}$, neighborhood distances $d_{12}(p, q)$ and $d_{18}(p, q)$ are taken into consideration. For better approximations of the exact Euclidean distance, the length of the moves from a point to its neighbors can be weighted according to some criteria, and the path of the minimal sum of weights called as *chamfer distances* are used [3, 25]. Let $\langle a, b, c \rangle$ denote the general *chamfer mask* for the FCC grid, where a and b are the weights assigned to all points in $N_{12}(p)$ and $N_6(p)$, respectively, weight c is assigned to all points which are $\sqrt{6}$ far from the central point 0 (see Fig. 3). For practical usage, we prefer integer weights. Some examples are presented in [7, 21], e.g.,

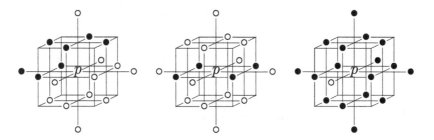

Fig. 2. Examples for simple and non-simple points in $(18, 12)$-pictures. Notice that central black point p is simple only in the left configuration because there are two and zero 12-connected white components in $N_{12}(p)$ in the middle and right situation, respectively. Hence, Condition 2 of Theorem 1 is violated in both cases.

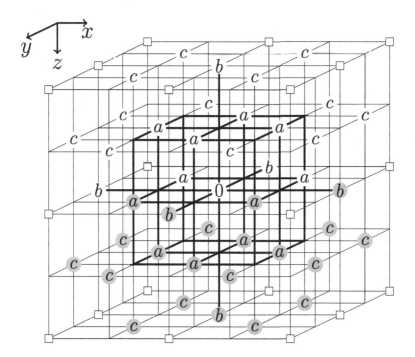

Fig. 3. Chamfer mask for distance transform on $\mathbb{F}$, in which the three weights $0 < a \leq b < c$ are taken into consideration. There are 42 examined positions in total, and they all fit into the $5 \times 5 \times 5$ local environment. Note that points depicted "□" are unused grid positions (according to the center point 0), and the unmarked elements of $\mathbb{Z}^3$ are not points in $\mathbb{F}$. Neighbors labeled letters with white and lightgray background are investigated in the forward and backward scan during distance transform, respectively.

the recommended $\langle 11, 16, 19 \rangle$, which was taken under detailed testing in Sect. 4. Note that distances d_{12} and d_{18} are equivalent to the $\langle 1, \infty, \infty \rangle$ and $\langle 1, 1, \infty \rangle$ chamfer distances, respectively. Computation of the chamfer DT requires two raster scans [7,21] (see Fig. 3), while the (non-errorfree) Euclidean DT takes four scans on the FCC grid [19]. Our testing explained in Sect. 4 is based on them.

3   The Proposed Thinning Algorithm

In this section, our distance-driven thinning algorithm is presented, see Algorithm 1, in which all border points with the same distance value are visited for possible deletion. The reported method extracts directly the centerline of the objects in $(18, 12)$ pictures. In order to preserve line branches, some geometric constraints must be applied. For this purpose, we retain either of two types of curve endpoints. Point $p \in B$ is an *endpoint of type E1*, if $|N_{18}(p) \cap B| = 1$, i.e., p has only one black 18-neighbor. A black point is said to be a *line point* if there are exactly two black points in its 18-neighborhood such that they are not 18-adjacent to each other. Point p is called an *endpoint of type E2*, if p is an endpoint of type $E1$ and its only black neighbor is a line point.

Algorithm 1: Distance-driven curve-thinning — DDCT

Input: picture $(\mathbb{F}, 18, 12, X)$ with the initial objects in it, distance d, and endpoint of type $\varepsilon \in \{E1, E2\}$

Output: picture $(\mathbb{F}, 18, 12, Y)$ containing the centerline of objects

```
   // Distance transform
1  DT ← computeDT(X, d)
   // Thinning
2  Y ← X
3  for k ← 1 to max(DT) do
4  │  repeat
5  │  │   D ← {p ∈ Y | DT(p) ≤ k, p is simple for Y and not an endpoint of type
   │  │       ε}
6  │  │   changed_any ← false
7  │  │   repeat
8  │  │   │   changed ← false
9  │  │   │   foreach p ∈ D do
10 │  │   │   │   if p is simple for Y and not an endpoint of type ε then
11 │  │   │   │   │   Y ← Y \ {p}
12 │  │   │   │   │   D ← D \ {p}
13 │  │   │   │   │   changed ← true
14 │  │   │   │   │   changed_any ← true
15 │  │   │   until changed = false
16 │  until changed_any = false
```

For the implementation, we use a computationally efficient scheme for thinning algorithms, which was introduced by Palágyi [14]. This method utilizes the fact that all thinning algorithms may delete only border points. Hence, we do not have to evaluate the deletion rules for interior points. All border points are stored in a list in the current picture. Thus, the repeated scans/traverses of the entire array (that stores the actual picture) can be avoided.

To ensure distance-driven thinning, all distinct distance values are collected into a set right after the distance transform. The *for* loop (see lines 3–16) iterates through this set in ascending order. As a consequence, the total number of iterations is equal to the number of (positive) unique distance values, which depends on the shape of objects and the chosen distance function.

4   Results

The reported algorithm was tested on several objects of different shapes. Five of them are shown in Fig. 4. We examined the reconstructibility of objects from their centerline (see Fig. 5, 6, 7, 8 and 9) by performing reverse distance transform [4, 18].

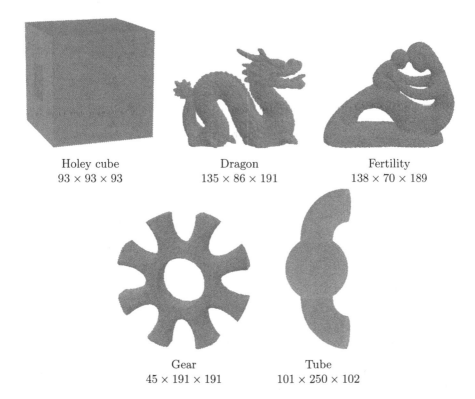

Holey cube
$93 \times 93 \times 93$

Dragon
$135 \times 86 \times 191$

Fertility
$138 \times 70 \times 189$

Gear
$45 \times 191 \times 191$

Tube
$101 \times 250 \times 102$

Fig. 4. The selected five test objects.

To make the difference between the applied parameters more visible, the produced centerlines with distinct line endpoint criteria are fusioned, where red, green and gray voxels belong to the $E1$, $E2$ and both produced centerlines with the indicated parameters, respectively. Furthermore, the transparent contour of the original object is displayed in order to verify the centeredness of the resulting centerlines. We can observe that preserving endpoints of type $E2$ leaves fewer branches compared to $E1$. Thus, the centerline with condition $E2$ gives a more concise representation of the object. However, the reconstructed object is usually less compact because of the loss of details of its shape. We got the worst quality of reconstruction in the case of the holey cube, because it is the least tubular object without any curved surface. Hence, centerline is not a suitable shape descriptor for it. Since sequential thinning algorithms are sensitive to the visiting order of border points, they may produce asymmetric skeletons, just like criterion $E1$ from the holey cube or the tube.

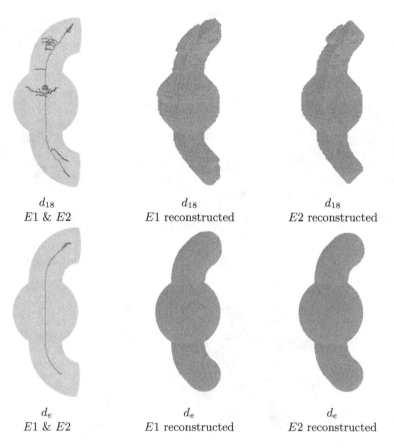

d_{18}
$E1$ & $E2$

d_{18}
$E1$ reconstructed

d_{18}
$E2$ reconstructed

d_e
$E1$ & $E2$

d_e
$E1$ reconstructed

d_e
$E2$ reconstructed

Fig. 5. Produced centerlines of the tube by using d_{18} and Euclidean distance (left) and the reconstructed objects (middle and right).

Table 1. Evaluation of Algorithm 1 on objects showed in Fig. 4. The investigated distance functions are listed in ascending order of their geometric accuracy.

Test object	Distance function	Endpoint type	Running time (millisec)	Object reconstruction (%)
Holey cube	d_{18}	$E1$	135	48.5
		$E2$	130	47.5
	d_{12}	$E1$	139	40.7
		$E2$	139	32.6
	$\langle 11, 16, 19 \rangle$	$E1$	177	57.9
		$E2$	176	51.8
	d_e	$E1$	299	58.8
		$E2$	289	58.0
Dragon	d_{18}	$E1$	122	73.3
		$E2$	120	65.7
	d_{12}	$E1$	139	82.6
		$E2$	137	77.1
	$\langle 11, 16, 19 \rangle$	$E1$	177	86.4
		$E2$	170	80.8
	d_e	$E1$	265	88.0
		$E2$	257	82.6
Fertility	d_{18}	$E1$	111	70.1
		$E2$	110	64.4
	d_{12}	$E1$	123	83.5
		$E2$	122	78.4
	$\langle 11, 16, 19 \rangle$	$E1$	163	87.3
		$E2$	159	82.3
	d_e	$E1$	238	88.5
		$E2$	232	81.1
Gear	d_{18}	$E1$	130	53.4
		$E2$	127	38.6
	d_{12}	$E1$	142	81.3
		$E2$	141	75.0
	$\langle 11, 16, 19 \rangle$	$E1$	164	82.1
		$E2$	159	76.2
	d_e	$E1$	310	80.3
		$E2$	294	73.0
Tube	d_{18}	$E1$	209	62.0
		$E2$	204	55.8
	d_{12}	$E1$	230	73.0
		$E2$	223	66.8
	$\langle 11, 16, 19 \rangle$	$E1$	305	86.7
		$E2$	296	86.0
	d_e	$E1$	520	94.2
		$E2$	512	93.6

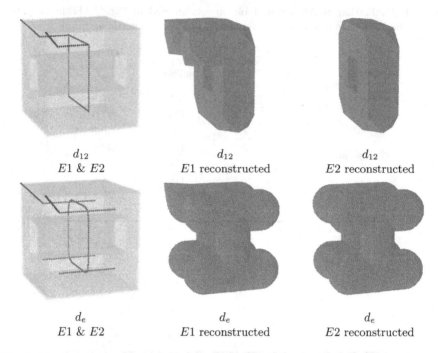

d_{12}
$E1$ & $E2$

d_{12}
$E1$ reconstructed

d_{12}
$E2$ reconstructed

d_e
$E1$ & $E2$

d_e
$E1$ reconstructed

d_e
$E2$ reconstructed

Fig. 6. Produced centerlines of the holey cube by using d_{12} and Euclidean distance (left) and the reconstructed objects (middle and right).

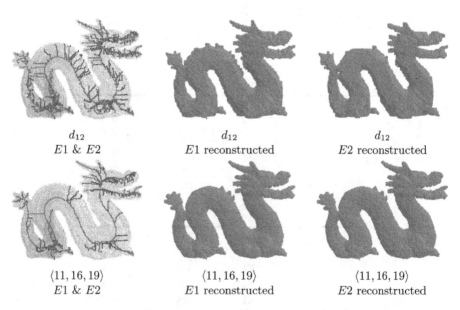

d_{12}
$E1$ & $E2$

d_{12}
$E1$ reconstructed

d_{12}
$E2$ reconstructed

$\langle 11, 16, 19 \rangle$
$E1$ & $E2$

$\langle 11, 16, 19 \rangle$
$E1$ reconstructed

$\langle 11, 16, 19 \rangle$
$E2$ reconstructed

Fig. 7. Produced centerlines of the dragon by using d_{12} and a chamfer distance (left) and the reconstructed objects (middle and right).

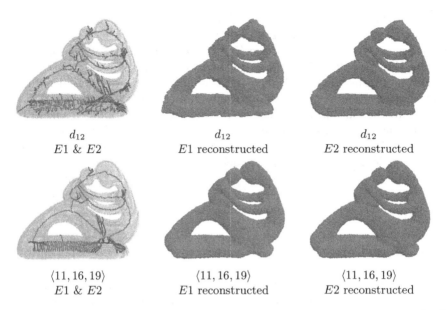

d_{12}
$E1$ & $E2$

d_{12}
$E1$ reconstructed

d_{12}
$E2$ reconstructed

$\langle 11, 16, 19 \rangle$
$E1$ & $E2$

$\langle 11, 16, 19 \rangle$
$E1$ reconstructed

$\langle 11, 16, 19 \rangle$
$E2$ reconstructed

Fig. 8. Produced centerlines of the 'fertility' by using d_{12} and a chamfer distance (left) and the reconstructed objects (middle and right).

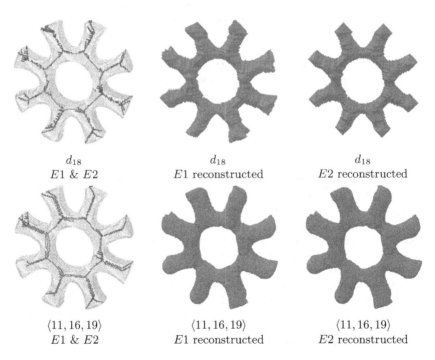

d_{18}
$E1$ & $E2$

d_{18}
$E1$ reconstructed

d_{18}
$E2$ reconstructed

$\langle 11, 16, 19 \rangle$
$E1$ & $E2$

$\langle 11, 16, 19 \rangle$
$E1$ reconstructed

$\langle 11, 16, 19 \rangle$
$E2$ reconstructed

Fig. 9. Produced centerlines of the gear by using d_{18} and a chamfer distance (left) and the reconstructed objects (middle and right).

We also measured the required time for the curve-thinning method to process. For this purpose, the `std::chrono` library was applied in the implementation written in C++ on a usual desktop (HP ProDesk 400 G4; 3.20 GHz Intel Core i5-6500; Windows 10 × 64). A detailed list can be found in Table 1. Note that just the distance transform and the iterative thinning process were considered here, file operations were not taken into account. Notice that the better the approximation to the Euclidean distance is, the more time the thinning takes. This happens due to the growing number of distinct values, which implies that the number of thinning iterations is increasing as well. Additionally, the Euclidean DT takes more raster scans than chamfer ones through the picture, which makes it computationally even more expensive. We can also observe that thinning with endpoint condition $E2$ takes less time because it leaves fewer border points to visit for possible deletion in the remaining iterations.

5   Conclusions

The proposed algorithm DDCT is the first curve-thinning method working on the FCC grid. Two types of line endpoints are introduced, and we showed that the obtained centerlines allow recovery of a significant subset of the object, especially in case of tubular-like models.

Future research will be devoted to adapting the presented results to $(12, 12)$ and $(12, 18)$ pictures and constructing parallel thinning algorithms on the FCC grid.

Acknowledgements. Project no. TKP2021-NVA-09 has been implemented with the support provided by the Ministry of Innovation and Technology of Hungary from the National Research, Development and Innovation Fund, financed under the TKP2021-NVA funding scheme.

References

1. Arcelli, C., di Baja, G.S., Serino, L.: Distance-driven skeletonization in voxel images. IEEE Trans. Pattern Anal. Mach. Intell. **33**(4), 709–720 (2011). https://doi.org/10.1109/TPAMI.2010.140
2. Biswas, R., Largeteau-Skapin, G., Zrour, R., Andres, E.: Digital objects in rhombic dodecahedron grid. Math. Morphol. Theory Appl. **4**(1), 143–158 (2020)
3. Borgefors, G.: Distance transformations in arbitrary dimensions. Comput. Vision Graph. Image Process. **27**(3), 321–345 (1984)
4. Borgefors, G., Nyström, I., di Baja, G.S.: Discrete skeletons from distance transforms in 2D and 3D. In: Siddiqi, K., Pizer, S.M. (eds.) Medial Representations. Computational Imaging and Vision, vol. 37, Springer, Dordrecht (2008). https://doi.org/10.1007/978-1-4020-8658-8_5
5. Čomić, L., Magillo, P.: Repairing 3D binary images using the FCC grid. J. Math. Imaging Vision **61**, 1301–1321 (2019)
6. Čomić, L., Magillo, P.: On hamiltonian cycles in the FCC grid. Comput. Graph. **89**, 88–93 (2020)

7. Fouard, C., Strand, R., Borgefors, G.: Weighted distance transforms generalized to modules and their computation on point lattices. Pattern Recogn. **40**(9), 2453–2474 (2007)
8. Gau, C.J., Kong, T.Y.: Minimal nonsimple sets of voxels in binary pictures on a face-centered cubic grid. Int. J. Pattern Recogn. Artif. Intell. **13**(4), 485–502 (1999)
9. Kittel, C.: Crystal structures. In: Kittel, C. (ed.) Introduction to Solid State Physics, 8th edn. Wiley, New York (2004)
10. Kong, T.Y.: On topology preservation in 2-D and 3-D thinning. Int. J. Pattern Recogn. Artif. Intell. **9**(5), 813–844 (1995)
11. Kong, T.Y., Rosenfeld, A.: Digital topology: Introduction and survey. Comput. Vision Graph. Image Process. **48**(3), 357–393 (1989)
12. Lam, L., Lee, S.W., Suen, C.Y.: Thinning methodologies - a comprehensive survey. IEEE Trans. Pattern Anal. Mach. Intell. **14**(9), 869–885 (1992)
13. Marchand-Maillet, S., Sharaiha, Y.M.: Binary Digital picture Processing: A Discrete Approach. Academic Press (2000). https://doi.org/10.1016/B978-0-12-470505-0.X5000-X
14. Palágyi, K.: A 3D fully parallel surface-thinning algorithm. Theor. Comput. Sci. **406**(1–2), 119–135 (2008)
15. Saha, P.K., Borgefors, G., di Baja, G.S.: A survey on skeletonization algorithms and their applications. Pattern Recogn. Lett. **76**(1), 3–12 (2016)
16. Saha, P.K., Borgefors, G., di Baja, G.S.: Skeletonization: Theory, Methods and Applications, 1st edn. Academic Press, Cambridge (2017)
17. Saito, T., Toriwaki, J.: A sequential thinning algorithm for three dimensional digital pictures using the Euclidean distance transformation. In: Proceedings 9th Scandinavian Conference on Picture Analysis, Uppsala, Sweden, pp. 507–516 (1995)
18. Sobiecki, A., Jalba, A., Telea, A.: Comparison of curve and surface skeletonization methods for voxel shapes. Pattern Recogn. Lett. **47**, 147–156 (2014)
19. Strand, R.: The euclidean distance transform applied to the FCC and BCC grids. In: Marques, J.S., Pérez de la Blanca, N., Pina, P. (eds.) IbPRIA 2005. LNCS, vol. 3522, pp. 243–250. Springer, Heidelberg (2005). https://doi.org/10.1007/11492429_30
20. Strand, R.: The face-centered cubic grid and the body-centered cubic grid: a literature survey. Technical Report 35, Centre for Image Analysis, Uppsala University, Uppsala, Sweden (2005)
21. Strand, R., Borgefors, G.: Distance transforms for three-dimensional grids with non-cubic voxels. Comput. Vision Image Underst. **100**(3), 294–311 (2005)
22. Strand, R., Nagy, B., Borgefors, G.: Digital distance functions on three-dimensional grids. Theor. Comput. Sci. **412**(15), 1350–1363 (2011)
23. Strand, R., Stelldinger, P.: Topology preserving marching cubes-like algorithms on the face-centered cubic grid. In: Proceedings 14th International Conference on Image Analysis and Processing (ICIAP), Modena, Italy, pp. 781–788 (2007)
24. Svensson, S.: Reversible surface skeletons of 3D objects by iterative thinning of distance transforms. In: Bertrand, G., Imiya, A., Klette, R. (eds.) Digital and Image Geometry. LNCS, vol. 2243, pp. 400–411. Springer, Heidelberg (2001). https://doi.org/10.1007/3-540-45576-0_24
25. Svensson, S., Borgefors, G.: Digital distance transforms in 3D images using information from neighbourhoods up to $5 \times 5 \times 5$. Comput. Vision Image Underst. **88**(1), 24–53 (2002). https://doi.org/10.1006/cviu.2002.0976

A New Lattice-Based Plane-Probing Algorithm

Jui-Ting Lu[(✉)] [ID], Tristan Roussillon [ID], and David Coeurjolly [ID]

Univ Lyon, CNRS, INSA Lyon, UCBL, LIRIS, UMR5205, Villeurbanne, France
jui-ting.lu@liris.cnrs.fr

Abstract. Plane-probing algorithms have become fundamental tools to locally capture arithmetical and geometrical properties of digital surfaces (boundaries of a connected set of voxels), and especially normal vector information. On a digital plane, the overall idea is to consider a local pattern, a triangle, that is expanded starting from a point of interest using simple probes of the digital plane with a predicate "Is a point x in the digital plane?". Challenges in plane-probing methods are to design an algorithm that terminates on a triangle with several geometrical properties: its normal vector should match with the expected one for digital plane (correctness), the triangle should be as compact as possible (acute or right angles only), and probes should be as close as possible to the source point (locality property). In addition, we also wish to minimize the number of iterations or probes during the computations. Existing methods provide correct outputs but only experimental evidence for these properties. In this paper, we present a new plane-probing algorithm that is theoretically correct on digital planes, and with better experimental compactness and locality than existing solutions. Additional properties of this new approach also suggest that theoretical proofs of the aforementioned geometrical properties could be achieved.

Keywords: Digital plane recognition · Plane-probing algorithm · Lattice reduction

1 Introduction

A digital surface is a quadrangular mesh that corresponds to the boundary of a union of regularly spaced unit cubes (voxels). We are interested in processing the geometry of such surfaces, for instance to recognize local elementary structures such as digital plane segments [1,4,5,14], or to estimate some differential quantities [2,3]. When performing such local computations, we usually need to capture local geometric properties of the surface around a given point. This can be done either by considering a fixed neighborhood, e.g., using a Euclidean ball with fixed radius, or by adapting such neighborhood to local geometric properties. Probing algorithms target the latter case by iteratively growing a pattern with update rules given from *probing* of the geometry. Plane-probing algorithms

---

This work has been partly funded by PARADIS ANR-18-CE23-0007-01 research grant.

É. Baudrier et al. (Eds.): DGMM 2022, LNCS 13493, pp. 366–381, 2022.
https://doi.org/10.1007/978-3-031-19897-7_29

analyze digital planes [10] without imposing a parameter that controls the size of the patch [6–9,11]. The key objective of these techniques is to exploit arithmetical and geometrical properties of the digital plane being *explored* in order to retrieve its unknown arithmetical parameters, e.g., its normal vector. When applied on generic non-planar (implicit or explicit) digital surfaces, outputs of plane-probing algorithms could be used to locally estimate the normal bundle of the surface, or could be a key ingredient for surface reconstruction [9].

Plane-probing algorithms can mainly be categorized into two types: tetrahedra-based plane-probing algorithms [8,9,11] and parallelepiped-based [6]. In this paper, we focus on tetrahedra-based plane-probing algorithms applied on a digital plane. Those algorithms update the three vertices of the tetrahedron base until it matches the normal of the digital plane. Meanwhile, the apex of the tetrahedron remains fixed (see Fig. 1).

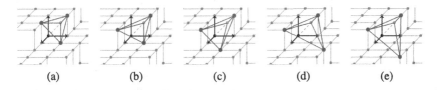

    (a)   (b)   (c)   (d)   (e)

Fig. 1. The evolution (from left to right) of a tetrahedra-based plane-probing algorithm for normal $(1, 2, 5)$.

Among existing approaches to update the tetrahedra vertices, we can mention the H-algorithm and the R-algorithm [9]. The main advantage of such approaches is their proximity to the source point. Indeed, the apex of the tetrahedron does not move, stays right above the starting point and always projects into the opposite face, i.e., the base, in the direction of the starting point (see Fig. 1 and [9, Lemma 4]). However, we do not have an upper bound of the probed area. A comparison of H-algorithm and R-algorithm is illustrated in Fig. 2, where only the triangles corresponding to the bases are drawn. The outputs of the two algorithms are identical, but H-algorithm probes a larger region than R-algorithm does. We also spot more obtuse triangles in H-algorithm's evolution. Furthermore, as stated in [12], our new algorithm also leads to additional theoretical results such as the minimality of the lattice generated by the last triangle. In this article, we mainly focus on the algorithmic sides of the new approach. The paper is divided into three parts: in Sect. 2, we recall some notations used in [11] and describe the general framework of plane-probing algorithms. We precisely describe and analyze our new algorithm in Sect. 3, whereas Sect. 4 is devoted to experimental results.

2   Plane-Probing Algorithm Variants

A standard and rational *digital plane* is an infinite digital set defined by a normal $\mathbf{N} \in \mathbb{Z}^3 \setminus \{\mathbf{0}\}$ and a shift value $\mu \in \mathbb{Z}$ as follows [10]:

$$\mathbf{P}_{\mu,\mathbf{N}} := \{\mathbf{x} \in \mathbb{Z}^3 \mid \mu \leq \mathbf{x} \cdot \mathbf{N} < \mu + \|\mathbf{N}\|_1\}.$$

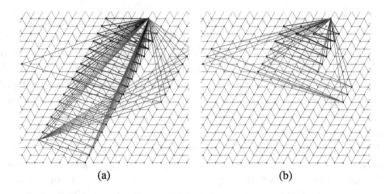

Fig. 2. The evolution for normal $(1, 73, 100)$ with H-algorithm (a) and R-algorithm (b). Every triangle of the evolution is superimposed. The initial triangle is black. The next ones are more and more blue while iterating. The last one is red. (Color figure online)

In this paper, we suppose w.l.o.g. that $\mu = 0$ and that the components of $\mathbf{N}$ are positive, i.e., $\mathbf{N} \in \mathbb{N}^3 \setminus \{\mathbf{0}\}$. Given a digital plane $\mathbf{P} \in \{\mathbf{P}_{0,\mathbf{N}} \mid \mathbf{N} \in \mathbb{N}^3 \setminus \{\mathbf{0}\}\}$ of unknown normal vector, a *plane-probing algorithm* computes the normal vector $\mathbf{N}$ of $\mathbf{P}$ by sparsely probing it with the predicate "is $\mathbf{x}$ in $\mathbf{P}$?" (InPlane($\mathbf{x}$) predicate hereafter). We describe below in a uniform way the algorithms H and R introduced in [9] as well as our new method (see also Algorithm 1).

Initialization. Let $(\mathbf{e}_0, \mathbf{e}_1, \mathbf{e}_2)$ be the canonical basis of $\mathbb{Z}^3$. We assume that a starting point $\mathbf{p}$ satisfies three conditions: (1) $\mathbf{p} \in \mathbf{P}$, (2) the apex $\mathbf{q} := \mathbf{p} + (1, 1, 1) \notin \mathbf{P}$ and (3) the initial triangle $\mathbf{T}^{(0)} := (\mathbf{v}_k^{(0)})_{k \in \mathbb{Z}/3\mathbb{Z}} \subset \mathbf{P}$, where $\mathbf{v}_k^{(0)} := \mathbf{q} - \mathbf{e}_k$ for all $k \in \mathbb{Z}/3\mathbb{Z}$ (see inset figure and Algorithm 1, line 1).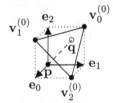

Neighborhood and Update Rule. At every step $i \in \mathbb{N}$, the triangle $\mathbf{T}^{(i)}$ is defined from updated vertices $\{\mathbf{v}_k^{(i)}\}_{k \in \mathbb{Z}/3\mathbb{Z}}$ and represents the current approximation of the plane $\mathbf{P}$. All algorithms update one vertex of $\mathbf{T}^{(i)}$ per iteration. That vertex is replaced by a point of $\mathbf{P}$ from a candidate set, also called *neighborhood* in [9]. To properly define distinct neighborhoods, we first define the following sets:

$$S_H := \{(\alpha, \beta) \in \{(1, 0), (0, 1)\}. \tag{1}$$
$$S_R := \{(\alpha, \beta) \in \{(1, \lambda), (\lambda, 1) \mid \lambda \in \mathbb{N}\}. \tag{2}$$
$$S_L := \{(\alpha, \beta) \in \mathbb{N}^2 \setminus (0, 0)\}. \tag{3}$$

Note that $S_H \subset S_R \subset S_L$. At every step i and for any $S \in \{S_H, S_R, S_L\}$, the neighborhood is now defined as follows:

$$\mathcal{N}_S^{(i)} := \left\{ \mathbf{v}_k^{(i)} + \alpha(\mathbf{q} - \mathbf{v}_{k+1}^{(i)}) + \beta(\mathbf{q} - \mathbf{v}_{k+2}^{(i)}) \mid k \in \mathbb{Z}/3\mathbb{Z}, (\alpha, \beta) \in S \right\}. \tag{4}$$

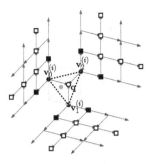

Fig. 3. Illustrations of the neighborhoods: $\mathcal{N}_{S_H}^{(i)}$ (black squares), $\mathcal{N}_{S_R}^{(i)}$ (white squares) and $\mathcal{N}_{S_L}^{(i)}$ includes every point on the lattices, excepted the triangle vertices.

See Fig. 3 for an illustration of the neighborhoods. The H-algorithm is based on $\mathcal{N}_{S_H}^{(i)}$, which looks like an Hexagon, whereas the R-algorithm is based on $\mathcal{N}_{S_R}^{(i)}$, which consists of Rays. In this paper, we propose a lattice-based algorithm, denoted by the letter L, for lattice, and which uses the largest neighborhood $\mathcal{N}_{S_L}^{(i)}$.

Let $\mathcal{H}_+^{(i)}$ be the half-space delimited by $\mathbf{T}^{(i)}$ and containing $\mathcal{N}_S^{(i)}$. In addition, let $\mathcal{B}(\mathbf{T}, \mathbf{x})$ be the closed ball defined by $\mathbf{T}^{(i)}$ and a fourth point $\mathbf{x}$ not in the plane passing by $\mathbf{T}^{(i)}$. As in [11], for any pair of points $\mathbf{x}, \mathbf{x}'$, not in the plane passing by $\mathbf{T}^{(i)}$, we say that $\mathbf{x}'$ is *closer* to $\mathbf{T}^{(i)}$ than $\mathbf{x}$, denoted $\mathbf{x}' \leq_{\mathbf{T}^{(i)}} \mathbf{x}$, if and only if $(\mathcal{B}(\mathbf{T}^{(i)}, \mathbf{x}') \cap \mathcal{H}_+^{(i)}) \subseteq (\mathcal{B}(\mathbf{T}^{(i)}, \mathbf{x}) \cap \mathcal{H}_+^{(i)})$. As $\leq_{\mathbf{T}^{(i)}}$ is reflexive and transitive, and since all pairs of points in $\mathcal{H}_+^{(i)}$ are comparable [12], it defines a total preorder.

The algorithms replace a vertex of $\mathbf{T}^{(i)}$ with a point of the set $\mathcal{N}_S^{(i)} \bigcap \mathbf{P}$ that is a closest one according to $\leq_{\mathbf{T}^{(i)}}$. More precisely, if $\mathcal{N}_S^{(i)} \bigcap \mathbf{P} \neq \emptyset$, there is at least an index $k \in \mathbb{Z}/3\mathbb{Z}$ and numbers $(\alpha, \beta) \in \mathbb{N}^2 \setminus (0, 0)$ such that

$$\forall \mathbf{x} \in \mathcal{N}_S^{(i)} \cap \mathbf{P}, \ \mathbf{v}_k^{(i)} + \alpha(\mathbf{q} - \mathbf{v}_{k+1}^{(i)}) + \beta(\mathbf{q} - \mathbf{v}_{k+2}^{(i)}) \leq_{\mathbf{T}} \mathbf{x}. \tag{5}$$

Note that the triple (k, α, β) may not be unique when several points are in a cospherical position. The update rule is then [9, Lemma 2]:

$$\begin{cases} \mathbf{v}_k^{(i+1)} := \mathbf{v}_k^{(i)} + \alpha(\mathbf{q} - \mathbf{v}_{k+1}^{(i)}) + \beta(\mathbf{q} - \mathbf{v}_{k+2}^{(i)}), \\ \mathbf{v}_{k+1}^{(i+1)} := \mathbf{v}_{k+1}^{(i)}, \\ \mathbf{v}_{k+2}^{(i+1)} := \mathbf{v}_{k+2}^{(i)}. \end{cases} \tag{6}$$

As shown in Algorithm 1, lines 5 to 7, Eqs. (5) and (6) are used to update the current triangle.

Termination. The algorithms terminate at a step n, when the neighborhood has an empty intersection with the plane, i.e., when $\mathcal{N}_S^{(n)} \bigcap \mathbf{P} = \emptyset$ (Algorithm 1, line 3). The number of steps, n, is less than or equal to $\|\mathbf{N}\|_1 - 3$ [9, Theorem 1], which is a tight bound reached for any normal of components $(1, 1, r)$ with $r \in \mathbb{N} \setminus \{0\}$.

If $\mathbf{p}$ is one of the least high points in $\mathbf{P}$, i.e., $\mathbf{p} \cdot \mathbf{N} = 0$, the vertices of $\mathbf{T}^{(n)}$ are among the highest ones in $\mathbf{P}$, i.e., $\forall k \in \mathbb{Z}/3\mathbb{Z}, \mathbf{v}_k^{(n)} \cdot \mathbf{N} = \|\mathbf{N}\|_1 - 1$ [9, Theorem 2]. A consequence is that $\mathbf{T}^{(n)}$ is aligned with $\mathbf{P}$. In other words, its normal is equal to $\mathbf{N}$ [9, Corollary 4]. In addition, one can deduce from the vertices of $\mathbf{T}^{(n)}$ a basis of the lattice $\{\mathbf{x} \mid \mathbf{x} \cdot \mathbf{N} = \|\mathbf{N}\|_1 - 1\}$ [9, Corollary 5].

Even if [9] only introduces the neighborhoods H and R, the above-mentionned results and their proofs are correct for parameters $(\alpha, \beta) \in S_L$ in the update rule and, as a consequence, for the newly introduced neighborhood $\mathcal{N}_{S_L}^{(i)}$ as well.

Our motivation for introducing such a neighborhood is linked to the compactness. Indeed, starting from an identical triangle, a point chosen by the L algorithm always lies in the circumscribed sphere that passes the point chosen by the H- or R-algorithm. Furthermore, it is proven that every circumscribed sphere of two consecutive triangles provided by the L-algorithm does not include any other points of the digital plane [12, Theorem 4]. In the next section, we show how to efficiently find a closest point in the L-neighborhood. The above-mentionned result will be crucial in Lemma 7.

Algorithm 1: Plane-probing algorithms H, R ([9]) and L (our method)

Input: The predicate InPlane := "Is a point $\mathbf{x} \in \mathbf{P}$?", a point $\mathbf{p} \in \mathbf{P}$ and the type of neighborhood $S \in \{S_H, S_R, S_L\}$ (see equations (1)–(4))

Output: A normal vector $\hat{\mathbf{N}}$ and a basis of the lattice $\{\mathbf{x} \mid \mathbf{x} \cdot \hat{\mathbf{N}} = \|\mathbf{N}\|_1 - 1\}$.

1   $\mathbf{q} \leftarrow \mathbf{p} + (1,1,1)$; $(\mathbf{v}_k^{(0)})_{k \in \mathbb{Z}/3\mathbb{Z}} \leftarrow (\mathbf{q} - \mathbf{e}_k)_{k \in \mathbb{Z}/3\mathbb{Z}}$;   // initialization

2   $i \leftarrow 0$;

3 **while** $\mathcal{N}_S^{(i)} \cap \{\mathbf{x} \mid \text{InPlane}(\mathbf{x})\} \neq \emptyset$ **do**

4   Let (k, α, β) be such that, for all $\mathbf{y} \in \mathcal{N}_S^{(i)} \cap \{\mathbf{x} \mid \text{InPlane}(\mathbf{x})\}$,

5      $\mathbf{v}_k^{(i)} + \alpha(\mathbf{q} - \mathbf{v}_{k+1}^{(i)}) + \beta(\mathbf{q} - \mathbf{v}_{k+2}^{(i)}) \leq_{\mathbf{T}^{(i)}} \mathbf{y}$;   // equation (5)

6      $\mathbf{v}_k^{(i+1)} \leftarrow \mathbf{v}_k^{(i)} + \alpha(\mathbf{q} - \mathbf{v}_{k+1}^{(i)}) + \beta(\mathbf{q} - \mathbf{v}_{k+2}^{(i)})$;   // equation (6)

7      $\forall l \in \mathbb{Z}/3\mathbb{Z} \setminus k, \mathbf{v}_l^{(i+1)} \leftarrow \mathbf{v}_l^{(i)}$;

8      $i \leftarrow i + 1$;

9   $B \leftarrow \{\mathbf{v}_0^{(i)} - \mathbf{v}_1^{(i)}, \mathbf{v}_1^{(i)} - \mathbf{v}_2^{(i)}, \mathbf{v}_2^{(i)} - \mathbf{v}_0^{(i)}\}$;

10 Let $\mathbf{b}_1$ and $\mathbf{b}_2$ be the shortest and second shortest vectors of B ;

11 **return** $\mathbf{b}_1 \times \mathbf{b}_2, (\mathbf{b}_1, \mathbf{b}_2)$;   // $\times$ denotes the cross product

---

3   The L-Algorithm

The most expensive task in Algorithm 1 is computing a point of $\mathcal{N}_S^{(i)} \cap \mathbf{P}$, which is *closest* according to $\leq_{\mathbf{T}^{(i)}}$ (see lines 4 and 5). A brute-force method would be computing the whole finite set $\mathcal{N}_S^{(i)} \cap \mathbf{P}$ and finding a point of that set closer than any others, which would require lots of probes. In practice, one does not need to probe so much, because one can safely discard a large part of $\mathcal{N}_S^{(i)} \cap \mathbf{P}$.

In this section, we focus on a step $i \in \{0, \dots, n\}$ and for the sake of simplicity, we drop the exponent (i) in the notations. Furthermore, we focus on the 2D lattice

$$\forall k \in \mathbb{Z}/3\mathbb{Z}, \ \mathcal{L}_k := \{\mathbf{v}_k + \alpha \mathbf{m}_{k+1} + \beta \mathbf{m}_{k+2} \mid (\alpha, \beta) \in S_L\},$$

where $\mathbf{m}_k := \mathbf{q} - \mathbf{v}_k$ for all $k \in \mathbb{Z}/3\mathbb{Z}$. We propose an algorithm (Algorithm 2) that selects a small and sufficient set of candidate points included in $\mathcal{L}_k$.

3.1   A Smaller Candidate Set

We introduce, in the first place, two general geometrical results which will be useful.

Lemma 1. *Let two non-zero vectors $\mathbf{u}, \mathbf{w} \in \mathbb{R}^3$ and a closed ball whose border passes through the origin o and the point $o + \mathbf{u} + \mathbf{w}$. If $\mathbf{u} \cdot \mathbf{w} \geq 0$, at least one of the two points $o + \mathbf{u}$ and $o + \mathbf{w}$ lies in the ball.*

Proof. We focus on the plane including $o, o + \mathbf{u}, o + \mathbf{w}$ (and $o + \mathbf{u} + \mathbf{w}$). In this plane, if $\mathbf{u} \cdot \mathbf{w} \geq 0$, one half of the disk of diameter $[o, o + \mathbf{u} + \mathbf{w}]$ contains $o + \mathbf{u}$, whereas the other contains $o + \mathbf{w}$. Furthermore, any other disk whose border passes through o and $o + \mathbf{u} + \mathbf{w}$ must include one of the previous halves, thus one of the two points. Since any ball whose border passes through o and $o + \mathbf{u} + \mathbf{w}$ covers such a disk, the result follows (see Fig. 4-(a)). □

Lemma 2. *Let a non-zero vector $\mathbf{u} \in \mathbb{R}^3$ and a closed ball whose border passes through the origin o and the point $o + \mathbf{u}$. No point $o + \delta \mathbf{u}$ such that $\delta > 1$ lies in the ball.*

Proof. The intersection between the ball and the ray starting from o in direction $\mathbf{u}$ is the segment $[o, o + \mathbf{u}]$, which is equal, by convexity, to the set $\{o + \delta' \mathbf{u}\}_{0 \leq \delta' \leq 1}$. The points $o + \delta \mathbf{u}$ such that $\delta > 1$ do not lie in that set and therefore do not lie in the ball. □

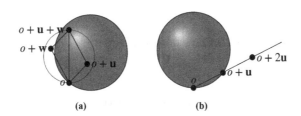

Fig. 4. Illustrations for (a) Lemma 3 and (b) Lemma 2.

An elementary application of the above lemmas is the following result:

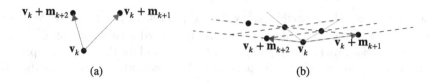

Fig. 5. Angle between $\mathbf{m}_{k+1}$ and $\mathbf{m}_{k+2}$: (a) when $\mathbf{m}_{k+1} \cdot \mathbf{m}_{k+2} \geq 0$, (b) when $\mathbf{m}_{k+1} \cdot \mathbf{m}_{k+2} < 0$ and $\mathbf{m}_{k+2} \cdot (\mathbf{m}_{k+1} + \mathbf{m}_{k+2}) < 0$. Here, we also have $(\mathbf{m}_{k+1} + \gamma\mathbf{m}_{k+2}) \cdot (\mathbf{m}_{k+1} + (\gamma + 1)\mathbf{m}_{k+2}) < 0$ with $\gamma = 1$ (see Lemma 4 and the green angle). (Color figure online)

Lemma 3. *For all $k \in \mathbb{Z}/3\mathbb{Z}$, let Λ_k be the set $\{\mathbf{v}_k + \alpha\mathbf{u} + \beta\mathbf{w} \mid (\alpha, \beta) \in S_L\}$, where $\mathbf{u}, \mathbf{w}$ are any two non-zero vectors of $\mathbb{Z}^3$ such that $\mathbf{v}_k + \mathbf{u}, \mathbf{v}_k + \mathbf{w} \in \mathcal{H}_+$. If $\mathbf{u} \cdot \mathbf{w} \geq 0$, we have either $\mathbf{v}_k + \mathbf{u} \leq_{\mathbf{T}} \mathbf{x}$ for all $\mathbf{x} \in \Lambda_k$ or $\mathbf{v}_k + \mathbf{w} \leq_{\mathbf{T}} \mathbf{x}$ for all $\mathbf{x} \in \Lambda_k$.*

Proof. Let us consider the ball $\mathcal{B}(\mathbf{T}, \mathbf{x})$ for a point $\mathbf{x} := \mathbf{v}_k + \alpha\mathbf{u} + \beta\mathbf{w}$, with $\alpha, \beta \geq 1$. Since $\alpha\mathbf{u} \cdot \beta\mathbf{w} \geq 0$, by Lemma 1, we know that either $\mathbf{v}_k + \alpha\mathbf{u}$ or $\mathbf{v}_k + \beta\mathbf{w}$ lies in $\mathcal{B}(\mathbf{T}, \mathbf{x})$. Let us assume w.l.o.g. that $\mathbf{v}_k + \alpha\mathbf{u} \in \mathcal{B}(\mathbf{T}, \mathbf{x})$, which means that $\mathbf{v}_k + \alpha\mathbf{u} \leq_{\mathbf{T}} \mathbf{x}$. By Lemma 2, we then conclude that $\mathbf{v}_k + \mathbf{u} \leq_{\mathbf{T}} \mathbf{v}_k + \alpha\mathbf{u} \leq_{\mathbf{T}} \mathbf{x}$. $\square$

Thanks to the previous lemma, if $\mathbf{m}_{k+1} \cdot \mathbf{m}_{k+2} \geq 0$, one can consider only two points of $\mathcal{L}_k$ (Fig. 5(a)). Otherwise, if both $\mathbf{m}_{k+1} \cdot (\mathbf{m}_{k+1} + \mathbf{m}_{k+2}) \geq 0$ and $(\mathbf{m}_{k+1} + \mathbf{m}_{k+2}) \cdot \mathbf{m}_{k+2} \geq 0$, one can again consider as few as three points: $\mathbf{v}_k + \mathbf{m}_{k+1}$, $\mathbf{v}_k + \mathbf{m}_{k+2}$ and $\mathbf{v}_k + \mathbf{m}_{k+1} + \mathbf{m}_{k+2}$. We now focus on the case where either $\mathbf{m}_{k+1} \cdot (\mathbf{m}_{k+1} + \mathbf{m}_{k+2})$ or $\mathbf{m}_{k+2} \cdot (\mathbf{m}_{k+1} + \mathbf{m}_{k+2})$ is strictly negative (see for instance Fig. 5(b)).

Lemma 4. *Let $\mathbf{u}, \mathbf{w}$ be two non-zero vectors in $\mathbb{Z}^3$. If there exists $\gamma \geq 1$ such that*

$$(\mathbf{u} + \gamma\mathbf{w}) \cdot (\mathbf{u} + (\gamma + 1)\mathbf{w}) < 0, \tag{7}$$

then γ is the unique integer greater than or equal to 1 that verifies

$$(\mathbf{u} + (\gamma + 1)\mathbf{w}) \cdot \mathbf{w} > 0 > (\mathbf{u} + \gamma\mathbf{w}) \cdot \mathbf{w}. \tag{8}$$

In this case, $\gamma = \left\lfloor \frac{-\mathbf{u} \cdot \mathbf{w}}{\|\mathbf{w}\|^2} \right\rfloor$.

Proof. We refer to Fig. 5(b) for an example where $\mathbf{u} = \mathbf{m}_{k+1}$ and $\mathbf{w} = \mathbf{m}_{k+2}$. By rewritting the left-hand side of (7) as $(\mathbf{u} + \gamma\mathbf{w}) \cdot ((\mathbf{u} + \gamma\mathbf{w}) + \mathbf{w})$ and developing, we get

$$\|\mathbf{u} + \gamma\mathbf{w}\|^2 + (\mathbf{u} + \gamma\mathbf{w}) \cdot \mathbf{w} < 0 \Rightarrow (\mathbf{u} + \gamma\mathbf{w}) \cdot \mathbf{w} < 0,$$

which is the right-hand side of (8). Similarly, by rewritting the left-hand side of (7) as $((\mathbf{u} + (\gamma + 1)\mathbf{w}) - \mathbf{w}) \cdot (\mathbf{u} + (\gamma + 1)\mathbf{w}) < 0$ and developing, we have

$$\|(\mathbf{u} + (\gamma + 1)\mathbf{w})\|^2 - \mathbf{w} \cdot (\mathbf{u} + (\gamma + 1)\mathbf{w}).$$

As this expression is strictly negative by (7), we obtain $(\mathbf{u} + (\gamma + 1)\mathbf{w}) \cdot \mathbf{w} > 0$, which is the left-hand side of (8). To end, by developing (8) and isolating the γ, we obtain $\gamma + 1 > \frac{-\mathbf{u} \cdot \mathbf{w}}{\|\mathbf{w}\|^2} > \gamma$, thus unicity.   □

Lemma 5. *Let $\mathbf{u}$, $\mathbf{w}$ be two non-zero vectors in $\mathbb{Z}^3$. If there exists $\gamma \geq 1$ verifying (7), then for all $c \in \{0, 1, \ldots, \gamma - 1\}$, $(\mathbf{u} + c\mathbf{w}) \cdot (\mathbf{u} + (c + 1)\mathbf{w}) > 0$.*

Proof. First, observe that for all $c \in \mathbb{N} \setminus \{0\}$,

$$(\mathbf{u} + (c - 1)\mathbf{w}) \cdot (\mathbf{u} + c\mathbf{w}) = (\mathbf{u} + c\mathbf{w}) \cdot (\mathbf{u} + (c + 1)\mathbf{w}) - 2\mathbf{w} \cdot (\mathbf{u} + c\mathbf{w}). \quad (9)$$

To determine the sign of $-2\mathbf{w} \cdot (\mathbf{u} + c\mathbf{w})$, note that we obviously have $c\mathbf{w}^2 < \gamma\mathbf{w}^2$ and, from the right-hand side of (8), $\gamma\mathbf{w}^2 < -\mathbf{u} \cdot \mathbf{w}$. As a result,

$$c\mathbf{w}^2 < -\mathbf{u} \cdot \mathbf{w} \Leftrightarrow \mathbf{w} \cdot (\mathbf{u} + c\mathbf{w}) < 0.$$

Since $-2\mathbf{w} \cdot (\mathbf{u} + c\mathbf{w}) > 0$, it is enough to show that the statement is true for $c = \gamma - 1$ because the result for the smaller values of c then follows by induction. By (8), we also have $\mathbf{w} \cdot (\mathbf{u} + \gamma\mathbf{w}) < 0 < (\mathbf{u} + \gamma\mathbf{w})^2$. Therefore,

$$2\mathbf{w} \cdot (\mathbf{u} + \gamma\mathbf{w}) < (\mathbf{u} + \gamma\mathbf{w})^2 + \mathbf{w} \cdot (\mathbf{u} + \gamma\mathbf{w}) = (\mathbf{u} + \gamma\mathbf{w}) \cdot (\mathbf{u} + (\gamma + 1)\mathbf{w}).$$

From this lower bound and replacing c by γ in (9), we finally obtain $(\mathbf{u} + (\gamma - 1)\mathbf{w}) \cdot (\mathbf{u} + \gamma\mathbf{w}) > 0$, which concludes the proof.   □

The two previous lemmas provide a set of lattice bases whose vectors form an acute angle. Indeed, with $\mathbf{u} = \mathbf{m}_{k+1}$ and $\mathbf{w} = \mathbf{m}_{k+2}$ and assuming that γ exists, we have $(\mathbf{m}_{k+1} + (\gamma + 1)\mathbf{m}_{k+2}) \cdot \mathbf{m}_{k+2} > 0$ (Lemma 4) and for all $c \in \{0, 1, \ldots, \gamma - 1\}$, $(\mathbf{m}_{k+1} + c\mathbf{m}_{k+2}) \cdot (\mathbf{m}_{k+1} + (c + 1)\mathbf{m}_{k+2}) > 0$ (Lemma 5). Then, it straightforwardly follows from Lemma 3 that the closest points in the set

$$\{\mathbf{v}_k + \mathbf{m}_{k+2}\} \cup \{\mathbf{v}_k + \mathbf{m}_{k+1} + c\mathbf{m}_{k+2} \mid c \in \{0, \ldots, \gamma + 1\}\}$$

are closer than any other points in the set

$$\mathcal{L}_k \setminus \{\mathbf{v}_k + \alpha(\mathbf{m}_{k+1} + \gamma\mathbf{m}_{k+2}) + \beta(\mathbf{m}_{k+1} + (\gamma + 1)\mathbf{m}_{k+2}) \mid \alpha, \beta \geq 1\}.$$

One part of $\mathcal{L}_k$ cannot be covered because $(\mathbf{m}_{k+1} + \gamma\mathbf{m}_{k+2}) \cdot (\mathbf{m}_{k+1} + (\gamma + 1)\mathbf{m}_{k+2}) < 0$. In order to cope with that problem, we simply recursively apply the previous results.

Definition 1. *For any pair of linearly independent non-zero vectors $(\mathbf{u}, \mathbf{w}) \in \mathbb{Z}^3 \times \mathbb{Z}^3$, we define a sequence of vector pairs $\Omega_{\mathbf{u},\mathbf{w}} = \{(\mathbf{u}_j, \mathbf{w}_j)\}_{j \geq 0}$ as follows:*

1. $\mathbf{u}_0 = \mathbf{u}$ *and* $\mathbf{w}_0 = \mathbf{w}$.
2. *For any $j \geq 0$, the pair $(\mathbf{u}_{j+1}, \mathbf{w}_{j+1})$ exists if and only if there exists $\gamma_j \geq 1$ such that*

$$(\mathbf{u}_j + \gamma_j \mathbf{w}_j) \cdot (\mathbf{u}_j + (\gamma_j + 1)\mathbf{w}_j) < 0, \quad (10)$$

then

$$\mathbf{u}_{j+1} = \mathbf{w}_j, \quad \mathbf{w}_{j+1} = \mathbf{u}_j + \gamma_j \mathbf{w}_j. \quad (11)$$

Definition 2 (Candidate set). *For* $k \in \mathbb{Z}/3\mathbb{Z}$ *and for any pair of vectors* $(\mathbf{u}, \mathbf{w})$ *in the set* $\{(\mathbf{m}_{k+1}, \mathbf{m}_{k+2}), (\mathbf{m}_{k+2}, \mathbf{m}_{k+1})\}$, *we define*

$$C_k := \bigcup_{(\mathbf{u}_j, \mathbf{w}_j) \in \Omega(\mathbf{u}, \mathbf{w})} \{\mathbf{v}_k + \mathbf{w}_j\} \cup \{\mathbf{v}_k + \mathbf{u}_j + c\mathbf{w}_j \mid c \in \{0, \ldots, \gamma_j + 1\}\}.$$

The finiteness of C_k stems from the finiteness of $\Omega_{\mathbf{u}, \mathbf{w}}$:

Lemma 6. *The sequence* $\Omega_{\mathbf{u}, \mathbf{w}} = \{(\mathbf{u}_j, \mathbf{w}_j)\}_{j \geq 0}$ *is finite.*

Proof. From (11), we have for any $j \geq 0$, $-\mathbf{u}_{j+1} \cdot \mathbf{w}_{j+1} = -\mathbf{w}_j \cdot (\mathbf{u}_j + \gamma_j \mathbf{w}_j)$. Developing the last expression, we obtain $-\mathbf{u}_j \cdot \mathbf{w}_j - \gamma_j \|\mathbf{w}_j\|^2$, which is strictly less than $-\mathbf{u}_j \cdot \mathbf{w}_j$. Therefore, the sequence of natural numbers $\{-\mathbf{u}_j \cdot \mathbf{w}_j\}_{j \geq 0}$ is strictly decreasing. Since, in addition, $-\mathbf{u}_j \cdot \mathbf{w}_j \geq \|\mathbf{w}_j\|^2$, while there exists $\gamma_j \geq 1$, the sequence $\Omega_{\mathbf{u}, \mathbf{w}}$ is finite. $\qquad\square$

3.2   Even Smaller Candidate Set

The set C_k described in the previous section is a union of subsets of aligned points. We show below that, for each subset, the last point is always closer than the other ones:

Lemma 7. *For any vectors* $\mathbf{u}, \mathbf{w} \in \mathcal{L}_k$ *such that there exists* $\gamma \geq 1$ *such that* $(\mathbf{u} + \gamma\mathbf{w}) \cdot (\mathbf{u} + (\gamma+1)\mathbf{w}) < 0$, *then for any* $0 \leq c \leq \gamma - 1$, *we have* $\mathbf{u} + \gamma\mathbf{w} \leq_{\mathbf{T}} \mathbf{u} + c\mathbf{w}$.

Proof. We assume w.l.o.g. that $k = 0$ and we use the notation $\delta_{\mathbf{T}}^0(\mathbf{x}, \mathbf{y})$ introduced in [11], where $\mathbf{x}$ and $\mathbf{y}$ are relative points of $\mathbb{Z}^3$ when considering $\mathbf{v}_0$ as origin. We recall that if $\mathbf{v}_0 + \mathbf{x} \in \mathcal{H}_+$, then $\mathbf{v}_0 + \mathbf{x} \leq_{\mathbf{T}} \mathbf{v}_0 + \mathbf{y} \Leftrightarrow \delta_{\mathbf{T}}^0(\mathbf{x}, \mathbf{y}) \geq 0$.

In order to show that for all $0 \leq c \leq \gamma - 1$, $\delta_{\mathbf{T}}^0(\mathbf{u} + \gamma\mathbf{w}, \mathbf{u} + c\mathbf{w}) \geq 0$, we use the following identity [11, equation (6)]:

$$\delta_{\mathbf{T}}^0(\mathbf{z}, \mathbf{z}' + \mathbf{z}'') = \delta_{\mathbf{T}}^0(\mathbf{z}, \mathbf{z}') + \delta_{\mathbf{T}}^0(\mathbf{z}, \mathbf{z}'') + (2\mathbf{z}' \cdot \mathbf{z}'') \det[\mathbf{m}_0 - \mathbf{m}_1, \mathbf{m}_0 - \mathbf{m}_2, \mathbf{z}]. \quad (12)$$

Indeed, as $c = \gamma - (\gamma - c)$, we obtain (with $\mathbf{z} = \mathbf{z}' = \mathbf{u} + \gamma\mathbf{w}$ and $\mathbf{z}'' = -(\gamma - c)\mathbf{w}$):

$$\delta_{\mathbf{T}}^0(\mathbf{u} + \gamma\mathbf{w}, \mathbf{u} + c\mathbf{w}) = \underbrace{\delta_{\mathbf{T}}^0(\mathbf{u} + \gamma\mathbf{w}, \mathbf{u} + \gamma\mathbf{w})}_{=0} + \underbrace{\delta_{\mathbf{T}}^0(\mathbf{u} + \gamma\mathbf{w}, -(\gamma - c)\mathbf{w})}_{\geq 0, \text{ see item } 1.}$$

$$- 2(\gamma - c)\ \underbrace{(\mathbf{u} + \gamma\mathbf{w}) \cdot \mathbf{w}}_{\leq 0 \text{ by Lemma } 4\ (8)}\ \underbrace{\det[\mathbf{m}_0 - \mathbf{m}_1, \mathbf{m}_0 - \mathbf{m}_2, \mathbf{u} + \gamma\mathbf{w}]}_{>0, \text{ see item } 2.}.$$

1. Let $\mathcal{H}_-$ be the half-space lying below the plane incident to $\mathbf{T}$. Let us set $\mathbf{x} := \mathbf{u} + \gamma\mathbf{w}$ and $\mathbf{y} := -(\gamma - c)\mathbf{w}$. By definition, $\mathbf{v}_0 + \mathbf{x} \in \mathcal{H}_+$ and $\mathbf{v}_0 + \mathbf{y} \in \mathcal{H}_-$. We have to prove that $\mathbf{v}_0 + \mathbf{x} \leq_{\mathbf{T}} \mathbf{v}_0 + \mathbf{y}$. Let $\mathbf{x}^\star$ be the closest point chosen for update. By definition, $\mathbf{x}^\star \leq_{\mathbf{T}} \mathbf{v}_0 + \mathbf{x}$, which implies that $(\mathcal{H}_- \cap \mathcal{B}(\mathbf{T}, \mathbf{v}_0 + \mathbf{x})) \subseteq (\mathcal{H}_- \cap \mathcal{B}(\mathbf{T}, \mathbf{x}^\star))$ (see Lemma 8 in appendix). Due to the above inclusion relation, since $\mathbf{v}_0 + \mathbf{y}$ is not in the interior of $\mathcal{B}(\mathbf{T}, \mathbf{x}^\star)$ [12, Theorem 4], $\mathbf{v}_0 + \mathbf{y}$ is not in the interior of $\mathcal{B}(\mathbf{T}, \mathbf{v}_0 + \mathbf{x})$ either, i.e., $\mathbf{v}_0 + \mathbf{x} \leq_{\mathbf{T}} \mathbf{v}_0 + \mathbf{y}$.

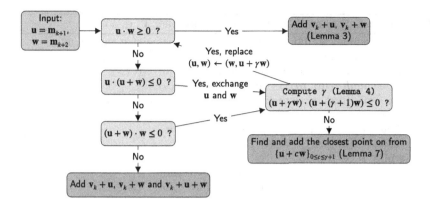

Fig. 6. Roadmap

2. For any $(\alpha, \beta) \in S_L$, $\det[\mathbf{m}_0 - \mathbf{m}_1, \mathbf{m}_0 - \mathbf{m}_2, \alpha\mathbf{m}_1 + \beta\mathbf{m}_2] = \alpha + \beta > 0$, because $\det[\mathbf{m}_0, \mathbf{m}_1, \mathbf{m}_2] = 1$ [9, Lemma 3]. Notably, $\det[\mathbf{m}_0 - \mathbf{m}_1, \mathbf{m}_0 - \mathbf{m}_2, \mathbf{u} + \gamma\mathbf{w}] > 0$. $\qquad\square$

Lemma 7 shows that the last point should be the closest. However, in the case where this last point is not in **P**, we can resort to a binary search as in [9, Algorithm 4].

3.3 Algorithm and Complexity

Figure 6 sums up the process of filtering the set $\mathcal{L}_k$. However, we have to discard the points that are not in **P**. For this purpose, we use the predicate InPlane in the whole procedure detailed in Algorithm 2. We set $\omega := \|\mathbf{N}\|_1$. The worst-case number of predicate calls is in $O(\omega)$ for the H-algorithm, $O(\omega \log \omega)$ for the R-algorithm [9] and $O(\omega)$ for the R^1-algorithm [11]. We give below a upper bound for the L-algorithm.

Theorem 1. *Algorithm 2 requires $O(\log \omega)$ calls to the predicate* InPlane: *"is* **x** *in* **P** *?".*

Proof. We consider the sequence of vectors $(\mathbf{u}_j, \mathbf{w}_j)_{0 \leq j \leq j_{max}}$. For any $j \geq 2$, if we rewrite the Eq. (11) with only $\mathbf{u}_{j-2}$, $\mathbf{u}_{j-1}$ and $\mathbf{u}_j$, we obtain the relation $\mathbf{u}_j = \mathbf{u}_{j-2} + \gamma_j \mathbf{u}_{j-1}$. We use the bar notation $\bar{\cdot}$ above any vector **x** to denote its height relative to **N**. Otherwise said, $\bar{\mathbf{x}} := \mathbf{x} \cdot \mathbf{N}$. Then, we have $\bar{\mathbf{u}}_j = \bar{\mathbf{u}}_{j-2} + \gamma_j \bar{\mathbf{u}}_{j-1} \geq \bar{\mathbf{u}}_{j-2} + \bar{\mathbf{u}}_{j-1}$ (because $\gamma_j \geq 1$ and $\bar{\mathbf{u}}_{j-1} \geq 0$ by recurrence). By induction, we have for all $2 \leq j \leq j_{max}$, $\bar{\mathbf{u}}_j \geq 2^{\lfloor \frac{j}{2} \rfloor}(\bar{\mathbf{u}}_0 + \bar{\mathbf{u}}_1)$, which leads to $j_{max} \in O(\log \omega)$, because the last point must be in **P**, i.e., $\bar{\mathbf{u}}_{j_{max}} \leq \omega$. Note that there is only one call to the predicate at each rank $2 \leq j \leq j_{max}$ (and at most four calls before), hence a total of $O(\log \omega)$ calls at the last rank. It remains to notice that the final search also requires at most $O(\log \omega)$ calls with an appropriate procedure such as [9, Algorithm 4]. $\qquad\square$

Algorithm 2: CREATECANDIDATELIST(InPlane, $\mathbf{T}$, $\mathbf{q}$, k)

Input: The predicate InPlane, the triangle $\mathbf{T}$, the point $\mathbf{q}$ and an index
$k \in \{0, 1, 2\}$

Output: A list $Cand_k$ of candidate points around vertex $\mathbf{v}_k$

1 Initialize $Cand_k$; $(\mathbf{m}_1, \mathbf{m}_2) \leftarrow (\mathbf{q} - \mathbf{v}_{k+1}, \mathbf{q} - \mathbf{v}_{k+2})$;

2 Add $\mathbf{v}_k + \mathbf{m}_1$ (resp. $\mathbf{v}_k + \mathbf{m}_2$) to $Cand_k$ if InPlane($\mathbf{v}_k + \mathbf{m}_1$) (resp.
InPlane($\mathbf{v}_k + \mathbf{m}_2$));

3 **if** InPlane($\mathbf{v}_k + \mathbf{m}_1$) *and* InPlane($\mathbf{v}_k + \mathbf{m}_2$) **then**

4      $(\mathbf{u}, \mathbf{w}) \leftarrow (\mathbf{m}_1, \mathbf{m}_2)$;

5   **while** $\mathbf{u} \cdot \mathbf{w} < 0$ **do**

6   **if** $\mathbf{u} \cdot (\mathbf{u} + \mathbf{w}) \leq 0$ *or* $\mathbf{w} \cdot (\mathbf{u} + \mathbf{w}) \leq 0$ **then**

7   **if** $\mathbf{u} \cdot (\mathbf{u} + \mathbf{w}) \leq 0$ **then**

8                  $(\mathbf{u}, \mathbf{w}) \leftarrow (\mathbf{w}, \mathbf{u})$;

9   Compute $\gamma = \left\lfloor \frac{-\mathbf{u} \cdot \mathbf{w}}{\|\mathbf{w}\|^2} \right\rfloor$;

10   **if** $(\mathbf{u} + \gamma\mathbf{w}) \cdot (\mathbf{u} + (\gamma + 1)\mathbf{w}) < 0$ **then**

11   **if** InPlane($\mathbf{u} + \gamma\mathbf{w}$) **then**

12   Add $\mathbf{v}_k + \mathbf{u} + \gamma\mathbf{w}$ to $Cand_k$;

13                      $(\mathbf{u}, \mathbf{w}) \leftarrow (\mathbf{w}, \mathbf{u} + \gamma\mathbf{w})$;

14   **else**

15   Find a closest point $\mathbf{x}^* \in \{\mathbf{v}_k + \mathbf{u} + c\mathbf{w}\}_{0 \leq c \leq \gamma+1}$ such that InPlane($\mathbf{x}^*$) and add it to $Cand_k$; **break**;

16   **else**

17   Find a closest point $\mathbf{x}^* \in \{\mathbf{v}_k + \mathbf{u} + c\mathbf{w}\}_{0 \leq c \leq \gamma-1}$ such that InPlane($\mathbf{x}^*$) and add it to $Cand_k$; **break**;

18   **else**

19   Add $\mathbf{v}_k + \mathbf{u} + \mathbf{w}$ to $Cand_k$ if InPlane($\mathbf{v}_k + \mathbf{u} + \mathbf{w}$); **break**;

20 **return** $Cand_k$;

A straightforward corollary is that the total number of predicate calls is in $O(\omega \log \omega)$ for the L-algorithm, because there are $O(\omega)$ steps (see Sect. 2) and $O(\log \omega)$ calls to the predicate at every step due to the use of Algorithm 2 (Theorem 1).

4 Experimental Results

Overall Performance. First of all, Fig. 7 compares the number of predicate calls for different plane-probing algorithms in a simple family of digital planes. The figure also shows the result of an optimized variant of the L-algorithm, denoted L-opt, that decreases the number of calls at each step by some values that are bounded by a constant. The points of the H-neighborhood are included in the L-neighborhood and are necessarily probed by our method. However, some of the points remain inside the H-neighborhood after the vertex update.

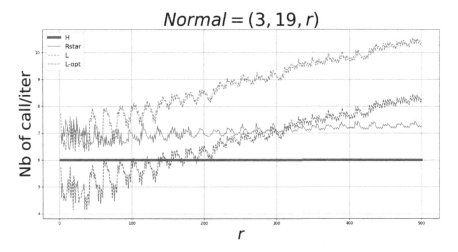

Fig. 7. Number of calls to predicate per iteration for $\mathbf{N} \in \{(3, 19, r), 1 \leq r \leq 500\}$.

Table 1. Statistics of plane-probing algorithms (on planes whose normal is in χ on the left or lying between $(1, 1, 1)$ and $(80, 80, 80)$ on the right). $\mathcal{N}_{call}^i$ denotes the number of calls to predicate at a step i and n is the number of steps. $C_{\mathbf{N}}$ denotes the number of points lying both in $\mathbf{P}$ and in $\mathcal{B}_{\mathbf{T}^{(i)}, \mathbf{T}^{(i+1)}}$, the closed ball that passes through the vertices of two consecutive triangles.

	n	$\mathcal{N}_{call}^i$		$\sum_{i=0}^{n-1} \mathcal{N}_{call}^i$		$\mathbf{N}$ s.t. $C_{\mathbf{N}} > 0$	$C_{\mathbf{N}}$	
alg.	avg.	avg.	max.	avg.	alg.	tot.	tot.	avg.
H	25.3756	6.00	6	152.25	H	247457	75235972	471.46
R	19.2534	17.73	25	271.31	R	90	424	2.44
R^1	19.2534	9.77	15	131.23	R^1	-	-	-
L	19.2529	12.03	21	144.85	L	0	0	0

Therefore, instead of probing all of them repeatedly at each iteration, one can use a cache so as not to probe twice the same point.

To provide more statistics, we have considered a large collection of implicit digital planes with normal vectors in a set χ with relatively prime components, in the range $(1, 1, 1)$ to $(200, 200, 200)$ ($|\chi| = 6578833$). For all variants of plane-probing algorithms, including our L-algorithm, we compare in Table 1 (left): the number n of steps, the number $\mathcal{N}_{call}^i$ of calls to the predicate per iteration and the total number of calls $\sum_{i=0}^{n-1} \mathcal{N}_{call}^i$. The results are obtained from a C++ implementation using the DGtal Library [13]. The numbers do not perfectly match with the table shown in [11] due to different implementation choices (*e.g.*, the ordering in case of co-spherical points).

In average, the L-algorithm requires a fewer number of steps to obtain the exact normal vector of the plane. We also remark that it usually examines fewer points at each step than the R-algorithm. However, it does not beat the R^1-

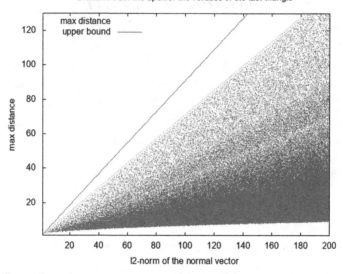

Fig. 8. The relation between maximum distance and the l_2-norm of normal vectors. Each green dot corresponds to the output of the L-algorithm for a given normal vector in χ. The theoretical upper bound is in blue. (Color figure online)

algorithm, the optimized version of R-algorithm, in terms of the number of calls to the predicate.

Locality. We wish to estimate the proximity of the probes to the initial vertex during the iterations. We define the *max distance $Dist_{max}$* of the last triangle $\mathbf{T}^{(n)}$ as $\max_k\{\|\mathbf{m}_k^{(n)}\|\}$. Since the last triangle obtained by the L-algorithm has only acute or right angles [12, Corollary 1], one can derive the following upper bound (see the last section of [12]):

$$\|Dist_{max}\| \leq \sqrt{\frac{2}{3}\|\mathbf{N}\|_2^2 + \frac{2}{\sqrt{3}}\|\mathbf{N}\|_2 + \frac{1}{\|\mathbf{N}\|_2^2}} . \tag{13}$$

In Fig. 8, we measure the max distance of the last triangle computed by the L-algorithm for all normals whose l_2-norm is less than 200 and compare them with the above theoretical bound. Both the theoretical bound and the bound given by experiments shows that the max distance is linear with respect to $\|\mathbf{N}\|_2$ and the thickness of the digital plane.

Compactness. For all $i \in \{0, \cdots, n-1\}$, let $\mathcal{B}_{\mathbf{T}^{(i)},\mathbf{T}^{(i+1)}}$ be the closed ball that passes through the vertices of two consecutive triangles. We tested for all normals in the set χ, that for the L-algorithm the sequence of radii of $\{\mathcal{B}_{\mathbf{T}^{(i)},\mathbf{T}^{(i+1)}}\}_{0 \leq i \leq n-1}$ is non-decreasing. This is not the case for H-algorithm

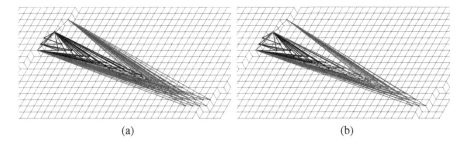

Fig. 9. The evolution of the algorithm R(a) and L(b) for the normal vector $(2, 5, 156)$.

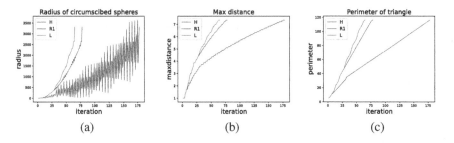

Fig. 10. Various measures for $\mathbf{N} = (198, 195, 193)$ per iteration during probings with the H, the R^1 and our L-algorithm: (from left to right) radius of $\mathcal{B}_{\mathbf{T}^{(i)}, \mathbf{T}^{(i+1)}}$, maximal distance to $\mathbf{q}$, and perimeters of the triangles.

nor R-algorithm. An example is shown in the Fig. 10. We also count the number of points in $\mathbf{P}$ and strictly inside the balls $\left\{ \mathcal{B}_{\mathbf{T}^{(i)}, \mathbf{T}^{(i+1)}} \right\}_{0 \leq i \leq n-1}$ for all normals of coprime coordinates between $(1, 1, 1)$ and $(80, 80, 80)$ (See Table 1, right). No points are found in any balls for the L-algorithm while 75235972 points are found for H-algorithm and 424 points for R-algorithm.

In particular, we observe for all vectors in χ that the steps of L-algorithm are included in the ones of R-algorithm. This implies that L-algorithm always needs fewer steps than the R-algorithm. For example, for the plane of normal vector $\mathbf{N} = (2, 5, 6)$ (see Fig. 9), the L-algorithm uses 40 steps while the R-algorithm uses 50 steps to find the exact normal vector. In Fig. 10, we also observe that the curves of L-algorithm stop earlier than other plane-probing algorithms.

5   Conclusion

In this paper, we present a new plane-probing algorithm, called L-algorithm, that takes into account more candidate points at each step than its predecessors. We also observe that the L-algorithm requires fewer steps than the R-algorithm and H-algorithm. In contrary, at each step, it needs to examine more candidate points to find the closest point in its neighborhood. Despite this downside, the point selected by the L-algorithm provides more interesting compactness features at every step. The circumspheres of consecutive triangles has non-decreasing radii and do not include any point in the plane. In other words, the L-algorithm creates a local 3D delaunay triangulation of the digital plane. We also proposed an optimization of the plane-probing algorithm to reduce the number of predicate calls. This improvement could be extended to the cases where we probe on a digital surface.

In the future, we wish to bound from above the distance of all vertices to the starting point in order to objectively measure the localness of plane-probing algorithms. Another perspective is to further optimize the L-algorithm to reduce its complexity to an amortized linear complexity.

Acknowledgements. We would like to thank the anonymous reviewers for having devoted their time and effort to their extensive and constructive feedback. It helps us considerably improve the content of this paper.

A   Inclusion Relation in $\mathcal{H}_-$ Used in Lemma 7

Lemma 8. *If* $\mathbf{y} \leq_{\mathbf{T}} \mathbf{x}$, *then* $(\mathcal{B}(\mathbf{T}, \mathbf{x}) \cap \mathcal{H}_-) \subseteq (\mathcal{B}(\mathbf{T}, \mathbf{y}) \cap \mathcal{H}_-)$.

Proof. For any pair $\mathbf{x}, \mathbf{y} \in \mathcal{H}_-$, we denote $\mathbf{y} \preceq_{\mathbf{T}} \mathbf{x}$ if and only if $(\mathcal{B}(\mathbf{T}, \mathbf{y}) \cap \mathcal{H}_-) \subseteq (\mathcal{B}(\mathbf{T}, \mathbf{x}) \cap \mathcal{H}_-)$. As for $\leq_{\mathbf{T}}$, note that $\preceq_{\mathbf{T}}$ is a total preorder. Let us consider now two points $\mathbf{x}' \in (\partial\mathcal{B}(\mathbf{T}, \mathbf{x}) \cap \mathcal{H}_-)$ and $\mathbf{y}' \in (\partial\mathcal{B}(\mathbf{T}, \mathbf{y}) \cap \mathcal{H}_-)$ (both points lie on the boundary of either $\mathcal{B}(\mathbf{T}, \mathbf{x})$ and $\mathcal{B}(\mathbf{T}, \mathbf{y})$ in $\mathcal{H}_-$). Note that, by construction, $\mathcal{B}(\mathbf{T}, \mathbf{y}') = \mathcal{B}(\mathbf{T}, \mathbf{y})$ and $\mathcal{B}(\mathbf{T}, \mathbf{x}') = \mathcal{B}(\mathbf{T}, \mathbf{x})$.

Since the relation $\preceq_{\mathbf{T}}$ is total, we have either $\mathbf{y}' \preceq_{\mathbf{T}} \mathbf{x}'$ or $\mathbf{x}' \preceq_{\mathbf{T}} \mathbf{y}'$. As the second case implies the remark statement by definition, we focus below on the first case. By definition, $\mathbf{y}' \preceq_{\mathbf{T}} \mathbf{x}'$ implies $(\mathcal{B}(\mathbf{T}, \mathbf{y}) \cap \mathcal{H}_-) \subseteq (\mathcal{B}(\mathbf{T}, \mathbf{x}) \cap \mathcal{H}_-)$. Since we assume $\mathbf{y} \leq_{\mathbf{T}} \mathbf{x}$, we also have $(\mathcal{B}(\mathbf{T}, \mathbf{y}) \cap \mathcal{H}_+) \subseteq (\mathcal{B}(\mathbf{T}, \mathbf{x}) \cap \mathcal{H}_+)$ by definition. If we take the union of both sides of the inclusion, we have $\mathcal{B}(\mathbf{T}, \mathbf{y}) \subseteq \mathcal{B}(\mathbf{T}, \mathbf{x})$. If $\mathcal{B}(\mathbf{T}, \mathbf{y}) = \mathcal{B}(\mathbf{T}, \mathbf{x})$, the overall remark statement is trivially true. If $\mathcal{B}(\mathbf{T}, \mathbf{y}) \subset \mathcal{B}(\mathbf{T}, \mathbf{x})$, we have a contradiction as both balls are constructed from the same triangle $\mathbf{T}$. $\square$

References

1. Charrier, E., Buzer, L.: An efficient and quasi linear worst-case time algorithm for digital plane recognition. In: Coeurjolly, D., Sivignon, I., Tougne, L., Dupont, F. (eds.) DGCI 2008. LNCS, vol. 4992, pp. 346–357. Springer, Heidelberg (2008). https://doi.org/10.1007/978-3-540-79126-3_31
2. Coeurjolly, D., Lachaud, J.-O., Levallois, J.: Integral based curvature estimators in digital geometry. In: Gonzalez-Diaz, R., Jimenez, M.-J., Medrano, B. (eds.) DGCI 2013. LNCS, vol. 7749, pp. 215–227. Springer, Heidelberg (2013). https://doi.org/10.1007/978-3-642-37067-0_19
3. Cuel, L., Lachaud, J.O., Mérigot, Q., Thibert, B.: Robust geometry estimation using the generalized voronoi covariance measure. SIAM J. Imaging Sci. **8**(2), 1293–1314 (2015)

4. Debled-Rennesson, I., Reveillès, J.: An incremental algorithm for digital plane recognition. In: Proceedings of Discrete Geometry for Computer Imagery, pp. 194–205 (1994)
5. Gérard, Y., Debled-Rennesson, I., Zimmermann, P.: An elementary digital plane recognition algorithm. Discret. Appl. Math. **151**(1), 169–183 (2005)
6. Lachaud, J.O., Meyron, J., Roussillon, T.: An optimized framework for plane-probing algorithms. J. Math. Imaging Vis. **62**, 718–736 (2020)
7. Lachaud, J.O., Provençal, X., Roussillon, T.: An output-sensitive algorithm to compute the normal vector of a digital plane. J. Theor. Comput. Sci. (TCS) **624**, 73–88 (2016)
8. Lachaud, J.O., Provençal, X., Roussillon, T.: Computation of the normal vector to a digital plane by sampling signicant points. In: 19th IAPR International Conference on Discrete Geometry for Computer Imagery, Nantes, France, April 2016
9. Lachaud, J.O., Provençal, X., Roussillon, T.: Two plane-probing algorithms for the computation of the normal vector to a digital plane. J. Math. Imaging Vis. **59**(1), 23–39 (2017)
10. Reveillès, J.P.: Géométrie Discrète, calculs en nombres entiers et algorithmique. Thèse d'etat, Université Louis Pasteur (1991)
11. Roussillon, T., Lachaud, J.-O.: Digital plane recognition with fewer probes. In: Couprie, M., Cousty, J., Kenmochi, Y., Mustafa, N. (eds.) DGCI 2019. LNCS, vol. 11414, pp. 380–393. Springer, Cham (2019). https://doi.org/10.1007/978-3-030-14085-4_30
12. Roussillon, T., Lu, J.T., Lachaud, J.O., Coeurjolly, D.: Delaunay property and proximity results of the L-algorithm. Research report, Université de Lyon, July 2022. https://hal.archives-ouvertes.fr/hal-03719592
13. The DGtal Project: DGtal (2010). https://dgtal.org
14. Veelaert, P.: Digital planarity of rectangular surface segments. IEEE Trans. Pattern Anal. Mach. Intell. **16**(6), 647–652 (1994)

Exact and Optimal Conversion of a Hole-free 2D Digital Object into a Union of Balls in Polynomial Time

Isabelle Sivignon[✉]

Univ. Grenoble Alpes, CNRS, Grenoble INP, GIPSA-Lab, 38000 Grenoble, France
isabelle.sivignon@gipsa-lab.grenoble-inp.fr

Abstract. This paper addresses the problem of converting a 2D digital object, *i.e.* a set S of points in $\mathbb{Z}^2$, into a finite union of balls $\mathcal{B}$ centered on $\mathbb{R}^2$, such that the digitization of $\mathcal{B}$ is exactly S and the cardinality of $\mathcal{B}$ is minimum. We prove that, for the specific case of 2D hole-free digital objects, there exists a greedy polynomial-time algorithm. The algorithm is based on the same principle as the simple greedy optimal algorithm for the interval cover problem. After bringing to light under which conditions the latter algorithm can be extended to tree-like structures, we show that such a structure can be defined for any hole-free 2D digital object, so that the extended algorithm applies.

Keywords: Digital object · Union of balls · Covering · Optimal algorithm · Arborescence

1 Introduction

Computer representation of shapes is a basic component to digitize, create, visualize or exchange models of physical objects. Different geometric models exist, either to represent the surface (B-rep, point clouds, triangle meshes) or the volume (tetrahedral meshes, digital objects, CSG models) of a solid shape. However, the model used to create or register a shape is not always the one tailored for subsequent processings or applications. Thus, the problem of converting one geometric model into another has been widely studied, for a variety of models. In particular, many provably good conversion algorithms have been designed to output a finite union of balls from other models, including point clouds, polygonal meshes or digital shapes. Indeed, being composed of very simple geometric shapes, finite union of balls are useful in a number of applications, for instance detection of collisions in computer graphics [6], or simulation of physical processes [12]. Various metrics can be used to measure the quality of the conversion such as the number of balls, or the difference in volume between the original model and the union of balls.

In this article, we consider the following problem:

É. Baudrier et al. (Eds.): DGMM 2022, LNCS 13493, pp. 382–394, 2022.
https://doi.org/10.1007/978-3-031-19897-7_30

Problem 1. Given a 2D digital object S, compute a finite union of balls $\mathcal{B}$ such that: $\mathcal{B}$ covers exactly the points of S (and no point of $\mathbb{Z}^2 \backslash S$), and the cardinality of $\mathcal{B}$ is minimum.

This problem is closely related to the more constrained problem where the balls of $\mathcal{B}$ must be centered in $\mathbb{Z}^2$, which is NP-hard [7]. It is also very close to the class of well-studied set cover problems that are also NP-hard [8]. The input of the set cover problem is a pair $(X, \mathcal{R})$, where X is a set of points (generally in $\mathbb{R}^n$) and $\mathcal{R}$ is a family of subsets of X called *ranges*. The problem is to find a minimum subset of $\mathcal{R}$ that covers all the points of X. In our problem, $X = S$ is a subset of $\mathbb{Z}^2$. However, the set of ranges $\mathcal{R}$ is not part of the input, but is constrained to be a set of balls centered on $\mathbb{R}^2$.

We show that, when S is a 4-connected digital object and $\mathbb{Z}^2 \backslash S = S^c$ has exactly one 8-connected component, the problem can be seen as a variant of the interval covering problem (1D set cover problem) for which an optimal greedy algorithm exists. The idea was introduced in [13] in the specific case of (δ, ε)-ball approximation problem: given a shape S, compute a finite union of balls included in the δ-dilation of S while covering its ε-erosion. It was shown that, while the general problem is NP-hard [4], a greedy optimal algorithm exists when the δ-dilation of S has a cycle-free medial axis [14].

In Sect. 2, we revisit the results of [13,14] in a more general context. We consider the case where the input is a generic set of ranges and exhibit sufficient conditions on this set to ensure that the greedy algorithm is optimal in this setting. Once the good tools and conditions have been defined, the proofs of termination and optimality unfold as in [13,14]. In Sect. 3, we show how to implement this algorithm to compute an exact and optimal conversion of a 2D hole-free digital object into a finite union of balls.

2   General Optimal Greedy Algorithm

2.1   Algorithm Specification

For the sake of simplicity, given a subset of ranges R we denote $\bigcup R = \bigcup_{r \in R} r$. We use the same vocabulary as in [13,14] in the broader context of sets of ranges. A covering of a set of ranges $\mathcal{R}$ is a subset of $\mathcal{R}$ that covers all the points in $\bigcup \mathcal{R}$. More formally,

Definition 1 (Covering). *Let $\mathcal{R}$ be a set of ranges, and R be a subset of $\mathcal{R}$. We say that R is a covering of $\mathcal{R}$ if $\bigcup R = \bigcup \mathcal{R}$.*

A covering R is said to be *minimal* if no range can be removed from R while keeping the covering property, and *minimum* if its cardinality is minimum among all possible coverings. In the following, we assume that $\mathcal{R}$ can be endowed with a partial order $\preceq$ such that the poset $(\mathcal{R}, \preceq)$ is anti-arborescent:

Definition 2 (Anti-arborescence [9]). *A poset $(V, \preceq)$ is anti-arborescent if:*

– *for all $v \in V$, the set of its successors $\{v' \in V, v \prec v'\}$ is totally ordered.*
– *for any two incomparable elements $v, v' \in V$, the predecessors of v and the predecessors of v' are pairwise incomparable.*

A range $r \in \mathcal{R}$ is said to be maximal (resp. minimal) in $R \subseteq \mathcal{R}$ if for all $r' \in R$, either $r' \preceq r$ (resp. $r' \succeq r$) or r' and r are incomparable. Given a range $r \in R$, we define the domain covered by ranges smaller than r : $C(\mathcal{R}, \preceq r) = (\bigcup_{r' \in \mathcal{R}, r' \prec r} r') \backslash r$. Similarly, we define the domain covered by ranges larger than or incomparable to r : $C(\mathcal{R}, \npreceq r) = (\bigcup_{r' \in \mathcal{R}, r' \npreceq r} r') \backslash r$. Remark that by definition, if $r_1 \preceq r_2$ then $C(\mathcal{R}, \preceq r_1) \cup r_1 \subseteq C(\mathcal{R}, \preceq r_2) \cup r_2$ and $C(\mathcal{R}, \npreceq r_1) \cup r_1 \supseteq C(\mathcal{R}, \npreceq r_2) \cup r_2$. It will also be useful later to extend these definitions to a set of ranges $R \subseteq \mathcal{R}$: $C(\mathcal{R}, \preceq R) = \bigcup_{r \in R} C(\mathcal{R}, \preceq r)$ and $C(\mathcal{R}, \npreceq R) = \bigcap_{r \in R} C(\mathcal{R}, \npreceq r)$.

Figure 1(c) illustrates these notations in the case of ranges being balls - $\mathcal{R} = \mathcal{B}$: a partial order $\preceq$ on the balls of $\mathcal{B}$ is depicted using arrows on the set of centers of the balls (in red). $(\mathcal{B}, \preceq)$ being an anti-arborescence, it has a root, indicated with a cross. The sets $C(\mathcal{B}, \preceq \theta)$ and $C(\mathcal{B}, \npreceq \theta)$ are depicted respectively in green and orange for a specific ball θ outlined in dashed gray.

Definition 3 (Partial covering). *Let $\mathcal{R}$ be a set of ranges, and R be a subset of $\mathcal{R}$. We say that R is a partial covering of $\mathcal{R}$ if it is a covering of $C(\mathcal{R}, \preceq R)$, i.e. $C(\mathcal{R}, \preceq R) \subseteq \bigcup R$.*

Definition 4 (Candidate range). *Let $R \subset \mathcal{R}$ be a partial covering of $\mathcal{R}$. A range $r \notin R$ is candidate to R if $R' = R \cup \{r\}$ is also a partial covering of $\mathcal{R}$ and $\bigcup R \subsetneq \bigcup R'$.*

A candidate range r with respect to R is said to be *maximal* if it is maximal in the set of candidate ranges. Algorithm 1 describes a greedy algorithm that computes a covering given a finite set of ranges $\mathcal{R}$. It uses the fact that, if $(\mathcal{R}, \preceq)$ is anti-arborescent, a topological ordering of the elements of $\mathcal{R}$ can be defined. The idea is pretty natural: considering ranges in topological order, if a range is critical for the set of uncovered points, then it is added to the covering.

Algorithm 1: GreedyCovering($\mathcal{R}, \preceq$)

Preconditions: $\mathcal{R}$ is finite, $(\mathcal{R}, \preceq)$ is an anti-arborescent poset

1 $R \leftarrow \emptyset$;
2 $U \leftarrow \bigcup \mathcal{R}$ (points of $\bigcup \mathcal{R}$ not in $\bigcup R$);
3 **for** $r \in \mathcal{R}$, *in topological order* **do**
4   **if** r *is a maximal candidate for* U **then**
5        $R \leftarrow R \cup \{r\}$;
6        $U \leftarrow U \backslash r$

7 **return** R

By definition of candidate range, and since Algorithm 1 only inserts candidate ranges to the computed covering, an invariant of Algorithm 1 is that R is always a partial covering of $\mathcal{R}$. The next section is dedicated to the proof of the fact that, provided that $\mathcal{R}$ fulfills two extra conditions, candidate ranges to non-empty subsets always exist (proving that Algorithm 1 terminates with a covering), and that Algorithm 1 computes a minimum-cardinal covering.

2.2   Correctness, Termination and Optimality of Algorithm 1

In the following, we prove that if the poset $(\mathcal{R}, \preceq)$ fulfills the two conditions below, Algorithm 1 terminates and computes a minimum covering :

Property (1) for any $r_1, r_2 \in \mathcal{R}$ such that $r_1 \cap r_2 \neq \emptyset$, for all $r_1 \prec r \prec r_2$, $r_1 \cap r_2 \subseteq r$.
Property (2) for $x \in \bigcup \mathcal{R}$, let $Cov(x, \mathcal{R}) = \{r \in \mathcal{R}, x \in r\}$; then $\forall x \in \bigcup \mathcal{R}$, $Cov(x, \mathcal{R})$ admits a greatest element that is called the *critical range* of x and is denoted $\mathrm{Crit}(x, \mathcal{R})$.

The proof of optimality requires several technical lemmas. These lemmas were stated and proven in [13,14] for a specific family of ranges. We show here that they are still valid when the set of ranges fulfills above properties. The proofs are in general very similar, and simply call properties (1) or (2) when necessary. Space being limited, we only provide the most relevant ones.

The first lemma shows that any range r separates the elements of $\bigcup \mathcal{R}$ into three disjoint subsets of elements: those before, those in, and those after.

Lemma 1 (Proposition 4.10 [13]). *Let $r \in \mathcal{R}$. For any $x \in \bigcup \mathcal{R}$, x belongs to one and only one of the three subsets r, $C(\mathcal{R}, \preceq r)$, $C(\mathcal{R}, \npreceq r)$.*

Proof. By definition, r is disjoint from $C(\mathcal{R}, \preceq r)$ and $C(\mathcal{R}, \npreceq r)$. Suppose now that there exists an element $x \in \bigcup \mathcal{R}$ such that $x \in C(\mathcal{R}, \preceq r) \cap C(\mathcal{R}, \npreceq r)$. Let $r^- \prec r$ such that $x \in r^- \backslash r$ and r^+ such that $r \prec r^+$ or r^+ and r are incomparable and $x \in r^+ \backslash r$. By definition, $x \notin r$ but r^- and r^+ are in $Cov(x, \mathcal{R})$. By property (2), $Cov(x, \mathcal{R})$ admits a greatest element $r_M = \mathrm{Crit}(x, \mathcal{R})$, i.e. $r^- \preceq r_M$, $r^+ \preceq r_M$ and $x \in r_M$. If $r_M = r^-$, then $r^+ \preceq r^- \prec r$, a contradiction. Thus r_M is a strict successor of r^-, as r. By Definition 2, they are comparable. If $r \prec r_M$, then by property (1), $r^- \cap r_M \subseteq r$, leading to a contradiction since $x \in r^- \cap r_M$. If $r_M \prec r$, then r is a successor of r_M which is either a successor of r^+ or r^+ itself. Then $r^+ \prec r$ which is a contradiction with the fact that either $r \prec r^+$ or r and r^+ are incomparable. $\square$

The following two lemmas were not stated as such in [13,14], but used in the proofs. Lemma 2 shows that, given a partial covering, there always exists a candidate.

Lemma 2. *Let $R \subseteq \mathcal{R}$ be a minimal covering of $\mathcal{R}$. Let $R_- \subsetneq R$ be a partial covering, and $R_+ = R \backslash R_-$. Then any range r_+ minimal in R_+ is candidate to R_-.*

The proof is similar to part of the proof of Lemma 4.27 [13] and calls Lemma 1 to assert that the points of $C(\mathcal{R}, \preceq r_+)$ are disjoint from $r_+ \cup C(\mathcal{R}, \npreceq r_+)$ and thus cannot be covered by ranges in R_+. Lemma 2 implies in particular that any range $r = min_{x \in (\bigcup \mathcal{R}) \backslash R} Crit(x, \mathcal{R})$ is a candidate to R (there may be several incomparable candidates). By definition of $Crit(x, \mathcal{R})$, any range $r' \succ r$ does not contain the point $p = \arg\min_{x \in (\bigcup \mathcal{R}) \backslash R} Crit(x, \mathcal{R})$, $p \in (\bigcup \mathcal{R}) \backslash R$, so that r is actually a maximal candidate to R.

Lemma 3. *Let $R \subseteq \mathcal{R}$ be a minimal covering of $\mathcal{R}$. Let $R_- \subsetneq R$ be a partial covering, and let r be a candidate to R_-. Then any range $r' \in \mathcal{R} \backslash R_-$ such that $r' \prec r$ is also a candidate to R_-.*

Proof. Suppose by contradiction that there exists a range $r' \prec r$ that is not a candidate to R_-. Then there exists a point $x \in C(\mathcal{R}, \preceq(R_- \cup \{r'\}))$ which is not in $R_- \cup \{r'\}$. If x were in $C(\mathcal{R}, \preceq R_-)$, it would be covered by R_- since R_- is a partial covering, a contradiction. So $x \notin C(\mathcal{R}, \preceq R_-)$, which implies $x \in C(\mathcal{R}, \preceq r')$. By definition of C, there exists a range $r'' \prec r'$ that contains x. If $x \in r$, then by Property (1), we get $x \in r'$, a contradiction. Thus $x \notin r$. By transitivity of $\prec$, we have $r'' \prec r$. Using the fact that $x \notin r$, and by definition of C, we have $x \in C(\mathcal{R}, \preceq r)$. Again by definition of C, we have $C(\mathcal{R}, \preceq r) \subseteq C(\mathcal{R}, \preceq(R_- \cup r))$. r being candidate to R_-, $C(\mathcal{R}, \preceq(R_- \cup r)) \subseteq \bigcup(R_- \cup \{r\})$, a contradiction. $\square$

Combining the previous lemmas, we can prove that, to complete a partial covering R_-, it is necessary to add a range that is smaller than or equal to a maximal candidate to R_-.

Proposition 1 (Lemma 4.27 [13]). *Let $R \subseteq \mathcal{R}$ be a minimal covering of $\mathcal{R}$. Let $R_- \subsetneq R$ be a partial covering, and let r be a maximal candidate to R_-. Then $R \backslash R_-$ contains a candidate range that is smaller than or equal to r.*

Theorem 1 (Theorem 10 [13]). *Let $\mathcal{R}$ be a finite set of ranges. Suppose that $\mathcal{R}$ can be endowed with a partial order $\preceq$ such that $(\mathcal{R}, \preceq)$ is an anti-arborescent poset, and fulfills Properties (1) and (2). Then, Algorithm 1 outputs a cardinal minimum covering of $\mathcal{R}$.*

The proofs of the proposition and of the theorem follow exactly the ones of Lemma 4.27 and Theorem 10 in [13]. The proof of Proposition 1 calls Lemmas 2 and 3, and the proof of Theorem 1 appllies Proposition 1 to replace one by one the ranges of any optimal covering by the ranges computed by Algorithm 1.

3   From a Digital Set to a Set of Ranges

In this section, we show how Algorithm 1 can be used to solve Problem 1. Here, ranges are balls. Given a digital object S, a set of balls fulfilling Theorem 1 hypothesis is defined. Moreover, this set is such that the result of Algorithm 1 is indeed a collection of balls of minimum cardinality that covers S exactly.

Let $S \subset \mathbb{Z}^2$ be a finite 4-connected digital object such that $S^c = \mathbb{Z}^2 \backslash S$ has one exactly 8-connected component. A digital ball b is a subset of $\mathbb{Z}^2$ for which there exists a ball $\mathscr{b}$ such that $\mathring{\mathscr{b}} \cap \mathbb{Z}^2 = b$, where $\mathring{\mathscr{b}}$ denotes the interior of $\mathscr{b}$. Otherwise said, if Dig denotes the Gauss digitization function, we have which $Dig(\mathscr{b}) = \mathring{\mathscr{b}} \cap \mathbb{Z}^2 = b$. In the following, we assume that balls $\mathscr{b}$ are open, so that $\mathring{\mathscr{b}} = \mathscr{b}$. The preimage of a digital ball b, denoted $Dig^{-1}(b)$ will be useful later on. A digital ball b is said to be valid for a digital object S if $b \subseteq S$. It is said to be maximal if there is no other valid digital ball containing it.

Given a digital object S, we aim at finding a set of ranges $\mathcal{B}$ that are (non empty) valid digital balls and such that $\bigcup \mathcal{B} = S$. Given a set of ranges as input, Algorithm 1 computes a minimum covering for this set of ranges. In order to obtain the minimum covering of a digital object S, the input set of ranges $\mathcal{B}$ must contain all maximal digital balls valid for S. For instance, taking the set of balls ouput by a distance transform of S is not enough to ensure optimality: indeed, all the balls of this set have a center in $\mathbb{Z}^2$, so that it misses all digital balls for which $Dig^{-1}(b)$ contains only balls of center not in $\mathbb{Z}^2$.

The next sections are dedicated to exhibiting a way to grasp the set of all valid maximal digital balls and showing that this set can be endowed with an anti-arborescent poset structure that fulfills sufficient properties (1) and (2).

3.1   Getting a Grip on Valid Maximal Digital Balls

The center of a ball $\mathscr{b}$ is denoted by $c(\mathscr{b})$. For $p \in \mathbb{Z}^2$, let pixel(p) be the unit square centered on p. For any ball $\mathscr{b}$ such that $c(\mathscr{b}) \in pixel(q), q \in S^c$, either $Dig(\mathscr{b}) = \emptyset$ or $Dig(\mathscr{b}) \cap S^c \neq \emptyset$. These balls do not contribute to the set of valid maximal digital balls and can be discarded. Consequently we define $\mathcal{S} = \bigcup_{p \in S}$ pixel(p) and restrict the study to this set. For $x \in \mathcal{S}$, let $\mathscr{b}^S(x)$ be the maximal ball centered in x such that $Dig(\mathscr{b}^S(x)) \subseteq S$. Note that by maximality, $\partial \mathscr{b}^S(x)$ contains at least one point of S^c. The following Lemma shows that any valid maximal digital ball has a ball in its preimage with at least two points of S^c on its boundary.

Lemma 4. *Let b be a valid maximal digital ball for S. Then there exists $\mathscr{b}$ such that $Dig(\mathscr{b}) = b$ and $|\partial \mathscr{b} \cap S^c| \geq 2$.*

Proof. Let $\mathscr{b}'$ be a ball such that $\mathring{\mathscr{b}}' \cap \mathbb{Z}^2 = b$. If $\partial \mathscr{b}' \cap S^c = \emptyset$, then we increase the radius of $\mathscr{b}'$ until $\mathscr{b}' = \mathscr{b}^S(c(\mathscr{b}'))$. $\partial \mathscr{b}'$ contains at least one point of S^c. Now we use a classical projection from a set of balls to the balls of the medial axis of a shape [10,11]. The shape considered here is the whole space $\mathbb{R}^2$ punctured by the discrete set S^c. In this simple case, the medial axis is simply the set of edges of the Voronoi diagram of S^c, i.e. $\partial \mathrm{Vor}(S^c)$. The projection is illustrated in Fig. 1: it associates to any ball $\mathscr{b}$ a ball $\pi(\mathscr{b})$ centered on $\partial \mathrm{Vor}(S^c)$ and such that $\mathscr{b} \subseteq \pi(\mathscr{b})$. This projection is well defined since S is finite (in particular, no half-space is void of points of S^c). Consider the ball $\pi(\mathscr{b}')$. If $Dig(\pi(\mathscr{b}')) \neq b$, we have a contradiction with the maximality of b, and otherwise, we have found a ball $\mathscr{b}$ such that $Dig(\mathscr{b}) = b$ and $|\partial \mathscr{b} \cap S^c| \geq 2$. $\square$

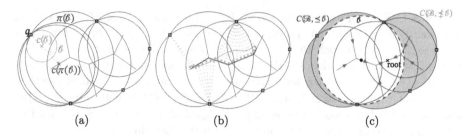

(a)   (b)   (c)

Fig. 1. (a) The projection $\pi(\mathscr{b})$ of $\mathscr{b}$ is defined from the center $c(\mathscr{b})$ and its closest point q in S^c. (b) Projection π is continuous on any continuous path: the continuous path in green is projected on the bolder dark green continuous subpath of the Voronoi diagram. Grey arrows represent the projection. (c) Illustration of a partial order (in red) on $\mathscr{B}$, and of the sets $C(\mathscr{B}, \preceq \mathscr{b})$ and $C(\mathscr{B}, \npreceq \mathscr{b})$. (Color figure online)

Consequently, for any valid maximal digital ball b, there exists a ball $\mathscr{b}$ in $Dig^{-1}(b)$ with $c(\mathscr{b}) \in \partial\mathrm{Vor}(S^c) \cap \mathcal{S}$. Note that: (i) all the balls $\mathscr{b}$ with $c(\mathscr{b})$ in this set are such that $Dig(\mathscr{b})$ is valid for S ; (ii) some balls $\mathscr{b}$ with $c(\mathscr{b})$ in this set may however be such that $Dig(\mathscr{b})$ is not maximal. In the following, we denote $\mathrm{Vor}^{\sqcap}(S) = \partial\mathrm{Vor}(S^c) \cap \bigcup \mathrm{pixel}(S)$ (see Fig. 2(a)), and we consider the set of balls $\mathscr{B} = \{\mathscr{b}^S(x), x \in \mathrm{Vor}^{\sqcap}(S)\}$. This set contains all the balls which digitization is a valid maximal digital ball for S.

3.2   Ordering Balls of $\mathscr{B}$

By construction, $\mathrm{Vor}^{\sqcap}(S)$ is a collection of segments.

Lemma 5. $\mathrm{Vor}^{\sqcap}(S)$ *is a geometric embedding of a tree in* $\mathbb{R}^2$.

Proof. Suppose that $\mathrm{Vor}^{\sqcap}(S)$ contains a cycle. This cycle is a Jordan curve, and since it is a subset of $\partial\mathrm{Vor}(S^c)$ it must contain a point of S^c in its interior. Moreover, this cycle is included in $\mathcal{S}$, which is an open polygon containing no point of S^c since S is 4-connected and S^c is 8-connected. A contradiction.   □

$\mathrm{Vor}^{\sqcap}(S)$ being a tree, it can be endowed with a partial order by picking any point on it as a root: indeed, it is enough to orient each edge/segment from the leaves to the root. This results in an oriented tree, denoted by $\mathcal{T}$, that defines a partial order $\leq_{\mathcal{T}}$ on the set (of centers) of balls $\mathscr{B}$ (see Fig. 1(c)). By construction, $(\mathscr{B}, \leq_{\mathcal{T}})$ is an anti-arborescent poset. Moreover, for any $p \in S$, the set of centers of the balls of $Cov(p, \mathscr{B}) = \{\mathscr{b} \in \mathscr{B}, p \in \mathscr{b}\}$ is a connected subset of $\mathrm{Vor}^{\sqcap}(S)$.

Lemma 6 (Lemma 4.9 [13]). *Let* $p \in S$. *If* $p \subseteq \mathring{\mathscr{b}}_1 \cap \mathring{\mathscr{b}}_2$, *then* $p \subseteq \mathring{\mathscr{b}}$ *for all* $\mathscr{b}$ *such that* $c(\mathscr{b})$ *is on the unique path* $\Gamma(\mathscr{b}_1, \mathscr{b}_2)$ *between* $c(\mathscr{b}_1)$ *and* $c(\mathscr{b}_2)$ *in* $\mathrm{Vor}^{\sqcap}(S)$.

The proof uses projection π defined in the previous section, together with the fact that $\mathrm{Vor}^{\sqcap}(S)$ is the geometric embedding of a tree.

This lemma implies that Property (1) is true for $\mathscr{B}$. It moreover implies that for all p, $Cov(p, \mathscr{B})$ admits a supremum according to the order $\mathcal{T}$. However, since the balls of $\mathscr{B}$ are open, these sets are open too (see illustration in Fig. 2(b)), except for points p that belong to the balls that are either the root or leaves of $\mathcal{T}$. A consequence is that, in general, $Cov(p, \mathscr{B})$ does not admit a greatest element, and $p \notin Dig(\sup_{\mathcal{T}} Cov(p, \mathscr{B}))$. This results in the following property:

Lemma 7. *For any $p \in S$ that does not belong to the root of $\mathcal{T}$, $\sup_{\mathcal{T}} Cov(p, \mathscr{B})$ either belongs to an open segment of $Vor^{\sqcap}(S)$ or, if it is a vertex, the balls of $Cov(p, \mathscr{B})$ are all in the same subtree of predecessors.*

Proof. Suppose that $\sup_{\mathcal{T}} Cov(p, \mathscr{B})$ is a vertex $v \in Vor^{\sqcap}(S)$, and, by contradiction, pick any ball of $Cov(p, \mathscr{B})$ in a first subtree, and another one in another subtree. Then the unique path between them goes through v, and the ball centered on v must contain p by Lemma 6 and thus be in $Cov(p, \mathscr{B})$. It cannot be the supremum of $Cov(p, \mathscr{B})$. □

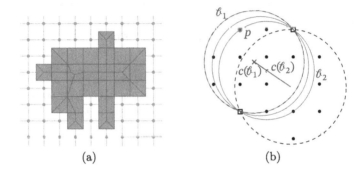

(a)   (b)

Fig. 2. (a) Cropped Voronoi diagram $Vor^{\sqcap}(S)$ for a set of pixels S depicted in grey. (b) $Cov(p, \mathscr{B})$ is an open set. Part of $Vor^{\sqcap}(S)$ is depicted in red : the centers of all the balls of $Cov(p, \mathscr{B})$ are on the blue segment, delimited by $c(\mathscr{b}_1)$ and $c(\mathscr{b}_2)$, but $\mathscr{b}_2$ does not belong to $Cov(p, \mathscr{B})$ since it contains p on its boundary. (Color figure online)

The set of ranges $\mathscr{B}$ does not fulfill property (2), which is required for Algorithm 1 to be valid. We turn to the set $\mathcal{B} = \{b \subseteq S, \exists \mathscr{b} \in \mathscr{B} \, Dig(\mathscr{b}) = b\}$ instead. Since $\mathcal{B}$ is finite, the sets $Cov(p, \mathcal{B}) = \{b \in \mathcal{B}, p \in b\}$ are also finite and are good candidates to admit a greatest element if equipped with a partial order. We show hereafter how to do this without explicitly computing the set $\mathcal{B}$.

3.3   Ordering Digital Balls of $\mathcal{B}$

Let the *representative* of b be $\text{Rep}(b) = \sup_T \{\mathscr{b} \in \mathscr{B}, \mathscr{b} \in Dig^{-1}(b)\}$. As seen before, usually, $Dig(\text{Rep}(b)) \neq b$. From the partial order $\mathcal{T}$ on $\mathscr{B}$, we define a partial order T on $\mathcal{B}$ as follows:

Definition 5. *Given two digital balls b_1 and b_2 of $\mathcal{B}$, $b_1 \leq_T b_2$ if:*

(1) either $b_1 = b_2$
(2) or $b_1 \neq b_2$ and
   (a) either $Rep(b_1) <_{\mathcal{T}} Rep(b_2)$
   (b) or $Rep(b_1) = Rep(b_2)$ and $Dig(Rep(b_2)) = b_2$.

Lemma 8. $(\mathcal{B}, \leq_T)$ *is a poset.*

Sketch of proof. Reflexivity follows directly from (1). Antisymmetry is shown by contradiction considering two cases: either $Rep(b_1) \neq Rep(b_2)$ and we get a contradiction by Definition 5 and definition of $\mathcal{T}$, or $Rep(b_1) = Rep(b_2)$ and we have a contradiction with unicity of $Dig(\mathcal{A})$ using Definition 5 (2)(b). To show transitivity, the case $b_1 = b_2$ or $b_2 = b_3$ is trivial. Otherwise, we distinguish the two cases $Rep(b_1) \neq Rep(b_2) \neq Rep(b_3)$ and $Rep(b_1) = Rep(b_2)$ and $Rep(b_2) \neq Rep(b_3)$ and conclude that $Rep(b_1) <_{\mathcal{T}} Rep(b_3)$ using the fact that $\mathcal{T}$ is a partial order. □

In order to prove that the poset $(\mathcal{B}, \leq_T)$ is anti-arborescent, we need two extra lemmas that express properties of the sets $Dig^{-1}(b)$. The following lemma, together with Lemma 6, moreover ensures that Property (1) is fulfilled for the set of ranges $\mathcal{B}$.

Lemma 9. *(i) For any $b \in \mathcal{B}$, $Dig^{-1}(b)$ is connected. (ii) For any $b, b' \in \mathcal{B}$, $b \neq b'$, $Dig^{-1}(b) \cap Dig^{-1}(b') = \emptyset$. As a consequence, we have (iii): let $\mathcal{b} \in Dig^{-1}(b)$ and $\mathcal{b}' \in Dig^{-1}(b')$ with $b \neq b'$: if $\mathcal{b} <_{\mathcal{T}} \mathcal{b}'$, then $\mathcal{b} \leq_{\mathcal{T}} Rep(b) \leq_{\mathcal{T}} \mathcal{b}' \leq_{\mathcal{T}} Rep(b')$.*

Proof. (i) follows from Lemma 6 since $Dig^{-1}(b) = \bigcap_{p \in b} Cov(p, \mathcal{B})$. (ii) is straightforward by unicity of the digitization. To prove (iii), note that $Rep(b)$ and $\mathcal{b}'$ are comparable since they are both successors of $\mathcal{b}$ and $\mathcal{T}$ is an anti-arborescence. Suppose by contradiction that $\mathcal{b}' \leq_{\mathcal{T}} Rep(b)$. Then $\mathcal{b}'$ is on the unique path between $\mathcal{b}$ and $Rep(b)$, a contradiction with (i) and (ii). □

Lemma 10. *Let $b_1, b_2 \in \mathcal{B}$, $b_1 \neq b_2$, and $\mathcal{b} = Rep(b_1) = Rep(b_2)$ such that $\mathcal{b} \notin Dig^{-1}(b_1)$ and $\mathcal{b} \notin Dig^{-1}(b_2)$. Then for all $\mathcal{b}_1 \in Dig^{-1}(b_1)$ and all $\mathcal{b}_2 \in Dig^{-1}(b_2)$, $\mathcal{b}_1$ ans $\mathcal{b}_2$ are incomparable.*

Proof. Suppose by contradiction that $\mathcal{b}_1 <_{\mathcal{T}} \mathcal{b}_2$. Thus $\mathcal{b}_1 \leq_{\mathcal{T}} Rep(b_1)$. Since $Dig^{-1}(b_2)$ cannot be empty, $Rep(b_2) \notin Dig^{-1}(b_2)$ implies that there exists $\mathcal{b}_2 \neq Rep(b_2)$ such that $\mathcal{b}_2 \in Dig^{-1}(b_2)$. Using Lemma 9, we get $\mathcal{b}_1 \leq_{\mathcal{T}} Rep(b_1) \leq_{\mathcal{T}} \mathcal{b}_2 \leq_{\mathcal{T}} Rep(b_2)$, which is a contradiction with $Rep(b_1) = Rep(b_2)$. □

Theorem 2. *The poset $(\mathcal{B}, \leq_T)$ is anti-arborescent.*

Sketch of proof. We first prove by contradiction that the successors of any $b \in \mathcal{B}$ are comparable, by considering three cases: $Rep(b) = Rep(b_1) = Rep(b_2)$, or $Rep(b) = Rep(b_1)$ and $Rep(b) <_{\mathcal{T}} Rep(b_2)$, or $Rep(b) <_{\mathcal{T}} Rep(b_1)$ and $Rep(b) <_{\mathcal{T}}$

Rep(b_2). In the first two cases, we have a direct contradiction with Definition 5. The third case is a little bit trickier and uses Lemma 10.

Next, we prove by contradiction that the predecessors of two incomparable balls b_1 and b_2 are also incomparable. Two cases are studied: if $b'_1 = b'_2$, then Rep(b'_1) = Rep(b'_2) and we use Lemma 9(iii) and the fact that $(\mathscr{B}, \mathfrak{I})$ is an anti-arborescence to get a contradiction; if $b'_1 \leq_T b'_2$ and Rep(b'_1) $\lesssim_{\mathfrak{I}}$ Rep(b'_2), we use again Lemma 9(iii) to get a contradiction.   □

It remains to prove that $Cov(p, \mathscr{B})$ admits a greatest element for all $p \in S$. To do so, we remark that the ball $\sup_{\mathfrak{I}} Cov(p, \mathscr{B})$ of a point $p \in S$ can be written as the maximum representative ball of $Cov(p, \mathscr{B})$.

$$
\begin{aligned}
\sup_{\mathfrak{I}} Cov(p, \mathscr{B}) &= \sup_{\mathfrak{I}} \{ \theta \in \mathscr{B}, p \in \theta \} \\
&= \sup_{\mathfrak{I}} \{ \theta \in \mathscr{B}, p \in Dig(\theta) \} \\
&= \sup_{\substack{\mathfrak{I} \\ b \in Cov(p, \mathscr{B})}} \{ \theta \in \mathscr{B}, Dig(\theta) = b \} \\
&= \sup_{\substack{\mathfrak{I} \\ b \in Cov(p, \mathscr{B})}} Rep(b) = \max_{\substack{\mathfrak{I} \\ b \in Cov(p, \mathscr{B})}} Rep(b)
\end{aligned}
\tag{1}
$$

The digital ball $b_{max}(p) \in \mathscr{B}$ that achieves the maximum in Eq. (1) is actually the critical ball Crit($p, \mathscr{B}$). In the last subsection, we show how to compute it.

Lemma 11. *Let $p \in S$, and $b_{max}(p) \in Cov(p, \mathscr{B})$ be such that Rep($b_{max}(p)$) = $\sup_{\mathfrak{I}} Cov(p, \mathscr{B})$. Then for any $b \in Cov(p, \mathscr{B})$, $b \leq_T b_{max}(p)$.*

Proof. Let $b \in Cov(p, \mathscr{B})$. If Rep($b$) $<_{\mathfrak{I}}$ Rep($b_{max}(p)$), by Definition 5, $b <_T b_{max}(p)$. The case Rep(b) $>_{\mathfrak{I}}$ Rep($b_{max}(p)$) is not possible by definition of $b_{max}(p)$. The case Rep(b) = Rep($b_{max}(p)$) remains. Since $Dig^{-1}(b)$ are connected and disjoint (Lemma 9), the only way for two balls b_1 and b_2 to have the same representative is when it is a vertex of the anti-arborescence. Then $Dig^{-1}(b_1)$ and $Dig^{-1}(b_2)$ belong to two different subtrees of this vertex, a contradiction with Lemma 7.   □

3.4   Computing Critical Balls

The first step is to find the edge of $Vor^{\sqcap}(S)$ Rep(Crit($p, \mathscr{B}$)) belongs to. It is convenient to note that each edge of $Vor^{\sqcap}(S)$ corresponds to balls of $\mathscr{B}$ that go through a pair of points of S^c. This edge can then be described as a parabolic pencil of circles [5,15] defined by two points of S^c and delimited by its two extremities. Each ball θ_λ of the pencil can be expressed as a convex combination of the two extremities, according to the following relation: $\forall p$, pow(p, θ_λ) = $(1 - \lambda)$pow(p, θ_1) + λpow(p, θ_2), where pow denotes the power of a point with respect to a ball $\theta(c, r)$ and is equal to pow(p, θ) = $d(c, p) - r^2$.

Given a topological order on the edges of $Vor^{\sqcap}(S)$, consider the edges $[\theta_1, \theta_2]$ in increasing order. If p belongs to $Dig(\theta_1)$ but not to $Dig(\theta_2)$, then Rep(Crit($p, \mathscr{B}$)) belongs to the edge $[\theta_1, \theta_2[$. Using the fact that $p \in$

$\partial\mathrm{Rep}(\mathrm{Crit}(p,\mathcal{B}))$, and the relation above, we can compute the value λ such that $\mathcal{B}_\lambda = \mathrm{Rep}(\mathrm{Crit}(p,\mathcal{B}))$ on the pencil $[\mathcal{B}_1,\mathcal{B}_2[$, as $\lambda = \frac{\mathrm{pow}(p,\mathcal{B}_1)}{\mathrm{pow}(p,\mathcal{B}_1)-\mathrm{pow}(p,\mathcal{B}_2)}$. For all $0 \le \lambda' < \lambda$, $Dig(\mathcal{B}_{\lambda'})$ contains p (see Fig. 3(a)). We look for a value $\lambda_{crit} < \lambda$ such that $Dig(\mathcal{B}_\lambda) \subset Dig(\mathcal{B}_{\lambda_{crit}})$. Such a value exist thanks to Lemma 7. For all the points $q \in Dig(\mathcal{B}_\lambda)\backslash Dig(\mathcal{B}_1)$, let $\mathcal{B}_{\lambda_q}$ be the ball of $[\mathcal{B}_1,\mathcal{B}_2[$ such that $q \in \partial\mathcal{B}_{\lambda_q}$. For all values $\mu > \lambda_q$, $q \in Dig(\mathcal{B}_\mu)$. By setting $\mu = \max_q\{\lambda_q\}$, we have that $\forall\mu' > \mu$, $Dig(\mathcal{B}_\lambda) \subset Dig(\mathcal{B}'_\mu)$.

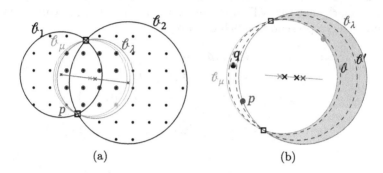

(a)   (b)

Fig. 3. (a) Computation of $\mathrm{Crit}(p,\mathcal{B})$: $\mathcal{B}_\lambda = \mathrm{Rep}(\mathrm{Crit}(p,\mathcal{B}))$; any ball between $\mathcal{B}_\mu$ and $\mathcal{B}_\lambda$ (see for instance the ball in gray) contains p and all the points of $Dig(\mathcal{B}_\lambda)$ (circled). (b) Illustration of the fact that the critical ball is not always maximal.

By setting λ_{crit} to any value strictly between μ and λ, we have $Dig(\mathcal{B}_{\lambda_{crit}}) \supset \{p\} \cup Dig(\mathcal{B}_\lambda)$ as desired. Note that, as mentioned before, $Dig(\mathcal{B}_{\lambda_{crit}})$ may not be maximal. Indeed, as illustrated in Fig. 3(b), by definition of $\mathcal{B}_\mu$ there is no point of S in the grey region. However, $Dig(\mathcal{B}_\mu)$ may contain points of S other than p, for instance point q in the figure. Thus, if we consider the two balls $\mathcal{B}$ and $\mathcal{B}'$ both between $\mathcal{B}_\mu$ and $\mathcal{B}_\lambda$, $Dig(\mathcal{B}') \subset Dig(\mathcal{B})$, so that $Dig(\mathcal{B}')$ is not maximal. As proven in the section before, this is not a problem: in the course of the algorithm, either q belongs to the subset not covered yet, and then the critical ball of q, which is equal to $Dig(\mathcal{B})$, is chosen, or q is already covered, and picking $Dig(\mathcal{B}')$ instead of $Dig(\mathcal{B})$ does not change anything.

4   Results

Algorithm 1 was implemented[1] using three open-source libraries: DGtal [2] to handle digital sets, CGAL [1] to compute $\mathrm{Vor}^\sqcap(S)$, and Boost Graph [3] to compute topological order on trees. A kernel with exact predicates and constructions was used to avoid rounding errors. As a conclusion, some results are presented in Fig. 4.

---

[1] https://github.com/isivigno/ConvertDigitalObjectToBalls.git.

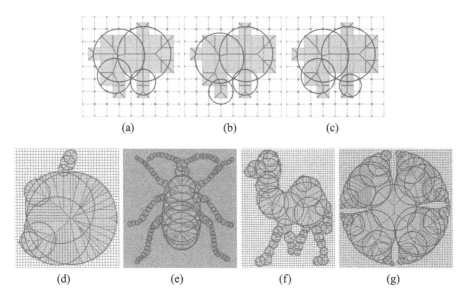

(a)   (b)   (c)

(d)   (e)   (f)   (g)

Fig. 4. (a-c) Three different optimal coverings of the same toy example with 4 balls, obtained using different roots ; (d-g) Results on images of the database MPEG7 CE Shape-1 Part B :(d) 9 balls (e) 113 balls (f) 40 balls (g) 36 balls.

References

1. CGAL: Computational Geometry Algorithms Library. https://cgal.org
2. DGtal: Digital Geometry Tools and Algorithms Library. https://dgtal.org
3. The Boost Graph Library. https://www.boost.org/doc/libs/1_78_0/libs/graph
4. Attali, D., Nguyen, T.B., Sivignon, I.: (δ, ε)-ball approximation of a shape: Definition and complexity. Discret. Comput. Geom. **61**, 595–625 (2019)
5. Boissonnat, J.D., Yvinec, M.: Géométrie algorithmique (1995)
6. Choi, A.R., Sung, M.Y.: Performance improvement of haptic collision detection using subdivision surface and sphere clustering. PLoS ONE **12**(9), 1–17 (2017)
7. Coeurjolly, D., Hulin, J., Sivignon, I.: Finding a minimum medial axis of a discrete shape is np-hard. Theoret. Comput. Sci. **406**(1), 72–79 (2008)
8. Cormen, T.H., Leiserson, C.E., Rivest, R.L., Stein, C.: Introduction to Algorithms, 3rd edn. The MIT Press, Cambridge (2009)
9. F. Fauvet, F. Menous, D.S.: Explicit linearization of one-dimensional germs through tree-expansions. Bull. de la SMF **146**(2), 241–285 (2018)
10. Lieutier, A.: Any open bounded subset of $\mathbb{R}^n$ has the same homotopy type as its medial axis. Comput. Aided Des. **36**(11), 1029–1046 (2004)
11. Matheron, G., Serra, J. (eds.): Image Analysis and Mathematical Morphology 2: Theoretical Advances, pp. 216–256. Academic Press (1988)
12. Mede, T., Chambon, G., Hagenmuller, P., Nicot, F.: A medial axis based method for irregular grain shape representation in dem simulations. Granular Matter **20**(16) (2018). https://doi.org/10.1007/s10035-017-0785-7
13. Nguyen, T.B.: Finite unions of balls with inner and outer margins. Ph.D. thesis, Université Grenoble Alpes (2018)

14. Nguyen, T.B., Sivignon, I.: Epsilon-covering: a greedy optimal algorithm for simple shapes. In: CCCG 2016–28th Canadian Conference on Computational Geometry (2016)
15. Schwerdtfeger, H.: Geometry of Complex Numbers: Circle Geometry, Moebius transformation, Non-euclidean Geometry. Courier Corporation (1979)

Density Functions of Periodic Sequences

Olga Anosova[iD] and Vitaliy Kurlin[✉][iD]

Department of Computer Science, University of Liverpool,
Liverpool L69 3BX, UK
{oanosova,vkurlin}@liv.ac.uk
http://kurlin.org

Abstract. This paper contributes to the emergent area of Periodic Geometry, which studies continuous spaces of solid crystalline materials (crystals) by new methods of metric geometry. Since crystal structures are determined in a rigid form, their strongest practical equivalence is rigid motion or isometry preserving inter-point distances. The most fundamental model of any crystal is a periodic set of points at all atomic centers. The previous work introduced an infinite sequence of density functions that are continuous isometry invariants of periodic point sets. These density functions turned out to be highly non-trivial even in dimension 1 for periodic sequences of points in the line. This paper fully describes the density functions of any periodic sequence and their symmetry properties. The explicit description confirms coincidences of density functions that were previously computed via finite samples.

Keywords: Periodic sequence · Isometry invariant · Density functions

1 Motivations for Density Functions of Periodic Sets

Motivated by applications to solid crystalline materials, the first paper [10] in the emergent area of Periodic Geometry rigorously stated the problem of designing continuous invariants and metrics for periodic point sets such as lattices.

The past work [5] introduced such continuous invariants for any periodic sets of points representing atoms in crystals. This point-set model is most fundamental for materials because nuclei of atoms are well-defined physical objects, while chemical bonds are not real sticks or strings but abstractly represent inter-atomic interactions depending on many thresholds for distances and angles.

Since crystal structures are determined in a rigid form, their most practical equivalence is *rigid motion* (a composition of translations and rotations) or *isometry* that maintains all inter-point distances and includes reflections [14].

Since atoms always vibrate at any finite temperature above absolute zero, X-ray diffraction patterns of the same material contain inevitable noise and lead to slightly different crystal structures determined at variable temperatures.

In the past, crystallography distinguished periodic structures by coarser isometry invariants such as symmetry groups, which are discontinuous under perturbations [14, Fig. 1]. To continuously quantify the similarity between near-duplicates among experimental and simulated structures, we need stronger isometry invariants that continuously change under perturbations [1, Problem 3].

© Springer Nature Switzerland AG 2022
E. Baudrier et al. (Eds.): DGMM 2022, LNCS 13493, pp. 395–408, 2022.
https://doi.org/10.1007/978-3-031-19897-7_31

The past work [5] introduced an infinite sequence of density functions $\psi_k[S](t)$ that are continuous isometry invariants of a periodic point set S as defined below. Let $\mathbb{R}^n$ be the n-dimensional Euclidean space, $\mathbb{Z}$ be the set of all integers.

Definition 1 (a lattice Λ, a unit cell U, a motif M, a periodic set $S = M + \Lambda$).

For a linear basis $v_1, \ldots, v_n$ of $\mathbb{R}^n$, a *lattice* is $\Lambda = \{\sum_{i=1}^{n} c_i v_i : c_i \in \mathbb{Z}\}$. The *unit cell* $U(v_1, \ldots, v_n) = \left\{\sum_{i=1}^{n} c_i v_i : c_i \in [0, 1)\right\}$ is the parallelepiped spanned by the basis. A *motif* $M \subset U$ is any finite set of points $p_1, \ldots, p_m \in U$. A *periodic point set* [14] is the Minkowski sum $S = M + \Lambda = \{u + v \mid u \in M, v \in \Lambda\}$. ∎

In dimension $n = 1$, a lattice is defined by any non-zero vector $v \in \mathbb{R}$, any periodic point set S is a periodic sequence $\{p_1, \ldots, p_m\} + |v|\mathbb{Z}$ of the period $|v|$.

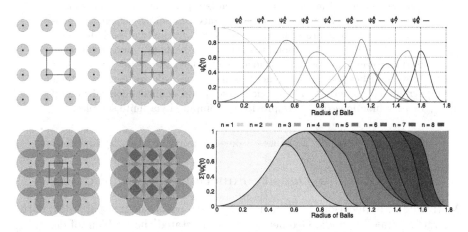

Fig. 1. Illustration of Definition 2 for the square lattice. **Left:** subregions $U_k(t)$ are covered by k disks for the radii $t = 0.25, 0.55, 0.75, 1$. **Right:** the nine density functions are above the corresponding *densigram* of accumulated functions $\sum_{i=1}^{k} \psi_i(t)$ [5, Fig. 2].

Definition 2 (density functions). Let a periodic set $S = \Lambda + M \subset \mathbb{R}^n$ have a unit cell U. For any integer $k \geq 0$, let $U_k(t)$ be the region within the cell U covered by exactly k closed balls with a radius $t > 0$ and centers at all points of S. The k-th *density function* is $\psi_k[S](t) = \mathrm{Vol}[U_k(t)]/\mathrm{Vol}[U]$. The *density fingerprint* is the sequence $\Psi[S] = \{\psi_k(t)\}_{k=0}^{+\infty}$, see [5, section 3] and Fig. 1, 2. ∎

The implementation [5] computes the density functions $\psi_k(t)$ at uniform radii t up to given upper bounds of t and k. This paper explicitly describes all density functions $\psi_k(t)$ for any periodic sequence $S \subset \mathbb{R}$ in Theorems 5 and 7. Theorem 8 proves the symmetry and periodicity of $\psi_k(t)$ in the variables t and k. Corollary 12 concludes that the 1st density function $\psi_1(t)$ distinguishes all non-isometric periodic sequences with distinct distances between motif points.

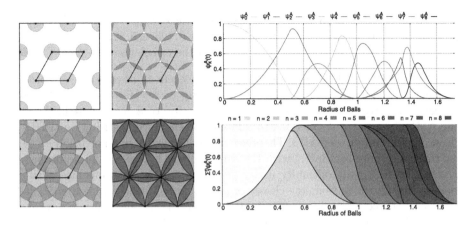

Fig. 2. Illustration of Definition 2 for the hexagonal lattice. **Left:** subregions $U_k(t)$ are covered by k disks for the radii $t = 0.25, 0.55, 0.75, 1$. **Right:** the nine density functions are above the corresponding *densigram* of accumulated functions $\sum_{i=1}^{k} \psi_i(t)$ [5, Fig. 2].

2   Past Work on Isometry Invariants of Periodic Sets

The strongest result about the density fingerprint $\Psi[S]$ is [5, Theorem 2] proving that any non-isometric periodic point sets in $\mathbb{R}^3$ have different sequences $\psi_k(t)$, though there was no simple upper bound for k. However, the density fingerprint turned out to be incomplete [5, section 5] for the periodic sequences below.

Example 3 (periodic sequences $S_{15}, Q_{15} \subset \mathbb{R}$). [14, Appendix B] discusses homometric periodic sets that can be distinguished by the recent invariant AMD (Average Minimum Distances) and not by inter-point distance distributions. The sets $S_{15} = \{0, 1, 3, 4, 5, 7, 9, 10, 12\} + 15\mathbb{Z}$, $Q_{15} = \{0, 1, 3, 4, 6, 8, 9, 12, 14\} + 15\mathbb{Z}$ have the period 15 and the unit cell $[0, 15]$ shown as a circle in Fig. 3.

These periodic sequences [6] are obtained as Minkowski sums $S_{15} = U + V + 15\mathbb{Z}$ and $Q_{15} = U - V + 15\mathbb{Z}$ for $U = \{0, 4, 9\}$ and $V = \{0, 1, 3\}$. The last picture

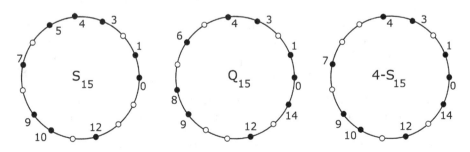

Fig. 3. Circular versions of the periodic sets S_{15}, Q_{15}. Distances are along round arcs.

in Fig. 3 shows the periodic set $4 - S_{15}$ isometric to S_{15}. Now the difference between Q_{15} and $4 - S_{15}$ is better visible: points $0, 1, 3, 4, 5, 12, 14$ are common, but points $6, 8, 9 \in Q_{15}$ are shifted to $7, 9, 10$ in the circular set $4 - S_{15}$. ∎

For rational-valued periodic sequences, [6, Theorem 4] proved that r-th order invariants (combinations of r-factor products) up to $r = 6$ are enough to distinguish such sequences up to a shift (a rigid motion of $\mathbb{R}$ without reflections). The AMD invariant was extended to a Pointwise Distance Distribution (PDD), whose generic completeness [13, Theorem 11] was proved in any dimension $n \geq 1$, but there are finite sets in $\mathbb{R}^3$ with the same PDD [11, Fig. S4]. In addition to the completeness and continuity under perturbations, applications also need a computable metric on isometry classes of periodic point sets. Such a metric was defined on the complete isoset invariant [1, section 7] but has only an approximate algorithm because of a minimization over infinitely many rotations.

This paper fully elucidates all density functions and their exact computation for any periodic sequence, leading to new problems at the end of Sect. 4.

3   A Description of Density Functions of Periodic Sequences

The key results of this section are Theorems 5 and 7 explicitly describing all density functions $\psi_k[S](t)$ for any periodic sequence $S \subset \mathbb{R}$ and $k \geq 0$. For convenience, scale any periodic sequence to period 1 so that $S = \{p_1, \ldots, p_m\} + \mathbb{Z}$. Since the expanding balls in $\mathbb{R}$ are growing intervals, volumes of their intersections linearly change in the variable radius t. Hence any density function $\psi_k(t)$ is piecewise linear and uniquely determined by *corner* points (a_j, b_j) where the gradient changes. Examples 4 and 6 explain how the density functions $\psi_k(t)$ are computed for the periodic sequence $S = \{0, \frac{1}{3}, \frac{1}{2}\} + \mathbb{Z}$, see all graphs in Fig. 4.

Example 4 (0-th density $\psi_0(t)$ for $S = \{0, \frac{1}{3}, \frac{1}{2}\} + \mathbb{Z}$). By Definition 2 $\psi_0(t)$ is the fractional length within the period interval $[0, 1]$ not covered by the intervals of radius t (length $2t$), which are the red intervals $[0, t] \cup [1 - t, 1]$, green dashed interval $[\frac{1}{3} - t, \frac{1}{3} + t]$ and blue dotted interval $[\frac{1}{2} - t, \frac{1}{2} + t]$. The graph of $\psi_0(t)$ starts from the point $(0, 1)$ at $t = 0$. Then $\psi_0(t)$ linearly drops to the point $(\frac{1}{12}, \frac{1}{3})$ at $t = \frac{1}{12}$ when a half of the interval $[0, 1]$ remains uncovered.

The next linear piece of $\psi_0(t)$ continues to the point $(\frac{1}{6}, \frac{1}{6})$ at $t = \frac{1}{6}$ when only $[\frac{2}{3}, \frac{5}{6}]$ is uncovered. The graph of $\psi_0(t)$ finally returns to the t-axis at the point $(\frac{1}{4}, 0)$ and remains there for $t \geq \frac{1}{4}$. The piecewise linear behavior of $\psi_0(t)$ can be briefly described via the *corner* points $(0, 1)$, $(\frac{1}{12}, \frac{1}{3})$, $(\frac{1}{6}, \frac{1}{6})$, $(\frac{1}{4}, 0)$. ∎

Theorem 5 extends Example 4 to any periodic sequence S and implies that $\psi_0(t)$ is uniquely determined by the ordered distances within a unit cell of S.

Theorem 5 (description of ψ_0). For any periodic sequence $S = \{p_1, \ldots, p_m\} + \mathbb{Z}$ with motif points $0 \leq p_1 < \cdots < p_m < 1$, set $d_i = p_{i+1} - p_i \in (0, 1)$, where $i = 1, \ldots, m$ and $p_{m+1} = p_1 + 1$. Put the distances in the increasing order

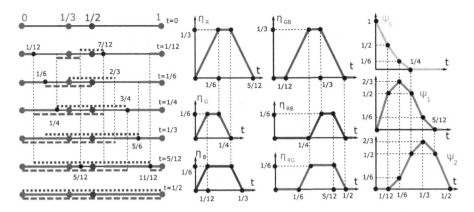

Fig. 4. Left: the periodic sequence $S = \{0, \frac{1}{3}, \frac{1}{2}\} + \mathbb{Z}$ with points of three colors. The growing intervals around the red point $0 \equiv 1$ (mod 1), green point $\frac{1}{3}$, blue point $\frac{1}{2}$ have the same color for various radii t. **Right**: the trapezoid functions η from Example 6.

$d_{[1]} \leq d_{[2]} \leq \cdots \leq d_{[m]}$. Then the 0-th density function ψ_0 is piecewise linear with the following (unordered) corners: $(0, 1)$ and $(\frac{1}{2}d_{[i]}, 1 - \sum_{j=1}^{i-1} d_{[j]} - (m - i + 1)d_{[i]})$ for $i = 1, \ldots, m$, so the last corner is $(\frac{1}{2}d_{[m]}, 0)$. If any corner points are repeated, e.g. when $d_{[i-1]} = d_{[i]}$, these corners are collapsed into one corner point. ∎

Proof. The function $\psi_0(t)$ measures the total length of subintervals in $[0, 1]$ that are not covered by growing intervals $[p_i - t, p_i + t]$, $i = 1, \ldots, m$. Hence $\psi_0(t)$ linearly decreases from the initial value $\psi_0(0) = 1$ except for m critical values of t where one of the intervals $[p_i, p_{i+1}]$ between successive points become completely covered and can not longer shrink. These critical radii t are ordered according to the distances $d_{[1]} \leq d_{[2]} \leq \cdots \leq d_{[m]}$. The first critical radius is $t = \frac{1}{2}d_{[1]}$, when the shortest interval $[p_i, p_{i+1}]$ of the length $d_{[1]}$ is covered by the intervals centered at p_i, p_{i+1}. At this moment, all m intervals cover the subregion of the length $md_{[1]}$. Then $\psi_0(t)$ has the first corner points $(0, 1)$ and $(\frac{1}{2}d_{[1]}, 1 - md_{[1]})$.

The second critical radius is $t = \frac{1}{2}d_{[2]}$, when the covered subregion has the length $d_{[1]} + (m - 1)d_{[2]}$, i.e. the next corner point is $(\frac{1}{2}d_{[2]}, 1 - d_{[1]} - (m-1)d_{[2]})$. If $d_{[1]} = d_{[2]}$, then both corner points coincide, so $\psi_0(t)$ will continue from the joint corner point. The above pattern generalizes to the i-th critical radius $t = \frac{1}{2}d_{[i]}$, when the covered subregion has the length $\sum_{j=1}^{i-1} d_{[j]}$ (for the finally covered intervals) plus $(m - i + 1)d_{[i]}$ (for the still growing intervals). For the final critical radius $t = \frac{1}{2}d_{[m]}$, the whole interval $[0, 1]$ is covered by the grown intervals because $\sum_{j=1}^{m} d_{[j]} = 1$. So the final corner point of $\psi_0(t)$ is $(\frac{1}{2}d_{[m]}, 0)$. □

Theorem 5 for the sequence $S = \{0, \frac{1}{3}, \frac{1}{2}\} + \mathbb{Z}$ gives the ordered distances $d_{[1]} = \frac{1}{6} < d_{[2]} = \frac{1}{3} < d_{[3]} = \frac{1}{2}$, which determine the corner points $(0, 1)$, $(\frac{1}{12}, \frac{1}{2})$, $(\frac{1}{6}, \frac{1}{6})$, $(\frac{1}{4}, 0)$ of the density function $\psi_0(t)$ in Fig. 4, see Example 4.

By Theorem 5 any 0th density function $\psi_0(t)$ is uniquely determined by the (unordered) set of lengths of intervals between successive points. Hence we can re-order these intervals without changing $\psi_0(t)$. For instance, the periodic sequence $Q = \{0, \frac{1}{2}, \frac{2}{3}\} + \mathbb{Z}$ has the same set of interval lengths $d_{[1]} = \frac{1}{6}$, $d_{[2]} = \frac{1}{3}$, $d_{[3]} = \frac{1}{2}$ as the periodic sequence $S = \{0, \frac{1}{3}, \frac{1}{2}\} + \mathbb{Z}$ in Example 4.

The above sequences S, Q are related by the mirror reflection $t \mapsto 1 - t$. One can easily construct many non-isometric sequences with $\psi_0[S](t) = \psi_0[Q](t)$. For any $1 \leq i \leq m-3$, the sequences $S_{m,i} = \{0, 2, 3, \ldots, i+2, i+4, i+5, \ldots, m+2\} + (m+2)\mathbb{Z}$ have the same interval lengths $d_{[1]} = \cdots = d_{[m-2]} = 1$, $d_{[m-1]} = d_{[m]} = 2$ but are not related by isometry (translations and reflections in $\mathbb{R}$) because the intervals of length 2 are separated by $i - 1$ intervals of length 1 in $S_{m,i}$.

Corollary 12 will prove that the 1st density function $\psi_1[S](t)$ uniquely determines a periodic sequence $S \subset \mathbb{R}$ in general position up to isometry of $\mathbb{R}$.

Example 6 (functions $\psi_k(t)$ for $S = \{0, \frac{1}{3}, \frac{1}{2}\} + \mathbb{Z}$). The 1st density function $\psi_1(t)$ can be obtained as a sum of the three *trapezoid* functions η_R, η_G, η_B, each measuring the length of a region covered by a single interval (of one color). The red intervals $[0, t] \cup [1 - t, 1]$ grow until $t = \frac{1}{6}$ when they touch the green interval $[\frac{1}{6}, \frac{1}{2}]$. So the length $\eta_R(t)$ of this interval linearly grows from the origin $(0, 0)$ to the corner point $(\frac{1}{6}, \frac{1}{3})$. For $t \in [\frac{1}{6}, \frac{1}{4}]$, the left red interval is shrinking at the same rate due to the overlapping green interval, while the right red interval continues to grow until $t = \frac{1}{4}$, when it touches the blue interval $[\frac{1}{4}, \frac{3}{4}]$. Hence the graph of $\eta_R(t)$ remains constant up to the corner point $(\frac{1}{4}, \frac{1}{3})$. After that $\eta_R(t)$ linearly returns to the t-axis at $t = \frac{5}{12}$. Hence the trapezoid function η_R has the piecewise linear graph through the corner points $(0, 0)$, $(\frac{1}{6}, \frac{1}{3})$, $(\frac{1}{4}, \frac{1}{3})$, $(\frac{5}{12}, \frac{1}{6})$.

The 2nd function $\psi_2(t)$ is the sum of the *trapezoid* functions $\eta_{GB}, \eta_{RG}, \eta_{RB}$, each measuring the length of a double intersection. For the green interval $[\frac{1}{3} - t, \frac{1}{3} + t]$ and the blue interval $[\frac{1}{2} - t, \frac{1}{2} + t]$, the graph of the trapezoid function $\eta_{GB}(t)$ is piecewise linear and starts at the point $(\frac{1}{12}, 0)$, where the intervals touch. The green-blue intersection interval $[\frac{1}{2} - t, \frac{1}{3} + t]$ grows until $t = \frac{1}{4}$, when $[\frac{1}{4}, \frac{7}{12}]$ touches the red interval on the left. At the same time $\eta_{GB}(t)$ is linearly growing to the point $(\frac{1}{4}, \frac{1}{3})$. For $t \in [\frac{1}{4}, \frac{1}{3}]$, the green-blue intersection interval becomes shorter on the left, but grows at the same rate on the right until $[\frac{1}{3}, \frac{2}{3}]$ touches the red interval $[\frac{2}{3}, 1]$. Then $\eta_{GB}(t)$ remains constant up to the point $(\frac{1}{3}, \frac{1}{3})$. For $t \in [\frac{1}{3}, \frac{1}{2}]$ the green-blue intersection interval is shortening from both sides. Finally, the graph of $\eta_{GB}(t)$ returns to the t-axis at $(\frac{1}{2}, 0)$, see Fig. 4. ∎

Theorem 7 extends Example 6 and proves that any $\psi_k(t)$ is a sum of trapezoid functions whose corners are explicitly described. We consider any index $i = 1, \ldots, m$ (of a point p_i or a distance d_i) modulo m so that $m + 1 \equiv 1 \pmod{m}$.

Theorem 7 (description of ψ_k, $k > 0$). For any sequence $S = \{p_1, \ldots, p_m\} + \mathbb{Z}$ with motif points $0 \leq p_1 < \cdots < p_m < 1$, set $d_i = p_{i+1} - p_i \in (0, 1)$, where $i = 1, \ldots, m$ and $p_{m+1} = p_1 + 1$. Any interval $[p_i - t, p_i + t]$ is projected to $[0, 1]$ modulo $\mathbb{Z}$. For $1 \leq k \leq m$, the density function $\psi_k(t)$ is the sum of m *trapezoid* functions $\eta_{k,i}$ with the corner points $(\frac{s}{2}, 0)$, $(\frac{d_{i-1}+s}{2}, d)$, $(\frac{s+d_{i+k-1}}{2}, d)$,

$(\frac{d_{i-1}+s+d_{i+k-1}}{2}, 0)$, where $d = \min\{d_{i-1}, d_{i+k-1}\}$, $s = \sum\limits_{j=i}^{i+k-2} d_j$, $i = 2, \ldots, m+1$.

If $k = 1$, then $s = 0$ is the empty sum. So $\psi_k(t)$ is determined by the unordered set of triples (d_{i-1}, s, d_{i+k-1}) whose first and last entries are swappable. ∎

Proof. For simplicity, we separately prove the case $k = 1$. The 1st density function $\psi_1(t)$ measures the total length of subregions covered by a single interval $[p_i - t, p_i + t]$. Hence $\psi_1(t)$ is the sum of the functions η_{1i}, each measuring the length of the subinterval of $[p_i - t, p_i + t]$ not covered by other such intervals.

Each function η_{1i} starts from $\eta_{1i}(0) = 0$ and linearly grows up to $\eta_{1i}(\frac{1}{2}d) = d$, where $d = \min\{d_{i-1}, d_i\}$, when the interval $[p_i - t, p_i + t]$ of the length $2t = d$ touches the growing interval centered at the closest of its neighbors $p_{i\pm1}$.

If (say) $d_{i-1} < d_i$, then the subinterval covered only by $[p_i - t, p_i + t]$ is shrinking on the left and is growing at the same rate on the right until it touches the growing interval centered at the right neighbor. During this period, when t is between $\frac{1}{2}d_{i-1}$ and $\frac{1}{2}d_i$, the trapezoid function $\eta_{1i}(t) = d$ remains constant.

If $d_{i-1} = d_i$, this horizontal piece collapses to one point in the graph of $\eta_{1i}(t)$. For $t \geq \max\{d_{i-1}, d_i\}$, the subinterval covered only by $[p_i - t, p_i + t]$ is shrinking on both sides until the intervals centered at $p_{i\pm1}$ meet at a mid-point between them for $t = \frac{d_{i-1}+d_i}{2}$. So the graph of η_{1i} has a trapezoid form with the corner points $(0, 0)$, $(\frac{d_{i-1}}{2}, d)$, $(\frac{d_i}{2}, d)$, $(\frac{d_{i-1}+d_i}{2}, 0)$.

In Example 6 for $S = \{0, \frac{1}{3}, \frac{1}{2}\} + \mathbb{Z}$, the distances $d_1 = \frac{1}{3}$, $d_2 = \frac{1}{6}$, $d_3 = \frac{1}{2} = d_0$ give $\eta_{11} = \eta_R$ with the corner points $(0, 0)$, $(\frac{1}{4}, \frac{1}{3})$, $(\frac{1}{6}, \frac{1}{3})$, $(\frac{5}{12}, 0)$ as in Fig. 4.

In the case $k > 1$, the k-th density function $\psi_k(t)$ measures the total length of k-fold intersections among m intervals $[p_i - t, p_i + t]$, $i = 1, \ldots, m$.

A k-fold intersection appears only when two intervals $[p_i - t, p_i + t]$ and $[p_{i+k-1} - t, p_{i+k-1} + t]$ overlap because their intersection is covered by the k intervals centered at k points $p_i < p_{i+1} < \cdots < p_{i+k-1}$. Since only k successive intervals can contribute to k-fold intersections, $\psi_k(t)$ becomes the sum of the functions $\eta_{k,i}$, each equal to the length of the subinterval of $[p_i - t, p_{i+k-1} + t]$ covered by exactly k intervals of the form $[p_j - t, p_j + t]$, $j = 1, \ldots, m$.

The above function $\eta_{k,i}(t)$ remains 0 until the radius $t = \frac{1}{2} \sum\limits_{j=i}^{i+k-2} d_j$ because $2t$ is the length between the points $p_i < p_{i+k-1}$. Then $\eta_{k,i}(t)$ is linearly growing until the k-fold intersection touches one of the intervals centered at the points p_{i-1}, p_{i+k}, which are left and right neighbors of p_i, p_{i+k-1}, respectively.

If (say) $d_{i-1} < d_{i+k-1}$, this critical radius is $t = \frac{1}{2} \sum\limits_{j=i-1}^{i+k-2} d_j = \frac{d_{i-1}+s}{2}$. The function $\eta_{k,i}(t)$ measures the length of the k-fold intersection $[p_{i+k-1} - t, p_i + t]$.

$$\eta_{k,i}(t) = (p_i + t) - (p_{i+k-1} - t) = 2t - (p_{i+k-1} - p_i) = (d_{i-1} + s) - s = d_{i-1}.$$

Then the k-fold intersection is shrinking on the left and is growing at the same rate on the right until it touches the growing interval centered at the right neighbor p_{i+k}. During this time, when t is between $\frac{1}{2} \sum\limits_{j=i-1}^{i+k-2} d_j$ and $\frac{1}{2} \sum\limits_{j=i}^{i+k-1} d_j$, the function $\eta_{k,i}(t)$ remains equal to d_{i-1}. If $d_{i-1} > d_{i+k-1}$, the last argument should include the smaller distance d_{i+k-1} instead of d_{i-1}. Hence we will use below the single value $d = \min\{d_{i-1}, d_{i+k-1}\}$ to cover both cases. If $d_{i-1} = d_i$, this horizontal piece collapses to one point in the graph of $\eta_{k,i}(t)$. The k-fold intersection within $[p_i, p_{i+k-1}]$ disappears when the intervals centered at p_{i-1}, p_{i+k} have the radius t equal to the half-distance $\frac{1}{2}\sum\limits_{j=i-1}^{i+k-1} d_j$ between p_{i-1}, p_{i+k}.

Then $\eta_{k,i}(t)$ is the trapezoid function with the expected four corner points expressed as $(\frac{s}{2}, 0)$, $(\frac{d_{i-1}+s}{2}, d)$, $(\frac{s+d_{i+k-1}}{2}, d)$, $(\frac{d_{i-1}+s+d_{i+k-1}}{2}, 0)$ for $s = \sum\limits_{j=i}^{i+k-2} d_j$ and $d = \min\{d_{i-1}, d_{i+k-1}\}$. These corners are uniquely determined by the triple (d_{i-1}, s, d_{i+k-1}), where the components d_{i-1}, d_{i+k-1} can be swapped. $\qquad\square$

In Example 6 for $S = \{0, \frac{1}{3}, \frac{1}{2}\} + \mathbb{Z}$, we have $d_1 = \frac{1}{3}$, $d_2 = \frac{1}{6}$, $d_3 = \frac{1}{2} = d_0$. For $k = 2$, $i = 2$, we get $d_{i-1} = d_1 = \frac{1}{3}$, $d_{i+k-1} = d_3 = \frac{1}{2}$, i.e. $d = \min\{d_1, d_3\} = \frac{1}{3}$, $s = d_2 = \frac{1}{6}$. Then $\eta_{22} = \eta_{GB}$ has the corner points $(\frac{1}{12}, 0)$, $(\frac{1}{4}, \frac{1}{3})$, $(\frac{1}{3}, \frac{1}{3})$, $(\frac{1}{2}, 0)$.

4   Symmetries, Computations, and Generic Completeness

Theorem 8 (symmetries of $\psi_k(t)$). For any periodic sequence $S \subset \mathbb{R}$ with a unit cell $[0, 1]$, we have the *periodicity* $\psi_{k+m}(t + \frac{1}{2}) = \psi_k(t)$ for any $k \geq 0$, $t \geq 0$, and the *symmetry* $\psi_{m-k}(\frac{1}{2} - t) = \psi_k(t)$ for $k = 0, \ldots, [\frac{m}{2}]$, and $t \in [0, \frac{1}{2}]$. $\qquad\blacksquare$

Proof. To prove $\psi_{m-k}(\frac{1}{2} - t) = \psi_k(t)$ for $k = 1, \ldots, [\frac{m}{2}]$, we establish a bijection between the triples of parameters that determined ψ_{m-k} and ψ_k in Theorem 7.

Take a triple (d_{i-1}, s, d_{i+k-1}) of ψ_k, where $s = \sum\limits_{j=i}^{i+k-2} d_j$ is the sum of $k - 1$ distances from d_{i-1} to d_{i+k-1} in the increasing (cyclic) order of distance indices. Under $t \mapsto \frac{1}{2} - t$, the corner points of trapezoid function $\eta_{k,i}$ map to

$$\left(\frac{1-s}{2}, 0\right), \quad \left(\frac{1-s-d_{i-1}}{2}, d\right), \quad \left(\frac{1-s-d_{i+k-1}}{2}, d\right), \quad \left(\frac{1-d_{i-1}-s-d_{i+k-1}}{2}, 0\right).$$

Notice that $\bar{s} = 1 - d_{i-1} - s - d_{i+k-1}$ is the sum of the $m - k - 1$ intermediate distances from d_{i+k-1} to d_{i-1} in the increasing (cyclic) order of indices.

The four corner points can be re-written with the above notation $\bar{s}$ as follows:

$$\left(\frac{d_{i-1}+\bar{s}+d_{i+k-1}}{2}, 0\right), \quad \left(\frac{\bar{s}+d_{i+k-1}}{2}, d\right), \quad \left(\frac{\bar{s}+d_{i-1}}{2}, d\right), \quad \left(\frac{\bar{s}}{2}, 0\right).$$

These resulting points are re-ordered corners of the trapezoid function $\eta_{m-k,i+k}$. Hence $\eta_{k,i}(\frac{1}{2} - t) = \eta_{m-k,i+k}(t)$. Taking the sum over all indices $i = 1, \ldots, m$, we get $\psi_k(\frac{1}{2} - t) = \psi_{m-k}(t)$. Figure 4 shows the symmetry $\psi_1(t) = \psi_2(\frac{1}{2} - t)$.

For periodicity, we compare ψ_k and ψ_{k+m} for $k \geq 0$. Any $(k+m)$-fold intersection should involve intervals centered at $k + m$ successive points of the sequence $S \subset \mathbb{R}$. Then we can find a period interval $[t, t+1]$ covering m of these points. By collapsing this interval to a single point, the $(k + m)$-fold intersection becomes

k-fold, but its fractional length within any period interval of length 1 remains the same. Since the radius t is twice smaller than the length of the corresponding interval, this collapse gives us $\psi_{k+m}(t + \frac{1}{2}) = \psi_k(t)$.

The symmetry $\psi_m(\frac{1}{2} - t) = \psi_0(t)$ follows from $\psi_m(\frac{1}{2} - t) = \psi_m(\frac{1}{2} + t)$. Indeed, any trapezoid of ψ_m has $s = 1 - d_{i-1}$. Since its four corners $(\frac{1-d_{i-1}}{2}, 0), (\frac{1}{2}, \frac{d_{i-1}}{2}), (\frac{1}{2}, \frac{d_{i-1}}{2}), (\frac{1+d_{i-1}}{2}, 0)$ are symmetric in $t = \frac{1}{2}$, then so is the sum ψ_m. □

Corollary 9 (computation of $\psi_k(t)$). Let $S, Q \subset \mathbb{R}$ be periodic sequences with at most m motif points. For $k \geq 1$, one can draw the graph of the k-th density function $\psi_k[S]$ in time $O(m^2)$. One can check in time $O(m^3)$ if $\Psi[S] = \Psi[Q]$. ∎

Proof. To draw the graph of $\psi_k[S]$ or evaluate the k-th density function $\psi_k[S](t)$ at any t, we first use the symmetry and periodicity from Theorem 8 to reduce k to the range $0, 1, \ldots, [\frac{m}{2}]$. In time $O(m \log m)$ we put the points from a unit cell U (scaled to $[0, 1]$ for convenience) in the increasing (cyclic) order $p_1, \ldots, p_m$. In time $O(m)$ we compute the distances $d_i = p_{i+1} - p$ between successive points.

For $k = 0$, we put the distances in the increasing order $d_{[1]} \leq \cdots \leq d_{[m]}$ in time $O(m \log m)$. By Theorem 5 in time $O(m^2)$, we write down the $O(m)$ corner points whose horizontal coordinates are the critical radii where $\psi_0(t)$ can change its gradient. We evaluate ψ_0 at every critical radius t by summing up the values of m trapezoid functions at t, which needs $O(m^2)$ time. It remains to plot the points at all $O(m)$ critical radii t and connect the successive points by straight lines, so the total time is $O(m^2)$. For any larger fixed index $k = 1, \ldots, [\frac{m}{2}]$, in time $O(m^2)$ we write down all $O(m)$ corner points from Theorem 7, which leads to the graph of $\psi_k(t)$ similarly to the above argument for $k = 0$.

To decide if the infinite sequences of density functions coincide. $\Psi[S] = \Psi[Q]$, by Theorem 8 it suffices to check only if $O(m)$ density functions coincide: $\psi_k[S](t) = \psi_k[Q](t)$ for $k = 0, 1, \ldots, [\frac{m}{2}]$. To check if two piecewise linear functions coincide, it remains to compare their values at all $O(m)$ critical radii t from the corner points in Theorems 5 and 7. Since these values were found in time $O(m^2)$ above, the total time for $k = 0, 1, \ldots, [\frac{m}{2}]$ is $O(m^3)$. □

To illustrate Corollary 9, Example 10 will justify that the periodic sequences S_{15} and Q_{15} in Fig. 3 have identical density fingerprints $\Psi[S_{15}] = \Psi[Q_{15}]$.

Example 10 (S_{15}, Q_{15} have equal density functions). To avoid fractions, we keep the unit cell $[0, 15]$ of the sequences S_{15}, Q_{15} because all quantities in Theorem 7 can be scaled up by factor 15. To conclude that $\psi_0[S_{15}] = \psi_0[Q_{15}]$, by Theorem 5 we check that S_{15}, Q_{15} have the same set of the ordered distances $d_{[i]}$ between successive points, which is shown in identical rows 3 of Tables 1 and 2.

To conclude that $\psi_1[S_{15}] = \psi_1[Q_{15}]$ by Theorem 7, we check that S_{15}, Q_{15} have the same set of unordered pairs (d_{i-1}, d_i) of distances between successive points. Indeed, Tables 1 and 2 have identical rows 5, where pairs are *lexicograpically* ordered for comparison: $(a, b) < (c, d)$ if $a < b$ or $a = b$ and $c < d$.

To conclude that $\psi_k[S_{15}] = \psi_k[Q_{15}]$ for $k = 2, 3, 4$, we compare the triples $(d_{i-1}, \mathbf{s}, d_{i+k-1})$ from Theorem 7 for S_{15}, Q_{15}. For $k = 2$ and $k = 3$, Tables 1

Table 1. Row 1: points p_i from the set S_{15} in Fig. 3. **Row 2**: the distances d_i between successive points of S_{15}. **Row 3**: the distances $d_{[i]}$ are in the increasing order. **Row 4**: the unordered set of these pairs determines the density function ψ_1 by Theorem 7. **Row 5**: the pairs are lexicographically ordered for comparison with row 5 in Table 2. **Rows 6, 8, 10**: the unordered sets of these triples determine the density functions ψ_2, ψ_3, ψ_4 by Theorem 7 for $k = 2, 3, 4$. **Rows 7,9,11**: the triples from rows 6,8,10 are ordered for comparison with corresponding rows 7, 9, 11 in Table 1, see Example 10.

p_i	0	1	3	4	5	7	9	10	12
$d_i = p_{i+1} - p_i$	1	2	1	1	2	2	1	2	3
ordered $d_{[i]}$	1	1	1	1	2	2	2	2	3
(d_{i-1}, d_i)	(3,1)	(1,2)	(2,1)	(1,1)	(1,2)	(2,2)	(2,1)	(1,2)	(2,3)
order (d_{i-1}, d_i)	(1,1)	(1,2)	(1,2)	(1,2)	(1,2)	(1,2)	(1,3)	(2,2)	(2,3)
$(d_{i-1}, \mathbf{d_i}, d_{i+1})$	(3,**1**,2)	(1,**2**,1)	(2,**1**,1)	(1,**1**,2)	(1,**2**,2)	(2,**2**,1)	(2,**1**,2)	(1,**2**,3)	(2,**3**,1)
order $(d_{i-1}, \mathbf{d_i}, d_{i+1})$	(1,**1**,2)	(1,**1**,2)	(2,**1**,2)	(2,**1**,3)	(1,**2**,1)	(1,**2**,2)	(1,**2**,2)	(1,**2**,3)	(1,**3**,2)
$(d_{i-1}, \mathbf{s}, d_{i+2})$	(3,**3**,1)	(1,**3**,1)	(2,**2**,2)	(1,**3**,2)	(1,**4**,1)	(2,**3**,2)	(2,**3**,3)	(1,**5**,1)	(2,**4**,2)
order $(d_{i-1}, \mathbf{s}, d_{i+2})$	(2,**2**,2)	(1,**3**,1)	(1,**3**,2)	(1,**3**,3)	(2,**3**,2)	(2,**3**,3)	(1,**4**,1)	(2,**4**,2)	(1,**5**,1)
$(d_{i-1}, \mathbf{s}, d_{i+3})$	(3,**4**,1)	(1,**4**,2)	(2,**4**,2)	(1,**5**,1)	(1,**5**,2)	(2,**5**,3)	(2,**6**,1)	(1,**6**,2)	(2,**6**,1)
order $(d_{i-1}, \mathbf{s}, d_{i+3})$	(1,**4**,2)	(1,**4**,3)	(2,**4**,2)	(1,**5**,1)	(1,**5**,2)	(2,**5**,3)	(1,**6**,2)	(1,**6**,2)	(1,**6**,2)

Table 2. Row 1: points p_i from the set Q_{15} in Fig. 3. **Row 2**: the distances d_i between successive points of Q_{15}. **Row 3**: the distances $d_{[i]}$ are in the increasing order. **Row 4**: the unordered set of these pairs determines the density function ψ_1 by Theorem 7b. **Row 5**: the pairs are lexicographically ordered for comparison with row 5 in Table 1. **Rows 6, 8, 10**: the unordered sets of these triples determine the density functions ψ_2, ψ_3, ψ_4 by Theorem 7 for $k = 2, 3, 4$. **Rows 7, 9, 11**: the triples from rows 6,8,10 are ordered for comparison with corresponding rows 7, 9, 11 in Table 2, see Example 10.

p_i	0	1	3	4	6	8	9	12	14
$d_i = p_{i+1} - p_i$	1	2	1	2	2	1	3	2	1
ordered $d_{[i]}$	1	1	1	1	2	2	2	2	3
(d_{i-1}, d_i)	(1,1)	(1,2)	(2,1)	(1,2)	(2,2)	(2,1)	(1,3)	(3,2)	(2,1)
ordered (d_{i-1}, d_i)	(1,1)	(1,2)	(1,2)	(1,2)	(1,2)	(1,2)	(1,3)	(2,2)	(2,3)
$(d_{i-1}, \mathbf{d_i}, d_{i+1})$	(1,**1**,2)	(1,**2**,1)	(2,**1**,2)	(1,**2**,2)	(2,**2**,1)	(2,**1**,3)	(1,**3**,2)	(3,**2**,1)	(2,**1**,1)
order $(d_{i-1}, \mathbf{d_i}, d_{i+1})$	(1,**1**,2)	(1,**1**,2)	(2,**1**,2)	(2,**1**,3)	(1,**2**,1)	(1,**2**,2)	(1,**2**,2)	(1,**2**,3)	(1,**3**,2)
$(d_{i-1}, \mathbf{s}, d_{i+2})$	(1,**3**,1)	(1,**3**,2)	(2,**3**,2)	(1,**4**,1)	(2,**3**,3)	(2,**4**,2)	(1,**5**,1)	(3,**3**,1)	(2,**2**,2)
order $(d_{i-1}, \mathbf{s}, d_{i+2})$	(2,**2**,2)	(1,**3**,1)	(1,**3**,2)	(1,**3**,3)	(2,**3**,2)	(2,**3**,3)	(1,**4**,1)	(2,**4**,2)	(1,**5**,1)
$(d_{i-1}, \mathbf{s}, d_{i+3})$	(1,**4**,2)	(1,**5**,2)	(2,**5**,1)	(1,**5**,3)	(2,**6**,2)	(2,**6**,1)	(1,**6**,1)	(3,**4**,2)	(2,**4**,1)
order $(d_{i-1}, \mathbf{s}, d_{i+3})$	(1,**4**,2)	(1,**4**,2)	(2,**4**,3)	(1,**5**,2)	(1,**5**,2)	(1,**5**,3)	(1,**6**,1)	(1,**6**,2)	(2,**6**,2)

and 2 have identical rows 7 and 9, where the triples are ordered for easier comparison as follows. If needed, we swap d_{i-1}, d_{i+k-1} to make sure that the first entry is not larger than the last. Then we order by the middle bold number **s**. Finally, we lexicographically order the triples with the same middle value s.

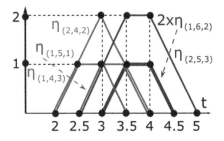 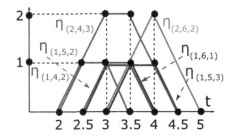

Fig. 5. The 4th-density function $\psi_4[S_{15}]$ includes the six trapezoid functions on the left, which are replaced by other six trapezoid functions in $\psi_4[Q_{15}]$ on the right, compare the last rows of Tables 1 and 2. However, the sums of these six functions are equal, which can be checked at critical radii: both sums of six functions have $\eta(2.5) = 2$, $\eta(3) = 5$, $\eta(3.5) = 6$, $\eta(4) = 4$, $\eta(4.5) = 1$. Hence the periodic sequences S_{15}, Q_{15} in Fig. 3 have identical density functions ψ_k for all $k \geq 0$, see details in Example 10.

Final rows 11 of Tables 1 and 2 look different for $k = 4$. More exactly, the rows share three triples $(1,4,2)$, $(1,5,2)$, $(1,6,4)$, but the remaining six triples differ. However, the density function ψ_4 is the *sum* of nine trapezoid functions. Figure 5 shows that these sums are equal for S_{15}, Q_{15}. Then the sequences S_{15}, Q_{15} have identical density functions ψ_k for $k = 0, 1, 2, 3, 4$, hence for all k by the symmetry and periodicity from Theorem 8. Figure 6 shows ψ_k, $k = 0, 1, \ldots, 9$. ∎

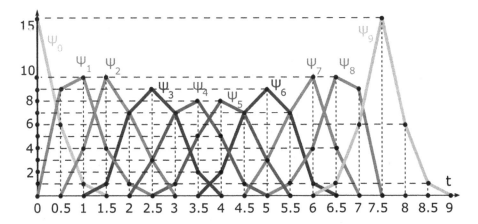

Fig. 6. The periodic sequences S_{15}, Q_{15} in Fig. 3 have identical density functions $\psi_k(t)$ for all $k \geq 0$. Both axes are scaled by factor 15. Theorem 8 implies the symmetry $\psi_k(\frac{15}{2} - t) = \psi_{9-k}(t)$, $t \in [0, \frac{15}{2}]$, and periodicity $\psi_9(t + \frac{15}{2}) = \psi_0(t)$, $t \geq 0$.

Recall that all indices i of distances d_i are considered modulo m.

Corollary 11 (*k*-th density ρ_k). For any periodic sequence $S = \{p_1, \ldots, p_m\} + \mathbb{Z}$ with inter-point distances $d_i = p_{i+1} - p_i$, where $i = 1, \ldots, m$ and $p_{m+1} = p_1 + 1$, the *k*-th *density* $\rho_k[S] = \int\limits_{-\infty}^{+\infty} \psi_k(t)dt$ defined as the area under the graph of $\psi_k(t)$ over $\mathbb{R}$ equals $\rho_k[S] = \frac{1}{2} \sum\limits_{i=1}^{m} d_{i-1}d_{i+k-1}$ for any $k > 0$ and $\rho_0[S] = \frac{1}{4} \sum\limits_{i=1}^{m} d_i^2$. ∎

Proof. By Theorem 7 for $k > 0$, each $\psi_k(t)$ is the sum of m trapezoid functions. Hence $\rho_k[S]$ equals the sum of the areas under the graphs of these trapezoids with corners $(\frac{s}{2}, 0)$, $(\frac{d_{i-1}+s}{2}, d)$, $(\frac{s+d_{i+k-1}}{2}, d)$, $(\frac{d_{i-1}+s+d_{i+k-1}}{2}, 0)$, where $d = \min\{d_{i-1}, d_{i+k-1}\}$. The area of each trapezoid is $A_i = \frac{d}{2}(\frac{d_{i-1}+d_{i+k-1}}{2} + |d_{i+k-1} - d_{i-1}|) - \frac{dD}{2}$, where $D = \max\{d_{i-1}, d_{i+k-1}\}$. Then $\rho_k = \sum\limits_{i=1}^{m} A_i = \frac{1}{2} \sum\limits_{i=1}^{m} d_{i-1}d_{i+k-1}$. Since $\psi_0(t) = 0$ for $t < 0$, ρ_0 is a half of the area $\rho_m = \frac{1}{2} \sum\limits_{i=1}^{m} d_i^2$ under $\psi_m(t)$ due to $\psi_m(\frac{1}{2} \pm t) = \psi_0(t)$ for $t \in [0, \frac{1}{2}]$ by Theorem 8, see Fig. 6. □

For $S = \{0, \frac{1}{3}, \frac{1}{2}\} + \mathbb{Z}$, Corollary 11 gives $\rho_0 = \frac{7}{72}$, $\rho_1 = \rho_2 = \frac{11}{12^2}$ as in Fig. 4.

Corollary 12 (generic completeness of ψ_1). Let $S \subset \mathbb{R}$ be a sequence with period 1 and m points $0 \leq p_1 < \cdots < p_m < 1$. The sequence S is called *generic* if $d_i = p_{i+1} - p_i$ are distinct, where $i = 1, \ldots, m$ and $p_{m+1} = p_1 + 1$. Then any generic S can be reconstructed from the 1st density function $\psi_1[S](t)$ up to isometry in $\mathbb{R}$. Hence $\psi_1(t)$ is a complete isometry invariant for all generic S. ∎

Proof. As always, one can scale a unit cell of S to the standard interval $[0, 1]$ as in Theorem 7. Hence, up to translation and reflection of $\mathbb{R}$, one can assume that $p_1 = 0 < p_2 < 1 = p_{m+1}$. It suffices to uniquely locate $p_2, \ldots, p_m \in (0, 1)$.

The 1st density function $\psi_1[S](t)$ is the sum of the trapezoid functions that have the initial gradient 2 and the corner points $(0, 0)$, $(\frac{d_{i-1}}{2}, d)$, $(\frac{d_i}{2}, d)$, $(\frac{d_{i-1}+d_i}{2}, 0)$, where $d = \min\{d_{i-1}, d_i\}$, $i = 1, \ldots, m$, all indices are modulo m.

Due to a cyclic order of inter-point distances d_i, one can assume that the minimum distance is $d_{[1]} = d_1$. For any $0 \leq t \leq d_1$, the function $\psi_1[S](t)$ is linearly increasing with the gradient $2m$. This gradient drops to $2m - 2$ at the first critical radius $t = \frac{d_1}{2}$, which differs from all other larger points $\frac{d_i}{2}$ and $\frac{d_{i-1}+d_i}{2}$ where the gradient of $\psi_1(t)$ changes. Then the first corner of $\psi_1(t)$ uniquely determines d_1 and the second point $p_2 = d_1$ of the sequence S.

At the radius $t = \frac{d_1}{2}$, subtracting from $\psi_1(t)$ the contribution $(m-1)d_1$ from other still growing $m-1$ trapezoid functions, we get the value $\frac{d_1+d_2}{2}$. So the first corner of $\psi_1(t)$ also determines the length $d_2 = p_3 - p_2$ of the second inter-point interval after $[p_1, p_2]$ of the length d_1, and the third point $p_3 = d_1 + d_2$.

Since we know both d_1, d_2, we can subtract from $\psi_1(t)$ the whole trapezoid function $\eta(t)$ with the above corners for $i = 2$ for all $t \in [0, 1]$. The resulting function $\tilde{\psi}_1(t)$ is the sum of $m - 1$ trapezoid functions depending on $m - 1$ inter-point distances $d_2, \ldots, d_m$. We continue analyzing $\tilde{\psi}_1(t)$ by looking at the first corner where its gradient drops from $2m - 2$ to $2m - 4$, which gives us another pair (d_{i-1}, d_i) of successive interval lengths, and so on. Since all distances d_i are distinct, the above pairs uniquely determine the ordered sequence $d_1, \ldots, d_m$ of all interval lengths, hence the points $p_2, \ldots, p_m \in (0, 1)$ of the sequence S. $\square$

The recent developments in Periodic Geometry include algorithms of metrics on periodic point sets [2,8], Lattice Isometry Spaces in dimension two [3,9] and three [4,7], Pointwise Distance Distributions [13], and applications to materials [12,15]. We thank all reviewers for helpful suggestions. This research was supported by the EPSRC grant Application-driven Topological Data Analysis', Royal Academy of Engineering Industrial Fellowship 'Data Science for Next Generation Engineering of Solid Crystalline Materials', and EPSRC New Horizons grant 'Inverse design of periodic crystals'

References

1. Anosova, O., Kurlin, V.: Introduction to periodic geometry and topology (2021). https://arxiv.org/abs/2103.02749
2. Anosova, O., Kurlin, V.: Algorithms for continuous metrics on periodic crystals (2022). https://arxiv.org/abs/2205.15298
3. Bright, M., Cooper, A.I., Kurlin, V.: Geographic-style maps for 2-dimensional lattices (2021). https://arxiv.org/abs/2109.10885
4. Bright, M., Cooper, A.I., Kurlin, V.: Welcome to a continuous world of 3-dimensional lattices (2021). https://arxiv.org/abs/2109.11538
5. Edelsbrunner, H., Heiss, T., Kurlin, V., Smith, P., Wintraecken, M.: The density fingerprint of a periodic point set. In: SoCG, vol. 189, pp. 32:1–32:16 (2021)
6. Grünbaum, F., Moore, C.: The use of higher-order invariants in the determination of generalized Patterson cyclotomic sets. Acta Cryst. A **51**, 310–323 (1995)
7. Kurlin, V.: A complete isometry classification of 3-dimensional lattices (2022). https://arxiv.org/abs/2201.10543
8. Kurlin, V.: A computable and continuous metric on isometry classes of high-dimensional periodic sequences (2022). https://arxiv.org/abs/2205.04388
9. Kurlin, V.: Mathematics of 2-dimensional lattices. arxiv:2201.05150 (2022)
10. Mosca, M., Kurlin, V.: Voronoi-based similarity distances between arbitrary crystal lattices. Cryst. Res. Technol. **55**(5), 1900197 (2020)
11. Pozdnyakov, S., et al.: Incompleteness of atomic structure representations. Phys. Rev. Let. **125**, 166001 (2020)
12. Ropers, J., et al.: Fast predictions of lattice energies by continuous isometry invariants. In: Proceedings of DAMDID 2021. https://arxiv.org/abs/2108.07233
13. Widdowson, D., Kurlin, V.: Resolving the data ambiguity for periodic crystals. Adva. Neural Inf. Proc. Syst. **35** (NeurIPS 2022). arXiv:2108.04798

14. Widdowson, D., et al.: Average minimum distances of periodic point sets. MATCH Commun. Math. Comput. Chem. **87**, 529–559 (2022)
15. Zhu, Q., et al.: Analogy powered by prediction and structural invariants. J. Am. Chem. Soc. **144**, 9893–9901 (2022). https://doi.org/10.1021/jacs.2c02653

Approximation of Digital Surfaces by a Hierarchical Set of Planar Patches

Jocelyn Meyron and Tristan Roussillon[✉]

Univ Lyon, INSA Lyon, CNRS, UCBL, LIRIS, UMR 5205,
69622 Villeurbanne, France
{jocelyn.meyron,tristan.roussillon}@liris.cnrs.fr

Abstract. We show that the plane-probing algorithms introduced in Lachaud et al. (J. Math. Imaging Vis., 59, 1, 23–39, 2017), which compute the normal vector of a digital plane from a starting point and a set-membership predicate, are closely related to a three-dimensional generalization of the Euclidean algorithm. In addition, we show how to associate with the steps of these algorithms generalized substitutions, i.e., rules that replace square faces by unions of square faces, to build finite sets of elements that periodically generate digital planes. This work is a first step towards the incremental computation of a hierarchy of pieces of digital plane that locally fit a digital surface.

Keywords: Digital planes · Multi-dimensional continued-fraction algorithms · Generalized substitutions · Plane-probing algorithms

1   Introduction

Digital geometry mainly deals with sets of discrete elements considered to be digitized versions of Euclidean objects. A digital surface may be seen as a mesh of unit square faces whose vertices have integer coordinates. A challenge is to decompose digital surfaces into patches, such as pieces of digital planes.

A digital plane has been analytically defined as a set of points with integer coordinates lying between two parallel planes. Given a finite point set, one can decide whether this set belongs to a digital plane or not in linear time using linear programming. For a review on digital planarity, see [7]. However, a linear programming solver does not help so much for the analysis of digital surfaces, because one does not know which point set should be tested to obtain patches that approximates the tangent plane of the underlying surface.

In order to cope with this problem, *plane-probing* algorithms have been developed [17,18]. Their main feature is to decide on-the-fly how to probe a given point set and locally align a triangle with it. However, if probing for points in a sparse way is perfect for digital planes, it is not enough for non-convex parts where the triangles may jump over holes or stab the digital surface. Therefore, that approach also requires to associate pieces of digital planes to the triangles and check whether they fit the digital surface or not.

This work has been funded by PARADIS ANR-18-CE23-0007-01 research grant.

© Springer Nature Switzerland AG 2022
É. Baudrier et al. (Eds.): DGMM 2022, LNCS 13493, pp. 409–421, 2022.
https://doi.org/10.1007/978-3-031-19897-7_32

In this paper, we make a first step towards the incremental generation of a hierarchy of pieces of digital plane, which can be used during the execution of a plane-probing algorithm (see Fig. 1). To do that, we take advantage of the combinatorial properties of digital planes.

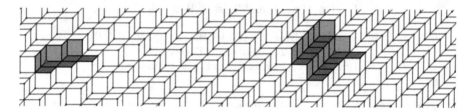

Fig. 1. Local approximation of a digital sphere of radius 63 by planar patches (in green) from two starting points (in blue). The implementation combines the plane-probing algorithm H [18] with the generation method of Sect. 3. (Color figure online)

The particular case of digital lines has been studied for a long time in different contexts and has led to many applications. One key result is that digital lines are hierarchical point sets whose structure is exactly described by the continued fraction expansion of their slope and strongly relies on the Euclidean algorithm. For a survey on digital straightness, see [15].

There has been much effort done in order to find similar results in three dimensions despite the lack of a canonical algorithm and the diversity of existing generalizations of the Euclidean algorithm. Some combinatorial results, involving symmetries, piece exchanges and flips, have been stated thanks to an appropriate representation of digital planes [16]. However, most of other related works depend on a multi-dimensional generalization of the Euclidean algorithm such as *Brun*, e.g., [4], *Jacobi-Perron*, e.g., [6], *Fully substractive*, e.g. [3], or a mix of several of them [13]. Those algorithms have been used to generate digital planes from a normal vector. There are two different but closely related construction schemes in the literature. The first one is based on union and translation of point sets [5,10,13]. It has been used mostly to construct the thinnest digital plane that is connected [3,5,8,9]. The second one is based on a description of standard digital planes as unions of square faces and uses rules that replace square faces by unions of square faces. Since the pioneer work of Arnoux and Ito [2], that formalism has been used for instance in [4–6,11,12]. Both construction schemes incrementally generate sets so that the current set will be included in the next one, but suffer from topological and geometrical limitations. This work is based on the second scheme that generates few elements in comparison with the first scheme.

In Sect. 2, we show how one can relate plane-probing algorithms to a three-dimensional generalization of the Euclidean algorithm. In Sect. 3, we introduce the generalized substitutions, describe our approach and examine several properties on the generated sets. Finally, the paper ends with some concluding remarks and perspectives.

2   Generalization of the Euclidean Algorithm

Let $\{e_1, \ldots, e_d\}$ be the canonical basis of $\mathbb{R}^d$. We denote $\mathbf{0}$ the origin and $\mathbf{1} = \sum_{i=1}^{d} e_i$ the vector with all coordinates equal to 1 whatever the dimension d. We are interested below in vectors with coprime positive integer coordinates, i.e., in the set $\mathcal{V}^d := \{(v_1, \ldots, v_d) \mid \forall i, v_i \in \mathbb{N} \setminus \{0\} \text{ and } \gcd(v_1, \ldots, v_d) = 1\}$.

2.1   Three-Dimensional Euclidean Algorithms

In its additive form, the Euclidean algorithm can simply be expressed as: "for a rational number represented by a pair of integers: subtract the smaller element to the larger one and repeat". It can be also expressed as a map $\Pi : \mathcal{V}^2 \to \mathcal{V}^2$ such that $\Pi(\mathbf{1}) = \mathbf{1}$ and

$$\text{for any } \left(\begin{smallmatrix} a \\ b \end{smallmatrix}\right) \in \mathcal{V}^2 \setminus \{\mathbf{1}\}, \ \Pi\left(\left(\begin{smallmatrix} a \\ b \end{smallmatrix}\right)\right) = \begin{cases} \left(\begin{smallmatrix} 1 & -1 \\ 0 & 1 \end{smallmatrix}\right)\left(\begin{smallmatrix} a \\ b \end{smallmatrix}\right) & \text{if } a > b, \\ \left(\begin{smallmatrix} 1 & 0 \\ -1 & 1 \end{smallmatrix}\right)\left(\begin{smallmatrix} a \\ b \end{smallmatrix}\right) & \text{otherwise.} \end{cases}$$

Note that for all $\mathbf{v} \in \mathcal{V}^2$, there exists a non-negative integer N such that N successive applications of Π returns $\mathbf{1}$, i.e., $\Pi^N(\mathbf{v}) = \mathbf{1}$. In dimension 2, it is clear to decide which number has to be subtracted to the other, whereas in dimension 3, i.e., for triplets, it is not the case, hence the diversity of existing generalizations. Several existing algorithms stick to a convention, which may lead to ambiguities or to a null coordinate in case of ties:

- *Brun* consists in subtracting the second largest entry to the largest one; $(1, 2, 2)$ may lead to either $(1, 0, 2)$ or to $(1, 2, 0)$.
- *Farey* consists in subtracting the smallest entry to the second largest one; $(1, 1, 2)$ may lead to either $(0, 1, 2)$ or to $(1, 0, 2)$. In both cases, the numbers cannot be reduced further.
- *Selmer* consists in subtracting the smallest entry to the largest one, $(1, 2, 2)$ may lead to either $(1, 1, 2)$ or to $(1, 2, 1)$. While the numbers are not all equal, no zero appears.

We generalize below that last algorithm with the following two definitions:

Definition 1. *Let $\mathcal{T}$ be the set of all permutations over $\{1, 2, 3\}$. For a permutation $\tau \in \mathcal{T}$, let $\mathbf{U}_{\tau(1),\tau(2),\tau(3)}$ be a 3×3 matrix having -1 at the intersection between the $\tau(1)$-th row and the $\tau(2)$-th column and the same entries as $\mathbf{I}_3$, i.e. the 3×3 identity matrix, elsewhere. Let $\mathcal{U}$ be the set $\{\mathbf{U}_{\tau(1),\tau(2),\tau(3)} \mid \tau \in \mathcal{T}\}$.*

$\mathbf{U}_{1,2,3}$	$\mathbf{U}_{1,3,2}$	$\mathbf{U}_{2,3,1}$	$\mathbf{U}_{2,1,3}$	$\mathbf{U}_{3,1,2}$	$\mathbf{U}_{3,2,1}$
$\begin{pmatrix} 1 & -1 & 0 \\ 0 & 1 & 0 \\ 0 & 0 & 1 \end{pmatrix}$	$\begin{pmatrix} 1 & 0 & -1 \\ 0 & 1 & 0 \\ 0 & 0 & 1 \end{pmatrix}$	$\begin{pmatrix} 1 & 0 & 0 \\ 0 & 1 & -1 \\ 0 & 0 & 1 \end{pmatrix}$	$\begin{pmatrix} 1 & 0 & 0 \\ -1 & 1 & 0 \\ 0 & 0 & 1 \end{pmatrix}$	$\begin{pmatrix} 1 & 0 & 0 \\ 0 & 1 & 0 \\ -1 & 0 & 1 \end{pmatrix}$	$\begin{pmatrix} 1 & 0 & 0 \\ 0 & 1 & 0 \\ 0 & -1 & 1 \end{pmatrix}$

Note that the elements of $\mathcal{U}$ are matrices with determinant 1.

Definition 2. *A three-dimensional Euclidean algorithm is a map* $\Pi : \mathcal{V}^3 \rightarrow \mathcal{V}^3$ *such that* $\Pi(\mathbf{1}) = \mathbf{1}$ *and for any* $\mathbf{w} \in \mathcal{V}^3 \setminus \{\mathbf{1}\}$*, there is a matrix* $\mathbf{U} \in \mathcal{U}$ *satisfying* $\Pi(\mathbf{w}) = \mathbf{U}\mathbf{w}$.

Note that Selmer is a three-dimensional Euclidean algorithm according to Definition 2.

The following proposition shows that the repeated application of Π always brings a vector $\mathbf{v} \in \mathcal{V}^3$ to $\mathbf{1}$, exactly as with the Euclidean algorithm in 2D.

Proposition 1. *Let* Π *be a three-dimensional Euclidean algorithm. For all* $\mathbf{v} \in \mathcal{V}^3$*, there exists a non-negative integer* N *such that* $\Pi^N(\mathbf{v}) = \mathbf{1}$ *and* $N \leq \|\mathbf{v}\|_1 - 3$.

Proof. For all $n \geq 0$ for which the coordinates of $\Pi^n(\mathbf{v})$ are not all equal, by definition of Π, there exists a matrix $\mathbf{U} \in \mathcal{U}$ such that $\Pi^{n+1}(\mathbf{v}) = \mathbf{U}\Pi^n(\mathbf{v}) \in \mathcal{V}^3$. Furthermore, we have $\|\Pi^{n+1}(\mathbf{v})\|_1 < \|\Pi^n(\mathbf{v})\|_1$ in that case.

If there is $N \geq 1$ for which the coordinates of $\Pi^N(\mathbf{v})$ are all equal, there is a strictly positive integer h such that $\Pi^N(\mathbf{v}) = h\mathbf{1}$. However, $\Pi^N(\mathbf{v}) = \mathbf{M}\mathbf{v}$, where $\mathbf{M}$ is a product of elementary matrices belonging to $\mathcal{U}$. It follows that $\mathbf{M}$ has determinant 1 and is therefore invertible. We have $\mathbf{M}\mathbf{v} = h\mathbf{1} \Leftrightarrow \mathbf{v} = h\mathbf{M}^{-1}\mathbf{1}$ and since the coordinates of $\mathbf{v}$ are coprime, we obtain $h = 1$ and $\Pi^N(\mathbf{v}) = \mathbf{1}$.

As a consequence, $(\|\Pi^n(\mathbf{v})\|_1)_{n=0,\ldots,N}$ is a strictly decreasing integer sequence from $\|\mathbf{v}\|_1$ to $\|\Pi^N(\mathbf{v})\|_1 = 3$, which concludes the proof. □

We focus now on the following finite sequences:

Definition 3. *A sequence of matrices* $(\mathbf{U}_n)_{0 \leq n \leq N}$ *is* valid *iff* $\mathbf{U}_0 = \mathbf{I}_3$ *and every* $\mathbf{U}_n$, $n \in \{1, \ldots, N\}$, *is in* $\mathcal{U}$. *In addition, a valid sequence* reduces *a vector* $\mathbf{v} \in \mathcal{V}^3$ *iff for all* $n \in \{0, \ldots, N-1\}$, $\mathbf{U}_n \cdots \mathbf{U}_0\mathbf{v} \in \mathcal{V}^3 \setminus \{\mathbf{1}\}$ *and* $\mathbf{U}_N \cdots \mathbf{U}_0\mathbf{v} = \mathbf{1}$. *In that case, we set* $\mathbf{v}_n := \mathbf{U}_n \cdots \mathbf{U}_0\mathbf{v}$ *and* $\mathbf{a}_n := \mathbf{U}_0^{-1} \cdots \mathbf{U}_n^{-1}\mathbf{1}$.

Let $\langle \cdot, \cdot \rangle$ stand for the usual scalar product on $\mathbb{R}^3$. Note that for all $\mathbf{x} \in \mathbb{R}^3$, $\langle \mathbf{x}, \mathbf{e}_i \rangle$ is equal to the i-th coordinate of $\mathbf{x}$. The following proposition shows several properties of the above-defined sequences:

Proposition 2. *We have* $\mathbf{a}_N = \mathbf{v}_0$ *and for each* $n \in \{0, \ldots, N\}$:

    (i) *for each* $i \in \{1, 2, 3\}$, $\langle \mathbf{U}_0^T \cdots \mathbf{U}_n^T \mathbf{e}_i, \mathbf{v}_0 \rangle$ *is the* i-*th coordinate of* $\mathbf{v}_n$,
    (ii) *for each* $i \in \{1, 2, 3\}$, $\langle \mathbf{U}_0^T \cdots \mathbf{U}_n^T \mathbf{e}_i, \mathbf{a}_n \rangle = 1$,
    (iii) *the differences* $(\mathbf{U}_0^T \cdots \mathbf{U}_n^T(\mathbf{e}_2 - \mathbf{e}_1), \mathbf{U}_0^T \cdots \mathbf{U}_n^T(\mathbf{e}_3 - \mathbf{e}_2))$ *form a basis of*
    the lattice $\{\mathbf{x} \in \mathbb{Z}^3 \mid \langle \mathbf{x}, \mathbf{a}_n \rangle = 0\}$.

Proof. By Definition 3, we have $\mathbf{v}_N = \mathbf{U}_N \cdots \mathbf{U}_0\mathbf{v}_0 = \mathbf{1}$, which is equivalent to $\mathbf{v}_0 = \mathbf{U}_N^{-1} \cdots \mathbf{U}_0^{-1}\mathbf{1} = \mathbf{a}_N$. For (i), $\langle \mathbf{U}_0^T \cdots \mathbf{U}_n^T \mathbf{e}_i, \mathbf{v}_0 \rangle = \langle \mathbf{e}_i, \mathbf{U}_n \cdots \mathbf{U}_0\mathbf{v}_0 \rangle = \langle \mathbf{e}_i, \mathbf{v}_n \rangle$. For (ii), $\langle \mathbf{U}_0^T \cdots \mathbf{U}_n^T \mathbf{e}_i, \mathbf{a}_n \rangle = \langle \mathbf{e}_i, \mathbf{U}_n \cdots \mathbf{U}_0\mathbf{a}_n \rangle = \langle \mathbf{e}_i, \mathbf{1} \rangle = 1$. Finally, for the last item, note that $\langle \mathbf{U}_0^T \cdots \mathbf{U}_n^T(\mathbf{e}_2 - \mathbf{e}_1), \mathbf{a}_n \rangle = 0$ and $\langle \mathbf{U}_0^T \cdots \mathbf{U}_n^T(\mathbf{e}_3 - \mathbf{e}_2), \mathbf{a}_n \rangle = 0$ by (ii). The fact that $(\mathbf{U}_0^T \cdots \mathbf{U}_n^T(\mathbf{e}_2 - \mathbf{e}_1), \mathbf{U}_0^T \cdots \mathbf{U}_n^T(\mathbf{e}_3 - \mathbf{e}_2))$ form a *basis* of the lattice $\{\mathbf{x} \in \mathbb{Z}^3 \mid \langle \mathbf{x}, \mathbf{a}_n \rangle = 0\}$ comes from the fact that the matrix $\mathbf{U}_0^T \cdots \mathbf{U}_n^T$ has determinant 1. □

Figure 2 illustrates the action of the matrices $\mathbf{U}_N \cdots \mathbf{U}_0$ and $\mathbf{U}_0^{-1} \cdots \mathbf{U}_N^{-1}$ in (a), and $\mathbf{U}_0^T \cdots \mathbf{U}_N^T$ in (b). Note that for each $n \in \{0, \dots, N\}$, $\mathbf{U}_0^T \cdots \mathbf{U}_n^T$ transforms the basis $\{\mathbf{e}_i \mid i \in \{1, 2, 3\}\}$ to $\{\mathbf{U}_0^T \cdots \mathbf{U}_n^T \mathbf{e}_i \mid i \in \{1, 2, 3\}\}$. In other words, it deforms an orthonormal basis to a basis that is more and more aligned with the plane of normal $\mathbf{v} = \mathbf{v}_0$ because the quantities $\{\langle \mathbf{U}_0^T \cdots \mathbf{U}_n^T \mathbf{e}_i, \mathbf{v}_0 \rangle \mid i \in \{1, 2, 3\}\}$ are smaller and smaller by Proposition 2, item (i).

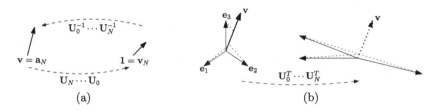

(a)   (b)

Fig. 2. Geometrical interpretation of the matrices $(\mathbf{U}_n)_{0 \le n \le N}$. In (b), we have implicitly represented the scalar projection of the basis vectors in the direction of $\mathbf{v}$ with the help of dotted lines. In the rightmost figure, the scalar products $\{\langle \mathbf{U}_0^T \cdots \mathbf{U}_N^T \mathbf{e}_i, \mathbf{v} \rangle \mid i \in \{1, 2, 3\}\}$ are equal to the coordinates of $\mathbf{v}_N$ and are thus all equal to 1.

2.2   Relation with Plane-Probing Algorithms

A *digital plane* is formally defined by a normal $\mathbf{v} \in \mathbb{Z}^3 \setminus \{\mathbf{0}\}$ and a position $\mu \in \mathbb{Z}$ as follows:

$$\mathcal{P}_{\mu, \mathbf{v}} := \{\mathbf{x} \in \mathbb{Z}^3 \mid \mu \le \mathbf{x} \cdot \mathbf{v} < \mu + \|\mathbf{v}\|_1\}. \tag{1}$$

In what follows, we assume w.l.o.g. that $\mathbf{v} \in \mathcal{V}^3$ and $\mu = 0$. Given a digital plane $\mathcal{P} \in \{\mathcal{P}_{0, \mathbf{v}} \mid \mathbf{v} \in \mathcal{V}^3\}$ of unknown normal vector a *plane-probing algorithm* computes the normal vector $\mathbf{v}$ of $\mathcal{P}$ by sparsely probing it with the predicate "is $\mathbf{x}$ in $\mathcal{P}$?".

We now describe the plane-probing algorithm H introduced in [18]. The state of the algorithm is a basis of three vectors, which can be stored in a 3×3 matrix, denoted by the letter $\mathbf{B}$, with the index of the step as a subscript. At initialization, $\mathbf{B}_0$ is set to identity. In order to explain how $\mathbf{B}_{n+1}$ is computed from $\mathbf{B}_n$ at a step $n \ge 0$ of the algorithm, let us introduce the following set of differences (see Fig. 3 for an illustration):

$$\mathcal{D}_n := \{\mathbf{B}_n(-\mathbf{e}_{\tau(1)} + \mathbf{e}_{\tau(2)}) \mid \tau \in \mathcal{T}\}.$$

If $\mathcal{P} \cap \{1 + \mathbf{d} \mid \mathbf{d} \in \mathcal{D}_n\} = \emptyset$ (see Fig. 3 on the right), the algorithm halts. Otherwise, there exists a permutation $\tau \in \mathcal{T}$ such that:

1. $1 + \mathbf{B}_n(-\mathbf{e}_{\tau(1)} + \mathbf{e}_{\tau(2)}) \in \mathcal{P}$ (see Fig. 3),
2. the sphere passing by $\{1 + \mathbf{B}_n \mathbf{e}_i \mid i \in \{1, 2, 3\}\}$ and the point $1 + \mathbf{B}_n(-\mathbf{e}_{\tau(1)} + \mathbf{e}_{\tau(2)})$ does not include in its interior any other point of $\mathcal{P} \cap \{1 + \mathbf{d} \mid \mathbf{d} \in \mathcal{D}_n\}$.

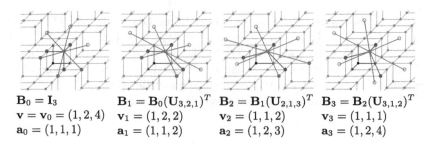

$$\mathbf{B}_0 = \mathbf{I}_3 \qquad \mathbf{B}_1 = \mathbf{B}_0(\mathbf{U}_{3,2,1})^T \quad \mathbf{B}_2 = \mathbf{B}_1(\mathbf{U}_{2,1,3})^T \quad \mathbf{B}_3 = \mathbf{B}_2(\mathbf{U}_{3,1,2})^T$$

$$\mathbf{v} = \mathbf{v}_0 = (1,2,4) \quad \mathbf{v}_1 = (1,2,2) \qquad \mathbf{v}_2 = (1,1,2) \qquad \mathbf{v}_3 = (1,1,1)$$

$$\mathbf{a}_0 = (1,1,1) \qquad \mathbf{a}_1 = (1,1,2) \qquad \mathbf{a}_2 = (1,2,3) \qquad \mathbf{a}_3 = (1,2,4)$$

Fig. 3. Execution of the algorithm H on a digital plane of normal vector $(1,2,4)$. For $n \in \{0,1,2,3\}$, the column vectors of $\mathbf{B}_n$ (resp. elements of $\mathcal{D}_n$) are depicted with blue (resp. red) arrows that points to $\mathbf{1}$. The end points are depicted with disks (resp. circles) if they belong (resp. do not belong) to the digital plane. (Color figure online)

From that permutation τ, we set $\mathbf{B}_{n+1} := \mathbf{B}_n(\mathbf{U}_{\tau(1),\tau(2),\tau(3)})^T$. In other words, the $\tau(1)$-th column of $\mathbf{B}_{n+1}$, i.e., $\mathbf{B}_{n+1}\mathbf{e}_{\tau(1)}$, is equal to the difference $\mathbf{B}_n(\mathbf{e}_{\tau(1)} - \mathbf{e}_{\tau(2)})$, while the other columns are identical in $\mathbf{B}_{n+1}$ and $\mathbf{B}_n$.

Note that there may be as much as three points in $\mathcal{P} \cap \{\mathbf{1} + \mathbf{d} \mid \mathbf{d} \in \mathcal{D}_n\}$ because $\mathbf{1}$ does not belong to $\mathcal{P}$ and $\mathcal{D}_n$ contains three distinct pairs of vectors of opposite sign. The in-sphere criterion generally provides a way of selecting one of those points and thus one elementary matrix. If several points are in a cospherical position, one can resort to a lexicographic order so that the algorithm is defined without any ambiguity.

That algorithm is very similar to a three-dimensional Euclidean algorithm, but does not exactly correspond to Definition 2, because two consecutive steps are not independent. However, it produces a sequence of matrices $(\mathbf{U})_{0 \leq n \leq N}$ such that $\mathbf{U}_0 = \mathbf{I}_3$ and $\mathbf{B}_n^T = \mathbf{B}_{n-1}^T\mathbf{U}_n$ for $n \in \{1, \ldots, N\}$. That sequence is not only valid by construction, but also reduces the normal vector $\mathbf{v}$ (Definition 3):

- $\forall n \in \{0, \ldots, N-1\}$, $\mathbf{U}_n \cdots \mathbf{U}_0\mathbf{v} \in \mathcal{V}^3 \setminus \{\mathbf{1}\}$ by item 1 of [18, Lemma 1],
- $\mathbf{U}_N \cdots \mathbf{U}_0\mathbf{v} = \mathbf{1}$ by [18, Theorem 2].

As a consequence, Proposition 2 also applies. For instance, $\mathbf{a}_N = \mathbf{v}$ [18, Corollary 4] means that the algorithm, which computes $\mathbf{a}_N$ only from a predicate "is $\mathbf{x}$ in $\mathcal{P}$?", is indeed able to retrieve the normal vector $\mathbf{v}$ of $\mathcal{P}$ (see also Fig. 3).

What makes the algorithm H very different from the classical three-dimensional Euclidean algorithms like Selmer is twofold: on one hand, it uses the past results for each new computation and on the other hand, it uses a geometrical criterion to decide which elementary matrix has to be applied. That is why less elongated patterns are obtained with that algorithm in the next section and for instance in Fig. 5. In addition, since it uses only a set-membership predicate, it can also be applied to arbitrary digital surfaces with slight changes as shown in Fig. 1 and [18, Section 5].

Lastly, note that an update step in the algorithm R, also introduced in [18], can be decomposed into elementary steps that are update steps in the algorithm H [17, Section 3.5]. Therefore, we have the same results for the algorithm R.

3   Pattern Generation with Generalized Substitutions

The goal of this section is to show how one can use a sequence of matrices that reduces a vector $\mathbf{v} \in \mathcal{V}^3$ to generate a piece of digital plane of normal $\mathbf{v}$.

The generation method is based on a description of standard digital planes as union of faces. For $\mathbf{x} \in \mathbb{Z}^3$ and $i \in \{1, 2, 3\}$, we define the *pointed face* of type i and origin $\mathbf{x}$ as the following subset of $\mathbb{R}^3$: $(\mathbf{x}, i^*) := \{\mathbf{x} + \sum_{j \neq i} \lambda_j \mathbf{e}_j, \lambda_j \in [0, 1]\}$.

$\mathbf{e}_3$

$\mathbf{e}_1$ [image] $\mathbf{e}_2$ shows $(\mathbf{0}, 1^*)$ in red, $(\mathbf{0}, 2^*)$ in green and $(\mathbf{0}, 3^*)$ in blue.

We will use the following notation for translations of faces: if $(\mathbf{x}, i^*)$ is a face and $\mathbf{y}$ is a vector, then $(\mathbf{x}, i^*) + \mathbf{y} := (\mathbf{x} + \mathbf{y}, i^*)$, which extends in a natural way to union of faces.

For a vector $\mathbf{v} \in \mathcal{V}^3$, a *stepped plane* is defined as an infinite set of pointed faces where each face $(\mathbf{x}, i^*)$ verifies $0 \leq \langle \mathbf{x}, \mathbf{v} \rangle < \langle \mathbf{e}_i, \mathbf{v} \rangle$ (see for instance [14, Definition 1.2.1]). That way, the points of a digital plane, as defined in the previous section, are the vertices of the faces of a stepped plane. By abuse of notation, we will use $\mathcal{P}_{0,\mathbf{v}}$ or simply $\mathcal{P}$ to denote both a digital plane and a stepped plane in the following.

3.1   Substitutions and Generalized Substitutions

In this subsection, we first recall the definition of generalized substitutions and then show how to use them to generate a stepped plane of normal $\mathbf{v}$.

We consider a 3-letter alphabet $\mathcal{A} := \{1, 2, 3\}$. A word is an element of the free monoid $\mathcal{A}^*$ generated by $\mathcal{A}$. The empty word is denoted by ϵ and the concatenation operation is denoted by $\cdot$ or is left implicit. A *substitution* σ over $\mathcal{A}$ is a non-erasing endomorphism of $\mathcal{A}^*$, completely defined by its image on the letters of $\mathcal{A}$ by the relation $\sigma(w \cdot w') = \sigma(w) \cdot \sigma(w')$. The *abelianization mapping* $l : \mathcal{A}^* \to \mathbb{N}^3$ is such that $l(w) = (|w|_1, |w|_2, |w|_3)$, where $|w|_i$ denotes the number of occurrences of the letter i in w. The *incidence matrix* $\mathbf{M}_\sigma$ of σ is the 3×3 matrix whose i-th column is $l(\sigma(i))$ for every $i \in \mathcal{A}$. We assume that all the substitutions we work with are unimodular, i.e., such that $\det(\mathbf{M}_\sigma) = \pm 1$.

Furthermore, we define the following set:

$$\mathcal{S}_\sigma^i := \{(s, j) \in \mathcal{A}^* \times \mathcal{A} \mid j \in \mathcal{A}, \ i \cdot s \text{ is a suffix of } \sigma(j)\}. \tag{2}$$

To obtain $\mathcal{S}_\sigma^i$, one splits the words $\{\sigma(j) \mid j \in \mathcal{A}\}$ at each occurrence of the letter i. For each decomposition, we keep j as well as the suffix s located just after i.

Example 1. For σ: $1 \mapsto 1, 2 \mapsto 21, 3 \mapsto 32$, $\mathcal{S}_\sigma^2 = \{(1, 2), (\epsilon, 3)\}$.

The *generalized substitution* of a pointed face $(\mathbf{x}, i^*)$ is [14, Definition 1.2.3]:

$$E_1^*(\sigma)(\mathbf{x}, i^*) := \bigcup_{(s,j) \in \mathcal{S}_\sigma^i} (\mathbf{M}_\sigma^{-1}(\mathbf{x} + l(s)), j^*). \tag{3}$$

Example 2. For σ: $1 \mapsto 1, 2 \mapsto 21, 3 \mapsto 32$, $E_1^*(\sigma)(\mathbf{0}, 2^*) = \{(\mathbf{M}_\sigma^{-1}\mathbf{e}_1, 2^*), (\mathbf{0}, 3^*)\}$.

We extend this definition to unions of faces: $E_1^*(\sigma)(\mathcal{F}_1 \cup \mathcal{F}_2) := E_1^*(\sigma)(\mathcal{F}_1) \cup E_1^*(\sigma)(\mathcal{F}_2)$.

This setting is just fine for our purpose, but note that generalized substitutions allow a more general setting by considering the free group – instead of the free monoid – generated by $\mathcal{A}$ (see for instance [1, 6, 12]).

One of the main results about generalized substitutions and stepped planes is that $E_1^*(\sigma)$ preserves stepped planes ([14, Proposition 1.2.4, item (3)] or [11, Theorem 1] for a proof). Indeed, for any substitution σ whose incidence matrix $\mathbf{M}_\sigma$ is unimodular,

$$E_1^*(\sigma)(\mathcal{P}_{0,\mathbf{v}}) = \mathcal{P}_{0,\mathbf{M}_\sigma^T \mathbf{v}}. \tag{4}$$

We focus now on specific finite sequences of substitutions defined below:

Definition 4. *A sequence of substitutions* $(\sigma_n)_{0 \leq n \leq N}$ *is admissible iff the sequence* $((\mathbf{M}_{\sigma_n}^T)^{-1})_{0 \leq n \leq N}$ *reduces the vector* $\mathbf{M}_{\sigma_0}^T \ldots \mathbf{M}_{\sigma_N}^T \mathbf{1}$ *(See Definition 3 to recall the notion of reduction). In that case, we also set* $\mathbf{a}_n := \mathbf{M}_{\sigma_0}^T \ldots \mathbf{M}_{\sigma_n}^T \mathbf{1}$ *for all* $n \in \{0, \ldots, N\}$ *in accordance with Definition 3 and the relation* $\mathbf{U}_n^{-1} = \mathbf{M}_{\sigma_n}^T$.

One can obtain 2^N admissible sequences of substitutions from any sequence of matrices that reduces a given vector. Indeed, since each matrix $\mathbf{U}$ of the sequence belongs to $\mathcal{U}$, there are exactly two substitutions σ such that $\mathbf{M}_\sigma = (\mathbf{U}^T)^{-1}$. For instance, the inverse $\begin{pmatrix} 1 & 0 & 0 \\ 0 & 1 & 0 \\ 0 & 1 & 1 \end{pmatrix}$ of the matrix $\begin{pmatrix} 1 & 0 & 0 \\ 0 & 1 & 0 \\ 0 & -1 & 1 \end{pmatrix}$ is the incidence matrix of the two substitutions: $1 \mapsto 1, 2 \mapsto 23, 3 \mapsto 3$ and $1 \mapsto 1, 2 \mapsto 32, 3 \mapsto 3$. This choice has an impact on the geometry and topology of the generated patterns (see Sect. 3.3).

For two substitutions σ', σ'', we denote by $\circ$ their composition: $(\sigma' \circ \sigma'')(w) := \sigma'(\sigma''(w))$. Note that $(\mathbf{M}_{\sigma' \circ \sigma''})^T = \mathbf{M}_{\sigma''}^T \mathbf{M}_{\sigma'}^T$ and $E_1^*(\sigma') \circ E_1^*(\sigma'') = E_1^*(\sigma'' \circ \sigma')$ [14, Proposition 1.2.4, item (1)]. To save space, we set $\sigma_{i \ldots 0} := \sigma_i \circ \cdots \circ \sigma_0$ for $1 \leq i \leq N$. The following theorem is a direct consequence of (4):

Theorem 1. *Let* $(\sigma_n)_{0 \leq n \leq N}$ *be an admissible sequence of substitutions (see Definition 4). For all* $n \in \{0, \ldots, N\}$, $E_1^*(\sigma_{n \ldots 0})(\mathcal{P}_{0,1}) = \mathcal{P}_{0,\mathbf{a}_n}$

Proof. Applying (4) and using the definition of $\mathbf{a}_n$, we get:

$$E_1^*(\sigma_{n \ldots 0})(\mathcal{P}_{0,1}) = \mathcal{P}_{0,\mathbf{M}_{\sigma_0}^T \ldots \mathbf{M}_{\sigma_n}^T \mathbf{1}} = \mathcal{P}_{0,\mathbf{a}_n}$$

$\square$

3.2   Generation Method and Properties of the Patterns

In this section, we do not apply generalized substitutions on the whole stepped plane $\mathcal{P}_{0,1}$ as in Theorem 1. We apply them only on the lower unit cube composed of the three pointed faces $(\mathbf{0}, 1^*)$, $(\mathbf{0}, 2^*)$ and $(\mathbf{0}, 3^*)$ because it periodically generates $\mathcal{P}_{0,1}$ and is also included in any stepped plane of normal $\mathbf{v} \in \mathcal{V}^3$. The result is a finite set of pointed faces that we call *pattern*.

Definition 5 (Pattern). *Let* $(\sigma_n)_{0 \leq n \leq N}$ *be an admissible sequence of substitutions (see Definition 4). Let* $\mathcal{W}_0$ *be the lower unit cube* $\cup_{i \in \mathcal{A}}(\mathbf{0}, i^*)$ *and for all* $n \in \{1, \ldots, N\}$, *let* $\mathcal{W}_n$ *be the image of* $\mathcal{W}_0$ *by* $E_1^*(\sigma_{n \ldots 0})$, *i.e.,* $\mathcal{W}_n := E_1^*(\sigma_{n \ldots 0})(\mathcal{W}_0)$.

Fortunately, there exists a way to incrementally generate $\mathcal{W}_n$ from $\mathcal{W}_{n-1}$ in the manner of a union-translation scheme. This is the process we use in practice.

Theorem 2. *Let* $(\sigma_n)_{0 \leq n \leq N}$ *be an admissible sequence of substitutions (see Definition 4). We have for all* $n \in \{1, \ldots, N\}$ *and for all* $i \in \mathcal{A}$,

$$E_1^*(\sigma_{n \cdots 0})(\mathbf{0}, i^*) = \bigcup_{(s,j) \in \mathcal{S}_{\sigma_n}^i} \left(\mathbf{M}_{\sigma_0}^{-1} \cdots \mathbf{M}_{\sigma_n}^{-1} l(s) + E_1^*(\sigma_{n-1 \cdots 0})(\mathbf{0}, j^*) \right).$$

The proof, based on (2) and (3), is given in appendix.

As shown in Fig. 4, for $n \geq 1$, σ_n describes how the parts of $\mathcal{W}_n$ relate to the ones of $\mathcal{W}_{n-1}$. As an example, let us consider $E_1^*(\sigma_{3 \cdots 1})(\mathbf{0}, 1^*)$, which is the red part of $\mathcal{W}_3$. It has been obtained as the union of two parts of $\mathcal{W}_2$: $E_1^*(\sigma_{2 \cdots 1})(\mathbf{0}, 1^*)$ (in red) and $E_1^*(\sigma_{2 \cdots 1})(\mathbf{0}, 3^*)$ (in blue), because letter 1 belongs to both $\sigma_3(1)$ and $\sigma_3(3)$ and is also in the last position, which means with no suffixes and thus no translations.

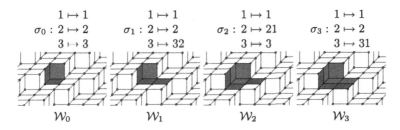

Fig. 4. The substitutions σ_1, σ_2 and σ_3 have been obtained from the reduction of $(1, 2, 4)$ using algorithm H (see also Fig. 3). The images of $(\mathbf{0}, 1^*)$, $(\mathbf{0}, 2^*)$ and $(\mathbf{0}, 3^*)$ by E_1^* are displayed respectively in red, green and blue. (Color figure online)

The following theorem gathers several properties of patterns:

Theorem 3. *Let* $(\sigma_n)_{0 \leq n \leq N}$ *be an admissible sequence of substitutions (see Definition 4). The following properties hold on the patterns defined in Definition 5:*

(i) $\forall n \in \{1, \ldots, N\}$, $\mathcal{W}_{n-1} \subset \mathcal{W}_n$ *and* $\forall n \in \{0, \ldots, N\}$, $\mathcal{W}_n \subset \mathcal{P}_{0, \mathbf{a}_N}$,
(ii) $\forall n \in \{1, \ldots, N\}$, $\mathcal{W}_n$ *periodically generates* $\mathcal{P}_{0, \mathbf{a}_n}$ *with period vectors:*

$$\mathbf{M}_{\sigma_0}^{-1} \cdots \mathbf{M}_{\sigma_n}^{-1}(\mathbf{e}_2 - \mathbf{e}_1) \ and \ \mathbf{M}_{\sigma_0}^{-1} \cdots \mathbf{M}_{\sigma_n}^{-1}(\mathbf{e}_3 - \mathbf{e}_2),$$

(iii) $\forall n \in \{0, \ldots, N\}$, $\forall i \in \{1, 2, 3\}$, $\mathcal{W}_n$ *has* $\langle \mathbf{e}_i, \mathbf{a}_n \rangle$ *faces of type* i.

Proof. (i) Since $\mathcal{W}_0 \subset \mathcal{P}_{0,1}$, Theorem 1 implies that $\mathcal{W}_N \subset \mathcal{P}_{0, \mathbf{a}_N}$. In addition, we have for all $n \in \{1, \ldots, N\}$:

$$\begin{aligned}
\mathcal{W}_n &= \cup_{i \in \mathcal{A}} E_1^*(\sigma_{n \cdots 0})(\mathbf{0}, i^*) &\text{(Definition 5)} \\
&= \cup_{i \in \mathcal{A}} \cup_{(s,j) \in \mathcal{S}_{\sigma_n}^i} \left(\mathbf{M}_{\sigma_0}^{-1} \cdots \mathbf{M}_{\sigma_n}^{-1} l(s) + E_1^*(\sigma_{n-1 \cdots 0})(\mathbf{0}, j^*) \right) &\text{(Theorem 2)} \\
&\supset \cup_{j \in \mathcal{A}} E_1^*(\sigma_{n-1 \cdots 0})(\mathbf{0}, j^*) = \mathcal{W}_{n-1} &\text{(Definition 5)},
\end{aligned}$$

where the inclusion comes from the trivial fact that for all $j \in \mathcal{A}$, $(\epsilon, j) \in \cup_{i \in \mathcal{A}} \mathcal{S}^i_{\sigma_n}$, i.e., the word $\sigma_n(j)$ ends with a letter $i \in \mathcal{A}$.

(ii) from (i), we have $\mathcal{W}_n \in \mathcal{P}_{0, \mathbf{a}_n}$ and from Proposition 2 (iii) (with the relation $\mathbf{U}_n^T = \mathbf{M}_{\sigma_n}^{-1}$), we have $\langle \mathbf{M}_{\sigma_0}^{-1} \cdots \mathbf{M}_{\sigma_n}^{-1}(\mathbf{e}_2 - \mathbf{e}_1), \mathbf{a}_n \rangle = 0$ and $\langle \mathbf{M}_{\sigma_0}^{-1} \cdots \mathbf{M}_{\sigma_n}^{-1}(\mathbf{e}_3 - \mathbf{e}_2), \mathbf{a}_n \rangle = 0$. Hence, for every $a, b \in \mathbb{Z}$, $\mathcal{W}_n + a\big(\mathbf{M}_{\sigma_0}^{-1} \cdots \mathbf{M}_{\sigma_n}^{-1}(\mathbf{e}_2 - \mathbf{e}_1)\big) + b\big(\mathbf{M}_{\sigma_0}^{-1} \cdots \mathbf{M}_{\sigma_n}^{-1}(\mathbf{e}_3 - \mathbf{e}_2)\big) \subset \mathcal{P}_{0, \mathbf{a}_n}$.

(iii) According to the definition of $E_1^\star$, Eq. (3), the number of faces of type j^0 in $E_1^*(\sigma)(\mathbf{0}, 1^\star)$ is equal to the number of pairs (s, j^0) in S_σ^1, which is equal to $|\sigma(j^0)|_1$, i.e., the number of occurrences of 1 in $\sigma(j^0)$. More generally, the number of faces of type j^0 in $\cup_{i \in \mathcal{A}} E_1^*(\sigma)(\mathbf{0}, i^\star)$ is equal to $|\sigma(j^0)|_1 + |\sigma(j^0)|_2 + |\sigma(j^0)|_3$, i.e., the number of letters in $\sigma(j^0)$. Similarly, the number of faces of type i in $\mathcal{W}_n$ is equal to the number of letters in $(\sigma_{n \cdots 0})(i)$ and is equal to

$$\langle l(\sigma_{n \cdots 0}), 1 \rangle = \langle \mathbf{M}_{\sigma_{n \cdots 0}} \mathbf{e}_i, 1 \rangle = \langle \mathbf{e}_i, (\mathbf{M}_{\sigma_{n \cdots 0}})^T 1 \rangle = \langle \mathbf{e}_i, \mathbf{M}_{\sigma_0}^T \cdots \mathbf{M}_{\sigma_n}^T 1 \rangle = \langle \mathbf{e}_i, \mathbf{a}_n \rangle.$$

$\square$

Theorem 3 shows that our method provides a hierarchical set of patterns, all included in a given stepped plane (i). The pattern of the highest level periodically generates the underlying stepped plane (ii) and is of minimal size (iii) because if one sum the normal vector of all its faces, we get exactly $\mathbf{a}_N$, i.e., the normal of the stepped plane, and one cannot expect to find a smaller pattern with the same normal. We discuss below two additional properties that we would like to have: shape compactness and connectivity.

3.3   Geometrical and Topological Issues

Our first remark is that the choice of the algorithm has a great impact on the shape of the pattern (see Fig. 5). In addition, the patterns generated using the algorithms H and R are much more compact. This is because the basis of the lattice $\{\mathbf{x} \in \mathbb{Z}^3 \mid \langle \mathbf{x}, \mathbf{a}_N \rangle = 0\}$ returned by the algorithm R (resp. algorithm H) is experimentally always (resp. almost always) reduced [18]. A first short-term perspective is to bound the distance of the pattern boundary to the origin in order to objectively compare the patterns generated by different algorithms. For the algorithms H and R, such bound may involve geometrical arguments based on the empty-circumsphere criterion.

Our second remark is about the connectivity of the patterns. There are no a priori guarantees that ensure that the patterns are vertex- or edge-connected. The connectivity of the last pattern is linked to the choice of substitutions, because one can associate two substitutions to one incidence matrix. Figure 6 and Fig. 7 show that one can end up with patterns of different topology, depending on which substitutions are used. We have experimentally noticed that there is always a sequence of substitutions among the 2^N admissible sequences that keep the pattern edge-connected (see Fig. 6). A second short-term perspective is to prove that such a connecting sequence indeed always exists and to design an algorithm that finds it. Even if there is a way of generating edge-connected patterns from specific sets of substitutions (see for instance [6]), the literature currently lacks general results we could directly reuse in our setting.

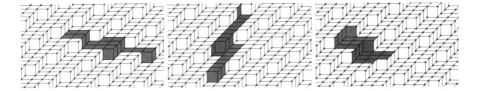

Fig. 5. Patterns of normal (2,6,15) generated by Brun, Selmer and H from left to right (same color convention as the one used in Fig. 4). In all cases, every substitution σ has been chosen so that for all $i \in \{1,2,3\}$, $\sigma(i)$ starts with i. Even if Brun does not correspond to Definition 2, we have included it for comparison, but keeping only the first red set because it ends on $(1,0,0)$ instead of $(1,1,1)$. (Color figure online)

$$\sigma_0 : \begin{matrix} 1 \mapsto 1 \\ 2 \mapsto 2 \\ 3 \mapsto 3 \end{matrix} \quad \sigma_1 : \begin{matrix} 1 \mapsto 1 \\ 2 \mapsto 2 \\ 3 \mapsto 32 \end{matrix} \quad \sigma_2 : \begin{matrix} 1 \mapsto 13 \\ 2 \mapsto 2 \\ 3 \mapsto 3 \end{matrix} \quad \sigma_3 : \begin{matrix} 1 \mapsto 1 \\ 2 \mapsto 21 \\ 3 \mapsto 3 \end{matrix} \quad \sigma_2' : \begin{matrix} 1 \mapsto 31 \\ 2 \mapsto 2 \\ 3 \mapsto 3 \end{matrix} \quad \sigma_3' = \sigma_3$$

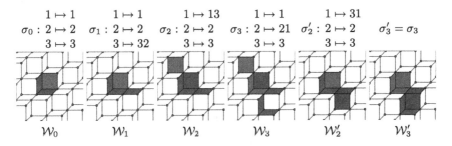

$$\mathcal{W}_0 \qquad \mathcal{W}_1 \qquad \mathcal{W}_2 \qquad \mathcal{W}_3 \qquad \mathcal{W}_2' \qquad \mathcal{W}_3'$$

Fig. 6. The substitutions have been obtained from the reduction of $(2,2,3)$ using the algorithm H (same color convention as the one used in Fig. 4). Using σ_2' instead of σ_2 can make the pattern edge-connected.

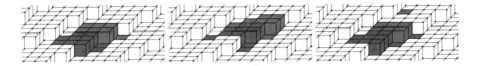

Fig. 7. Patterns of normal $(2,5,17)$ generated by the algorithm H. From left to right: connected pattern, not edge-connected, not vertex-connected.

4   Conclusion

We have introduced a three-dimensional version of the Euclidean algorithm that turns out to be closely related to plane-probing methods appearing in digital geometry. With the help of generalized substitutions, we have presented a way of generating hierarchical sets of pieces of digital planes. The patches of highest level periodically cover the underlying digital planes and are of limited size. We expect to obtain soon theoretical guarantees regarding the shape and connectivity of the generated patches. After having achieved this goal, we will use that generation method to improve the local analysis of digital surfaces using plane-probing.

A   Proofs

Proof (of Theorem 2).

$$E_1^*(\sigma_n \cdots 0)(\mathbf{0}, i^*) = E_1^*(\sigma_{n-1} \cdots 0)\big(E_1^*(\sigma_n)(\mathbf{0}, i^*)\big)$$

$$= E_1^*(\sigma_{n-1} \cdots 0)\left(\bigcup_{(s,j) \in \mathcal{S}_{\sigma_n}^i} (\mathbf{M}_{\sigma_n}^{-1} l(s), j^*) \right)$$

$$= \bigcup_{(s,j) \in \mathcal{S}_{\sigma_n}^i} E_1^*(\sigma_{n-1} \cdots 0)(\mathbf{M}_{\sigma_n}^{-1} l(s), j^*))$$

$$= \bigcup_{(s,j) \in \mathcal{S}_{\sigma_n}^i} ((\mathbf{M}_{\sigma_{n-1}} \cdots \mathbf{M}_{\sigma_0})^{-1}(\mathbf{M}_{\sigma_n})^{-1} l(s) + E_1^*(\sigma_{n-1} \cdots 0)(\mathbf{0}, j^*))$$

$$= \bigcup_{(s,j) \in \mathcal{S}_{\sigma_n}^i} ((\mathbf{M}_{\sigma_n} \cdots \mathbf{M}_{\sigma_0})^{-1} l(s) + E_1^*(\sigma_{n-1} \cdots 0)(\mathbf{0}, j^*)).$$

The second to last line comes from

$$E_1^*(\sigma_{n-1} \cdots 0)(\mathbf{x}, i^*) = (\mathbf{M}_{\sigma_{n-1}} \cdots \mathbf{M}_{\sigma_0})^{-1}\mathbf{x} + E_1^*(\sigma_{n-1} \cdots 0)(\mathbf{0}, i^*),$$

since $(\mathbf{M}_{\sigma_{n-1}} \cdots \mathbf{M}_{\sigma_0})^{-1}$ does not depend on the union in the definition of E_1^*, Eq. (3) (see also [14, Proposition 1.2.4, item (2)]).   □

References

1. Arnoux, P., Furukado, M., Harriss, E., Ito, S.: Algebraic numbers, free group automorphisms and substitutions on the plane. Trans. Amer. Math. Soc. **363**, 4651–4699 (2011)
2. Arnoux, P., Ito, S.: Pisot substitutions and Rauzy fractals. Bull. Belg. Math. Soc. Simon Stevin **8**(2), 181–208 (2001)
3. Berthé, V., Domenjoud, É., Jamet, D., Provençal, X.: Fully Subtractive algorithm, tribonacci numeration and connectedness of discrete planes. Res. Inst. Math. Sci. Lecture Note Kokyuroku Bessatu B **46**, 159–174 (2014)
4. Berthé, V., Fernique, T.: Brun expansions of stepped surfaces. Discret. Math. **311**(7), 521–543 (2011)
5. Berthé, V., Jamet, D., Jolivet, T., Provençal, X.: Critical connectedness of thin arithmetical discrete planes. In: Proceedings of DGCI, pp. 107–118 (2013)
6. Berthé, V., Lacasse, A., Paquin, G., Provençal, X.: A study of Jacobi-Perron boundary words for the generation of discrete planes. Theor. Comput. Sci. **502**, 118–142 (2013)
7. Brimkov, V., Coeurjolly, D., Klette, R.: Digital planarity - a review. Discret. Appl. Math. **155**(4), 468–495 (2007)
8. Domenjoud, E., Laboureix, B., Vuillon, L.: Facet connectedness of arithmetic discrete hyperplanes with non-zero shift. In: Proceedings of DGCI (2019)
9. Domenjoud, E., Provençal, X., Vuillon, L.: Facet connectedness of discrete hyperplanes with zero intercept: the general case. In: Proceedings of DGCI, pp. 1–12 (2014)

10. Domenjoud, E., Vuillon, L.: Geometric palindromic closure. Unif. Distrib. Theory **7**(2), 109–140 (2012)
11. Fernique, T.: Multidimensional Sturmian sequences and generalized substitutions. Int. J. Found. Comput. Sci. **17**, 575–600 (2006)
12. Fernique, T.: Generation and recognition of digital planes using multi-dimensional continued fractions. Pattern Recogn. **42**(10), 2229–2238 (2009)
13. Jamet, D., Lafrenière, N., Provençal, X.: Generation of digital planes using generalized continued-fractions algorithms. In: Proceedings of DGCI, pp. 45–56 (2016)
14. Jolivet, T.: Combinatorics of Pisot substitutions. Ph.d. thesis, Université Paris Diderot, University of Turku (2013)
15. Klette, R., Rosenfeld, A.: Digital straightness - a review. Discret. Appl. Math. **139**(1–3), 197–230 (2004)
16. Labbé, S., Reutenauer, C.: A d-dimensional extension of christoffel words. Discret. Comput. Geom. **54**(1), 152–181 (2015)
17. Lachaud, J.O., Meyron, J., Roussillon, T.: An optimized framework for plane-probing algorithms. J. Math. Imaging Vis. **62**, 718–736 (2020)
18. Lachaud, J.O., Provençal, X., Roussillon, T.: Two plane-probing algorithms for the computation of the normal vector to a digital plane. J. Math. Imaging Vis. **59**, 23–39 (2017)

Introduction to Discrete Soft Transforms

Bastien Laboureix[1]($\boxtimes$), Eric Andres[2]($\boxtimes$), and Isabelle Debled-Rennesson[1]($\boxtimes$)

[1] Université de Lorraine, CNRS, LORIA, 54000 Nancy, France
{bastien.laboureix,isabelle.debled-rennesson}@loria.fr
[2] Université de Poitiers, XLIM UMR CNRS 7252,
86962 Chasseneuil-Futuroscope, France
eric.andres@univ-poitiers.fr

Abstract. In this paper a new class of discrete transforms of discrete straight segments (DSS), called *Discrete Soft Transforms* is introduced. The soft transformation of a segment consists in moving a single discrete point at each step while keeping the segment property. We propose the soft rotation and soft translation of a segment and extend these results to the soft translation of a tree.

Keywords: Discrete straight segment · Discrete tree · Soft translation · Soft rotation · Discrete soft transform

1 Introduction

Commonly transformations in the discrete domain come in the form of discretized continuous transformations. The concern with such an approach is that it is difficult to obtain natural properties such as bijectivity or preservation geometric features or topology. An alternative approach promoted by the discrete geometry community is to consider transformations directly in the discrete domain. This led to interesting results on bijectivity or topology preservation for discrete rigid motions [1,6]. Such results are difficult to obtain and even more difficult to extend because, as global transformations on the whole domain, local properties are not easy to guarantee. In this paper we propose to introduce a new class of discrete transforms called *discrete soft transform*. The idea is to decompose a discrete transform into a sequence of atomic steps where only one discrete point at a time is moved/added/suppressed while maintaining a set of given properties. The recently, proposed morphing method by Lama Tarsissi and al. [9] that preserves the convexity of discrete objects by adding/removing one point at each step can be seen as an example of discrete soft transform.

As a proof of concept, we propose a discrete soft rotation and translation of discrete straight segments (DSS). The discrete lines have interesting arithmetical and combinatoric properties studied for a long time (see [3] for an historical review). The soft transformation of a segment consists in moving a single discrete point at each step while keeping the segment property. We present shortly the soft rotation of a DSS, but the focus of the paper is on the soft translation.

É. Baudrier et al. (Eds.): DGMM 2022, LNCS 13493, pp. 422–435, 2022.
https://doi.org/10.1007/978-3-031-19897-7_33

The soft translation can be seen as akin to a subpixel translation method when compared to the discretized continuous translation. After proving a fundamental result on the movable points in a DSS, we propose an algorithm for the soft translation of a DSS and the soft translation of tree embeddings where vertices are embedded as discrete points and edges as DSS. The next step will be to propose a soft translation for graphs of DSS and thus to the soft translation of segmented images.

This paper is decomposed into five parts, the first being this introduction. In section two, we recall basic notions about discrete lines and segments. Then, in the third section, the discrete soft tranform on a DSS is introduced and a fundamental result about movable points in a DSS is proved. In section four, we propose a first simple algorithm for soft rotations of a DSS before focusing on the soft translation of a DSS. There are in particular an important distinction that has to be made between the soft translation of a DSS of slope between 0 and 1 and a DSS of slopes greater or equal to 1. An extension to a set of DSS that embed a tree concludes Sect. 4. In the last section, we conclude and present perspectives for this work.

2   Preliminaries

Let $\{e_1, e_2\}$ denote the canonical basis of the 2-dimensional Euclidean vector space. In this paper we are dealing with discrete points in $\mathbb{Z}^2$. Two points $p, q \in \mathbb{Z}^2$ are said to be 4-neighbours iff $\|p - q\|_1 = 1$, and said to be 8-neighbours iff $\|p - q\|_\infty = 1$. A Digital Straight Line (DSL for short) $\mathcal{D}(a, b, \mu)$ of integer characteristics (a, b, μ) is the set of digital points $(x, y) \in \mathbb{Z}^2$ such that $0 \leqslant ax - by + \mu < \omega$ where $\omega = \max(|a|, |b|)$ and $\gcd(a, b) = 1$. These DSL are 8-connected and called naive DSL [8]. The slope of the DSL is the fraction $\frac{a}{b}$ (when $b \neq 0$). The value μ is sometimes called the translation constant or the off-set. The value $R(a, b, \mu)(x, y) = ax - by + \mu$ is called the remainder of the DSL. A DSL can also be defined as the integer points of a strip delimited by the lower leaning line $ax - by + \mu = \omega - 1$ and the upper leaning line $ax - by + \mu = 0$ [2]. Upper (resp. Lower) leaning points are the digital points of the DSL lying on the upper (resp. lower) leaning lines. A weakly exterior point is a point of a DSL that verifies $ax - by + \mu = -1$ (in this case we speak also of a weakly upper exterior point) or $ax - by + \mu = \omega$ (in this case we speak also of a weakly lower exterior point) [2]. We note $Inf(S)$ (resp. $Sup(S)$) the set of lower (resp. Upper) leaning points of S.

A digital straight line segment (DSS) $\mathcal{D}(a, b, \mu, E_0, E_1)$ is a finite 8-connected subset of the DSL $\mathcal{D}(a, b, \mu)$ with the end-points E_0 and E_1. We speak of horizontal (resp. vertical) segments, segments in the first octant (resp. second octant) with slope $0 \leqslant \frac{a}{b} < 1$ (resp. $\frac{a}{b} \geq 1$). For an horizontal line S, for a given x, there is one and only one $y = \lfloor \frac{ax+\mu}{b} \rfloor$ such that $(x, y) \in S$. Equivalently, in vertical lines, for a given y there is only one x.

For a given DSS $S = \mathcal{D}(a, b, \mu, E_0, E_1)$, there exists an infinite number of parameters (a', b', μ') such that $S = \mathcal{D}(a', b', \mu', S_0, S_1)$.

For instance, $\mathcal{D}(5, 8, 0, (0, 0), (11, 6)) = \mathcal{D}(8, 13, 1, (0, 0), (11, 6))$. There exists however only one set of parameters with minimal b, called *minimal parameters* [2].

3   Soft Transform

Our goal is to apply transforms to discrete structures by maintaining geometric properties and a form of *continuity*. If continuity has a topological definition for classical geometric structures, we must define a notion of *continuous transform* in the discrete case, which we call *soft transform*.

Definition 1. *Two DSS S and S' are said to be neighbours iff they can be described as $X \cup \{p\}$ and $X \cup \{q\}$ where X is a set of points and p, q are 4-neighbours.*
We note this relation as $S \leftrightarrow S'$.

Definition 2 (DSS Soft Transformation). *Considering a function $f : \mathbb{Z}^2 \to \mathbb{Z}^2$, and DSS S, a soft transformation of S into $f(S)$ is a sequence of segments $S_0...S_N$ such that:*

- $S_0 = S$
- $S_N = f(S)$
- $\forall i \in [\![0, N - 1]\!], S_i \leftrightarrow S_{i+1}$

For $S \leftrightarrow S'$, the transform from one to the other DSS is called *atomic soft transform*.

This definition can be extended to a set of DSS or more generally to a digital shape with some caveats: the soft transform of a set of segments can of course be handled as the independent separate soft transform of each segment, but in general, this is not what is expected. Segments may share vertices, and thus, for what follows, we consider graphs formed by digital straight segments and add an additional constraint which is the preservation of the overall structure of the graph. As for a digital shape, a shape can always be decomposed into a set of DSS, but this decomposition is not unique, and therefore the soft transform as applied on DSS graphs depends on this decomposition.

All our algorithms are implented. See for a couple of examples that go a beyond the present paper: https://imgur.com/a/j81nI6f.

3.1   Movable Points in a DSS

The goal of this paper is to define a notion of soft translation on a DSS, and introduce a first notion of soft rotation. The first question that is answered in this first subsection is the question of which points of a DSS can be moved while remaining a DSS:

Theorem 3 (Movable points in a DSS). *In a DSS S with minimal parameters of slope between 0 and 1, the points which can move up (resp. down) while keeping the segment property are exactly the extremal lower (resp. upper) leaning points. The number of these points can be one or two.*

By extremal, we mean the leaning points closests to the end-points of the DSS.

Proof. Let us concentrate for this proof on points that can move up while preserving the DSS property: the goal is to characterize all the points p such that $S \cup \{p + e_2\} \setminus \{p\}$ is still a DSS. The proof is similar for points that move down.

Let us consider a segment S with end-points E and E' and of analytical equation $0 \leqslant ax - by + \mu < b$ (with a, b, μ minimal).

Let $I \stackrel{\text{def}}{=} (x_I, y_I)$ be a point of S. We denote $S' \stackrel{\text{def}}{=} S \cup \{I + e_2\} \setminus \{I\}$.

Let us show S' is a DSS iff I is a lower extremal leaning point. The proof is similar for upper leaning points.

$\Rightarrow$: Let us reason by contradiction: Two cases are possible.

Case 1: Let us suppose that I is not a lower leaning point. This means that $R(a, b, \mu)(I) \leqslant b - 2$. So $R(a, b, \mu)(I + e_2) \leqslant -2$. Therefore $I + e_2$ is strongly exterior to S and S' is not a DSS [2].

Case 2: Let us now suppose that I is indeed a lower leaning point, but not extremal. Let us then call A and B the extremal lower leaning points of S. This means that I is a point of the continuous segment $[AB]$. Since $A, B \in S'$, by convexity, $I \in S'$ which is not.

$\Leftarrow$: Let us illustrate the proof of Theorem 3 with Fig. 1. In Fig. 1, the segment S is represented in blue. Let us extend the segment S upto an upper leaning point $P \stackrel{\text{def}}{=} (x_P, y_P)$ such that $\|P - I\|_\infty \geqslant b$ (green part of the segment): we have then a cover-segment of S denoted S_0.

Since P is an upper leaning point and an end-point of S_0, by [2], $P - e_2$ is slightly exterior to S_0, therefore $S'_0 \stackrel{\text{def}}{=} S_0 \cup \{P - e_2\} \setminus \{P\}$ is a DSS.

Moreover, S'_0 admits as lower extremal leaning points I and $P - e_2$.

Let us denote a', b', μ' the parameters of S'_0, and Π (resp. Π') the restriction of S_0 (resp. S'_0) to the abscissa interval $[x_P, x_I - 1]$.

This way Π (in green) and Π' (in magenta) form respectively a period of S_0 and of S'_0 because their lengths are b'.

One can extend Π' upto E (in red) by periodicity (since $\|P - I\|_\infty \geqslant b$), in a segment containing I and who is a cover-segment of S.

In the same way, one can extend Π upto E' (in yellow) in a segment containing $I + e_2$ and who is a cover segment of S'. Therefore S' is a segment. $\square$

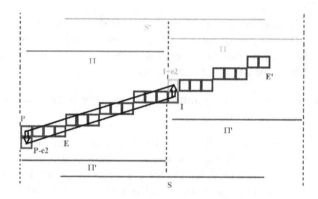

Fig. 1. Upward translation of a lower leaning point I

3.2   Computing the New Parameters of a DSS

In [2], I. Debled-Rennesson explained how to compute the minimal parameters of a DSS when adding a new point at the end of the DSS. In our case, the problem is slightly different since we move one point inside a DSS. Let us consider a DSS S with parameters (a, b, μ), of slope smaller than 1 defined on the interval $x \in [\![x_0, x_1]\!]$ and let us suppose that $m(x_m, y_m) \in S$ is the lower leaning point closest to x_1 (the rightmost movable point), with S' of parameters (a', b', μ') the resulting DSS after the move m to $m + e_2$. M is defined as the upper leaning point in S closest to x_1: $x_m = a^{-1}(b - 1 - \mu) + b \lfloor \frac{x_1 - a^{-1}(b-1-\mu)}{b} \rfloor$ and $y_m = \lfloor \frac{ax_m + \mu}{b} \rfloor$. M is a lower leaning point of S and $M + e_2$ is a upper leaning point in the new DSS. Let us consider the upper leaning point $u(x_u, y_u)$ that lies before with $x_u = -a^{-1}\mu + \lfloor \frac{x_0 - 1 + a^{-1}\mu}{b} \rfloor$ and $y_u = \lfloor \frac{ax_u + \mu}{b} \rfloor$. The parameters of the new DSS S' are given by $(b', a') = m + e_2 - u$ and $\mu = -a'x_u + b'y_u$. Now, nothing guarantees that these parameters are minimal: this means that if we want to iterate, we can use the results proposed in [5,7] to determine the minimal parameters of the new DSS, which can be made in time $O(\ln(n))$, where n is the length of the segment.

3.3   A Simple Exemple: The Soft Rotation of a DSS

In this section, we present an exemple of soft transform of a DSS that will be proved useful in future works: the soft rotation of a DSS. We consider only rotation in the first octant, the others can be obtained by symmetry. Note that our notion of rotation does not change the number of points but changes the length of the continuous segment (from n to $n\sqrt{2}$).

Problem 4. *Considering an integer $n \in \mathbb{N}$, how can we soft transform the flat segment $S_0 \stackrel{def}{=} \{(k, 0) \mid k \in [\![0, n-1]\!]\}$ into the diagonal segment $S_N \stackrel{def}{=} \{(k, k) \mid k \in [\![0, n-1]\!]\}$?*

Algorithm: Direct Soft Rotation
Input: the length n of the considered flat segment
Output: a sequence of moves from S_0 to S_N

$S \leftarrow S_0$;
while $S \neq S_N$ **do**
   | $p \leftarrow$ lower leaning point with greatest abscissa in S;
   | replace p by $p + e_2$ in S;
   | compute the new parameters of the DSS;
end

The reverse rotation can be easily obtained in a similar way. With the previous subsections, we have all we need to make this algorithm work and prove its correction. The overall complexity of an atomic soft rotation is $O(\ln(n))$. As there is a quadratic number of points to move, the complexity for the whole process is $O(n^2 \ln(n))$.

4   Soft Translations of Discrete Straight Line Segments

The goal is henceforth to deal with the soft translation of segments. Let us first note that we only consider here the upward translation by one position. Other translations can be obtained by accumulation and symmetries.

Problem 5. *Given a discrete Straight segement S, how to compute the soft translation from S to $S + e_2$?*

By translation and symmetry, we can limit our study to segments of the first quater with an end-point in $(0,0)$. We need however to differentiate the cases of horizontal segments (of slope between 0 and 1) from vertical segments (of slope greater than 1).

4.1   Horizontal Segments

In this section we are considering horizontal segments, i.e. of slope smaller than 1. The points of the DSS S are (k, y_k) for $k \in [\![0, n-1]\!]$, where n is the length of the segment. We propose an algorithm allowing to translate this segment simply by shifting the offset value μ.

Definition 6 (Primitive of a DSS). *A DSS S defined by $0 \leqslant ax - by + \mu < b$ is said to be primitive iff it contains exactly b points.*

Proposition 7. *Let S be a primitive DSS defined by $0 \leqslant ax - by + \mu < b$, then S admits a unique lower leaning point p and the segments $S \cup \{p + e_2\} \setminus \{p\}$ is defined by $0 \leqslant ax - by + \mu + 1 < b$.*

Proof. Since $\gcd(a, b) = 1$ and S is primitive, $x \mapsto ax + \mu$ is bijective over $\mathbb{Z}/b\mathbb{Z}$. Since S is primitive, the segment is of length b, therefore there exists a unique $p = (x, y) \in S$ such that $ax - by + \mu = b - 1$. The point p is a lower leaning point and all the points of the DSS $S \cup \{p + e_2\} \setminus \{p\}$ verify $0 \leqslant ax - by + \mu + 1 < \omega$.   □

Proposition 8 (Cover Segment). *For each DSS S of length n, there exists a primitive cover segment S' of length $\leqslant 3n$ such that $S \subset S'$. In addition, the parameters of S' can be effectively computed in $O(\ln(n))$.*

Proof. Let us consider a DSS S defined by $0 \leqslant ax - by + \mu < b$ with $b < n$. We are looking for parameters a', b' relative primes such that $\frac{a'}{b'} - \frac{a}{b} = \frac{1}{bb'}$ with $b' \in [\![2n, 3n]\!]$.

Such a couple exists (a', b') since $\frac{a'}{b'} - \frac{a}{b} = \frac{a'b - ab'}{bb'}$.

We consider $b' \in [\![2n, 3n]\!]$ such that $ab' \equiv -1 (b)$ (which exists since $\gcd(a, b) = 1$), and then a' such that $a'b - ab' = 1$.

And so, $\frac{a'}{b'} - \frac{a}{b} = \frac{1}{bb'}$.

For the offset, we consider μ' such that $\frac{\mu}{b} \leqslant \frac{\mu'}{b'} < \frac{\mu}{b} + \frac{1}{b'}$.

We call S' the DSS defined by $0 \leqslant a'x - b'y + \mu' < b'$ of length b'.

Let's consider $(x, y) \in S$. Then $y = \lfloor \frac{ax + \mu}{b} \rfloor$.

$$\frac{a'x + \mu'}{b'} - \frac{ax + \mu}{b} = x \left(\frac{a'}{b'} - \frac{a}{b} \right) + \frac{\mu'}{b'} - \frac{\mu}{b}.$$

This quantity is positive and inferior to $\frac{x}{bb'} + \frac{1}{b'} \leqslant \frac{x}{2bn} + \frac{1}{2n} < \frac{1}{2b} + \frac{1}{2b} = \frac{1}{b}$.

Therefore $y \leqslant \frac{ax + \mu}{b} \leqslant \frac{a'x + \mu'}{b'} < \frac{ax + \mu + 1}{b} \leqslant y + 1$.

By definition of the integer part, we obtain $y = \lfloor \frac{a'x + \mu'}{b'} \rfloor$ therefore $(x, y) \in S'$. □

We can then apply the soft translation algorithm on the cover segment S' of a DSS S, which allows us to perform the soft translation of S.

Algorithm: Horizontal DSS Soft Translation

Input: a segment S

Output: a sequence of moves from S to $S + e_2$

$n \leftarrow |S|$;

$S' \leftarrow$ a primitive cover segment of S of length $n' \leqslant 3n$;

$a', b', \mu' \leftarrow$ parameters of S';

Compute a'^{-1} in $\mathbb{Z}/b'\mathbb{Z}$;

for j *from* 0 *to* $n' - 1$ **do**

    $k \leftarrow -a'^{-1}(\mu' + j + 1)$ in $\mathbb{Z}/b'\mathbb{Z}$;

    if $k < n$ **then**

        replace p of abscissa k by $p + e_2$ in S;

    end

end

Theorem 9. *The soft translation of the horizontal DSS S is correct and has a complexity of $O(n)$ with n the length of S.*

Proof. At step j, we consider $k = -a'^{-1}(\mu' + j + 1)$ in $\mathbb{Z}/b'\mathbb{Z}$.

Henceforth, $0 \leqslant a'k + \mu + j = -1$ in $\mathbb{Z}/b'\mathbb{Z}$. Therefore, if $k < n$, the point (k, y_k) of S verifies $a'k - b'y_k + \mu + j = -1$ since $k < n' = b'$ because S' is primitive.

The point of abscissa k can therefore be moved in S', and thus in S.

As for the complexity, the length of S' is linearly proportional to the length of S and constructing S' requires a simple computation in $\mathbb{Z}/b'\mathbb{Z}$ once. The loop at each iteration has a constant time and thus operates in $O(n') = O(n)$.   □

The soft translation algorithm simply modifies the offset in the parameters. The slope of the segment is invariant during the translation.

4.2   Vertical Segments

In this section we are considering vertical segments, i.e. of slope greater or equal to 1. Let us note that we need here a new operation to handle the upward soft translation. For instance, in Fig. 2, we have a DSS of length 5 defined by $x - 2$ and $1 \leqslant y \leqslant 5$. No point can move upward without superposition or deconnection. To handle such situations, we propose another approach with two new operations for the soft translation of segments:

- Add a point over the point of highest ordinate (Fig. 2b.)
- Remove the point of lowest ordinate (Fig. 2c.).

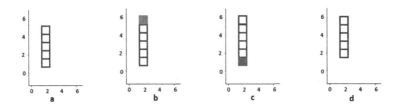

Fig. 2. A new operation for vertical segments

We propose a new algorithm that allows to create a soft translation of vertical segments in the general case by using these two operations.

Definition 10 (Pillar point). *Let S be a segment of the second octant admitting a cover segment S' defined by the parameters $0 \leqslant a'x - b'y + \mu' < a'$.*
The point $(x, y) \in S$ is called a pillar of S iff $a' - b' \leqslant a'x - b'y + \mu' < a'$.

Fig. 3. Soft translation of a vertical segment

It is easy to see that a point (x, y) of a DSS is a pillar iff $(x - 1, y - 1) \in S$ (except for the lower end-point) since $a' - b' \leqslant a'x - b'y + \mu' < a'$ means that $0 \leqslant a'(x - 1) - b'(y - 1) + \mu' < b'$. The idea of the algorithm is to translate to the left the set of pillar points, by decreasing remainder, and to manage the end-points via the operations of addition and suppression: indeed, for a pillar (x, y), $(x - 1, y - 1) + e_2 = (x, y) - e_1$. There is a special case that needs to be considered when the upper end-point is also a pillar point. Figure 3 illustrates the general idea of the algorithm. The red discrete point is removed. The green discrete point is added and all the discrete points with arrows are pillars and they are moved to the left.

Note that our algorithm can change the number of points of the DSS when we add or remove a discrete point.

Theorem 11. *The soft translation algorithm of a vertical segment S is correct (in the sense that is indeed a discrete soft translation) and has a time complexity of $O(n \ln(n))$ with n the length of S.*

Proof. The segment at the end of the algorithm is $S + e_2$ because the only points that move, beyond the removal/addition of the end-points, are the pillars and the property of a pillar (x, y) is to take the place of the DSS point $(x - 1, y - 1) + e_2$ as we have already mentioned. A point (x, y) that is neither pillar nor an end-point means that $(x, y - 1)$ belongs to S, which means that $(x, y - 1) + e_2 = (x, y)$. What we need to show now is that the set of points remains a DSS when moving pillar points and handling end-points. There are three cases for a pillar point p in the algorithm:

- p is not an end-point: identical to the horizontal DSS soft translation with a $90°$ rotation. The set of points remains thus a DSS.
- p is the lower end-point: a DSS where you remove an end-point remains a DSS.
- p is the upper end-point: here, the proof is slightly less direct and is presented in what follows.

Let us consider that the next pillar p to treat is the upper end-point. Let us call S_1 the current DSS before moving p. The point p may be moved upwards at position $p + e_2$ or $p + e_1 + e_2$.

However p is a pillar and therefore it has to be moved to the left, with $S_2 \stackrel{\text{def}}{=} S_1 \cup \{p - e_1\} \setminus \{p\}$ a DSS. This means $p + e_1 + e_2$ is not an option anymore, therefore S_2 can only be prolonged with $p + e_2$, and thus $S_1 \cup \{p + e_2\}$ is a DSS.

As for the complexity, it is easy to see that computing the pillar points is linear in the length of S and that ordering them in decreasing remainder order leads to the given complexity. $\qquad\qquad\qquad\qquad\qquad\qquad\qquad\qquad\qquad\qquad$ □

Algorithm: Vertical DSS Soft Translation

Input: a segment S
Output: a sequence of moves from S to $S + e_2$

$S' \leftarrow$ a primitive cover segment of S of size $\leqslant 3n$;
$a', b', \mu' \leftarrow$ parameters of S';
$P \leftarrow$ set of the pillar points of S in decreasing order of remainder $a'x - b'y + \mu'$;
while P *is not empty* **do**
$\quad$| $\quad p \leftarrow$ first element of P;
$\quad$| $\quad$ **if** p *is not an end-point* **then**
$\quad$| $\quad$ | $\quad$ replace p by $p - e_1$ in S;
$\quad$| $\quad$ | $\quad$ remove p from the set P
$\quad$| $\quad$ **end**
$\quad$| $\quad$ **if** p *is the lower end-point* **then**
$\quad$| $\quad$ | $\quad$ remove p from S;
$\quad$| $\quad$ | $\quad$ remove p from the set P
$\quad$| $\quad$ **end**
$\quad$| $\quad$ **if** p *is the upper end-point* **then**
$\quad$| $\quad$ | $\quad$ **if** *the lower end-point hasn't been removed yet* **then**
$\quad$| $\quad$ | $\quad$ | $\quad$ remove the lower end-point from S;
$\quad$| $\quad$ | $\quad$ | $\quad$ remove the lower end-point from the set P if it is a pillar.
$\quad$| $\quad$ | $\quad$ **end**
$\quad$| $\quad$ | $\quad$ Add a point $p + e_2$ to S;
$\quad$| $\quad$ | $\quad$ replace p by $p - e_1$ in S;
$\quad$| $\quad$ | $\quad$ remove p from the set P
$\quad$| $\quad$ **end**
end
Remove the lower end-point if it hasn't been done yet;
Add a point over the upper end-point if it hasn't been done yet.

4.3   Soft Translation of DSS Trees

In imaging, discrete straight line segments come rarely alone, that's why we propose here a first extension in the form of a soft DSS tree translation algorithm. A finite DSS tree Γ can be seen as an embedding ρ of a finite tree (Σ, A) in $\mathbb{Z}^2$, where vertices of Σ become points in $\mathbb{Z}^2$ and edges of A DSSs. Note that a DSS tree does not necessarily have to be planar here.

The idea is to move each DSS of the tree independently, however, contrary to a single DSS, there are end-points to consider. An end-point can be shared by a number of segments, and in order to ensure that each set of points corresponding to a segment remains a DSS during the set of atomic transforms and that the global structure of the tree is preserved, a shared end-point can only move once. If we move the end-points independently for each segment, we may loose the DSS tree structure. This means that the points of a DSS S can move until we reach the shared end-point E of the DSS (in a tree there is one and only one if the tree is not composed of only one segment which we can suppose here). The end-point E, and the remaining points of S, can only move when it is the turn of E to move in every other DSS S_i that shares E as end-point.

Algorithm: Soft DSS Tree Translation

Input: A DSS tree Γ
Output: A sequence of moves from Γ to $\Gamma + e_2$

$\Sigma, A, \rho \leftarrow$ Tree and embedding of Γ ;
while Γ *has not completely moved* **do**
    for $S \in \rho(A)$ *not totally translated* **do**
        $p \leftarrow$ next point of S to move ;
        if p *is not an end-point of* S **then**
           | move p ;
        end
        else
           $V(p) \leftarrow \{S \in \rho(A) \mid p$ is an end-point of $S\}$;
           if $\forall S \in V(p)$, *the next point to move in S is p* **then**
              | move p ;
           end
        end
    end
end

Theorem 12. *If Γ is a DSS tree with at least two vertices then the algorithm terminates, is correct and has a complexity* $O\left(\sum_{S \in \rho(A)} |S| \ln |S| \right)$.

The algorithm is correct for an empty tree but not for a tree with only one vertex, but then it means translating a structure with just one discrete point. In case of only one segment, we can use the soft DSS translation algorithms even if formally the tree algorithm still holds.

Proof. The partial correction of the algorithm follows from the correction of the soft DSS translation algorithms.

Let us prove the termination by recurrence on $|\Sigma|$: if the tree has only two vertices then Γ is a DSS and we use the DSS translation algorithm which ends.

Let us suppose that $|\Sigma| \geqslant 3$. G admits a leaf s attached to G by and edge $a = \{s, s'\}$. We denote $q \overset{\text{def}}{=} \rho(s)$ and $q' = \rho(s')$. We denote $S \overset{\text{def}}{=} \rho(A)$ with end-points q and q'.

Let us assume that the algorithm does not end and let us consider the state in which it is blocked. Let p be the next point to move in the DSS S. Since p can not be moved upwards, p is necessarily an end-point of S. Moreover, $p \neq q$ because $V(q) = \{S\}$ therefore for all $S_0 \in V(q)$, p is the next point to move is S_0. Therefore $p = q'$. This proves that the algorithms does not end on the discrete tree induced by $\Sigma \setminus \{s\}$, which contradicts our recurrence hypothesis. $\square$

An important property of this algorithm is that it preserves the slopes of the cover segments of the tree. Note that our algorithm can change the number of points in the tree when we add or remove points in vertical DSS.

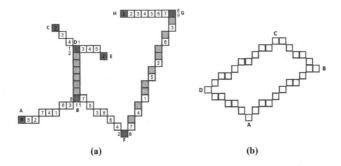

(a)   (b)

Fig. 4. (a) A DSS tree. The numbers represent the atomic movement order. The red discrete points are the vertices (that have multiple ordre movements, number next to the vertex). The discrete points in gray do not move. (b) A cyclic DSS graph that does not terminate with a naive approach (Color figure online)

Figure 4a illustrates the translation of a DSS tree. The numbers give the theoretical movement inside each DSS. The absolute movement order depends on the order with which you treat each DSS. If each DSS was treated independently, each discrete point with 1 would move, then those with value 2, etc. For trees, if the moving point is a end-point, the DSS is placed on hold until it is its turn to move for each of the other DSS sharing the end-point.

For example, we can start by translating the DSS $[BF]$. So we first consider the end-point B, however this point can't move because it has to wait its turn for segment $[AB]$. Meanwhile, other points in the segment $[AB]$ may move (because they are not end-points and are placed before in the order). After these moves (from 1 to 7 in $[AB]$), the next point to move in $[AB]$ is now the end-point B. Now, all the segments which share B as an end-point agree to move B so it can be done. We can continue the process of $[BF]$: the next point to move is F, which is an end-point so we have to wait that $[FG]$ agrees also to move F, etc. The theorem ensure us that the process finishes.

As we can see, there is no interlocking because there are no cycles in trees. It is different for DSS graphs. In Fig. 4b we see a polygon and here the proposed algorithm does not work: A moves before B, B before C, C before D and D before A. There is an interlocking. There is a way to solve this problem that we will present in a forthcoming paper.

5   Conclusion

In this paper we introduced Discrete Soft Transforms and showcase the soft rotation and soft translation of discrete straight segments. A discrete soft transform as a set of atomic transforms where at each step only one discrete point is modified while preserving some properties. For a discrete straight segment, we show that at most two points that can be moved in a given direction to stay a segment. After a brief presentation of soft rotation of a segment, we focused on soft translation of segments which required distinct algorithm for different slopes. The results on the soft translation of a segment are extended to a discrete tree formed of discrete segments. Som examples of soft rotation and translation of DSS can be seen at the following repositery: https://imgur.com/a/j81nI6f. At the end of the paper, we show that there may be interlocking when we have cycles. A solution which involves a new method of soft translation will be presented in a forthcoming paper. This paper opens up many questions. A first direct question is how to adapt these methods to standard discrete line segments. This would be particularly interesting for the soft translation of segmented images by moving the discrete inter-pixel boundary between the regions [4]. The extension to the soft translation of 3D/nD planes or hyperplanes is another interesting next step. Finally, we could develop our work on soft rotations and imagine other type of transforms such as soft homothety or general continuous transformations.

References

1. Andres, E.: The quasi-shear rotation. In: Miguet, S., Montanvert, A., Ubéda, S. (eds.) DGCI 1996. LNCS, vol. 1176, pp. 307–314. Springer, Heidelberg (1996). https://doi.org/10.1007/3-540-62005-2_26
2. Debled-Rennesson, I., Reveilles, J.-P.: A linear algorithm for segmentation of digital curves. IJPRAI **09**(04), 635–662 (1995)
3. Klette, R., Rosenfeld, A.: Digital straightness - a review. Discret. Appl. Math. **139**(1–3), 197–230 (2004)
4. Kovalevsky, V.A.: Finite topology and image analysis. In: Hawkes, P.W., (ed.) Image Mathematics and Image Processing, Volume 84 of Advances in Electronics and Electron Physics, pp. 197–259. Academic Press (1992)
5. Lachaud, J.-O., Said, M.: Two efficient algorithms for computing the characteristics of a subsegment of a digital straight line. Discret. Appl. Math. **161**(15), 2293–2315 (2013)
6. Ngo, P., Passat, N., Kenmochi, Y., Debled-Rennesson, I.: Geometric preservation of 2d digital objects under rigid motions. J. Math. Imaging Vis. **61**(2), 204–223 (2019). https://doi.org/10.1007/s10851-018-0842-9

7. Ouattara, J.S.D., Andres, E., Largeteau-Skapin, G., Zrour, R., Tapsoba, T.M.Y.: Remainder approach for the computation of digital straight line subsegment characteristics. Discret. Appl. Math. **183**, 90–101 (2015)
8. Reveillès, J.P.: Calcul en nombres entiers et algorithmique. Thèse d'état. Université Louis Pasteur, Strasbourg, France (1991)
9. Tarsissi, L., Coeurjolly, D., Kenmochi, Y., Romon, P.: Convexity preserving contraction of digital sets. In: Palaiahnakote, S., Sanniti di Baja, G., Wang, L., Yan, W.Q. (eds.) ACPR 2019. LNCS, vol. 12047, pp. 611–624. Springer, Cham (2020). https://doi.org/10.1007/978-3-030-41299-9_48

On the Validity of the Two Raster Sequences Distance Transform Algorithm

Édouard Thiel[(⊠)]

Aix Marseille Univ, Université de Toulon, CNRS, LIS, Marseille, France
Edouard.Thiel@univ-amu.fr
https://pageperso.lis-lab.fr/~edouard.thiel/

Abstract. This paper examines the validity of the two raster sequences distance transform algorithm, originally given by Rosenfeld and Pfaltz for the distance d_4, then extended to the weighted distances by Montanari and Borgefors. We show that the convergence in two passes does not hold for all chamfer masks, and we prove that the chamfer norm condition is a sufficient condition of validity for the algorithm.

Keywords: Discrete geometry · Distance transforms · Weighted distances · Chamfer norms

1 Introduction

Given a binary image A composed of shape points and background points, a Distance Transform (DT) of A is a copy where each shape point is labelled to its distance from the nearest background point. Both the computation and the result properties depend on the considered distance function. The computation of a DT is generally a global operation, which can be quite expensive; however for some distance functions there are very efficient algorithms based on local operations, using sequential or parallel approaches.

DTs have been extensively studied and have played an important role in Discrete Geometry and Image Analysis since the late 1960s. In the founding paper [1], Rosenfeld and Pfaltz introduced the notion of DT, and presented a two raster sequences DT algorithm in 2D for the direct neighbourhood distance d_4. They also proved that for any given local transformation on an image, the sequential and parallel approaches are mathematically equivalent. Following that, the notion of weighted (or chamfer) distances has emerged in [2–4] together with a rather straightforward extension of the DT algorithm.

We recall some definitions and hypotheses from [6]. A *weighting* $(\vec{v}, w)$ is a displacement $\vec{v} \neq \vec{0}$ associated to a weight $w > 0$. A *chamfer mask* $\mathcal{M}$ is a non-empty set of weightings, such that the set of displacements contains at least a basis of the image points (reachability), and such that $\forall(\vec{v}, w) \in \mathcal{M}$, $(-\vec{v}, w) \in \mathcal{M}$ (central-symmetry). Two points P and Q are $\mathcal{M}$-adjacent if there exists $(\vec{v}, w) \in \mathcal{M}$ such that $\vec{PQ} = \vec{v}$. Two points P and Q are $\mathcal{M}$-connected

© Springer Nature Switzerland AG 2022
É. Baudrier et al. (Eds.): DGMM 2022, LNCS 13493, pp. 436–446, 2022.
https://doi.org/10.1007/978-3-031-19897-7_34

if there exists a path of $\mathcal{M}$-adjacent points joining them, that is, a sequence of distinct points $P_0 = P$, $P_1, \ldots, P_k = Q$ with P_i a $\mathcal{M}$-neighbour of P_{i-1}, $1 \le i \le k$. The cost of the path is the sum of the weights of the displacements. The weighted distance $d_{\mathcal{M}}(P, Q)$ is the cost of a path having minimal cost:

$$d_{\mathcal{M}}(P, Q) = \min \left\{ \sum \lambda_i w_i \ : \ \sum \lambda_i \vec{v}_i = \vec{PQ}, \ (\vec{v}_i, w_i) \in \mathcal{M}, \ \lambda_i \in \mathbb{Z}_+ \right\}. \quad (1)$$

Equivalently, we can consider the weighted geometric graph (V, G), where the set of vertices V corresponds to the image points, and the set of edges G is defined as follows: each vertex $P \in V$ is connected to its $\mathcal{M}$-neighbours $P + \vec{v}$ by an edge having the weight w, $\forall (\vec{v}, w) \in \mathcal{M}$ s.t. $P + \vec{v} \in V$. The weighted distance $d_{\mathcal{M}}$ is then the intrinsic distance of this weighted graph, and always has the properties of a metric (positive definite, symmetric and triangular) since the graph is non oriented and the weights are strictly positive by hypothesis.

Let us go back to the origins. The first weighted distances d_4 and d_8 were presented in [1]; their mask correspond respectively to the 4- and 8-neighbourhood in $\mathbb{Z}^2$, each displacement having the weight 1; they coincide with the norms $\ell_1(\vec{x}) = |x_1| + \ldots + |x_n|$ and $\ell_\infty(\vec{x}) = \max(|x_1|, \ldots, |x_n|)$ in $\mathbb{Z}^n$.

In [2], Montanari introduced a family of weighted distances in $\mathbb{Z}^2$, where a mask $\mathcal{M}_k$ is the set of the displacements $\vec{v}(x, y)$ in the $(2k + 1) \times (2k + 1)$ neighbourhood (i.e. $-k \le x, y \le k$), such that (x, y) is visible (from the origin), i.e. $\gcd(x, y) = 1$. The weight of any displacement $\vec{v}(x, y)$ is its Euclidean length $\sqrt{x^2 + y^2}$. The distance values obtained $d_{\mathcal{M}_k}$ are no longer integers, but can give a good approximation of the Euclidean distance d_E (depending on k). The two raster sequences DT algorithm is extended to the masks $\mathcal{M}_k$ and the convergence in two passes is shown.

The weighted distances using integer weights, or chamfer distances, have then been popularized for $\mathbb{Z}^2$ and $\mathbb{Z}^n$ by Borgefors in [3,4]. The merits of several masks and weights are discussed so as to approximate d_E in an efficient manner, and some conditions are given to choose the weights in order to establish direct distances formulas. The two raster sequences DT algorithm is presented in $\mathbb{Z}^n$. But the problem is that the convergence in two passes is not actually shown; and if we look closer, it cannot be deduced from the Rosenfeld and Pfaltz or Montanari proofs for all chamfer masks.

For these reasons, we propose to study the validity of the DT according to the mask, see some counter-examples, and give a sufficient condition of convergence.

The remainder of the paper is organized as follows: the Sect. 2 first recalls the principle of the parallel and sequential DT algorithm for d_4 and d_8; we then examine in Sect. 3 the original proof of [1], by completing it with a missing hypothesis; the Sect. 4 presents an adaptation of the sequential DT algorithm for chamfer masks in $\mathbb{Z}^n$, in order to check the number of passes necessary for the convergence; in Sect. 5, we study a counter-example which shows that the convergence does not always hold in two passes; after that in Sect. 6 we show that the sequential DT algorithm always converges in two passes when using chamfer norms, and we conclude in Sect. 7.

2   Distance Transformations in $\mathbb{Z}^2$ for d_4 and d_8

Let $A = (a_{i,j})$ be an input image, where $a_{i,j}$ denotes the value of the point at row i $(1 \leq i \leq m)$ and column j $(1 \leq j \leq n)$; the foreground points have value 1 and the background points 0. Given a chamfer mask $\mathcal{M}$, the goal is to compute the DT $D = (d_{i,j})$ where $d_{i,j}$ is the distance $d_{\mathcal{M}}$ to the set of 0's (supposed non-empty) of A. For any weighting $(\vec{v}, w) \in \mathcal{M}$ we denote by (v_i, v_j) the coordinates of $\vec{v}$.

Here is the naive parallel algorithm to compute DT. At step 0, let B^0 be a copy of A, where the 1's are set to ∞, or a sufficient large value. We compute for each step $k > 0$ the image $B^k = (b^k_{i,j})$, where

$$b^k_{i,j} = \min \left\{ b^{k-1}_{i+v_i,j+v_j} + w \; : \; (\vec{v}, w) \in \mathcal{M}, \; \begin{array}{l} 1 \leq i + v_i \leq m, \\ 1 \leq j + v_j \leq n \end{array} \right\}. \tag{2}$$

The process is repeated until no point value changes. The number of iterations is bounded by the maximal number of displacements in a minimal $\mathcal{M}$-path, and can be quite large.

The same method can be processed in an iterative manner on a single image B. The order in which we compute the $b_{i,j}$ is arbitrary, and the convergence rate can be greatly increased by a clever choice of the order. The sequential DT algorithm of Rosenfeld and Pfaltz takes advantage of this idea, an converges in only two raster sequences on the image. Here is their original algorithm, presented in [1] for the distance d_4.

The forward scan processes the image row by row in the raster sequence $a_{1,1}, \ldots, a_{1,n}, a_{2,1}, \ldots, a_{2,n}, \ldots, a_{m,1}, \ldots, a_{m,n}$; the backward scan processes the points in the reverse order. During the forward scan the function f_1 is applied on A to obtain the image B, then during the backward scan the function f_2 is applied on B to get the image C. These functions are defined by:

$$
\begin{aligned}
f_1 : \quad b_{i,j} &= 0 && \text{if } a_{i,j} = 0, \\
&= \min (b_{i-1,j} + 1, b_{i,j-1} + 1) && \text{if } a_{i,j} = 1 \text{ and } (i,j) \neq (1,1), \\
&= \mu && \text{if } a_{i,j} = 1 \text{ and } (i,j) = (1,1); \\
f_2 : \quad c_{i,j} &= \min (b_{i,j}, c_{i+1,j} + 1, c_{i,j+1} + 1).
\end{aligned}
$$

The value μ is chosen to be an unattainable distance value in the image, e.g. $m + n$ (in the paper) or $+\infty$, and is set as an initialization for the top left point $(1,1)$. The min's are only evaluated on the neighbours inside the image; an alternative option is to consider the value μ for neighbours who are outside the image.

The algorithm can be easily adapted to d_8 by adding the indirect neighbours $(i-1, j-1)$ and $(i-1, j+1)$ in the min for f_1, and $(i+1, j-1)$, $(i+1, j+1)$ in the min for f_2.

Figure 1 shows an example with d_4 and Fig. 2 with d_8. For simplicity, we have considered in the min's that $\mu + x = \mu, \forall x \geq 0$.

1	1	1	1	1
1	1	0	1	1
1	1	1	1	1

$\xrightarrow{f_1}$

μ	μ	μ	μ	μ
μ	μ	0	1	2
μ	μ	1	2	3

$\xrightarrow{f_2}$

3	2	1	2	3
2	1	0	1	2
3	2	1	2	3

Fig. 1. Two raster sequences DT algorithm for d_4 on a 3×5 image.

1	1	1	1	1
1	1	0	1	1
1	1	1	1	1

$\xrightarrow{f_1}$

μ	μ	μ	μ	μ
μ	μ	0	1	2
μ	1	1	1	2

$\xrightarrow{f_2}$

2	1	1	1	2
2	1	0	1	2
2	1	1	1	2

Fig. 2. Two raster sequences DT algorithm for d_8 on a 3×5 image.

3   Original Proof for the Two Raster Sequences DT

The original proof in [1, Sect. 4.2] is rather compact; we will develop it and show that there was a missing hypothesis. The proof is constructed by induction for d_4 in $\mathbb{Z}^2$; the goal is to show that after applying f_1 and f_2, the obtained image C satisfies $C = D$ (using the notations of Sect. 2).

On the base case it is noted that if $a_{i,j} = 1$ and a direct neighbour inside the image is 0, evidently $c_{i,j} = 1$, and conversely.

The original induction hypothesis is: suppose for a given $k > 1$ that

$$c_{i,j} = d_{i,j} \quad \forall i, j \text{ s.t. } d_{i,j} < k. \tag{3}$$

Hence $\forall i, j$ we have

$$d_{i,j} < k \quad \Rightarrow \quad c_{i,j} = d_{i,j}; \tag{4}$$

but this does not exclude the existence of cases such as

$$d_{i,j} \geq k \quad \text{and} \quad c_{i,j} < k. \tag{5}$$

In fact, for the rest of the proof, we will have to exclude these cases in two places. The (extended) induction hypothesis has thus to be: suppose for a given $k > 1$ that

$$c_{i,j} = d_{i,j} \quad \forall i, j \text{ s.t. } d_{i,j} < k \text{ or } c_{i,j} < k. \tag{6}$$

We therefore further assumed that

$$c_{i,j} < k \quad \Rightarrow \quad c_{i,j} = d_{i,j}. \tag{7}$$

Remark. By (4) we have $d_{i,j} < k \Rightarrow c_{i,j} < k$, thus

$$c_{i,j} \geq k \quad \Rightarrow \quad d_{i,j} \geq k; \tag{8}$$

moreover, by (7) we have $c_{i,j} < k \Rightarrow d_{i,j} < k$, so $d_{i,j} \geq k \Rightarrow c_{i,j} \geq k$; hence

$$c_{i,j} \geq k \quad \Leftrightarrow \quad d_{i,j} \geq k. \tag{9}$$

We continue the induction by studying the case where $c_{i,j} = k$. By (8) we have $d_{i,j} \geq k$. If $d_{i,j} = k$ then $c_{i,j} = d_{i,j}$ and the proof is done. Let us suppose that $c_{i,j} = k$ and $d_{i,j} > k$. By definition of d_4, since $d_{i,j} > k$, the four direct neighbours are $\geq k$:

$$
\begin{array}{c|c|c}
 & d_{i-1,j} \geq k & \\
\hline
d_{i,j-1} \geq k & d_{i,j} > k & d_{i,j+1} \geq k \\
\hline
 & d_{i+1,j} \geq k &
\end{array} \tag{10}
$$

Thanks to the extended hypothesis, we have by (9)

$$
\begin{aligned}
d_{i,j+1} \geq k &\implies c_{i,j+1} \geq k, \\
d_{i+1,j} \geq k &\implies c_{i+1,j} \geq k,
\end{aligned} \tag{11}
$$

hence during the computation of $c_{i,j}$ by f_2 in backward sequence we have

$$
c_{i,j} = \min \begin{cases} b_{i,j} \\ c_{i,j+1} + 1 & (\geq k+1), \\ c_{i+1,j} + 1 & (\geq k+1) \end{cases} \tag{12}
$$

thus $c_{i,j} = k \implies b_{i,j} = k$. However, when calculating $b_{i,j}$ by f_1 in forward sequence we have applied

$$
b_{i,j} = \min \begin{cases} b_{i-1,j} + 1 \\ b_{i,j-1} + 1 \end{cases}, \tag{13}
$$

thus $b_{i,j} = k \implies b_{i-1,j} = k-1$ or $b_{i,j-1} = k-1$. Suppose that the former holds, that is $b_{i-1,j} = k-1$. During the calculation of $c_{i-1,j}$ by f_2 we have

$$
c_{i-1,j} = \min \begin{cases} b_{i-1,j} & (= k-1) \\ c_{i-1,j+1} + 1 \\ c_{i,j} + 1 \end{cases} \tag{14}
$$

therefore $c_{i-1,j} \leq k-1$; but $d_{i-1,j} \geq k$ by (10) so $d_{i-1,j} \neq c_{i-1,j}$, in contradiction with the extended hypothesis since $c_{i-1,j} < k$. $\qquad\square$

This proof can be easily extended for d_8 by adding the four indirect neighbours in the min's. More generally, the algorithm and the proof can be extended in $\mathbb{Z}^n$ for the distances d_1 and d_∞ induced by the ℓ_1 and ℓ_∞ norms.

It should be noted that the algorithm can also be adapted to chamfer masks in $\mathbb{Z}^n$ (see [4]), but we will show further with a counter-example that the convergence in two scans is not always guaranteed for any chamfer mask. At the proof level, we can see that this proof cannot be extended either, because the inequations are performed on neighbours of (i,j) only, and they use the fact that the distance values are consecutive integers.

4   Sequential DTs for Chamfer Masks in $\mathbb{Z}^n$

We present an adaptation of the sequential DT in $\mathbb{Z}^n$ which is a bit hardened to handle counter-examples masks.

The masks need to be split in two parts for the forward and backward scans. Using coordinates $p = (x_1, \ldots, x_n) \in \mathbb{Z}^n$, let us consider the forward raster sequence in ascending order by nested loops for x_n, for x_{n-1}, $\ldots$, for x_1.

The half-space visited by the loops after the origin in the raster sequence is $\mathcal{H}^n = \cup_{1 \le k \le n} \{ p : x_n = 0, \ldots, x_{k+1} = 0, x_k > 0 \}$. For instance, the half-space $\mathcal{H}^3$ is $\{ p : x_3 = 0, x_2 = 0, x_1 > 0 \} \cup \{ p : x_3 = 0, x_2 > 0 \} \cup \{ p : x_3 > 0 \}$. Given a chamfer mask $\mathcal{M} = \{ (\vec{v}, w) : \vec{v} \in \mathbb{Z}^n \}$, we define the half-mask $\mathcal{M}^h = \{ (\vec{v}, w) \in \mathcal{M} : \vec{v} \in \mathcal{H}^n \}$. During the sequential DT, the forward raster sequence will then use the half-mask $\mathcal{M} \setminus \mathcal{M}^h$, whereas the backward one will use $\mathcal{M}^h$.

The computation of one sequential DT scan is presented in Fig. 3, for convenience in Python language in $\mathbb{Z}^2$. The source code and examples are online in [9]. To extend the function in higher dimension it is sufficient to add coordinates and loops for the additional dimensions.

```
1   def compute_one_DT_scan (img, half_mask, scan_num) :
2       forward = scan_num % 2 == 1
3       if forward :
4           i_start = 0 ; i_end = img.m                      # 0 to m-1
5           j_start = 0 ; j_end = img.n ; step = 1           # 0 to n-1
6       else :
7           i_start = img.m-1 ; i_end = -1                   # m-1 to 0
8           j_start = img.n-1 ; j_end = -1 ; step = -1       # n-1 to 0
9       changed = False
10      for i in range (i_start, i_end, step) :
11          for j in range (j_start, j_end, step) :
12              if img.mat[i][j] == 0 : continue
13              min_w = -1 if scan_num == 1 else img.mat[i][j]
14              for p_i, p_j, p_w in half_mask.weightings :
15                  q_i = i - p_i*step ; q_j = j - p_j*step
16                  if not img.is_inside (q_i, q_j) : continue
17                  if img.mat[q_i][q_j] == -1 : continue
18                  q_w = img.mat[q_i][q_j] + p_w
19                  if min_w == -1 or q_w < min_w : min_w = q_w
20              if img.mat[i][j] != min_w : changed = True
21              img.mat[i][j] = min_w                        # can be -1
22      return changed
```

Fig. 3. Computation of one sequential DT scan in $\mathbb{Z}^2$ with $\mu = -1$.

The input and output image is img. The coordinates are $0 \le$ i $<$ img.m for x_2 (or y) and $0 \le$ j $<$ img.n for x_1 (or x); the point values are accessed by img.mat[i][j]. The method img.is_inside(i,j) returns True if the coordinates are inside the image. The parameter half_mask stores the $\mathcal{M}^h$ weightings as a list of tuples. The direction of the scan (forward or backward) is deduced from the scan number scan_num line 2. The loop step value is also used line 15 to compute the displacements of the half mask for the current scan direction.

The function is written with the special value $\mu = -1$. It indicates, as a forbidden distance value, the non currently propagated distance values in the image, and needs a test to handle the min's. Using signed pixel values, we find it more handy than choosing an arbitrary large integer to simulate $+\infty$.

The computation of the DT in two raster sequences is done by calling twice the function `compute_one_DT_scan` with the scan number, see the function `compute_sequential_DT_in_two_scans` in Fig. 4.

```
1   def compute_sequential_DT_in_two_scans (img, half_mask) :
2       compute_one_DT_scan (img, half_mask, 1)
3       compute_one_DT_scan (img, half_mask, 2)
4
5   def compute_sequential_DT_multi_scans (img, half_mask) :
6       scan_num = 1
7       while True :
8           if compute_one_DT_scan (img, half_mask, scan_num) :
9               scan_num += 1
10          else : break
11      return scan_num
```

Fig. 4. Sequential DT algorithms in $\mathbb{Z}^2$.

As for the parallel DT computation, the sequential DT can be performed scan by scan until no point value changes (all paths are propagated and convergence is reached). For this purpose, the function `compute_one_DT_scan` returns a boolean value `changed`, which is used to stop the loop in the function `compute_sequential_DT_multi_scans` in Fig. 4.

5   Counter-Example for the Two Raster Sequences DT

We present now a simple counter-example, which shows that the convergence of the DT in only two raster sequences does not hold for all chamfer masks.

One can imagine any kind of mask, see for instance [6, p. 42] for a gallery. In the literature, the most common category of studied masks are grid-symmetrical (8-symmetrical in $\mathbb{Z}^2$, 48- in $\mathbb{Z}^3$, $(2^n n!)$- in $\mathbb{Z}^n$). The weightings are chosen in the first octant (also called generator) $0 \leq x_n \leq \ldots \leq x_1$, then the grid symmetries are performed to populate the mask. For efficiency, the displacements are usually chosen among the visible points, because for a weighting $(\vec{v}, w)$, each period $O + \lambda\vec{v}$ is expected to get the distance value λw from O, if the mask has the good properties (see further), so adding $(\lambda\vec{v}, \lambda w)$ in the mask is useless.

In $\mathbb{Z}^2$, the first visible points in the first octant are denoted by $\mathbf{a} = (0, 1)$ (still using coordinates in the order (x_2, x_1)), $\mathbf{b} = (1, 1)$, $\mathbf{c} = (1, 2)$, $\mathbf{d} = (1, 3)$, $\mathbf{e} = (2, 3)$, etc. A grid-symmetrical mask constituted by a set of weightings $(\mathbf{v}, w)$ where $\mathbf{v}$ is a visible point is denoted by $\langle(\mathbf{v}, w), \ldots\rangle$. For instance, the mask for

d_4 is denoted by $\langle(\mathbf{a},1)\rangle$, the mask for d_8 is $\langle(\mathbf{a},1),(\mathbf{b},1)\rangle$, the mask for the chamfer distance 5,7,11 [4] is $\langle(\mathbf{a},5),(\mathbf{b},7),(\mathbf{c},11)\rangle$, and so on.

To find counter-examples it is sufficient to choose some displacements, loop on several weights, and compute the DTs on images of several sizes, where all points have value 1, except one point which has value 0 in the centre of the image. For each trial we can compare the results for the parallel algorithm and those of compute_sequential_DT_in_two_scans, or run the function compute_sequential_DT_multi_scans and check if it returns a number of scans > 3. See the program checkWDT.py in [9].

We have found a very simple counter-example for any image size larger then 3×3: this is the mask $\langle(\mathbf{c},1)\rangle$, also known as the Knight distance [5]. The Fig. 5 shows the full mask and the two half masks.

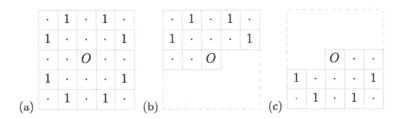

Fig. 5. Mask $\langle(\mathbf{c},1)\rangle$ around the origin O: (a) full, (b) forward, (c) backward mask.

The mask $\langle(\mathbf{c},1)\rangle$ is a chamfer mask because the basis vector $(0,1)$ can be obtained using the symmetrical displacements of $\mathbf{c}$, by $(-1,\;2) + (-1,2) + (2,1) = (0,1)$, and the same by symmetry for $(1,0)$.

The Fig. 6 shows the parallel passes for a 3×4 image; 6 passes are necessary to reach the correct DT values. See the program showWDT.py in [9].

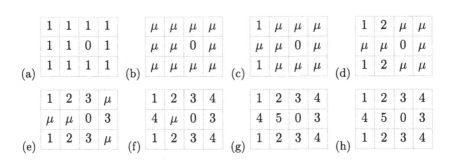

Fig. 6. Parallel DT for $\langle(\mathbf{c},1)\rangle$: (a) original image, (b) initialization, (c–h) passes 1–6.

On Fig. 7 we can see that the raster sequences DT algorithm also needs 6 passes: 5 to converge and the sixth to detect no changes and stop.

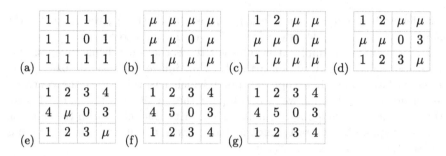

Fig. 7. Sequential DT for $\langle(\mathbf{c},1)\rangle$: (a) original image, (b–g) passes 1–6, (b,d,f) forward passes, (c,e,g) backward passes.

The following remarks can be made for the mask $\langle(\mathbf{c},1)\rangle$ taken as a counter-example for the convergence in two passes:

- The two raster sequences DT presented as the Knight Transform (KT) in [5] therefore does not converge in the general case.
- The necessary sequential passes number depends on the image size, and may decrease a little bit when the size grows.
- The passes number does not depend on the chosen $\mathbf{c}$ weight.
- We can replace $\mathbf{c}$ by any visible point $(1, 2k)$, $k \geq 1$ and still get a chamfer mask, since $k(-1, -2k) + k(-1, 2k) + (2k, 1) = (-2k, 0) + (2k, 1) = (0, 1)$.

6   Validity Holds for Chamfer Norms

A metric d in $\mathbb{Z}^n$ induces a discrete norm g defined by $g(q - p) = d(p, q)$ if g satisfies the property of homogeneity over $\mathbb{Z}$ [6, Sect. 2.2.3]:

$$\forall \vec{x} \in \mathbb{Z}^n,\ \forall \lambda \in \mathbb{Z},\ g(\lambda \vec{x}) = |\lambda|\, g(\vec{x}). \tag{15}$$

A chamfer norm is a discrete norm induced by a chamfer mask.

For instance, the masks $\langle(\mathbf{a},1)\rangle$ for d_4, $\langle(\mathbf{a},1),(\mathbf{b},1)\rangle$ for d_8, $\langle(\mathbf{a},3),(\mathbf{b},4)\rangle$ and $\langle(\mathbf{a},5),(\mathbf{b},7),(\mathbf{c},11)\rangle$ all induce chamfer norms, but $\langle(\mathbf{c},1)\rangle$ clearly not (no homogeneity: let $P = (0,1)$, then $d(O, P) = 3$ and $d(O, 2.P) = 2 \neq 2.d(O, P)$).

The chamfer norms have remarkable properties: they allow to completely characterize the geometry of the distance balls, to give direct distance formulas, and to determine the structure of minimal paths. Given a chamfer mask $\mathcal{M}$, we call rational ball of $\mathcal{M}$ the set

$$\mathcal{B}_{\mathcal{M}}^{\mathbb{Q}} = \mathrm{conv}\left(\frac{\vec{v}}{w} : (\vec{v}, w) \in \mathcal{M} \right). \tag{16}$$

The rational ball is a convex polyhedron, whose geometry is the same as the distance balls up to a scale factor.

The conditions for being a chamfer norm in $\mathbb{Z}^n$ are established in [6, Sect. 4.3.4] and [7, Sect. 4.3.2]: a chamfer mask $\mathcal{M}$ induces a discrete norm if and only if it exists a triangulation of $\mathcal{B}_{\mathcal{M}}^{Q}$ in unimodular cones of apex O. Now suppose that $\mathcal{M}$ induces a discrete norm and let $\mathcal{C}$ be such a cone, then $\mathcal{C}$ is bounded by a subset of n weightings of $\mathcal{M}$, denoted by $\mathcal{M}|_{\mathcal{C}} = \{ (\vec{v}_i', w_i'), 1 \leq i \leq n \}$; moreover, for each point P in $\mathcal{C}$, there is a minimal path from O to P which is a linear combination $\lambda_1 \vec{v}_1' + \ldots + \lambda_n \vec{v}_n'$, $\lambda_i \in \mathbb{Z}_+$ of displacements from $\mathcal{M}|_{\mathcal{C}}$, and whose intermediate points are all included in $\mathcal{C}$.

Proposition 1. *Let $\mathcal{M}$ be a chamfer norm mask, then the two raster sequences DT algorithm provides the correct DT values for $d_{\mathcal{M}}$.*

Proof. Let P be a feature point currently evaluated during a raster sequence, and Q a closest background point. Consider the unimodular cone $\mathcal{C}$ of apex P which contains a minimal $\mathcal{M}$-path $\mathcal{P}$ from P to Q, and the set $\mathcal{M}|_{\mathcal{C}}$ of weightings which are bounding $\mathcal{C}$. Then $\mathcal{P}$ is a sequence of distinct points $P_0 = P$, P_1, ..., $P_k = Q$ with P_i a $\mathcal{M}|_{\mathcal{C}}$-neighbour of P_{i-1}, $1 \leq i \leq k$.

The cone $\mathcal{C}$ is either (a) contained in the half-space $P - \mathcal{H}^n = \{ P - \overrightarrow{OX} : X \in \mathcal{H}^n \}$ (the points before P in the forward scan), see Fig. 8; (b) in the half-space $P + \mathcal{H}^n$ (the points before P in the backward scan); or (c) intersects both half-spaces.

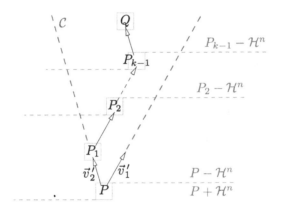

Fig. 8. Case (a) for the proof of proposition 1, here in $\mathbb{Z}^2$.

In the case (a) each P_i is contained in the half-space $P_{i-1} - \mathcal{H}^n$, $1 \leq i \leq k$, so during the forward scan, each P_i is evaluated before P_{i-1}. As P_{k-1} is an $\mathcal{M}|_{\mathcal{C}}$-neighbour of $P_k = Q$, the min computation will give the correct associated weight value in the DT for P_{k-1}, an so on from P_{k-1} to P_0.

In the case (b), the same reasoning can be made using $P_{i-1} + \mathcal{H}^n$ during the backward scan.

In case (c), if $Q \in P - \mathcal{H}^n$, then a minimal path can be chosen such that all the path points are included in $\mathcal{C} \cap (P - \mathcal{H}^n)$, so we can revert to case (a); the same for $Q \in P + \mathcal{H}^n$ and case (b). □

7 Conclusion and Future Work

In this paper, we have improved the proof of [1] for d_1 and d_∞, and proposed a hardened raster sequence DT algorithm for the chamfer masks. We have shown with a counter-example that the convergence does not always hold in two passes for all chamfer masks, and we have proved in proposition 1 that the two raster sequences DT algorithm provides the correct distance values for any chamfer norm.

It can be pointed out that the norm condition is sufficient but non necessary. For instance, the algorithm holds for the following non-norm chamfer masks: $\langle(\mathbf{a}, 1), (\mathbf{b}, 1), (\mathbf{c}, 1)\rangle$, $\langle(\mathbf{a}, 1), (\mathbf{b}, 3), (\mathbf{c}, 2)\rangle$, $\langle(\mathbf{a}, 2), (\mathbf{b}, 3), (\mathbf{c}, 4)\rangle$, $\langle(\mathbf{a}, 1), (\mathbf{c}, 1)\rangle$, $\langle(\mathbf{a}, 2), (\mathbf{c}, 3)\rangle$.

In future works, it would be interesting to investigate if necessary conditions could be established on non-norm chamfer masks, to predict the number of passes for their convergence, and also to study the convergence for the reverse distance transform. This work on weighted distances might be extended on semi-regular grids, or other families of weighted geometric graphs. One could finally relate this work to ns-weighted distances, of which weighted distances are a special case [7,8].

References

1. Rosenfeld, A., Pfaltz, J.: Sequential operations in digital picture processing. J. ACM **13**(4), 471–494 (1966)
2. Montanari, U.: A method for obtaining skeletons using a quasi-euclidean distance. J. ACM **15**, 600–624 (1968)
3. Borgefors, G.: Distance transformations in arbitrary dimensions. Comput. Vis. Graph. Image Process. **27**, 321–345 (1984)
4. Borgefors, G.: Distance transformations in digital images. Comput. Vis. Graph. Image Process. **34**, 344–371 (1986)
5. Das, P.P., Chatterji, B.N.: Knight's distance in digital geometry. Pattern Recogn. Lett. **7**, 215–226 (1988)
6. Thiel, E.: Géométrie des distances de chanfrein. Hábilitation à Diriger des Recherches, Université de la Méditerranée, Aix-Marseille, 2 Déc 2001. https:// pageperso.lis-lab.fr/~edouard.thiel/hdr/
7. Normand, N.: Projections et distances discrètes. Université de Nantes (Nov, Habilitation à Diriger des Recherches (2012)
8. Strand, R., Normand, N.: Distance transform computation for digital distance functions. Theoret. Comput. Sci. **448**, 80–93 (2012)
9. Annex with source code. https://pageperso.lis-lab.fr/~edouard.thiel/DGMM2022/

Learning Based Morphology
to Mathematical Morphology

MorphoActivation: Generalizing ReLU Activation Function by Mathematical Morphology

Santiago Velasco-Forero[(✉)] and Jesús Angulo

MINES Paris, PSL University, Centre for mathematical morphology (CMM),
Paris, France
{santiago.velasco,jesus.angulo}@mines-paristech.fr

Abstract. This paper analyses both nonlinear activation functions and spatial max-pooling for Deep Convolutional Neural Networks (DCNNs) by means of the algebraic basis of mathematical morphology. Additionally, a general family of activation functions is proposed by considering both max-pooling and nonlinear operators in the context of morphological representations. Experimental section validates the goodness of our approach on classical benchmarks for supervised learning by DCNN.

Keywords: Matheron's representation theory · Activation function · Mathematical morphology · Deep learning

1 Introduction

Artificial neural networks were introduced as mathematical models for biological neural networks [24]. The basic component is a *linear perceptron* which is a linear combination of weights with biases followed by a nonlinear function called *activation function*. Such components (usually called a *layer*) can then be concatenated eventually leading to very complex functions named *deep artificial neural networks (DNNs)* [7]. Activation function can also be seen as an attached function between two layers in a neural network. Meanwhile, in order to get the learning in a DNNs, one needs to update the weights and biases of the neurons on the basis of the error at the output. This process involves two steps, a *Back-Propagation* from prediction error and a Gradient Descent Optimization to update parameters [7]. The most famous activation function is the Rectified Linear Unit (ReLU) proposed by [27], which is simply defined as $\text{ReLU}(x) = \max(x, 0)$. A clear benefit of ReLU is that both the function itself and its derivatives are easy to implement and computationally inexpensive. However, ReLU has a potential loss during optimization because the gradient is zero when the unit is not active. This could lead to cases where there is a gradient-based optimization algorithm that will not adjust the weights of a unit that was never initially activated. An approach purely computational motivated to alleviate potential problems caused by the hard zero activation of ReLU, proposed a leaky

© Springer Nature Switzerland AG 2022
É. Baudrier et al. (Eds.): DGMM 2022, LNCS 13493, pp. 449–461, 2022.
https://doi.org/10.1007/978-3-031-19897-7_35

ReLU activation [18]: LeakyReLU$(x) = \max(x, .01x)$. A simple generalisation is the Parametric ReLU proposed by [11], defined as PReLU$_\beta(x) = \max(x, \beta x)$, where $\beta \in \mathbb{R}$ is a learnable parameter. In general, the use of *piecewise-linear functions* as activation function has been initially motivated by neurobiological observations; for instance, the inhibiting effect of the activity of a visual-receptor unit on the activity of the neighbouring units can be modelled by a line with two segments [10]. On the other hand, for the particular case of structured data as images, a translation invariant DNN called Deep Convolutional Neural Networks (DCNN) is the most used architecture. In the conventional DCNN framework interspersed convolutional layers and pooling layers to summarise information in a hierarchical structure. The common choice is the pooling by a maximum operator called *max-pooling*, which is particularly well suited to the separation of features that are very sparse [3].

As far as these authors know, that morphological operators have been used in the context of DCNNs following the paradigm of replacing lineal convolutions by non-linear morphological filters [13,15,26,31,34], or hybrid variants between linear and morphological layers [14,30,32,33]. Our contribution is more in the sense of [5] where the authors show favourable results in quantitative performance for some applications when seeing the max-pooling operator as a dilation layer. However, we go further to study both nonlinear activation and max-pooling operators in the context of morphological representation theory of nonlinear operators. Finally, in the experimental section, we compare different propositions in a practical case of training a multilayer CNNs for classification of images in several databases.

2   ReLU Activation and Max-Pooling are Morphological Dilations

2.1   Dilation and Erosion

Let us consider a complete lattice $(\mathcal{L}, \leq)$, where and $\bigvee$ and $\bigwedge$ are respectively its supremum and infimum. A lattice operator $\psi : \mathcal{L} \to \mathcal{L}$ is called *increasing operator* (or isotone) is if it is order-preserving, i.e., $\forall X, Y, X \leq Y \implies \psi(X) \leq \psi(Y)$. Dilation δ and erosion ε are lattice operators which are increasing and satisfy

$$\delta\left(\bigvee_{i \in J} X_i\right) = \bigvee_{i \in J} \delta(X_i); \quad \varepsilon\left(\bigwedge_{i \in J} X_i\right) = \bigwedge_{i \in J} \varepsilon(X_i).$$

Dilation and erosion can be then composed to obtain other operators [12]. In this paper, we also use morphological operators on the lattice of functions $\mathcal{F}(\mathbb{R}^n, \bar{\mathbb{R}})$ with the standard partial order $\leq$. The sup-convolution and inf-convolution of function f by structuring function g are given by

$$(f \oplus g)(x) = \delta_g(f)(x) := \sup_{y \in \mathbb{R}^n} \{f(x-y) + g(y)\}, \tag{1}$$

$$(f \ominus g)(x) = \varepsilon_g(f)(x) := \inf_{y \in \mathbb{R}^n} \{f(x+y) - g(y)\}. \tag{2}$$

2.2   ReLU and Max-Pooling

Let us now consider the standard framework of one-dimensional[1] signals on DCNNs where any operator is applied on signals $f(x)$ supported on a discrete grid subset of $\mathbb{Z}$. The *ReLU* activation function [27] applied on every pixel x of an image f is defined as

$$\text{ReLU}(f(x)) := \max(0, f(x)). \tag{3}$$

The *Max-pooling operator* of *pooling size* R and strides K, maps an image f of n pixels onto an image of $n' := \lfloor \frac{n-R}{K} + 1 \rfloor$ by taking the local maxima in a neighbour of size R, and moving the window K elements at a time, skipping the intermediate locations:

$$\text{MaxPool}_R(f)(x) = \delta_R^{\text{MaxPool}}(f)(x) := \bigvee_{y \in W_R(x)} \{f(K \cdot x - y)\}. \tag{4}$$

where $W_R(y) = 0$ if y belongs to the neighbour of size R centred in x and $-\infty$ otherwise. There are other operations in DCNN which use the maximum operation as main ingredient, namely the *Maxout layer* [8] and the *Max-plus layer (morphological perceptron)* [4,37].

From the definition of operators, it is straightforward to prove the following proposition

Proposition 1. *ReLU activation function and max-pooling are dilation operators on the lattice of functions.*

Proof. Using the standard partial ordering $\leq$, we note that both ReLU and max-pooling are increasing:

$$f \leq g \implies \text{ReLU}(f) \leq \text{ReLU}(g); \quad \delta_R^{\text{MaxPool}}(f) \leq \delta_R^{\text{MaxPool}}(g).$$

They commute with supremum operation

$$\text{ReLU}(f \vee g) = \text{ReLU}(f) \vee \text{ReLU}(g); \quad \delta_R^{\text{MaxPool}}(f \vee g) = \delta_R^{\text{MaxPool}}(f) \vee \delta_R^{\text{MaxPool}}(g).$$

These two operators are both also *extensive*, i.e., $f \leq \delta(f)$. ReLU is also idempotent, i.e., $\text{ReLU}(\text{ReLU}(f)) = \text{ReLU}(f)$. Then ReLU is both a dilation and a closing. **Remark 1: Factoring activation function and pooling.** The composition of dilations in the same complete lattice can often be factorized into a single operation. One can for instance define a nonlinear activation function and pooling dilation as

$$\delta_{R;\alpha}^{\text{ActPool}}(f)(x) := \bigvee_{y \in W_R(x)} \{\max(0, f(K \cdot x - y) + \alpha)\},$$

where W denotes a local neighbour, usually a square of side R. Note that that analysis does not bring any new operator, just the interpretation of composed nonlinearities as a dilation.

---

[1] The extension to d-dimensional functions is straightforward.

Remark 2: Positive and negative activation function, symmetric pooling. More general ReLU-like activation functions also keep a negative part. Let us consider the two parameters $\beta^+, \beta^- \in \mathbb{R}$, we define (β^+, β^-)-$ReLU$ as

$$\text{ReLU}_{\beta^+,\beta^-}(f)(x) := \begin{cases} \beta^+ f(x) & \text{if } f(x) > 0 \\ \beta^- f(x) & \text{if } f(x) \leq 0 \end{cases}$$

In the case when $\beta^+ \geq \beta^-$, one has

$$\text{ReLU}_{\beta^+,\beta^-}(f)(x) = \max\left(\beta^- f(x), \beta^+ f(x)\right). \tag{5}$$

Note that the Leaky ReLU [18] corresponds to $\beta^+ = 1$ and $\beta^- = 0.01$. The Parametric ReLU [11] takes $\beta^+ = 1$ and $\beta^- = \theta$ learned along with the other neural-network parameters. More recently [17] both β^+ and β^- are learned in the *ACtivateOrNot* (ACON) activation function, where a softmax is used to approximate the maximum operator.

Usually in CNNs, the *max-pooling* operator is used after activation, i.e., $\delta_R^{\text{MaxPool}}(\text{ReLU}_{\beta^+,\beta^-}(f))$, which is spatially enlarging the positive activation and removing the negative activation. It does not seem coherent with the goal of using the pooling to increase spatial equivariance and hierarchical representation of information. It is easy to *"fix"* that issue by using a more symmetric pooling based on taking the positive and negative parts of a function. Given a function f, it can be expressed in terms of its positive f^+ and negative parts f^-, i.e., $f = f^+ - f^-$, with $f^+(x) = \max(0, f(x))$ and $f^-(x) = \max(0, -f(x))$, where both f^+ and f^- are non-negative functions. We can now define a *positive and negative max-pooling*. The principle is just to take a max-pooling to each part and recompose, i.e.,

$$\delta_R^{\text{MaxPool}^+_-}(f)(x) = \delta_R^{\text{MaxPool}}(f^+)(x) - \delta_R^{\text{MaxPool}}(f^-)(x) \tag{6}$$
$$= \delta_R^{\text{MaxPool}}(\max(0, f))(x) + \varepsilon_R^{\text{MinPool}}(\min(0, f))(x).$$

We note that (6) is *self-dual* and related to the dilation on an inf-semilattice [16]. However, in the general case of (6) by learning both β^-, β^+,

$$\delta_{\beta^+,\beta^-,R}^{\text{MaxPool}^+}(f)(x) = \delta_R^{\text{MaxPool}}(\max(0, \beta^- f))(x) + \varepsilon_R^{\text{MinPool}}(\min(0, \beta^+ f))(x) \tag{7}$$

is not always self-dual.

3   Algebraic Theory of Minimal Representation for Nonlinear Operators and Functions

In the following section, we present the main results about representation theory of nonlinear operators from Matheron [23], Maragos [20] and Bannon-Barrera [1] (MMBB).

3.1   MMBB Representation Theory on Nonlinear Operators

Let us consider a translation-invariant (TI) increasing operator Ψ. The domain of the functions considered here is either $E = \mathbb{R}^n$ or $E = \mathbb{Z}^n$, with the additional condition that we consider only closed subsets of E. We consider first the set operator case applied on $\mathcal{P}(E)$ and functions $f : E \to \mathbb{R} \cap \infty$.

Kernel and Basis Representation of TI Increasing Set Operators. The kernel of the TI operator Ψ is defined as the following collection of input sets [23]: $\mathrm{Ker}(\Psi) = \{A \subseteq E : \mathbf{0} \in \Psi(A)\}$, where $\mathbf{0}$ denotes the origin of E.

Theorem 1 (Matheron (1975) [23]). Consider set operators on $\mathcal{P}(E)$. Let $\Psi : \mathcal{P}(E) \to \mathcal{P}(E)$ be a TI increasing set operator. Then

$$\Psi(X) = \bigcup_{A \in \mathrm{Ker}(\Psi)} X \ominus A = \bigcap_{B \in \mathrm{Ker}(\bar{\Psi})} X \oplus \check{B}.$$

where the dual set operator is $\bar{\Psi}(X) = [\Psi(X^c)]^c$ and $\check{B}$ is the transpose structuring element.

The kernel of Ψ is a partially ordered set under set inclusion which has an infinity number of elements. In practice, by the property of absorption of erosion, that means that the erosion by B contains the erosions by any other kernel set larger than B and it is the only one required when taking the supremum of erosions. The morphological basis of Ψ is defined as the minimal kernel sets [20]:

$$\mathrm{Bas}(\Psi) = \{M \in \mathrm{Ker}(\Psi) : [A \in \mathrm{Ker}(\Psi) \text{ and } A \subseteq M] \implies A = M\}.$$

A sufficient condition for the existence of $\mathrm{Bas}(\Psi)$ is for Ψ to be an upper semi-continuous operator. We also consider closed sets on $\mathcal{P}(E)$.

Theorem 2 (Maragos (1989) [20]). *Let* $\Psi : \mathcal{P}(E) \to \mathcal{P}(E)$ *be a TI, increasing and upper semi-continuous set operator*[2] *. Then*

$$\Psi(X) = \bigcup_{M \in \mathrm{Bas}(\Psi)} X \ominus M = \bigcap_{N \in \mathrm{Bas}(\bar{\Psi})} X \oplus \check{N}.$$

Kernel and Basis Representation of TI Increasing Operators on Functions. Previous set theory was extended [20] to the case of mappings on functions $\Psi(f)$ and therefore useful for signal or grey-scale image operators. We focus on the case of closed functions f, i.e., its epigraph is a closed set. In that case, the

---

[2] Upper semi-continuity meant with respect to the hit-miss topology. Let (X_n) be any decreasing sequence of sets that converges monotonically to a limit set X, *i.e.*, $X_{n+1} \subseteq X_n \forall n$ and $X = \cap_n X_n$; that is denoted by $X_n \downarrow X$.
An increasing set operator Φ on $\mathcal{F}(E)$ is upper semi-continuous if and only if $X_n \downarrow X$ implies that $\Phi(X_n) \downarrow \Phi(X)$.

dual operator is $\bar{\Psi}(f) = -\Psi(-f)$ and the transpose function is $\check{f}(x) = f(-x)$. Let

$$\mathrm{Ker}(\Psi) := \{f : \Psi(f)(\mathbf{0}) \geq 0\}$$

be the kernel of operator Ψ. As for the TI set operators, a basis can be obtained from the kernel functions as its minimal elements with respect to the partial order $\leq$, i.e.,

$$\mathrm{Bas}(\Psi) := \{g \in \mathrm{Ker}(\Psi) : [f \in \mathrm{Ker}(\Psi) \text{ and } f \leq g] \implies f = g\}.$$

This collection of functions can uniquely represent the operator.

Theorem 3 (Maragos (1989) [20]). *Consider an upper semi-continuous operator Ψ acting on an upper semi-continuous function[3] f. Let $Bas(\Psi) = \{g_i\}_{i \in I}$ be its basis and $Bas(\bar{\Psi}) = \{h_j\}_{j \in J}$ the basis of the dual operator. If Ψ is a TI and increasing operator then it can be represented as*

$$\Psi(f)(x) = \sup_{i \in I} (f \ominus g_i)(x) = \sup_{i \in I} \inf_{y \in \mathbb{R}^n} \{f(x+y) - g_i(y)\} \tag{8}$$

$$= \inf_{j \in J} (f \oplus \check{h}_j)(x) = \inf_{j \in J} \sup_{y \in \mathbb{R}^n} \{f(x-y) + \check{h}_j(y)\} \tag{9}$$

The converse is true. Given a collection of functions $\mathcal{B} = \{g_i\}_{i \in I}$ such that all elements of it are minimal in $(\mathcal{B}, \leq)$, the operator $\Psi(f) = \sup_{i \in I}\{f \ominus g_i\}$ is a TI increasing operator whose basis is equal to $\mathcal{B}$.

For some operators, the basis can be very large (potentially infinity) and even if the above theorem represents exactly the operator by using a full expansion of all erosions, we can obtain an approximation based on smaller collections or truncated bases $\mathcal{B} \subset \mathrm{Bas}(\Psi)$ and $\bar{\mathcal{B}} \subset \mathrm{Bas}(\bar{\Psi})$. Then, from the operators $\Psi_l(f) = \sup_{g \in \mathcal{B}}\{f \ominus g\}$ and $\Psi_u(f) = \inf_{h \in \bar{\mathcal{B}}}\{f \oplus \check{h}\}$ the original Ψ is bounded from below and above, i.e., $\Psi_l(f) \leq \Psi(f) \leq \Psi_u(f)$. Note also that in the case of a non minimal representation by a subset of the kernel functions larger than the basis, one just gets a redundant still satisfactory representation.

The extension to TI *non necessarily increasing mappings* was presented by Bannon and Barrera in [1], which involves a supremum of an operator involving an erosion and an anti-dilation. This part of the Matheron-Maragos-Bannon-Barrera (MMBB) theory is out of the scope of this paper.

---

[3] A function $f : \mathbb{R}^n \to \bar{\mathbb{R}}$ is *upper semi-continuous* (u.s.c) (resp. lower semi-continuous (l.s.c.)) if and only if, for each $x \in \mathbb{R}^m$ and $t \in \bar{\mathbb{R}}$, $f(x) < t$ (resp. $f(x) > t$) implies that $f(y) < t$ (resp. $f(y) < t$) for all in some neighbourhood of x. Similarly, f is u.s.c. (resp. l.s.c.) if and only if all its level sets are closed (resp. open) subsets of $\mathbb{R}^n$. A function is continuous iff is both u.s.c and l.s.c.

3.2   Max-Min Representation for Piecewise-Linear Functions

Let us also remind the fundamental results from the representation theory by Ovchinnikov [28,29] which is rooted in a Boolean and lattice framework and therefore related to the MMBB theorems. Just note that here we focus on a representation for functions and previously it was a representation of operators on functions. Let f be a smooth function on a closed domain $\Omega \subset \mathbb{R}^n$. We are going to represent it by a family of affine linear functions g_t which are tangent hyperplanes to the graph of f. Namely, for a point $t \in \Omega$, one defines

$$g_t(x) = \langle \nabla f(t), x - t \rangle + f(t), \quad x \in \Omega, \tag{10}$$

where $\nabla f(t)$ is the gradient vector of f at t. We have the following general result about the representation of piecewise-linear (PL) functions as max-min polynomial of its linear components.

Theorem 4 ([9] [2] [29]). *Let f be a PL function on a closed convex domain $\Omega \subset \mathbb{R}^n$ and $\{g_1 = \beta_1 x + \alpha_1, \cdots, g_d = \beta_d x + \alpha_d\}$ be the set of the d linear components of f, with $\beta_i, \alpha_i \in \mathbb{R}^n$. There is a family $\{K_i\}_{i \in I}$ of subsets of set $\{1, \cdots, d\}$ such that*

$$f(x) = \bigvee_{i \in I} \bigwedge_{j \in K_i} g_j(x), \quad x \in \Omega. \tag{11}$$

Conversely, for any family of distinct linear functions $\{g_1, \cdots, g_d\}$ the above formula defines a PL function.

The expression is called a max-min (or lattice) polynomial in the variable g_i. We note that a PL function f on Ω is a "selector" of its components g_i, i.e., $\forall x \in \Omega$ there is an i such that $f(x) = g_i(x)$. The converse is also true, with functions $\{g_i\}$ linearly ordered over Ω [29].

Let us also mention that from this representation we can show that a PL function is representable as a difference of two concave (equivalently, convex)-PL functions [29]. More precisely, let note $h_i(x) = \inf_{j \in K_i} g_j(x)$, with h_i a concave function. We are reminded that sums and minimums of concave functions are concave. One have $h_i = \sum_k h_k - \sum_{k \neq i} h_k$, therefore

$$f = \bigvee_{i \in I} h_i = \sum_k h_k - \bigwedge_{i \in I} \sum_{k \neq i} h_k.$$

4   Morphological Universal Activation Functions

Using the previous results, we can state the two following results for the activation function and the pooling by increasing operators. Additionally, a proposed layer used in the experimental section is formulated.

4.1   Universal Representation for Activation Function and Pooling

Proposition 2. *Any piecewise-linear activation function* $\sigma : \mathbb{R} \to \mathbb{R}$ *can be universally expressed as*

$$\sigma(x) = \bigwedge_{j \in J} \left[\bigvee_{i \in I} \left\{ \beta_i^j x + \alpha_i^j \right\} \right] = \bigwedge_{j \in J} p_j(x) \qquad (12)$$

where $p_j = \bigvee_{i \in I} \left\{ \beta_i^j x + \alpha_i^j \right\}$ *is a PL convex function.*

Proposition 3 (Pooling). *Any increasing pooling operator* $\pi : \mathbb{R}^n \to \mathbb{R}^{n'}$ *can be universally expressed as*

$$\pi(f)(x) = \bigwedge_{j \in J} \left[\delta_{b_j}(f)(K \cdot x) \right], \qquad (13)$$

where $\{b_j\}_{j \in J}$ *is a family of structuring functions defining by transpose the basis of the dual operator to* π.

In both cases, there is of course a dual representation using the maximum of erosions. The dilation operator of type $z \mapsto \bigvee_i [\beta_i z + \alpha_i]$ plays a fundamental role in multiplicative morphology [12].

Remark: Tropical Polynomial Interpretation. The max-affine function $p_j = \bigvee_{i \in I} \left\{ \beta_i^j z + \alpha_i^j \right\}$ is a tropical[4] polynomial such that in that geometry, the degree of the polynomial corresponds to the number of pieces of the PL convex function. The set of such polynomials constitutes the semiring $\mathbb{R}_{\max}$ of tropical polynomials. Tropical geometry in the context of lattice theory and neural networks is an active area of research [22] [21] [25], however those previous works have not considered the use of minimal representation of tropical polynomials as generalised activation functions.

Remark: Relationships to Other Universal Approximation Theorems. These results on universal representation of layers in DCNN are related to study the capacity of neural networks to be universal approximators for smooth functions. For instance, both maxout networks [8] and max-plus networks [37] can approximate arbitrarily well any continuous function on a compact domain. The proofs are based on the fact that [35] continuous PL functions can be expressed as a difference of two convex PL functions, and each convex PL can be seen as the maximum of affine terms.

Tropical formulation of ReLU networks has shown that a deeper network is exponentially more expressive than a shallow network [36]. To explore the expressiveness of networks with our universal activation function and pooling layer respect to the deepness of DCNN is therefore a fundamental relevant topic for future research.

---

[4] *Tropical geometry* is the study of polynomials and their geometric properties when addition is replaced with a minimum operator and multiplication is replaced with ordinary addition.

4.2   MorphoActivation Layer

We have now all the elements to justify why in terms of universal representation theory of nonlinear operators ReLU and max-pooling can be replaced by a more general nonlinear operator defined by a morphological combination of activation function, dilations and downsampling, using a max-plus layer or its dual.

More precisely, we introduce two alternative architectures of the MorphoActivation layer (Activation and Pooling Morphological Operator) $f \mapsto \Psi^{\text{Morpho}}$: $\mathbb{R}^n \to \mathbb{R}^{n'}$ either by composition $[\pi \circ \sigma(f)](x)$ or $[\sigma \circ \pi(f)](x)$ as follows:

$$\Psi_1^{\text{Morpho}}(f) = \bigwedge_{1 \leq j \leq M} \left\{ \delta_{R,b_j}^{\texttt{MaxPool}} \left(\bigvee_{1 \leq i \leq N} (\beta_i^j f + \alpha_i^j) \right) \right\}, \tag{14}$$

$$\Psi_2^{\text{Morpho}}(f) = \bigwedge_{1 \leq i \leq N} \left\{ \bigvee_{1 \leq j \leq M} \left(\beta_i^j \delta_{R,b_i}^{\texttt{MaxPool}}(f) + \alpha_i^j \right) \right\}, \tag{15}$$

where

$$\begin{cases} \delta_{R,b_j}^{\texttt{MaxPool}}(f)(x) = \delta_{b_j}(f)(R \cdot x), \quad \text{with} \\ \\ \delta_{b_j}(f)(x) = (f \oplus b_j)(x) = \bigvee_{y \in W} \{f(x-y) + b_j(y)\} \end{cases}$$

In the context of an end-to-end learning DCNN, the parameters β_j, α_j and structuring functions b_j are learnt by backpropagation [34]. The learnable structuring functions $\mathbf{b}_j$ play the same role as the kernel in the convolutions. Note that one can have $R = 1$, the pooling does not involve downsampling. We note that in a DCNN network the output of each layer T^k is composed of the affine function $x \mapsto \mathbf{W}^k x + \mathbf{b}^k$, where $\mathbf{W}^k$ is the weight matrix (convolution weights in a CNN layer) and $\mathbf{b}^k$ the bias, and the activation function σ, i.e., $T^k = \sigma \left(\mathbf{W}^k T^{k-1} + \mathbf{b}^k \right)$, where σ is acting elementwise. Using our general activation (12), we obtain that

$$T^k = \bigwedge_{j \in J} \left[\bigvee_{i \in I} \left\{ \beta_i^{jk} \mathbf{W}^k T^{k-1} + \beta_i^{jk} \mathbf{b}^k + \alpha_i^{jk} \right\} \right],$$

and therefore the bias has two terms which are learnt. We propose therefore to consider in our experiments that $\mathbf{b}^k$ is set to zero since its role will be replaced by learning the α_i^{jk}.

5   Experimental Section

Firstly, to illustrate the kind of activation functions that our proposition can learn, we use the MNIST dataset as a ten class supervised classification problem and an architecture composed of two convolutional layers and a dense layer for reducing to the number of classes. The activation functions that we optimise by stochastic gradient descent have as general form $\min(\max(\beta_0 x +$

$\alpha_0, \beta_1 x + \alpha_1, \alpha_2), \alpha_3)$, which corresponds to (14) and (15) where $R = 1$, i.e., without pooling. We have initialised all the activation functions to be equal to $\max(\min(\text{ReLU}(x), 6), -6)$ as it is illustrated in Fig. 1(left). The accuracy of this network without any training is 14%. Surprisingly when one optimises[5] *only* the parameters of activation functions the network accuracy increases to the acceptable performance of 92.38% and a large variability of activation functions are found Fig. 1(center). This is a way to assess the expressive power[6] of the parameter of the activation as it has been proposed in [6]. Additionally, an adequate separation among classes is noted by visualising the projection to two-dimensional space of the last layer via the t-SNE [19] algorithm. Of course, a much better accuracy $(98, 58\%)$ and inter-class separation is obtained by optimising all the parameters of the network Fig. 1(right).

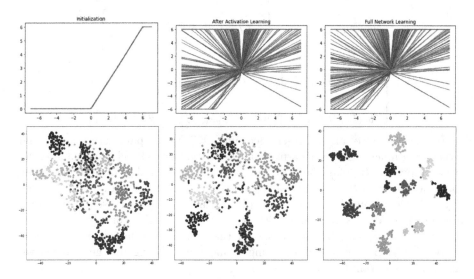

Fig. 1. First Row: Left: Random Initialisation with (14%) of accuracy on the test set, We use a simplified version of proposed activation $\min(\max(\beta_0 x + \alpha_0, \beta_1 x + \alpha_1, \alpha_2), \alpha_3)$, with initialisation $\max(\min(\text{ReLU}(x), 6), -6)$ Centre: Training only activation functions (92.38%), Right: Training Full Network (98,58%). Second Row: t-SNE visualisation of last layer is the 10-classes MNIST prediction for a CNN.

Secondly, we compare the performance of (6), (7), (14) and (15) following the common practice and train all the models using a training set and report the standard top-one error rate on a testing set. We use as architecture a classical two-layer CNN (without bias for (14) and (15)) with 128 filters of size (3×3) per layer, and a final dense layer with dropout. After each convolution the different

---

[5] We use ADAM optimizer with a categorical entropy as loss function, a batch size of 256 images and a learning rate of 0.001.

[6] The expressive power describes neural networks ability to approximate functions.

propositions are used to both produce a nonlinear mapping and reduce spatial dimension via pooling stride of two. As a manner of comparison, we include the case of a simple `ReLU` activation followed by a `MaxPool` with stride two. The difference in top-one error rate on a testing set is reported in Table 1 for CIFAR10, CIFAR100 and Fashion-MNIST databases. These quantitative results shown in propositions (6) and (7) do not seem to improve the performance in the explored cases. Additionally, (15) performs better than (14), and it improves the accuracy in comparison with our baseline in all the considered databases.

Table 1. Relative difference with respect to our baseline (`ReLU` followed by a `MaxPool`). Architecture used is a CNN with two layers. ADAM optimiser with an early stopping with patience of ten iterations. Only Random Horizontal Flip has been used as image augmentation technique for CIFARs. The results are the average over three repetitions of the experiments.

	Fashion MNIST			CIFAR10			CIFAR100		
`MaxPool(ReLU)`	93.11			78.04			47.57		
Self-dual Relu in (6)	−2.11			−20.12			−31.14		
(7)	−0.95			−1.75			−4.39		
MorphoActivation in (14)	$N=2$	$N=3$	$N=4$	$N=2$	$N=3$	$N=4$	$N=2$	$N=3$	$N=4$
$M=2$	−0.06	−0.05	−0.1	−0.42	0.02	−0.02	0.44	0.7	0.4
$M=3$	−0.14	−0.14	−0.06	−0.57	−0.4	−0.35	0.56	0.49	0.61
$M=4$	−0.02	−0.08	−0.01	0.05	−0.62	−0.5	0.41	0.35	0.73
MorphoActivation in (15)	$N=2$	$N=3$	$N=4$	$N=2$	$N=3$	$N=4$	$N=2$	$N=3$	$N=4$
$M=2$	0.04	−0.16	−0.12	1.84	2.02	1.49	3.31	3.5	3.45
$M=3$	0.08	0.09	**0.12**	2.39	1.96	1.82	3.48	3.55	**3.86**
$M=4$	−0.02	0.09	−0.03	**2.49**	2.25	2.13	3.47	3.73	3.58

6   Conclusions and Perspectives

To the best of our knowledge, this is the first work where nonlinear activation functions in deep learning are formulated and learnt as max-plus affine functions or tropical polynomials. We have also introduced an algebraic framework inspired from mathematical morphology which provides a general representation to integrate the nonlinear activation and pooling functions.

Besides more extended experiments on the performance on advanced DCNN networks, our next step will be to study the expressivity power of the networks based on our morphological activation functions. The universal approximation theorems for ReLU networks would just be a particular case. We conjecture that the number of parameters we are adding on the morphological activation can provide a benefit to get more efficient approximations of any function with the same width and depth.

Acknowledgments. This work has been supported by Fondation Mathématique Jacques Hadamard (FMJH) under the PGMO-IRSDI 2019 program. This work was granted access to the Jean Zay supercomputer under the allocation 2021-AD011012212R1.

References

1. Banon, G.J.F., Barrera, J.: Minimal representations for translation-invariant set mappings by mathematical morphology. SIAM J. Appl. Math. **51**(6), 1782–1798 (1991)
2. Bartels, S.G., Kuntz, L., Scholtes, S.: Continuous selections of linear functions and nonsmooth critical point theory. Nonlinear Anal. Theory Methods Appl. **24**(3), 385–407 (1995)
3. Boureau, Y.L., Ponce, J., LeCun, Y.: A theoretical analysis of feature pooling in visual recognition. In: ICML, pp. 111–118 (2010)
4. Charisopoulos, V., Maragos, P.: Morphological perceptrons: geometry and training algorithms. In: Angulo, J., Velasco-Forero, S., Meyer, F. (eds.) ISMM 2017. LNCS, vol. 10225, pp. 3–15. Springer, Cham (2017). https://doi.org/10.1007/978-3-319-57240-6_1
5. Franchi, G., Fehri, A., Yao, A.: Deep morphological networks. Pattern Recogn. **102**, 107246 (2020)
6. Frankle, J., Schwab, D.J., Morcos, A.S.: Training batchnorm and only batchnorm: On the expressive power of random features in CNNs. In: ICLR (2021)
7. Goodfellow, I., Bengio, Y., Courville, A.: Deep learning. MIT press (2016)
8. Goodfellow, I., Warde-Farley, D., Mirza, M., Courville, A., Bengio, Y.: Maxout networks. In: ICML, pp. 1319–1327 (2013)
9. Gorokhovik, V.V., Zorko, O.I., Birkhoff, G.: Piecewise affine functions and polyhedral sets. Optimization **31**(3), 209–221 (1994)
10. Hartline, H.K., Ratliff, F.: Inhibitory interaction of receptor units in the eye of limulus. J. Gen. Physiol. **40**(3), 357–376 (1957)
11. He, K., Zhang, X., Ren, S., Sun, J.: Delving deep into rectifiers: surpassing human-level performance on imagenet classification. In: IEEE ICCV, pp. 1026–1034 (2015)
12. Heijmans, H.J.A.M.: Theoretical aspects of gray-level morphology. IEEE Trans. Pattern Anal. Mach. Intell. **13**(06), 568–582 (1991)
13. Hermary, R., Tochon, G., Puybareau, É., Kirszenberg, A., Angulo, J.: Learning grayscale mathematical morphology with smooth morphological layers. J. Math. Imaging Vis. **64**, 736–753 (2022). https://doi.org/10.1007/s10851-022-01091-1
14. Hernández, G., Zamora, E., Sossa, H., Téllez, G., Furlán, F.: Hybrid neural networks for big data classification. Neurocomputing **390**, 327–340 (2020)
15. Islam, M.A., et al.: Extending the morphological hit-or-miss transform to deep neural networks. IEEE Trans. NNs Learn. Syst. **32**(11), 4826–4838 (2020)
16. Keshet, R.: Mathematical morphology on complete semilattices and its applications to image processing. Fund. Inform. **41**(1–2), 33–56 (2000)
17. Ma, N., Zhang, X., Liu, M., Sun, J.: Activate or not: learning customized activation. In: IEEE CVPR, pp. 8032–8042 (2021)
18. Maas, A.L., Hannun, A.Y., Ng, A.Y., et al.: Rectifier nonlinearities improve neural network acoustic models. In: Proceedings ICML, vol. 30, p. 3 (2013)
19. van der Maaten, L., Hinton, G.: Visualizing data using t-SNE. J. Mach. Learn. Res. **9**(11) (2008)

20. Maragos, P.: A representation theory for morphological image and signal processing. IEEE Trans. Pattern Anal. Mach. Intell. **11**(6), 586–599 (1989)
21. Maragos, P., Charisopoulos, V., Theodosis, E.: Tropical geometry and machine learning. Proc. IEEE **109**(5), 728–755 (2021)
22. Maragos, P., Theodosis, E.: Multivariate tropical regression and piecewise-linear surface fitting. In: ICASSP, pp. 3822–3826. IEEE (2020)
23. Matheron, G.: Random Sets and Integral Geometry. Wiley, Hoboken (1974)
24. McCulloch, W.S., Pitts, W.: A logical calculus of the ideas immanent in nervous activity. Bull. Math. Biophys. **5**(4), 115–133 (1943). https://doi.org/10.1007/BF02478259
25. Misiakos, P., Smyrnis, G., Retsinas, G., Maragos, P.: Neural network approximation based on hausdorff distance of tropical zonotopes. In: ICLR, pp. 0–8 (2022)
26. Mondal, R., Dey, M.S., Chanda, B.: Image restoration by learning morphological opening-closing network. MM-Theory Appl. **4**(1), 87–107 (2020)
27. Nair, V., Hinton, G.E.: Rectified linear units improve restricted boltzmann machines. In: ICML (2010)
28. Ovchinnikov, S.: Boolean representation of manifolds functions. J. Math. Anal. Appl. **263**, 294–300 (2001)
29. Ovchinnikov, S.: Max-min representations of piecewise linear functions. Beiträge Algebra Geom. **43**, 297–302 (2002)
30. Pessoa, L.F., Maragos, P.: Neural networks with hybrid morphological/rank/linear nodes: a unifying framework with applications to handwritten character recognition. Pattern Recogn. **33**(6), 945–960 (2000)
31. Ritter, G.X., Sussner, P.: An introduction to morphological neural networks. In: 13th International Conference on Pattern Recognition, vol. 4, pp. 709–717. IEEE (1996)
32. Sussner, P., Campiotti, I.: Extreme learning machine for a new hybrid morphological/linear perceptron. Neural Netw. **123**, 288–298 (2020)
33. Valle, M.E.: Reduced dilation-erosion perceptron for binary classification. Mathematics **8**(4), 512 (2020)
34. Velasco-Forero, S., Pagès, R., Angulo, J.: Learnable EMD based on mathematical morphology. SIAM J. Imag. Sci. **15**(1), 23–44 (2022)
35. Wang, S.: General constructive representations for continuous piecewise-linear functions. IEEE Trans. Circu. Syst. I **51**(9), 1889–1896 (2004)
36. Zhang, L., Naitzat, G., Lim, L.H.: Tropical geometry of deep neural networks. In: International Conference on Machine Learning, pp. 5824–5832. PMLR (2018)
37. Zhang, Y., Blusseau, S., Velasco-Forero, S., Bloch, I., Angulo, J.: Max-plus operators applied to filter selection and model pruning in neural networks. In: Burgeth, B., Kleefeld, A., Naegel, B., Passat, N., Perret, B. (eds.) ISMM 2019. LNCS, vol. 11564, pp. 310–322. Springer, Cham (2019). https://doi.org/10.1007/978-3-030-20867-7_24

Logarithmic Morphological Neural Nets Robust to Lighting Variations

Guillaume Noyel$^{(\boxtimes)}$ (ID), Emile Barbier-Renard (ID), Michel Jourlin (ID),
and Thierry Fournel (ID)

Laboratoire Hubert Curien, UMR CNRS 5516, Université Jean Monnet,
42 000 Saint-Etienne, France
{guillaume.noyel,michel.jourlin,fournel}@univ-st-etienne.fr,
emile.barbier.renard@etu.univ-st-etienne.fr

Abstract. Morphological neural networks allow to learn the weights of
a structuring function knowing the desired output image. However, those
networks are not intrinsically robust to lighting variations in images with
an optical cause, such as a change of light intensity. In this paper, we
introduce a morphological neural network which possesses such a robustness to lighting variations. It is based on the recent framework of Logarithmic Mathematical Morphology (LMM), i.e. Mathematical Morphology defined with the Logarithmic Image Processing (LIP) model. This
model has a LIP additive law which simulates in images a variation of the
light intensity. We especially learn the structuring function of a LMM
operator robust to those variations, namely: the map of LIP-additive
Asplund distances. Results in images show that our neural network verifies the required property.

Keywords: Morphological neural nets · Logarithmic Image
Processing · Logarithmic Mathematical Morphology · Robustness to
lighting variations · Functional Asplund metric

1 Introduction

Deep learning [8] based on convolutional neural networks (CNN) [16] has
emerged as a methodology to learn a model of the data in order to perform
a classification or a regression task [9]. During the training phase, the model
parameters are learnt by minimising a loss between a given truth and the model
prediction. In parallel to CNN, several morphological neural nets have been
defined and studied. First, fully connected morphological neural nets (where the
output depends on all the input pixels) have been defined in [6] and more recently
by Charisopoulos et al. [5], Mondal et al. [21] and Zhang et al. [35]. Second, Barrera et al. defined morphological neural nets in sliding windows [2] (where the
output only depends on the input pixels in the window). Moreover, deep morphological networks have also been defined by using either approximations of the

Supported by Lyon Informatics Federation, France, through the "FakeNets" project.

morphological operations [1,15,17,18,20,29,31], or exact morphological operations [7,23]. Deep morphological networks have been used e.g. for classification in hyperspectral images [23], image de-hazing or de-raining [19], image denoising [7]. A morphological network has a constant additive invariance [32]. In classical neural networks, a CNN was designed to have a shift invariance [4] and a neural net has a contrast invariance based on quaternion local phase [22].

However, morphological neural networks are not intrinsically robust to real lighting variations. The analysis of images presenting such variations is a challenging task that can occur in many settings [13,27,34]: industry, traffic control, underwater vision, face recognition, large public health databases, etc. In this paper, we propose a morphological neural network which is robust to such lighting variations due to a change of light intensity or of camera exposure-time.

Such a neural net is based on a metric, namely the functional Asplund metric [11] which presents this robustness property. This metric is defined with the *Logarithmic Image Processing* (LIP) model [12,13] which models those lighting variations. As the LIP model is based on a famous optical law, namely the *Transmittance Law*, we shall introduce in this way *Physics* in those neural nets. In addition, the maps of Asplund distances between an image and a reference template, the probe, are related to Mathematical Morphology [27]. We shall see that they are especially related to the newly introduced framework of *Logarithmic Mathematical Morphology* [24,25].

2 Background

2.1 Logarithmic Image Processing

The LIP model is defined for an image f acquired by transmission and, as it is consistent with the human vision [3,13], it can also be used for images acquired by reflection. In this model, the light is passing through a semi-transparent medium and is captured by the sensor. The resulting image f is a function defined on a domain $D \subset \mathbb{R}^2$ with values lying in the interval $[0, M[\subset \mathbb{R}$. It is important to note that the LIP greyscale is inverted with respect to the usual grey scale. 0 corresponds to the white extremity, when all the light passes through the medium. M is the black extremity, when no light is passing. For images digitised on 8-bits, M is always equal to $2^8 = 256$.

According to the *transmittance law*, the transmittance $T_{f \triangle g}$ of the superimposition $f \triangle g$ of two media which generate the images f and g, is equal to the product of the transmittances T_f and T_g of each image: $T_{f \triangle g} = T_f \cdot T_g$. The transmittance T_f of any medium generating the image f is equal to $T_f = 1 - f/M$. From both previous equations, the LIP-addition law is deduced:

$$f \triangle g = f + g - f \cdot g/M. \tag{1}$$

As the addition $f \triangle f$ may be written as $2 \triangle f$, the LIP-multiplication of an image f by a real number λ is expressed as:

$$\lambda \triangle f = M - M (1 - f/M)^\lambda. \tag{2}$$

When $\lambda = -1$, the LIP-negative function $\triangle f = -1 \triangle f$ can be defined, as well as the LIP-difference $f \triangle g$ between two images f and g. They are expressed as follows:

$$\triangle f = (-f)/(1 - f/M), \tag{3}$$

$$f \triangle g = (f - g)/(1 - g/M). \tag{4}$$

It can be noticed that $f \triangle g$ is an image (i.e. $f \triangle g \geq 0$) if and only if $f \geq g$.

The LIP model has a *strong mathematical property*. Let $\mathcal{F}_M =]-\infty, M[^D$ be the set of real functions defined on the domain D and whose values are less or equal than M. Let $\mathcal{I} = [0, M[^D$ be the set of images. The set $(\mathcal{F}_M, \triangle, \triangle)$ is a *real vector space* and the set $(\mathcal{I}, \triangle, \triangle)$ is its *positive cone*.

The LIP model also possesses a *strong physical property*. The LIP-negative values $\triangle f$, where $f \geq 0$, acts as light intensifiers. Those values can therefore be used to compensate the image attenuation in scenes captured with a low lighting. In particular, the LIP-addition of a positive constant to an image simulates the effect of a decrease of the light intensity or a decrease of the camera exposure-time. The resulting image is therefore darker than the original one. In an equivalent way, the LIP-subtraction of a positive constant from an image, simulates an increase and the resulting image becomes brighter.

2.2   Logarithmic Mathematical Morphology

Logarithmic Mathematical Morphology (LMM) was introduced by Noyel in [24, 25]. LMM consists of defining morphological operations [10, 30] in the LIP framework. In LMM, the dilation $\delta_b^{\triangle}$ and the erosion $\epsilon_b^{\triangle}$ are defined in the lattice $(\mathcal{F}_M, \leq)$. Let f and $b \in \overline{\mathcal{F}}_M$ be two functions, where $\overline{\mathcal{F}}_M = [-\infty, M]^D$. The function $b : D \mapsto]-\infty, M[$ is chosen as the structuring function, which implies that outside the domain $D_b \subset D$, all its values are equal to $-\infty$: $\forall x \in D \setminus D_b$, $b(x) = -\infty$. Both mappings $\delta_b^{\triangle}$ and $\varepsilon_b^{\triangle}$ are named *logarithmic-dilation* and *logarithmic-erosion*, respectively. They are defined by:

$$\delta_b^{\triangle}(f)(x) = \vee \{f(x - h) \triangle b(h), h \in D\} \tag{5}$$

$$\varepsilon_b^{\triangle}(f)(x) = \wedge \{f(x + h) \triangle b(h), h \in D\}. \tag{6}$$

In the case of ambiguous expressions, the following conventions will be used: $f(x-h) \triangle b(h) = -\infty$ when $f(x-h) = -\infty$ or $b(h) = -\infty$, and $f(x+h) \triangle b(h) = M$ when $f(x + h) = M$ or $b(h) = -\infty$.

Both operations form an adjunction, i.e. for all $f, g \in \mathcal{F}_M$, $\delta_b^{\triangle}(g) \leq f \Leftrightarrow g \leq \varepsilon_b^{\triangle}(f)$. As a consequence, the composition $\gamma_b^{\triangle} = \delta_b^{\triangle} \circ \varepsilon_b^{\triangle}$ is an *opening* and the composition $\varphi_b^{\triangle} = \varepsilon_b^{\triangle} \circ \delta_b^{\triangle}$ is a *closing*. LMM operations are adaptive to lighting variations modelled by the LIP-additive law. We shall see that at least an operation which is robust to those lighting variations can be defined in the LMM framework.

The logarithmic-dilation $\delta_b^{\triangle}$ and the logarithmic-erosion $\varepsilon_b^{\triangle}$, which are defined in the lattice $(\mathcal{F}_M, \leq)$, are related to the usual functional dilation δ_b

(or $\oplus$) and erosion ε_b (or $\ominus$), which are defined in the lattice of real functions $(\mathbb{R}^D, \leq)$. These usual dilation and erosion are defined, for all $x \in D$, by:

$$\delta_b(f)(x) = (f \oplus b)(x) = \vee \{f(x-h) + b(h), h \in D\} \qquad (7)$$

$$\varepsilon_b(f)(x) = (f \ominus b)(x) = \wedge \{f(x+h) - b(h), h \in D\}. \qquad (8)$$

Such relations are based on the isomorphism $\xi : \overline{\mathcal{F}}_M \mapsto \overline{\mathbb{R}}^D$ and its inverse $\xi^{-1} : \overline{\mathbb{R}}^D \mapsto \overline{\mathcal{F}}_M$, which are expressed by $\xi(f) = -M \ln(1 - f/M)$ and $\xi^{-1}(f) = M(1 - \exp(-f/M))$ [14]. The relations between the logarithmic-operations $\delta_b^{\triangle}$, $\varepsilon_b^{\triangle}$ and the usual functional erosion are the following ones [24,25]:

$$\delta_b^{\triangle}(f) = \xi^{-1}\left(\delta_{\xi(b)}[\xi(f)]\right) = \xi^{-1}\left[\xi(f) \oplus \xi(b)\right] \qquad (9)$$

$$\varepsilon_b^{\triangle}(f) = \xi^{-1}\left(\varepsilon_{\xi(b)}[\xi(f)]\right) = \xi^{-1}\left[\xi(f) \ominus \xi(b)\right], \qquad (10)$$

where f and $b \in \mathbb{R}^D$. These relations are not only important from a theoretical point of view but also from a practical point of view. Indeed, they facilitate the programming of the logarithmic operations by using the usual morphological operations which exist in numerous image analysis software.

In Fig. 1, in an image f, the logarithmic-erosion $\varepsilon_b^{\triangle}(f)$ and dilation $\delta_b^{\triangle}(f)$ are compared to the usual erosion $\varepsilon_b(f)$ and dilation $\delta_b(f)$. For the logarithmic operations, the amplitude of the structuring function b varies according to the amplitude of the image f; whereas, for the usual morphological operations, the amplitude of b does not change. In Fig. 1a, when the image intensity is close to the maximal possible value $M = 256$, the intensity of the logarithmic-dilation of f, $\delta_b^{\triangle}(f)$, is less or equal than M, because the structuring function becomes flat; whereas the intensity of the usual dilation of f, $\delta_b(f)$, is greater than M.

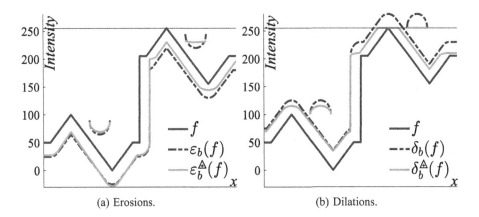

(a) Erosions.   (b) Dilations.

Fig. 1. In an image f, comparison between functional MM and LMM for the (a) erosions $\varepsilon_b(f)$, $\varepsilon_b^{\triangle}(f)$ and (b) dilations $\delta_b(f)$, $\delta_b^{\triangle}(f)$. (a) and (b) For both signal peaks, the structuring function b is represented (after an horizontal translation) for the erosions $\varepsilon_b(f)$, $\varepsilon_b^{\triangle}(f)$ and the dilations $\delta_b(f)$, $\delta_b^{\triangle}(f)$. (a) Erosions. (b) Dilations.

2.3   The LIP-Additive Asplund Metric

The LIP-additive Asplund metric was defined by Jourlin [11].

Definition 1 (LIP-additive Asplund metric). *Let f and $g \in \mathcal{F}_M$ be two functions. One of them, e.g. g, is chosen as a probing function. Both following numbers are then defined by: $c_1 = \inf \{c, f \leq c \triangle g\}$ and $c_2 = \sup \{c, c \triangle g \leq f\}$, where c lies within the interval $]-\infty, M[$. The LIP-additive Asplund metric $d_{asp}^{\triangle}$ is defined by*

$$d_{asp}^{\triangle}(f, g) = c_1 \triangle c_2. \tag{11}$$

Property 1 (Robustness to lighting variations [11]). Importantly, this metric is invariant under lighting changes modelled by a LIP-addition of a constant:

$$\forall k \in]-\infty, M[, \quad d_{asp}^{\triangle}(f, g) = d_{asp}^{\triangle}(f \triangle k, g) \text{ and } d_{asp}^{\triangle}(f, f \triangle k) = 0. \tag{12}$$

Those changes correspond to a modification of the light intensity or of the camera exposure-time.

2.4   Learning the Structuring Function in Morphological Operations

For machine learning in Mathematical Morphology (MM), the functions are defined in discrete grids of $\mathbb{Z}^2$; $f : D \mapsto \mathbb{R}$ and $b : D_b \mapsto \mathbb{R}$, where $D_b \subset D \subset \mathbb{Z}^2$. Let the cardinal of the set D_b be equal to $2n+1$. The bidimensional functions can always be represented as unidimensional arrays, e.g. by concatenating their rows. The structuring function are therefore written as follows: $b = \{b_{-n}, \ldots, b_n\}$. The dilation δ_b and the erosion ε_b layers can be expressed by [7,19]:

$$\delta_b(f)(x) = \max_{i \in [[-n,n]]} \{f(x - i) + b_i\} \tag{13}$$

$$\varepsilon_b(f)(x) = \min_{i \in [[-n,n]]} \{f(x + i) - b_i\} . \tag{14}$$

An image f is passed through a dilation layer or an erosion layer that gives an output equal to $\widehat{g}$ (i.e. $\widehat{g} = \delta_b(f)$ or $\widehat{g} = \varepsilon_{b(f)}$). The weights of the structuring function b are learnt so that the *loss* $L(b) = L(g, \widehat{g})$ between the output $\widehat{g}$ of the neural net layer and the desired output g, is minimised. The loss minimisation is performed by a *stochastic gradient descent* algorithm [8], which requires to compute the derivative of each weight to the loss:

$$\frac{\partial L}{\partial b_i} = \sum_x \frac{\partial \widehat{g}(x)}{\partial b_i} \frac{\partial L}{\partial \widehat{g}(x)} = \sum_x \nabla \widehat{g}(x) \frac{\partial L}{\partial \widehat{g}(x)}. \tag{15}$$

Let us denote $i^{x*} = \arg\max_{i \in [[-n,n]]}\{f(x - i) + b_i\}$ or $i^{x*} = \arg\min_{i \in [[-n,n]]}\{f(x + i) - b_i\}$, the index for which the dilation or the erosion takes its value. For the dilation, the gradient $\nabla \widehat{g}$ is equal to :

$$\nabla \widehat{g}(x) = \begin{cases} 1 & \text{if } \widehat{g}(x) = f(x - i^{x*}) + b_{i^{x*}} \\ 0, & \text{otherwise} \end{cases} \tag{16}$$

and for the erosion to

$$\nabla \widehat{g}(x) = \begin{cases} -1 & \text{if } \widehat{g}(x) = f(x + i^{x*}) - b_{i^{x*}} \\ 0, & \text{otherwise.} \end{cases} \tag{17}$$

The structuring function is therefore updated as : $b(x) = b(x) - \alpha \partial L / \partial b(x)$.

3   Maps of LIP-additive Asplund Distances

3.1   Definition

Definition 2 (Map of LIP-additive Asplund distances). *Let $f \in \mathcal{F}_M$ be a function and $b \in]-\infty, M[^{D_b}$ a probe, where $D_b \subset D$. The map of Asplund distances is the mapping $Asp_b^{\triangle} : \mathcal{F}_M \mapsto \mathcal{I}$ defined by:*

$$Asp_b^{\triangle} f(x) = d_{asp}^{\triangle}(f_{|D_b(x)}, b), \tag{18}$$

where $f_{|D_b(x)}$ is the restriction of f to the neighbourhood $D_b(x)$ centred on $x \in D$. The LIP addition $\triangle$ makes the map of distances robust to contrast variations due to exposure-time changes: $\forall c \in] \infty, M[, Asp_b^{\triangle}(f \triangle c) = Asp_b^{\triangle}(f)$.

3.2   Link with Logarithmic Mathematical Morphology

The map of Asplund distances is related to Mathematical Morphology (MM) [26,27]. Noyel as shown in [25] that it is specifically related to LMM as follows.

Proposition 1. *Let $f \subset \overline{\mathcal{F}}_M$ be a function and $b \in \overline{\mathcal{F}}_M$ be a structuring function, where for all $x \in D_b$, $D_b \subset D$, $b(x) > -\infty$. The map of Asplund distances between the function f and the structuring function b is equal to:*

$$Asp_b^{\triangle} f = \delta_{\triangle \overline{b}}^{\triangle}(f) \triangle \varepsilon_b^{\triangle}(f). \tag{19}$$

In the case of ambiguous expressions, the following conventions will be used: $Asp_b^{\triangle} f(x) = M$ when $\delta_{\triangle \overline{b}}^{\triangle}(f)(x) = M$ or $\varepsilon_b^{\triangle}(f)(x) = -\infty$, and $Asp_b^{\triangle} f(x) = 0$ when $\delta_{\triangle \overline{b}}^{\triangle}(f)(x) = \varepsilon_b^{\triangle}(f)(x)$.

4   Neural Net of Map of Asplund Distances

From Eqs. (19), (9), (10) and knowing that $\xi(\triangle b) = -\xi(b)$ and $\xi(f \triangle g) = -\xi(f) - \xi(g)$, one deduces that:

$$Asp_b^{\triangle} f = \xi^{-1} \left[\delta_{-\xi(\overline{b})} \xi(f) - \varepsilon_{\xi(b)} \xi(f) \right]. \tag{20}$$

We then create a *map of Asplund distance layer $Asp_b^{\triangle}$*, where we apply this layer to the input image f in order to give an output $\widehat{g} = Asp_b^{\triangle}(f)$. The structuring function b is learnt so as to minimise a loss $L(b) = L(g, \widehat{g})$ between $\widehat{g}$ and a desired output $g = Asp_{b_r}^{\triangle}(f)$, where b_r is a reference structuring function. The goal is to learning b in order to discover b_r.

In morphological neuron implementations [7], learning b is equivalent to learn the weight matrix $W \in \mathbb{R}^{A \times B}$, where $W(x) = b(x)$, for all $x \in D_b$ and $W(x) = -\infty$, otherwise. $A \times B$ is the window size in pixels of $D_W \subset D$. However, in Eq. (19), in the term $\delta_{\triangle \overline{b}}^{\triangle}$, we also expect that for all $x \notin D_b$, $\triangle \overline{W}(x) = -\infty$, which is not compatible with the current definition of W.

We therefore introduce a definition of b relying on two learnt kernels: the height kernel W_h and the mask kernel $W_m \in \mathbb{R}^{A \times B}$. First, W_h corresponds to the height-map of the probe satisfying $W_h(x) = b(x)$, for all $x \in D_b$. Second, W_m characterises the definition domain of the probe $D_b = \{x \in D_{W_m} \mid W_m(x) > 0\}$. We then rewrite Eq. (20) as:

$$Asp_b^{\triangle} f = \xi^{-1}\left[\delta_{b_{dil}}(\xi(f)) - \varepsilon_{b_{ero}}(\xi(f))\right], \tag{21}$$

where, $\forall x \in D_{W_m}$:

$$b_{dil}(x) = \begin{cases} -\xi(\overline{W_h})(x) & \text{if } \overline{W_m}(x) > 0 \\ -\infty & \text{otherwise} \end{cases}, b_{ero}(x) = \begin{cases} \xi(W_h)(x) & \text{if } W_m(x) > 0 \\ -\infty & \text{otherwise} \end{cases}. \tag{22}$$

In order to ensure that the gradient descent is smooth, a soft-binarisation function $\chi : \mathbb{R}^{A \times B} \mapsto]0,1[^{A \times B}$, such as the sigmoid $\chi(v) = 1/(1 + \exp(-v))$, is applied to the mask kernel W_m instead of a threshold. We therefore define $V = \chi(W_m) \in]0,1[^{A \times B}$ as the soft-mask of the probe in the window of size $A \times B$. Because V is not a binary mask, $-\infty$ cannot be used when computing b_{ero} or b_{dil}. As f is an image, we have for all $x \in D$, $f(x) \in [0, M-1]$. This implies that $\xi(f(x)) \in [0, \xi(M-1)]$. A bottom value $\perp = -\xi(M-1)$ is chosen such that $\xi(f)(x) - \perp \geq \xi(M-1)$. This implies that $\xi(f)(x) + \perp \geq \xi(0)$. We then define the approximations $\tilde{b}_{dil}$ and $\tilde{b}_{ero}$ of b_{dil} and b_{ero} as follows:

$$\tilde{b}_{dil} = -\xi(\overline{W_h}) \cdot \overline{V} + \perp \cdot (1 - \overline{V}) \tag{23}$$

$$\tilde{b}_{ero} = \xi(W_h) \cdot V + \perp \cdot (1 - \overline{V}). \tag{24}$$

From Eq. (21), (23), (24), an expression of $\widehat{g}$ is deduced:

$$\widehat{g} = \xi^{-1}\left[\delta_{\tilde{b}_{dil}}(\xi(f)) - \varepsilon_{\tilde{b}_{ero}}(\xi(f))\right]. \tag{25}$$

It only contains components with derivatives and which can be used in the back-propagation algorithm. In practice, W_h and W_m are both initialised as null matrices.

Remark 1. In order to push the weights of the kernel W_h away from zero, one might introduce a mean Gaussian or a mean squared Gaussian, as a regularisation function of W_m. This idea will be explored in future works.

5   Illustration and Results

We have illustrated our LMM network by using the Fashion MNIST dataset composed of a training set of 60 000 images and a test set of 10 000 images [33]. Each image is digitised with a 8-bit greyscale and has a size of 28 × 28 pixels.

The goal was to learn a structuring function (or probe) b (represented by both matrices W_h and W_m) so as to discover a reference structuring function b_r. This reference structuring function was defined as follows, for all $x \in D_{W_{b_r}}$, where $D_{W_{b_r}} \subset D$ and W_{b_r} is a matrix of size 7 × 7:

$$b_r(x) = \begin{cases} h(x) & \text{if } h(x) \geq 0, \text{ where } h(x) = -\beta\sqrt{7}\|x\|^2 \triangle c \\ -\infty & \text{otherwise.} \end{cases} \quad (26)$$

Let $D_{b_r} = \{x \in D_{W_{b_r}} \mid h(x) \geq 0\}$ be the domain of the probe. The mask kernel $W_{m,r}$ and the height kernel $W_{h,r}$ of the reference structuring function b_r are defined by $W_{m,r} = \mathbf{1}_{D_{b_r}}$ and $\forall x \in D_{W_{b_r}}$, $W_{h,r}(x) = b_r(x)$, if $x \in D_{b_r}$ and $W_{h,r}(x) = 0$, otherwise. $\mathbf{1}_{D_{b_r}}$ is the indicator function of the set D_{b_r}.

By varying the parameters $\beta \in \{0.2, 0.4, \ldots, 1.2\}$ and $c \in \{10, 25, \ldots, 250\}$, a total of 102 reference structuring functions b_r were generated. With those b_r, the desired outputs $g - Asp_{b_r}^{\triangle}(f)$ (i.e. a ground-truth) were computed in both train and test datasets. In the train set, the weights of b were learnt by minimising a loss $L(g, \widehat{g})$. In the test set, the *map of Asplund distance layer* $\widehat{g} = Asp_b^{\triangle}(f)$ was applied to the images f with the learnt structuring function b. The average of a validation metric Val_m was computed between the estimated outputs $\widehat{g}$ and the ground-truth g. For the loss L and the validation metric Val_m, we used the mean square error MSE or the LIP-mean square error $LIPMSE$ [28]:

$$MSE(g, \widehat{g}) = \frac{1}{P} \sum_{i=1}^{P} [g_i - \widehat{g}_i]^2 \quad (27)$$

$$LIPMSE(g, \widehat{g}) = \frac{M^2}{P} \sum_{i=1}^{P} \left[\ln \left(\frac{M - g_i}{M - \widehat{g}_i} \right) \right]^2, \quad (28)$$

where P is the number of pixels of g. The results of the validation metrics are shown in Table 1. In order to verify the robustness to lighting variations of our neural network, we have performed two other experiments with the same train set, but two additional test sets. We had therefore three test sets. i) The first test set is the initial test set. ii) The second test set is composed of the images of the first test set which were darkened by LIP-adding to them a constant of 100. iii) The third test set is composed of the images of the first test set which were brightened by LIP-subtracting from them a constant of 100. In Table 1, one can notice that the averaged validation metrics between the three test sets are similar with a residual difference less than 1.2×10^{-6} grey levels. Our neural network is therefore robust to lighting variations which are modelled by the LIP-addition of a constant.

Table 1. Comparison of the validations metrics Val_m in three test sets: average, standard deviation and absolute average differences with the ground truth of the 1st test set. Parameters: 15 epochs, Adam optimiser, learning rate $\alpha = 0.5$, batch size: 20.

Test sets	Metrics	Averages	Std dev.	Abs. av. diff.
1st test set	MSE	9.740×10^{-5}	0.673×10^{-3}	
(initial, f)	$LIPMSE$	6.528×10^{-4}	2.580×10^{-3}	
2nd test set	MSE	9.740×10^{-5}	0.673×10^{-3}	0
(dark, $f \triangle 100$)	$LIPMSE$	6.529×10^{-4}	2.581×10^{-3}	7×10^{-8}
3rd test set	MSE	9.740×10^{-5}	0.673×10^{-3}	1×10^{-10}
(bright, $f \triangle 100$)	$LIPMSE$	6.516×10^{-4}	2.576×10^{-3}	1.2×10^{-6}

The error $E_{pr}(W_h, W_{h,r})$ between the height kernels of the learnt probe b and of the reference probe b_r is defined as follows:

$$E_{pr}(W_h, W_{h,r}) = \frac{1}{A.B} \left[\min_k \sum_{x \in D_{b_r}} (W_{h,r}(x) - (W_h(x) \triangle k))^2 + \sum_{x \notin D_{b_r}} W_h^2(x) \right] \tag{29}$$

It takes into account that the map of Asplund distances is invariant under the LIP-addition of a constant to its probe. Table 2 shows the average errors between the kernels W_m and W_h of the learnt structuring function b and the kernels $W_{m,r}$ and $W_{h,r}$ of the reference structuring function b_r, in the train set. One can notice that the average MSE between the soft mask kernels W_m and $W_{m,r}$ is very small, with a value in $1e - 05$. The average error between the height kernels W_h and $W_{h,r}$ has a value in 1e-04 grey levels. This means that the learnt kernels W_m and W_h are similar to the reference kernels $W_{m,r}$ and $W_{h,r}$. The learnt probe b is therefore similar to the reference probe b_r.

Table 2. In the train set, comparison of the average errors between the kernels W_m and W_h of the learnt probe b and the kernels $W_{m,r}$, $W_{h,r}$ of the reference probe b_r. A total of 102 probes b have been learnt.

Kernels	Errors	Averages	Std dev.
Heights of the probe W_h	E_{pr}	6.89×10^{-5}	4.50×10^{-4}
Soft mask of the probe W_m	MSE	2.23×10^{-4}	7.74×10^{-4}

Figure 2 shows an image f of the test set (Fig. 2a). This image was darkened $f \triangle 100$ (Fig. 2b) or brightened $f \triangle 100$ (Fig. 2c). The ground-truth g was computed with the map of Asplund distances with the reference probe b_r: $g = Asp_{b_r}^{\triangle} f$ (Fig. 2d). The predictions $\hat{g}$ of the maps of Asplund distances were made with the learnt structuring function b: (i) in the initial image: $Asp_b^{\triangle} f_1$ (Fig. 2e), (ii) in the darkened images: $Asp_b^{\triangle}(f_1 \triangle 100)$ (Fig. 2f) and (iii) in the brightened image: $Asp_b^{\triangle}(f_1 \triangle 100)$ (Fig. 2g). These images show that there is no noticeable differences between the predictions $Asp_b^{\triangle} f$, $Asp_b^{\triangle}(f \triangle 100)$, $Asp_b^{\triangle}(f \triangle 100)$ and the ground-truth $g = Asp_{b_r}^{\triangle} f$ and that the predictions are robust to lighting variations which are modelled by the LIP-addition or the LIP-subtraction of a constant. Figure 3 shows that the height kernels W_{h,b_r} of the reference probe b_r and W_h of the learnt probe b are similar. The mask kernels W_{m,b_r} and W_m are also similar.

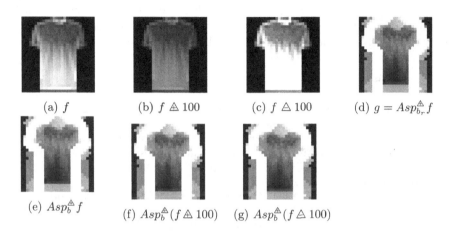

(a) f (b) $f \triangle 100$ (c) $f \triangle 100$ (d) $g = Asp_{b_r}^{\triangle} f$

(e) $Asp_b^{\triangle} f$ (f) $Asp_b^{\triangle}(f \triangle 100)$ (g) $Asp_b^{\triangle}(f \triangle 100)$

Fig. 2. (a) Image f coming from the Fashion MNIST test dataset. (b) $f \triangle 100$ darkened image. (c) $f \triangle 100$ brightened image. (d) $Asp_{b_r}^{\triangle} f$: ground-truth g, i.e. map of Asplund distances with the reference probe b_r. Predictions $\widehat{g}$, i.e. maps of Asplund distances with the learnt probe b: (e) in the initial image $Asp_h^{\triangle} f$; (f) in the darkened image $Asp_b^{\triangle}(f \triangle 100)$; (g) in the brightened image $Asp_b^{\triangle}(f \triangle 100)$.

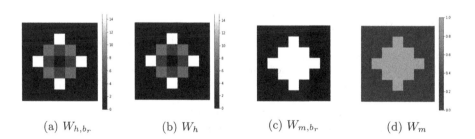

(a) W_{h,b_r} (b) W_h (c) W_{m,b_r} (d) W_m

Fig. 3. In the inverted grey scale, height kernels (a) W_{h,b_r} of the reference probe b_r and (b) W_h of the learnt probe b. Mask kernels (c) W_{m,b_r} of the reference probe b_r and (d) W_m of the learnt probe b. As W_m is a soft mask, a color scale was used.

6   Conclusion

We have introduced a logarithmic morphological neural network which is robust to real lighting variations. Those variations are modelled by the LIP-addition of a constant and they are caused by a change in the light intensity or in the camera exposure-time. Such a neural net is based on the functional Asplund metric defined with the LIP-additive law. In the future, by combining several logarithmic morphological layers, we will define neural nets for numerous practical applications where the light is uncontrolled. We will also study neural nets for the LIP-multiplicative Asplund metric, which is invariant under changes of opacity. Such changes are modelled by the LIP-multiplication by a scalar.

References

1. Aouad, T., Talbot, H.: Binary morphological neural network (2022). https://doi.org/10.48550/ARXIV.2203.12337
2. Barrera, J., Dougherty, E.R., Tomita, N.S.: Automatic programming of binary morphological machines by design of statistically optimal operators in the context of computational learning theory. Journal of Electronic Imaging 6(1), 54–67 (1997). https://doi.org/10.1117/12.260010
3. Brailean, J., Sullivan, B., Chen, C., Giger, M.: Evaluating the EM algorithm for image processing using a human visual fidelity criterion. In: ICASSP 1991. pp. 2957–2960 vol 4 (1991). https://doi.org/10.1109/ICASSP.1991.151023
4. Chaman, A., Dokmanić, I.: Truly shift-invariant convolutional neural networks. In: CVPR 2021. pp. 3772–3782 (2021). https://doi.org/10.1109/CVPR46437.2021.00377
5. Charisopoulos, V., Maragos, P.: Morphological perceptrons: Geometry and training algorithms. In: Lect Notes Comput Sc. vol. 10225, pp. 3–15. Springer, Cham (2017). https://doi.org/10.1007/978-3-319-57240-6_1
6. Davidson, J.L., Hummer, F.: Morphology neural networks: An introduction with applications. Circuits, Systems and Signal Processing 12(2), 177–210 (1993). https://doi.org/10.1007/BF01189873
7. Franchi, G., Fehri, A., Yao, A.: Deep morphological networks. Pattern Recognition 102, 107246 (2020). https://doi.org/10.1016/j.patcog.2020.107246
8. Goodfellow, I., Bengio, Y., Courville, A.: Deep Learning. MIT Press (2016), https://www.deeplearningbook.org
9. Hastie, T., Tibshirani, R., Friedman, J.: The Elements of Statistical Learning. Springer, New York, NY, 2 edn. (2009). https://doi.org/10.1007/978-0-387-84858-7
10. Heijmans, H., Ronse, C.: The algebraic basis of mathematical morphology I. Dilations and erosions. Comput. Vision Graphics and Image Process. 50(3), 245–295 (Jun 1990). https://doi.org/10.1016/0734-189X(90)90148-O
11. Jourlin, M.: Chapter three - metrics based on logarithmic laws. In: Logarithmic Image Processing: Theory and Applications, Adv. Imag. Electron Phys., vol. 195, pp. 61–113. Elsevier (2016). https://doi.org/10.1016/bs.aiep.2016.04.003
12. Jourlin, M., Pinoli, J.: Logarithmic image processing: The mathematical and physical framework for the representation and processing of transmitted images. Adv. Imag. Electron Phys., vol. 115, pp. 129–196. Elsevier (2001). https://doi.org/10.1016/S1076-5670(01)80095-1

13. Jourlin, M.: Logarithmic Image Processing: Theory and Applications, Adv. Imag. Electron Phys., vol. 195. Elsevier (2016)
14. Jourlin, M., Pinoli, J.C.: Image dynamic range enhancement and stabilization in the context of the logarithmic image processing model. Signal Process. **41**(2), 225–237 (1995). https://doi.org/10.1016/0165-1684(94)00102-6
15. Kirszenberg, A., Tochon, G., Puybareau, É., Angulo, J.: Going beyond p-convolutions to learn grayscale morphological operators. In: Lect Notes Comput Sc. vol. 12708, pp. 470–482. Springer, Cham (2021). https://doi.org/10.1007/978-3-030-76657-3_34
16. LeCun, Y., Bengio, Y., Hinton, G.: Deep learning. Nature **521**(7553), 436–444 (2015). https://doi.org/10.1038/nature14539
17. Masci, J., Angulo, J., Schmidhuber, J.: A learning framework for morphological operators using counter-harmonic mean. In: Lect Notes Comput Sc. vol. 7883, pp. 329–340. Springer, Berlin, Heidelberg (2013). https://doi.org/10.1007/978-3-642-38294-9_28
18. Mellouli, D., Hamdani, T.M., Ayed, M.B., Alimi, A.M.: Morph-CNN: A morphological convolutional neural network for image classification. In: Lect Notes Comput Sc. vol. 10635, pp. 110–117. Springer, Cham (2017). https://doi.org/10.1007/978-3-319-70096-0_12
19. Mondal, R., Dey, M.S., Chanda, B.: Image restoration by learning morphological opening-closing network. Mathematical Morphology - Theory and Applications **4**(1), 87–107 (2020). https://doi.org/10.1515/mathm-2020-0103
20. Mondal, R., Mukherjee, S.S., Santra, S., Chanda, B.: Morphological Network: How Far Can We Go with Morphological Neurons? (2019). https://doi.org/10.48550/arxiv.1901.00109
21. Mondal, R., Santra, S., Chanda, B.: Dense morphological network: An universal function approximator. CoRR (2019), https://arxiv.org/abs/1901.00109
22. Moya-Sánchez, E.U., Xambó-Descamps, S., Sánchez Pérez, A., Salazar-Colores, S., et al.: A bio-inspired quaternion local phase CNN layer with contrast invariance and linear sensitivity to rotation angles. Pattern Recognition Letters **131**, 56–62 (2020). https://doi.org/10.1016/j.patrec.2019.12.001
23. Nogueira, K., Chanussot, J., Mura, M.D., Santos, J.A.D.: An introduction to deep morphological networks. IEEE Access **9**, 114308–114324 (2021). https://doi.org/10.1109/ACCESS.2021.3104405
24. Noyel, G.: Logarithmic mathematical morphology: A new framework adaptive to illumination changes. Lect Notes Comput Sc, vol. 11401, pp. 453–461. Springer (2019). https://doi.org/10.1007/978-3-030-13469-3_53
25. Noyel, G.: Morphological and logarithmic analysis of large image databases. Dissertation of accreditation to supervise research, Université de Reims Champagne-Ardenne, France (Jun 2021), https://tel.archives-ouvertes.fr/tel-03343079
26. Noyel, G., Jourlin, M.: Double-sided probing by map of Asplund's distances using logarithmic image processing in the framework of mathematical morphology. Lect Notes Comput Sc, vol. 10225, pp. 408–420. Springer (2017). https://doi.org/10.1007/978-3-319-57240-6_33
27. Noyel, G., Jourlin, M.: Functional asplund metrics for pattern matching, robust to variable lighting conditions. Image Anal. Stereol. **39**(2), 53–71 (2020). https://doi.org/10.5566/ias.2292
28. Pinoli, J.C.: Metrics, scalar product and correlation adapted to logarithmic images. Acta Stereologica 11(2), 157–168 (1992), https://popups.uliege.be:443/0351-580x/index.php?id=1901

29. Saeedan, F., Weber, N., Goesele, M., Roth, S.: Detail-preserving pooling in deep networks. In: CVPR 2018. pp. 9108–9116 (2018). https://doi.org/10.1109/CVPR.2018.00949
30. Serra, J.: Image Analysis and Mathematical Morphology, vol. 1. Academic, Orlando, FL, USA (1982)
31. Shen, Y., Zhong, X., Shih, F.Y.: Deep morphological neural networks. CoRR (2019), https://arxiv.org/abs/1909.01532
32. Velasco-Forero, S., Pagès, R., Angulo, J.: Learnable empirical mode decomposition based on mathematical morphology. SIAM J. Imaging Sci. **15**(1), 23–44 (2022). https://doi.org/10.1137/21M1417867
33. Zalando: Fashion MNIST. https://www.kaggle.com/datasets/zalando-research/fashionmnist (2017)
34. Zhang, W., Zhao, X., Morvan, J.M., Chen, L.: Improving shadow suppression for illumination robust face recognition. IEEE T Pattern Anal **41**(3), 611–624 (2019). https://doi.org/10.1109/TPAMI.2018.2803179
35. Zhang, Y., Blusseau, S., Velasco-Forero, S., Bloch, I., Angulo, J.: Max-plus operators applied to filter selection and model pruning in neural networks. In: Lect Notes Comput Sc. vol. 11564, pp. 310–322. Springer, Cham (2019). https://doi.org/10.1007/978-3-030-20867-7_24

Author Index

Printed in the United States
by Baker & Taylor Publisher Services